Wind Power in Power Systems

ROYAL INSTITUTE OF TECHNOLOGY

Electric Power Systems

http://www.ets.kth.se/ees

Wind Power in Power Systems

Edited by

Thomas Ackermann
Royal Institute of Technology
Stockholm, Sweden

John Wiley & Sons, Ltd

Other Wiley Editorial Offices

John Wiley & Sons Inc., 111 River Street, Hoboken, NJ 07030, USA

Jossey-Bass, 989 Market Street, San Francisco, CA 94103-1741, USA

Wiley-VCH Verlag GmbH, Boschstr. 12, D-69469 Weinheim, Germany

John Wiley & Sons Australia Ltd, 33 Park Road, Milton, Queensland 4064, Australia

John Wiley & Sons (Asia) Pte Ltd, 2 Clementi Loop #02-01, Jin Xing Distripark, Singapore 129809

John Wiley & Sons Canada Ltd, 22 Worcester Road, Etobicoke, Ontario, Canada M9W 1L1

Library of Congress Cataloging in Publication Data

Wind power in power systems / edited by Thomas Ackermann.
    p. cm
  Includes bibliographical references and index.
  ISBN 0-470-85508-8 (cloth : alk. paper)
  1. Wind power plants. 2. Wind power. I. Ackermann, Thomas. II. Title.

  TK1541.W558 2005
  621.31′2136—dc22

                                                        2004018711

British Library Cataloguing in Publication Data

A catalogue record for this book is available from the British Library

ISBN 0-470-85508-8

Typeset in 10/12pt Times by Integra Software Services Pvt. Ltd, Pondicherry, India
Printed and bound in Great Britain by Antony Rowe Ltd, Chippenham, Wiltshire
This book is printed on acid-free paper responsibly manufactured from sustainable forestry in which at least
two trees are planted for each one used for paper production.

To Moana, Jonas and Nora

Contents

Contributors

Thomas Ackermann has a *Diplom Wirtschaftsingenieur* (MSc in Mechanical Engineering combined with an MBA) from the Technical University Berlin, Germany, an MSc in Physics from Dunedin University, New Zealand, and a PhD from the Royal Institute of Technology in Stockholm, Sweden. In addition to wind power, his main interests are related to the concept of distributed power generation and the impact of market regulations on the development of distributed generation in deregulated markets. He has worked in the wind energy industry in Germany, Sweden, China, USA, New Zealand, Australia and India. Currently, he is a researcher with the Royal Institute of Technology (KTH) in Stockholm, Sweden, and involved in wind power education at KTH and the University of Zagreb, Croatia, via the EU TEMPUS program. He is also a partner in Energynautics.com, a consulting company in the area of sustainable energy supply. Email: Thomas.Ackermann@ieee.com.

Vladislav Akhmatov has an MSc (1999) and a PhD (2003) from the Technical University of Denmark. From 1998 to 2003 he was with the Danish electric power company NESA. During his work with NESA he developed dynamic wind turbine models and carried out power system stability investigations, using mainly the simulation tool PSS/E^{TM}. He combined his PhD with work on several consulting projects involving Danish wind turbine manufacturers on grid connection of wind farms in Denmark and abroad. Specifically, he participated in a project regarding power system stability investigations in connection with the grid connection of the Danish offshore wind farm at Rødsand/ Nysted (165 MW). He demonstrated that blade angle control can stabilise the operation of the wind farm during grid disturbances. This solution is now applied in the Rødsand/ Nysted offshore wind farm. In 2003 he joined the Danish transmission system operator in Western Denmark, Eltra. His primary work is dynamic modelling of wind turbines in the simulation tool Digsilent Power-Factory, investigations of power system stability and projects related to the Danish offshore wind farm at Horns Rev (160 MW). In 2002 he received the Angelo Award, which is a Danish award for exceptional contributions to

Wind Power in Power Systems   Edited by T. Ackermann
© 2005 John Wiley & Sons, Ltd   ISBN: 0-470-85508-8 (HB)

the electric power industry, for 'building bridges between the wind and the electric power industries'. He has authored and co-authored a number of international publications on dynamic wind turbine modelling and power system stability. Email: vla@eltra.dk.

E. Ian Baring-Gould graduated with a master's degree in mechanical engineering from the University of Massachusetts Renewable Energy Research Laboratory in the spring of 1995, at which point he started working at the National Renewable Energy Laboratory (NREL) of the USA. Ian's work at NREL has focused on two primary areas: applications engineering for renewable energy technologies and international assistance in renewable energy uses. His applications work concentrates on innovative uses of renewable energies, primarily the modelling, testing and monitoring of small power systems, end-use applications and large diesel plant retrofit concepts. International technical assistance has focused on energy development for rural populations, including the design, analysis and implementation of remote power systems. Ian continues to manage and provide general technical expertise to international programs, focusing on Latin America, Asia and Antarctica. Ian also sits on IEA and IEC technical boards, is an editor for *Wind Engineering* and has authored or co-authored over 50 publications. His graduate research centred on the Hybrid2 software hybrid, power system design, code validation and the installation of the University's 250 kW ESI-80 wind turbine. Email: ian_baring_gould@nrel.gov.

Sigrid M. Bolik graduated in 2001 with a master's degree in electrical engineering (Diplom) from the Technical University Ilmenau in Germany. Currently, she works for Vestas Wind Systems A/S in Denmark and also on her PhD in cooperation with Aalborg University and Risø. Her research focuses on modelling induction machines for wind turbine applications and developing wind turbine models for research in specific abnormal operating conditions. Email: s.bolik@web.de.

Thomas Bopp is currently a research associate at the Electrical Energy and Power System Research Group at UMIST, UK. His main research interests are power system protection as well as power system economics and regulation. Email: T.Bopp@umist.ac.uk.

S. W. H. (Sjoerd) de Haan received his MSc degree in applied physics from the Delft University of Technology, the Netherlands, in 1975. In 1995 he joined the Delft University of Technology as associate professor in power electronics. His research interest is currently mainly directed towards power quality conditioning (i.e. the development of power electronic systems for the conditioning of the power quality in the public electricity network). Email: s.w.h.dehaan@ewi.tudelft.nl.

Predrag Djapić is currently a research associate at the Electrical Energy and Power System Research Group at UMIST, UK. His main research interests are power system planning and operation of distribution networks. Email: P.Djapic@umist.ac.uk.

Peter Borre Eriksen received an MSc degree in engineering from the Technical University of Denmark (DTU) in 1975. From 1980 until 1990 his work focused on the environmental consequences of power production. Between 1990 and 1998 he was employed in the System Planning Department of the former Danish utility ELSAM. In 1998, he joined Eltra, the independent transmission system operator of western Denmark. In 2000, he became head of Eltra's Development Department. Peter Borre Eriksen is the author of numerous technical papers on system modelling. Email: pbe@eltra.dk.

Bernhard Ernst is an electrical engineer and has a master's degree (Diplom) in measurement and control from the University of Kassel, Germany. In 1994, still a student, he joined ISET. In 2003, he completed at ISET a PhD on the prediction of wind power. Bernhard Ernst has contributed to numerous publications on the subject of the integration of wind energy into energy supply. Email: bernie.ernst@web.de.

Anca D. Hansen received her PhD in modelling and control engineering from the Technical University of Denmark (DTU) in 1997. In 1998 she joined the Wind Energy Department of Risø National Laboratory. Her work and research interests focus on dynamic modelling and the control of wind turbines as well as on the interaction of wind farms with the grid. As working tools she uses the dynamic modelling and simulation tools Matlab and Digsilent Power Factory. Her major contribution is the electromechanical modelling of active stall wind turbines and recently of a pitch-controlled variable-speed wind turbine with a doubly fed induction generator. She has also modelled PV modules and batteries. Email: anca.daniela.hansen@risoe.dk.

Carl Hilger received a BSc in electrical engineering from the Engineering Academy of Denmark and a general philosophy diploma as well as a bachelor of commerce degree. In 1966 he joined Brown Boveri, Switzerland, as an electrical engineer and later the Research Institute for Danish Electric Utilities (DEFU). In 1978 he became sectional engineer in the Planning Department of Elsam (the Jutland-Funen Power Pool). Between 1989 and 1997 he was executive secretary at Elsam and after that at Eltra, the independent transmission system operator in the western part of Denmark. In 1998, he was appointed head of the Operation Division at Eltra. Carl Hilger is a member of Eurelectric Working Group SYSTINT and Nordel's Operations Committee. Email: carl@hilger.dk.

Ritva Hirvonen has MSc and PhD degrees in electrical engineering from Helsinki University of Technology and an MBA degree. She has broad experience regarding power systems, transmission and generators. She has worked for the power company Imatran Voima Oy and transmission system operator Fingrid as a power system specialist and at VTT Technical Research Centre of Finland as research manager in the energy systems area. Her current position is head of unit of Natural Gas and Electricity Transmission for the Energy Market Authority (EMA) and she is actively involved in research and teaching at the Power Systems Laboratory of Helsinki University of Technology. Email: Ritva.Hirvonen@Energiamarkkinavirasto.fi.

Hannele Holttinen has MSc (Tech) and LicSc (Tech) degrees from Helsinki University of Technology. She has acquired broad experience regarding different aspects of wind energy research since she started working for the VTT Technical Research Centre of Finland in 1989. In 2000–2004 she worked mainly on her PhD on 'Effects of Large Scale Wind Power Production on the Nordic Electricity System', with Nordic Energy Research funding. Email: Hannele.Holttinen@vtt.fi.

Nick Jenkins is a professor of electrical energy and power systems at UMIST, UK. His research interests are in the area of sustainable energy systems including renewable energy and its integration in electricity distribution and transmission networks. Email: N.Jenkins@umist.ac.uk.

W. L. (Wil) Kling received an MSc degree in electrical engineering from the Technical University of Eindhoven in 1978. Currently, he is a part-time professor at the Electric Power Systems Laboratory of Delft University of Technology. His expertise lies in the area of planning and operating power systems. He is involved in scientific organisations, such as IEEE. He is also the Dutch representative in the Cigré Study Committee C1 'System Development and Economics'. Email: w.l.kling@ewi.tudelft.nl.

Hans Knudsen received a MScEE from the Technical University of Denmark in 1991. In 1994 he received an industrial PhD, which was a joint project between the Technical University of Denmark and the power companies Elkraft, SK Power and NESA. He then worked in the in the Transmission Planning Department of the Danish transmission and distribution company NESA and focused on network planning, power system stability and computer modelling, especially on modelling and simulation of HVDC systems and wind turbines. In 2001, he joined the Danish Energy Authority, where he works on the security of supply and power system planning. Email: HKN@ENS.dk.

Åke Larsson received in 2000 a PhD from Chalmers University of Technology, Sweden. His research focused on the power quality of wind turbines. He has broad experience in wind power, power quality, grid design, regulatory requirements, measurements and evaluation. He also participated in developing new Swedish recommendations for the grid connection of wind turbines. Currently, he works for Swedpower. Email: ake.larsson@swedpower.com.

Christer Liljegren has a BScEE from Thorildsplan Technical Institute, Sweden. He worked with nuclear power at ASEA, Vattenfall, with different control equipment, mainly concerning hydropower, and at Cementa factory working with electrical industrial designing. In 1985, he joined Gotland Energiverk AB (GEAB) and in 1995 became manager engineer of the electrical system on Gotland. He was project manager of the Gotland HVDC-Light project. In 2001, Christer Liljegren started his own consulting company, Cleps Electrical Power Solutions AB (CLEPS AB), specialising in technical and legal aspects of distributed power generation, especially wind turbines and their connection to the grid. He has been involved in developing guidelines and recommendations for connecting distributed generation in Sweden. Email: chl@cleps.se.

Eva Centeno López received an MSc degree in electrical engineering from Universidad Pontificia Comillas in Madrid, Spain, in 2001, and a master's degree at the Royal

Institute of Technology, Stockholm, Sweden, in 2000. She then worked at Endesa, Madrid, Spain, at the Department of Electrical Market. Currently, she works at the Swedish Energy Agency in Eskilstuna, Sweden. Email: Eva.Centeno@stem.se.

Per Lundsager started working full-time with wind energy in 1975, including R&D, assessment, planning, implementation and evaluation of energy systems and concepts, for wind energy and other renewables. Between 1984 and 1993 he was head of the wind diesel development programme at Risø National Laboratory. As senior consultant he has been advisor to the national wind energy centres in the USA, Canada, Finland, Denmark, Russia, Estonia, Poland, Brazil, India and Egypt, regarding projects, programmes and strategies. He has also been manager and/or participant in projects and studies in the USA, Canada and Europe, including Greenland, Eastern Europe, Africa and Asia. Email: per.lundsager@risoe.dk.

Matthias Luther received a PhD in the field of electrical switchgear devices from the Technical University of Braunschweig, Germany. In 1993, he joined PreussenElektra AG, Germany. He was the project manager of various European network studies, mainly concerning system stability. Between 1998 and 2000 he was in charge of network development and customer services at the Engineering and Sales Department of PreussenElektra Netz. Presently, Matthias Luther is head of network planning at E.ON Netz GmbH, Bayreuth, Germany. He is member of several national and international institutions and panels. Email: Matthias.luther@eon-energie.com.

Julija Matevosyan (Sveca) received a BSc degree in electrical engineering from Riga Technical University, Latvia, in 1999. From 1999 to 2000 she worked as a planning engineer in the Latvian power company Latvenergo. She received an MSc in electrical engineering from the Royal Institute of Technology, Stockholm, Sweden, in 2001. She is currently working at the Royal Institute of Technology towards a PhD on the large-scale integration of wind power in areas with limited transmission capability. Email: julija@ekc.kth.se.

Poul Erik Morthorst has a MEcon from the University of Århus and is a senior research specialist in the Systems Analysis Department at Risø National Laboratory. He joined this institute in 1978. His work has focused on general energy and environmental planning, development of long-term scenarios for energy, technology and environmental systems, evaluation of policy instruments for regulating energy and environment and the assessment of the economics of renewable energy technologies, especially wind power. He has participated in a large number of projects within these fields and has extensive experience in international collaboration. Email: p.e.morthorst@risoe.dk.

Jørgen Nygård Nielsen received a BScEE from the Engineering College of Sønderborg, Denmark, in 1984. From 1984 to 1988 he worked on developing digital control systems and designing software for graphical reproduction systems. Between 1988 and 1994 he was a lecturer at the College of Chemical Laboratory and Technician Education, Copenhagen. In 1996, he received an MScEE from the Technical University of Denmark and in 2000 an industrial PhD, a joint project between the Technical

University of Denmark, the Institute for Research and Development of the Danish Electric Utilities, Lyngby, Denmark, and Electricité de France, Clamart, France. In 2000 he joined the Department of Transmission and Distribution Planning of the Danish transmission and distribution company NESA. He works on general network planning, power system stability and the development of wind turbine simulation models. Email: JON@NESA.dk

Jonas Persson received an MSc degree in electrical engineering from Chalmers University of Technology, Göteborg, Sweden, in 1997 and a Tech. Lic. degree in electric power systems from the Royal Institute of Technology, Stockholm, Sweden, in 2002. He joined ABB, Västerås, Sweden, in 1995 where he worked on the development of the power system simulation software Simpow. In 2004 he joined STRI, Ludvika, Sweden, where he develops and teaches Simpow. Currently, he also works at the Royal Institute of Technology in Stockholm, Sweden, towards a PhD on bandwidth-reduced linear models of noncontinuous power system components. Email: Jonas.Persson@stri.se.

Henk Polinder received in 1992 an MSc degree in electrical engineering and in 1998 a PhD, both from the Delft University of Technology. Currently, he is an associate professor at the Electrical Power Processing Laboratory at the same university, where he gives courses on electrical machines and drives. His main research interest is generator systems in renewable energy, such as wind energy and wave energy. Email: h.polinder@ewi.tudelft.nl.

Uwe Radthe was born in 1948. He received the Doctor degree in Power Engineering from the Technical University of Dresden in 1980. In 1990 he joined PreussenElektra and worked in the network planning department. He was involved in international system studies as project manager and specialist of high voltage direct current transmission systems. From 2000 until 2003 he worked for E.ON Netz, responsible for system integration of renewable energy, especially wind power generation.

Harold M. Romanowitz is president and chief operating officer of Oak Creek Energy Systems Inc. and a registered professional engineer. He holds a BScEE from Purdue University and an MBA from the University of California at Berkeley. He has been involved in the wind industry in California since 1985 and received the AWEA Technical Achievement Award in 1991 for his turnaround work at Oak Creek. He has been directly involved in efforts to improve the Tehachapi area grid over this time, including the achievement of a better understanding of the impacts of induction machines and improved VAR support. In 1992–93 he designed and operated a 2.88 MW 17 280 KWH battery storage system directly integrated with wind turbines to preserve a firm capacity power purchase agreement. He was a manufacturer of engineered industrial drive systems for many years, produced the first commercial regenerative thyristor drives in the USA and WattMiser power recovery drives. He has extensive experience with dynamic systems, including marine main propulsion (10 MW), large material handling robots, container and bulk-handling cranes, large pumps and coordinated process lines. Email: hal@rwitz.net.

Fritz Santjer received an MSc (Diplom) in electrical engineering from the University of Siegen, Germany, in 1989. In 1990 he joined the German Wind Energy Institute (DEWI) where he works on grid connection and the power quality of wind turbines and wind farms and on standalone systems. In 2000 he became head of the Electrical Systems Group in DEWI. He has performed commercial power quality and grid protection measurements in many different countries in Europe, South America and Asia. He is an assessor for the MEASNET power quality procedure and is involved in national and international working groups regarding guidelines on power quality and the grid connection of wind turbines. He lectures at national and international courses. He was involved in various European research projects concerning grid connection and power quality of wind turbines, standalone systems and simulations of wind turbines and networks. Email: f.santjer@dewi.de.

J. G. (Han) Slootweg received an MSc degree in electrical engineering from Delft University of Technology, the Netherlands, in 1998. The topic of his MSc thesis was modelling magnetic saturation in permanent-magnet linear machines. In December 2003 he obtained a PhD from the Delft University of Technology. His thesis was on 'Wind Power; Modelling and Impact on Power System Dynamics'. He also holds an MSc degree in business administration from the Open University of the Netherlands. His MSc thesis focuses on how to ensure and monitor the long-term reliability of electricity networks from a regulator's perspective. Currently, he works with Essent Netwerk B.V. in the Netherlands. Email: han.slootweg@essent.nl.

Lennart Söder received MSc and PhD degrees in electrical engineering from the Royal Institute of Technology, Stockholm, Sweden, in 1982 and 1988, respectively. He is currently a professor in electric power systems at the Royal Institute of Technology. He works with projects concerning deregulated electricity markets, distribution systems, protection systems, system reliability and integration of wind power. Email: lennart. soder@ets.kth.se.

Robert Steinberger-Wilckens received a physics degree in 1985 on the simulation of passive solar designs. In 1993 he completed a PhD degree on the subject of coupling geographically dispersed renewable electricity generation to electricity grids. In 1985 he started an engineering consultancy PLANET (Planungsgruppe Energie und Technik) in Oldenburg, Germany, of which he became a full-time senior manager in 1993. His work has focused on complex system design and planning in energy and water supply, energy saving, hydrogen applications, building quality certificates and in wind, solar and biomass projects. In 1999–2000 he developed the hydrogen filling station EUHYFIS, funded within the CRAFT scheme of the EU. In 2002 he joined the Forschungszentrum Jülich as project manager for fuel cells. He is currently head of solid oxide fuel cell development at the research centre. Email: r.steinberger@planet-energie.de.

Poul Sørensen has an MSc in electrical engineering (1987). He joined the Wind Energy Department (VEA) of Risø National Laboratory in 1987 and now is a senior scientist there. Initially, he worked in the areas of wind turbine structural and aerodynamic modelling. Now, his research focuses on the interaction between wind energy and power systems, with special interest in modelling and simulation. He has been project manager on a number of research projects in the field. The modelling involves electrical aspects as well as aeroelasticity and turbulence modelling. Poul Sørensen has worked for several years on power quality issues, with a special focus on flicker emission from wind turbines, and has participated in the work on the IEC 61400-21 standard for the measurement and assessment of power quality characteristics for wind turbines. Email: poul.e.soerensen@risoe.dk.

Goran Strbac is a professor of electrical power engineering at UMIST, UK. His research interests are in the area of power system analysis, planning and economics and in particular in the technical and commercial integration of distributed generation in the operation and development of power systems. Email: G.Strbac@umist.ac.uk.

John Olav Giæver Tande received his MSc in electrical engineering from the Norwegian Institute of Science and Technology in 1988. After graduating he worked at the Norwegian Electric Power Research Institute (EFI) and then, from 1990 to 1997, he worked at Risø National Laboratory in Denmark. After this he returned to SINTEF Energy Research (formerly EFI), where he is currently employed. Throughout his career, his research has been focused on the electrical engineering aspects of wind power. He has participated in several international studies, including convening an IEC working group on preparing an international standard on the measurement and assessment of the power quality characteristics of grid-connected wind turbines, and is the operating agent representative of IEA Annex XXI: Dynamic Models of Wind Farms for Power System Studies (2002–2005). Email: john.o.tande@sintef.no.

Wilhelm R. Winter received an MSc and a PhD in power engineering from the Technical University of Berlin in 1995 and 1998, respectively. In 1995 he joined Siemens and worked in the department for protection development and in the system planning department. He was involved in large-system studies including stability calculations, HVDC and FACTS optimisations, modal analysis, transient phenomena, real-time simulation and renewable energy systems. He was responsible for the development of the NETOMAC Eigenvalue Analysis program. In 2000 he started working at E.ON Netz, and is responsible for system dynamics and the integration of large-scale wind power. Email: Wilhelm.Winter@eon-energie.com.

Abbreviations

A

ABB	Asea Brown Boveri
AC	Alternating current
AEC	Aeroelastic Code
AFC	Alkaline fuel cell
AM	Active management
ANN	Artificial neural network
ATC	Available transfer capacity
ATP	Alternative Transient Program
AWEA	American Wind Energy Association
AWPT	Advanced Wind Power Prediction Tool
AWTS	Atlantic Wind Test Site (Canada)

B

BEM	Blade element momentum (method)
BJT	Bipolar junction transistor

C

CAD	Computer-aided design
CANWea	Canadian Wind Energy Association
CA-OWEA	Concerted Action on Offshore Wind Energy in Europe
CBA	Cost–benefit analysis
CEC	California energy Commission
CEDRL	CANMET Energy Diversification Research Laboratory
CENELEC	Comité Européen de Normalisation Electrotechnique
CF	Capacity factor
CGH_2	Compressed gaseous hydrogen
CHP	Combined heat and power (also known as co-generation)
CIGRÉ	Conseil International des Grands Réseaux Électriques
COE	Cost of energy

Wind Power in Power Systems   Edited by T. Ackermann
© 2005 John Wiley & Sons, Ltd   ISBN: 0-470-85508-8 (HB)

CP	Connection point
CRES	Centre for Renewable Energy Sources
CSC	Current source converter

D

DANIDA	Danish International Development Agency
DC	Direct current
DEFU	Research Institute for Danish Electric Utilities (also translated as Danish Utilities Research Association)
DEWI	German Wind Energy Institute
DFIG	Doubly fed induction generator
DG	Distributed generation
DKK	Danish Crowns
DMI	Danish Meteorological Institute
DNC	Distribution network company
DR	Distributed resources
DRE	Distributed renewable energy
DRES	Distributed renewable energy systems
DS	Distribution system
DSB	Demand-side bidding
DSM	Demand-side management
DTU	Technical University of Denmark
DWD	Deutscher Wetterdienst (German Weather Service)

E

EDF	Electricité de France
EEG	Renewable Energy Sources Act (Germany)
EFI	Electric Power Research Institute (Norway)
EHV	Extra high voltage
EMTP	Electromagnetic transients program
EPS	Ensemble Prediction System
ER	Engineering recommendation
ESB	Electricity Supply Board (Republic of Ireland)
ESBNG	Electricity Supply Board National Grid (Republic of Ireland)
ETR	Engineering technical report
EU	European Union
EU-15	European Union 15 Member States
EUHYFIS	European Hydrogen Filling Station
EWEC	European Wind Energy Conference

F

F	Filter
FACTS	Flexible AC transmission systems
FC	Fuel cell
FGW	Fördergesellschaft Windenergie (Germany)

G

GC	General curtailment
GEAB	Gotlands Energi AB
GEB	Gujarat Electricity Board
GEDA	Gujarat Energy Development Agency
GIS	Geographical information system
GSP	Grid supply point
GTO	Gate turn-off thyristor

H

HFF	High-frequency filter
HHV	Higher heating value
HIRLAM	High Resolution Limited Area Model
HPP	Hydro power plant
HS	High-speed (shaft)
HV	High-voltage
HVAC	High-voltage alternating-current
HVDC	High-voltage direct-current
HVG	High-voltage generator

I

IC	Installed capacity
IEC	International Electrotechnical Commission
IEEE	Institute of Electrical and Electronic Engineers
IG	Induction generator
IGBT	Insulated gate bipolar transistor
IGCT	Integrated gate commutated thyristor
IM	Induction motor
IMM	Department of Informatics and Mathematical Modelling (Technical University of Denmark)
IREQ	Insitut de Recherche D'Hydro-Québec
IRL	Ireland
IRR	Internal rate of return
ISET	Institüt für Solare Energieversorgnungstechnik
ISO	Independent system operator (also commonly used for the International Organisation for Standardisation, Geneva)
IVS	Instantaneous value simulation

K

KTH	Royal Institute of Technology, Stockholm, Sweden

L

LCC	Line-commutated converter
LF	Load flow
LHV	Lower heating value
LH_2	Liquid hydrogen

LM	Local Model (Lokal-Modell)
LOEE	Loss of energy expectation
LOLE	Loss of load expectation
LOLP	Loss of load probability
LS	Low-speed (shaft)
LV	Low-voltage
LYSAN	Linear System analysis (program module)

M

MARS	Market Simulation Tool (Eltra)
MASS	Mesoscale Atmospheric Simulation System
MCFC	Molten carbonate fuel cell
MOB	Man overboard (boot)
MOS	Model output statistics
MOSFET	Metal oxide semiconductor field effect transistor
MSEK	Million Swedish crowns (krona)
MV	Medium-voltage

N

NERC	North American Electricity Reliability Council
NETA	New Electricity Trading Arrangement (UK)
NFFO	The Non-Fossil Fuel Obligation (UK)
NL	The Netherlands
NOIS	Nordic Operational Information System
NPV	Net present value
NREA	New and Renewable Energy Agency (Egypt)
NREL	National Renewable Energy Laboratory (USA)
NTC	Net transmission capacity
NTP	Normal temperature and pressure
NWP	Numerical weather prediction

O

OLTC	On-load tap-changing (transformer)
OM&R	Operation, maintenance and repair
OPF	Optimal power flow
OSIG	OptiSlip^{TM} induction generator (OptiSlip^{TM} is a registered trademark of Vestas Wind Systems A/S)
OSS	Offshore substation

P

PAFC	Phosphoric acid fuel cell
PAM	Pulse amplitude modulated
PAS	Publicly available specification
PCC	Point of common coupling
PDF	Probability density function
PE	Power exchange

PEFC	Polymer electrolyte fuel cell
PF	Power factor
PG&E	Pacific Gas and Electric (Company)
PI	Proportional–integral (controller)
PMG	Permanent magnet generator
PMSG	Permanent magnet synchronous generator
PPA	Power purchase agreement
PPP	Power purchase price
PQ	Power quality
PSDS	Power system dynamics simulation
PSLF	Load Flow Program (GE)
PSS/E^{TM}	Power System Simulator for Engineers [PSS/E^{TM} is a registered trademark of Shaw Power Technologies Inc. (PTI)]
PTC	Production Tax Credit (USA)
PTI	Shaw Power Technologies Inc.
p.u.	Per unit
PURPA	Public Utility Regulatory Policies Act (USA)
PV	Photovoltaic
PWM	Pulse-width modulated; pulse-width modulation
PX	Power exchange

R

RAL	Rutherford Appleton Laboratory (UK)
REC	Renewable energy credit
REE	Red Eléctrica de España
RERL-UMASS	Renewable Energy Research Lab at UMass Amherst (USA)
RES	Renewable energy source(s)
RMS	Root mean square
RMSE	Root mean square error
RO	Reverse osmosis
RPM	Rounds per minute; rotations per minute
RPS	Renewable portfolio standard
RWTH	Rheinisch-Westfälische Technische Hochschule Aachen, Germany (Institute of Technology of the Land North Rhine-Westphalia)

S

SAFT	SAFT Batteries SA
SC	Short-circuit
SCADA	Supervisory control and data acquisition
SCE	Southern California Edison (Company)
SCIG	Squirrel cage induction generator
SCR	Short-circuit ratio
SG	Synchronous generator
SIL	Surge impedance loading
SIMPOW	Simulation of Power Systems
SOFC	Solid oxide fuel cell

SRG	Switch reluctance generator
STATCOM	Static VAR compensator
STATCON	Static VAR converter
STD	Standard deviation
STP	Standard temperature and pressure
SVC	Static VAR compensator
SvK	Svenka Kraftnät (Swedish transmission system operator)
SW	Switch

T

TDC	Transmission duration curve
TFG	Transverse flux generator
THD	Total harmonic distortion
TL	Transmission limit
TNEB	Tamil Nadu Electricity Board
TSO	Transmission system operator
TSP	Transient stability program
Type A	Fixed-speed wind turbine, with asynchronous induction generator directly connected to the grid, with or without reactive power compensation (see also Section 4.2.3)
Type B	Limited variable-speed wind turbine with variable generator rotor resistance (see also Section 4.2.3)
Type C	Variable-speed wind turbine with doubly-fed asynchronous induction generators and partial-load frequency converter on the rotor circuit (see also Section 4.2.3)
Type D	Full variable-speed wind turbine, with asynchronous or synchronous induction generator connected to the grid through a full-load frequency converter (see also Section 4.2.3)

U

UCTE	Union pour la Coordination du Transport d'Electricité (Union for the Coordination of Transmission of Electricity, formerly UCPTE)
UK	United Kingdom
UK MESO	UK Meteorological Office Meso-scale Model

V

VDEW	Verband der Elektrizitätswirtschaft (German Electricity Association)
VRR	Variable rotor resistance
VSC	Voltage source converter

W

WASP	Wind Atlas Analysis and Application Program
WD	Wind–diesel
WETEC	Wind Economics and TEchnology (USA)
WF	Wind farm
WPDC	Wind power production duration curve

WPMS   Wind Power Management System
WPP   Wind power production
WPPT   Wind Power Prediction Tool
WRIG   Wound rotor induction generator
WRSG   Wound rotor synchronous generator
WT   Wind turbine
WTG   Wind turbine generator

X
XLPE   Polyethylen insulation

Notation

Note: this book includes contributions from different authors who work in different fields (i.e. electrical and mechanical engineering and others). Within each of these fields, certain variables may be used for different concepts (e.g. the variable R can denote resistance and also radius, or it can represent the specific gas constant for air). It has been the editor's intention to reduce multiple definitions for one symbol. However, sometimes there will be different denotations because some variables are commonly used within different engineering disciplines. This also means that similar concepts may not be denoted with the same variable throughout the entire book.

English Symbols

A

a	Subscribed level of power
$\hat{A}_g$	Amplitude of wind speed gust
$\hat{A}_r$	Amplitude of wind speed ramp
A_R	Area of wind turbine rotor; area through which wind flows

C

c	Numerical coefficient
c_P	Power coefficient (of a wind turbine rotor)
$\cos\varphi$	Power factor
$c(\psi_k)$	Flicker coefficient for continuous operation as a function of network impedance angle
$c_c(\psi_k)$	Flicker emission factor during normal operation
$c(\Psi_k, V_a)$	Flicker coefficient for continuous operation as a function of network impedance angle and annual average wind speed
C	Capacitor; capacitance; DC-link capacitor
$C(x)$	Cost function
C_{IC}	Installed capacity
C_P	Power coefficient

Wind Power in Power Systems Edited by T. Ackermann
© 2005 John Wiley & Sons, Ltd ISBN: 0-470-85508-8 (HB)

D

d	Steady-state voltage change as a percentage of nominal voltage
D	Shaft damping constant; load level
D_n	Harmonic interference for each individual harmonic n

E

e_R	Real part of transformer impedance
e_X	Imaginary part of transformer impedance
E_{Plti}	Long-term flicker emission limit
E_{Psti}	Short-term flicker emission limit
E_1	Voltage source of doubly fed induction generator (DFIG) rotor converter or permanent magnet generator (PMG) converter
$|E_1|$	Magnitude of voltage source of DFIG rotor converter or PMG generator converter
$E_{1\alpha}$	Active component of voltage source of DFIG rotor converter or PMG generator converter
$E_{1\beta}$	Reactive component of voltage source of DFIG rotor converter or PMG generator converter
E_2	Voltage source of grid-side converter
$E_{2\alpha}$	Active component of voltage source of grid-side converter
$E_{2\beta}$	Reactive component of voltage source of grid-side converter
E_c	Expected cost
E_{load}	Average primary electrical load
E_{wind}	Wind energy output
$E(X)$	Expected value of quantity X

F

f	Frequency
$f_{free-fixed}$	Eigenfrequency of a free–fixed shaft system
$f_{free-free}$	Eigenfrequency of a free–free shaft system
f_{ps}	Pitch angle controller sample frequency
f_{ss}	Rotor speed controller sample frequency
$f(x)$	Gaussian probability function
$f_X(x)$	Probability mass function for variable X
$F(x)$	Gaussian distribution function
F_{cap}	Capacity factor
F_N	Power frequency

G

g	Scale-parameter of Weibull distribution; gravity constant
G	Generation

H

h	Wind turbine hub height; harmonic order
h_{sea}	Height above sea level
H	Inertia constant

H_g	Inertia constant of the induction generator
H_{gen}	Generator rotor inertia constant
H_m	Inertia constant of generator (mechanical)
H_{wr}	Inertia constant of wind turbine

I

i	Current
i_n	Harmonic current of order n
$i_{n,k}$	Harmonic current of order n from source k
$i(t)$	Current as a function of time
$\vec{i}_s^r$	Stator current in the rotor reference frame
$\vec{i}_r^r$	Rotor current in the rotor reference frame
$\vec{i}_s^s$	Stator current in the stator reference frame
I	Current; complex current
I_1	Rotor current of DFIG or generator current of PMG
$I_{1\alpha}$	Active rotor current of DFIG or active generator current of PMG
$I_{1\alpha,\,Ref}$	Desired active rotor current of DFIG or desired active generator current of PMG
$I_{1\beta}$	Reactive rotor current of DFIG or reactive generator current of PMG
$I_{1\beta,\,Ref}$	Desired reactive rotor current of DFIG or desired reactive generator current of PMG
I_2	Current of grid-side converter
$I_{2\alpha}$	Active current of grid-side converter
$I_{2\alpha,\,Ref}$	Desired active current of grid-side converter
$I_{2\beta}$	Reactive current of grid-side converter
$I_{2\beta,\,Ref}$	Desired reactive current of grid-side converter
I_h	Maximum harmonic current
I_L	Phase current
I_M	Maximum current amplitude
I_{Max}	Current-carrying capacity
I_n	Rated current
I_R	Generator rotor current
I_S	Generator stator current
I_z	Capacitor current

J

J_1	Charging DC current
J_2	Discharging DC current
J_{gen}	Generator rotor inertia
J_{turb}	Wind turbine inertia

K

k	Shaft stiffness; shape parameter of the Weibull distribution
k_{base}	Base value of shaft stiffness, for use in a per unit system
k_f	Flicker step factor
k_{HS}	High-speed shaft stiffness in per unit

k_{LS}	Low-speed shaft stiffness in per unit
k_{tot}	Total shaft stiffness in per unit
$k_{\mathrm{f}}(\psi_k)$	Flicker step factor
k_i	Inrush current factor
$k_{i\Psi}(\Psi_k)$	Grid-dependent switching current factor
k_{spill}	Spilled wind energy in percent of wind energy production
$k_{\mathrm{u}}(\psi_k)$	Voltage change factor
K	Shaft torsion constant
K_{p}	Pitch angle controller constant
K_{HS}	High-speed shaft stiffness
K_{LS}	Low-speed shaft stiffness
K_{s}	Shaft stiffness
K_{v}	Voltage controller constant

L

l	Turbulence length scale; transmission line length
L	Inductance
L_{fd}	Field inductance
L_{m}	Mutual inductance
$L_{\sigma r}$	Rotor leakage inductance
$L_{\sigma s}$	Stator leakage inductance

M

m_{G}	Mean value of Gaussian distribution
m_x	Mean inflow (hydro power)
m_y	Mean inflow (renewable power source)

N

n	Number of points
n_{gear}	Gear ratio
n_{pp}	Number of pole pairs
N	Maximum number of switchings
N_{10}	Maximum number of one type of switchings within a 10-minute period
N_{120}	Maximum number of one type of switchings within a 120-minute period
N_{wt}	Number of wind turbines

P

p	Number of generator poles; pressure
$p(t)$	Power as a function of time
P	Active power; wind power production; probability
P_2	Electric power of grid-side converter
$P_{0.2}$	Maximum measured power (0.2-second average value)
P_{60}	Maximum measured power (60-second average value)
P_{base}	Power base value for use in a per unit system
P_{D}	Power spectral density; power consumption
P_{Dt}	Power spectral density of turbulence

P_G	Additional required power production; generator electric power
$P_G{}^{curt}$	Curtailed active power
$P_G{}^{max}$	Maximum generated power
$P_{G,Ref}$	Desired generator electric power
P_L	Power losses on the line; active power of load
P_{load}	Instantaneous primary electrical load
P_{lt}	Long-term flicker disturbance factor
P_m	Mean power production
P_{mc}	Maximum permitted power
P_{MECH}	Mechanical power of the wind turbine
P_n	Rated power of wind turbine
P_o	Natural load of the line
P_O	Power of moving air mass; standard sea level atmospheric pressure
P_R	Active power at the receiving end of the line; rated power
P_{REF}	Generator electric active power reference
P_S	Power delivered from kinetic energy stored in the rotating mass (turbine, shaft and rotor)
P_{st}	Short-term flicker disturbance factor
P_T	Power delivered by turbine
P_{TL}	Transmission limit
$P_{TOC}(t)$	Power at time t on the transmission duration curve
P_W	Wind power production
P_{wind}	Instantaneous wind power output
P_{WIND}	Power of the wind within the rotor swept area of the wind turbine
$P_{WPDC}(t)$	Power at time t on the wind power production duration curve
P_{wt}	Power extracted from the wind
P_x	Active power of system component x
P_{year}	Yearly mean price
$P(X = x)$	Probability that variable X is equal to x

Q

Q	Reactive power
Q_2	Reactive power of grid-side converter
$Q_{2,Ref}$	Desired reactive power of grid-side converter
Q_C	Capacitor reactive power; reactive power of compensation device
Q_{Comp}	Reactive power of compensation device
Q_G	Generator reactive power
$Q_{G,Ref}$	Desired generator reactive power
Q_{import}	Reactive power absorbed by the network
Q_L	Reactive power of load
Q_n	Rated reactive power of a wind turbine
Q_{REF}	Generator electric reactive power reference
Q_x	Reactive power of system component x

R

r	Length along blade (local radius), measured from the hub ($r = 0$)
r^r	Rotor winding resistance
r^s	Stator winding resistance
r_{xy}	Cross-correlation
R	Resistance; blade length; wind turbine radius; specific gas constant for air; radius of a cylinder
R_2	Resistance of smoothing inductor
R_{fd}	Field resistance
R_{gas}	Gas constant
R_k	Short-circuit ratio at the point of connection
R_r	Rotor resistance
R_s	Stator resistance

S

s	Slip; complex frequency
S	Apparent power; complex power
S_{base}	Base power, in per unit system
S_n	Rated apparent power of a wind turbine
S_{park}	Nominal apparent power of wind farm
S_{rG}	Generator rated apparent power
S_k	Short circuit power
S_N	Nominal three-phase power of the induction generator
S_T	Transformer power capacity
$\overline{S}, \overline{S}(t)$	Space vector
$S_a(t)$	Instantaneous (momentaneous) value of a quantity for phase a
$S_b(t)$	Instantaneous (momentaneous) value of a quantity for phase b
$S_c(t)$	Instantaneous (momentaneous) value of a quantity for phase c
$S_d, S_d(t)$	Direct axis of a space vector $\overline{S}(t)$
$S_q, S_q(t)$	Quadrature axis of a space vector $\overline{S}(t)$
$\overline{S}_r, \overline{S}_r(t)$	Space vector $\overline{S}(t)$ referred to the rotor reference frame
$S_{r,a}$	Instantaneous (momentaneous) value of a quantity for phase a in rotor reference frame
$S_{r,b}$	Instantaneous (momentaneous) value of a quantity for phase b in rotor reference frame
$S_{r,c}$	Instantaneous (momentaneous) value of a quantity for phase c in rotor reference frame
S_{rd}	Real part of $\overline{S}_r$
S_{rq}	Imaginary part of $\overline{S}_r$
$\overline{S}_s, \overline{S}_s(t)$	Space vector $\overline{S}(t)$ referred to the stator reference frame
$S_{s,d}$	Real part of $\overline{S}_s$
$S_{s,q}$	Imaginary part of $\overline{S}_s$

T

t	Time
$t \circ_C$	Temperature

t_{TL}	Number of hours over which the transmission limit is exceeded
T	Time; torque; temperature
T_1	Cost for subscription per kW per year
T_2	Price of excess power per kW per year
T_{base}	Base value of torque, in per unit system
$T_{damping}$	Damping torque of the shaft
T_e	Torque of generator (electrical)
T_{eg}	End time of wind speed gust
T_{el}	Electrical air gap torque of the induction generator
T_{er}	End time of wind speed ramp
T_h	Time period
T_m	Mechanical torque produced by the wind turbine
T_{MECH}	Mechanical torque of the wind turbine
T_p	Duration of voltage variation caused by a switching operation
T_{sg}	Start time of wind speed ramp
T_{shaft}	Incoming torque from the shaft connecting the induction generator with the wind turbine
$T_{torsion}$	Elasticity torque of the shaft
T_{wr}	Torque of wind turbine

U

u	Voltage; integration variable
$u(t)$	Voltage as a function of time
u_N	Nominal voltage
$\bar{u}_r^s$	Stator voltage in the rotor reference frame
$\bar{u}_r^r$	Rotor voltage in the rotor reference frame
$\bar{u}_s^s$	Stator voltage in the stator reference frame
U	Voltage
U_1	Fixed voltage at the end of the power system
U_2	Terminal voltage
$U_{2\alpha}$	Terminal voltage magnitude
$U_{2\alpha,Ref}$	Desired terminal voltage magnitude
U_{DC}	DC-link voltage
$U_{DC,Ref}$	Desired DC-link voltage
U_h	Harmonic voltage
U_i	Voltage of bus i
U_i^{min}	Minium voltage at bus i
U_{LL}	Phase-to-phase voltage
U_M	Maximum voltage amplitude
U_{max}	Maximum voltage
U_{min}	Minium voltage
U_N	Root mean square value of the phase-to-phase voltage of a machine
U_n	Nominal phase-to-phase voltage
U_R	Generator rotor voltage; voltage at the receiving end
U_{REF}	Generator stator voltage reference

U_S	Generator stator voltage; voltage at the sending end
U_t	Terminal voltage

V

v	Wind speed
v_a	Annual average wind speed
v_{ci}	Cut-in wind speed
v_R	Rated wind speed
v_t	Rotor tip speed
v_w	Wind speed at hub height of the power system
$v_w(t)$	Wind speed at time t
v_{wa}	Average value of wind speed
v_{wg}	Gust component of wind speed
$v_{wr}(t)$	Ramp component of wind speed
v_{wt}	Turbulence component of wind speed
V	Wind speed
V_{rel}	Relative wind speed
V_{tip}	Speed of the blade tip
V_{WIND}	Wind speed
$V(x)$	Value of water inflow

W

W	Yearly energy production
W_{spill}	Spilled energy
W_W	Energy generated by wind farm

X

x	Power transmission; water inflow
x_l^r	Leakage reactance of the rotor
x_l^s	Leakage reactance of the stator
x^m	Mutual reactance between the stator and rotor windings
x^r	Rotor self-reactance
x^s	Stator self-reactance
X	Reactance; power flow; cable length
X_2	Reactance of smoothing inductor
X_D	Direct axis reactance
X_Q	Quadrupol axis reactance

Y

Y	Generated wind power
Y_e	Lumped admittance of long transmission line

Z

z	Altitude above sea level
z_0	Roughness length
Z	Impedance; desired power transmission

Z_c	Transmission line surge impedance
Z_e	Lumped impedance of long transmission line
Z_l	Load impedance
Z_L	Line impedance
Z_{LD}	Load impedance

Greek symbols

α	Exponent; reference angle
β	Pitch angle; reference angle
β_{const}	Fixed blade angle
β_{ref}	Blade reference angle
γ	Torsional displacement between shaft ends
ΔP	Differences between consecutive production values
$d\beta/dt$	Pitch speed
θ	Pitch angle
θ_{base}	Base value of pitch angle, for use in per unit system
$\theta_{base,el}$	Electrical base angle
θg	Generator rotor angle
φ	Angle between terminal voltage and current
φ_{LD}	Load angle
$\cos\varphi_{LD}$	Load power factor
λ	Tip-speed ratio
λ_{opt}	Tip-speed ratio corresponding to the optimal rotor speed
μ	Average
ρ	Air density
ρ_{AIR}	Air density
ρ_{xy}	Correlation between x and y
$\rho(z)$	Air density as a function of altitude
σ	Standard deviation of Gaussian distribution; total leakage factor
σ_{load}	Standard deviation of load time series
σ_m	Position of wind turbine
σ_{total}	Standard deviation of net load time series
σ_{wind}	Standard deviation of wind power production time series
τ_V	Characteristic time constant of induced velocity lag
φ	Angle of incidence; phase angle
φ_1	Phase angle of voltage source of DFIG rotor converter or PMG converter
$\cos\varphi_{min}$	Expected minimum power factor at full load
ψ	Flux linkage
ψ_{base}	Base flux
ψ_k	Network impedance phase angle
ψ_{pm}	Amount of flux of the permanent magnets mounted on the rotor that is coupled to the stator winding
ψ_{rd}^s	Direct axis of the stator flux in rotor reference frame
ψ_{rq}^s	Quadrature axis of the stator flux in rotor reference frame

ψ^{r}_{rd}	Direct axis of the rotor flux in rotor reference frame
ψ^{r}_{rq}	Quadrature axis of the rotor flux in rotor reference frame
$\underline{\psi}^{\mathrm{s}}_{s}$	Stator flux in the stator reference frame
$\underline{\psi}^{\mathrm{r}}_{s}$	Stator flux in the rotor reference frame
$\underline{\psi}^{\mathrm{r}}_{r}$	Rotor flux in the rotor reference frame
ω	Angular frequency; angular speed
ω_{base}	Base value of rotational speed, in per unit system
$\omega_{\mathrm{base, el}}$	Electrical base angular speed
ω_{g}	Speed of machine
ω_{G}	Generator rotor speed
$\omega_{\mathrm{G, Ref}}$	Disired generator rotor speed
ω_{gen}	Generator rotor rotational speed
ω_{m}	Angular speed of the wind turbine; angular frequency of generator (mechanical)
ω_{N}	Angular speed (e.g. $2\pi f_{N}$)
ω_{turb}	Turbine rotational speed
$\omega_{\mathrm{turb, opt}}$	Optimal turbine rotational speed
ω_{wr}	Angular frequency of wind turbine

Units

SI Units

Basic unit	Name	Symbol
Length	meter	m
Mass	kilogram	kg
Time	second	s
Electric current	ampere	A
Temperature	kelvin	K

SI-derived units

Unit	Name	Unit Symbol
Area	square meter	m^2
Volume	cubic meter	m^3
Speed or velocity	meter per second	m/s
Acceleration	meter per second squared	m/s^2
Density	kilogram per cubic meter	kg/m^3
Specific volume	cubic meter per kilogram	m^3/kg
Current density	ampere per square meter	A/m^2
Magnetic field strength	ampere per meter	A/m

Derived units with special names and symbols

Unit	Name	Unit Symbol	In SI Units
Frequency	hertz	Hz	s^{-1}
Force	Newton	N	$m\,kg\,s^{-2}$
Pressure	pascal	Pa	$m^{-1}\,kg\,s^{-2}$
Energy, work, quantity of heat	joule	J	$m^2\,kg\,s^{-2}$
Power	watt	W	$m^2\,kg\,s^{-3}$

Wind Power in Power Systems Edited by T. Ackermann
© 2005 John Wiley & Sons, Ltd ISBN: 0-470-85508-8 (HB)

Electromotive force	volt	V	$m^2\,kg\,s$
Apparent power	volt ampere	VA	$m^2\,kg\,s^{-3}$
Reactive power	var	var	$m^2\,kg\,s^{-3}$
Capacitance	farad	F	$m^{-2}\,kg^{-1}\,s^4\,A^2$
Electric resistance	ohm	Ω	$m^2\,kg\,s^{-3}\,A^{-2}$
Electric conductance	siemens	S	$m^{-2}\,kg^{-1}\,s^3\,A^2$
Magnetic flux	weber	Wb	$m^2\,kg\,s^{-2}\,A^{-1}$
Magnetic flux density	tesla	T	$kg\,s^{-2}\,A^{-1}$
Inductance	henry	H	$m^2\,kg\,s^{-2}\,A^{-2}$

SI prefixes

Prefix	Symbol	Value	Prefix	Symbol	Value
Atto	a	10^{-18}	Kilo	k	10^3
Femto	f	10^{-15}	Mega	M	10^6
Pico	p	10^{-12}	Giga	G	10^9
Nano	n	10^{-9}	Tera	T	10^{12}
Micro	μ	10^{-6}	Peta	P	10^{15}
Milli	m	10^{-3}	Exa	E	10^{18}

1

Introduction

Thomas Ackermann

Wind energy is gaining increasing importance throughout the world. This fast development of wind energy technology and of the market has large implications for a number of people and institutions: for instance, for scientists who research and teach future wind power, and electrical engineers at universities; for professionals at electric utilities who really need to understand the complexity of the positive and negative effects that wind energy can have on the power system; for wind turbine manufacturers; and for developers of wind energy projects, who also need that understanding in order to be able to develop feasible, modern and cost-effective wind energy projects.

Currently, five countries – Germany, USA, Denmark, India and Spain – concentrate more than 83 % of worldwide wind energy capacity in their countries. Here, we also find most of the expertise related to wind energy generation and its integration into the power system in those countries. However, the utilisation of this renewable source of power is fast spreading to other areas of the world. This requires the theoretical knowledge and practical experience accumulated in the current core markets of wind energy to be transferred to actors in new markets. A main goal of this book is to make this knowledge available to anybody interested and/or professionally involved in this area.

The utilisation of wind energy has a tradition of about 3000 years, and the technology has become very complex. It involves technical disciplines such as aerodynamics, structural dynamics and mechanical as well as electrical engineering. Over past years a number of books on aerodynamics and the mechanical design of wind power have been published. There is, however, no general publication that discusses the integration of wind power into power systems. This books aims to fill this gap.

I first realised the need for such a book in 1998, shortly after arriving at the Royal Institute of Technology in Stockholm. There I met Lawrence Jones who wrote his PhD

Wind Power in Power Systems Edited by T. Ackermann
© 2005 John Wiley & Sons, Ltd ISBN: 0-470-85508-8 (HB)

on high-voltage direct-current (HVDC) technology. We had long discussions on possible applications of HVDC technology for offshore wind farms. The more we discussed, the more questions there were. As a result, in 2000 we organised a workshop on the topic of 'HVDC Transmission Networks for Offshore Wind Farms'. This workshop turned out to be a successful forum for the discussion of this subject, resulting in the decision to hold workshops on the same subject in 2001 and 2002. The discussions during these workshops became broader and so did the subject of the workshop. Hence, in 2003 the workshop was entitled the 'Fourth International Workshop on the Large-Scale Integration of Wind Power and Transmission Networks for Offshore Wind Farms'. That time, the co-organiser was Eltra, the transmission system operator of Western Denmark, and 175 participants from academia and industry attended the workshop.[1]

During the workshops it became clear that the subject of wind power in power systems met an increased general interest. In order to satisfy this interest, the initial idea was simply to summarise the papers from the workshop. This turned out to be more complicated than initially assumed. Designing a publication that can be of interest to a wider readership, including professionals in the industry, authorities and students, was not easy. Another challenge was to keep the content to a large extent consistent. Finally, I wanted to include not only papers from the workshop but also contributions from other authors who are renowned researchers in this field.

The final version of the book now comprises four parts. Part A aims to present basic theoretical background knowledge. Chapter 2 gives a brief overview of the historical development and current status of wind power, and Chapter 3 provides a brief introduction to wind power in power systems. Here, Chapter 4, which was written by Anca-Daniela Hansen, is central to the entire book as it presents an overview of current wind turbine designs. Throughout the book, the authors refer to wind turbine designs (types A, B, C and D) from Chapter 4 and do not describe them in the individual chapters again. In addition, this part of the book presents power quality standards (Chapter 5), power quality measurements (Chapter 6), network interconnection standards (Chapter 7) as well as a general discussion of power system requirements regarding wind power (Chapter 8) and of the value that wind power contributes to a power system (Chapter 9).

Part B showcases practical international experience regarding the integration of wind power. It starts with contributions from Eltra, the transmission system operator (TSO) in Western Denmark (Chapter 10) and the German TSO E.on Netz (Chapter 11). These

─────────────

[1] Thanks to the members of the International Advisory Committee of the first four workshops: Göran Andersson (Swiss Federal Institute of Technology, Zurich, Switzerland), Gunnar Asplund (ABB, Sweden), Peter Christensen (NVE, Denmark), Paul Gardner (Garrad Hassan , Siegfried Heier (University of Kassel, ISET, Germany), Hans Knudsen (Danish Energy Agency), Lawrence Jones (University of Washington, Seattle, USA), James Manwell (University of Massachusetts, Amherst, USA), Patrice Noury (Alstom, France) and Lennart Söder (Royal Institute of Technology, Stockholm, Sweden).

Thanks also to the following persons, who helped with or supported the organisation of the workshop in various ways over the years: Peter Bennich, Lillemor Hyllengren, Lawrence Jones, Valery Knyazkin, Magnus Lommerdal, Jonas Persson, Julija Matevosyan, Lennart Söder, Erik Thunberg (at the time all members of the Department of Electrical Engineering, Royal Institute of Technology, Stockholm, Sweden), Jari Ihonen (Royal Institute of Technology, Stockholm, Sweden), Lawrence Jones (University of Washington, Seattle, USA), Jens Hobohm (Prognos AG, Germany), Ralf Leutz (at that time with the Tokyo University of Agriculture and Technology) and Peter Børre Eriksen, Gitte Agersæk and John Eli Nielsen (all at Eltra, Denmark).

are the TSOs that probably have to deal with the largest wind power penetration worldwide. The discussion of the situation in California (Chapter 12) and on the Swedish island of Gotland (Chapter 13) focuses on the integration of comparatively simple wind turbines. Practical experience from wind power in isolated systems is presented in Chapter 14 and from developing countries such as India in Chapter 15. A more general discussion on practical experience regarding power quality and wind power is presented in Chapter 16. Part B also includes chapters on current issues regarding wind power forecasting (Chapter 17). Finally, we present economic issues that have arisen in the integration of wind power in the deregulated electricity industry (Chapter 18).

Part C discusses future concepts related to an increasing penetration level of wind power in power systems. The issues cover voltage control (Chapter 19), transmission congestion (Chapter 20) and the active management of distribution systems (Chapter 21). Additionally, this part discusses transmission solutions for offshore wind farms (Chapter 22) and the use of hydrogen as an alternative means of transporting wind power (Chapter 23).

Finally, Part D shows how dynamic modelling is used to study the impact of the large-scale integration of wind power. As a start, general wind power modelling issues are presented and discussed (Chapter 24). This is followed by chapters on low-order models (Chapter 25) and high-order models (Chapter 26) for wind turbines as well as on the full verification of dynamic wind turbine models (Chapter 27). The impact of wind power on power system dynamics is presented in Chapter 28, and the last chapter of the book (Chapter 29) discusses aggregated wind turbine models that represent a whole wind farm rather than a single wind turbine.

Owing to the large number of contributors it has not always been possible to avoid overlaps between chapters. Even though I have tried to limit them, the existing overlaps show that there may be diverging opinions regarding individual subjects. The careful reader will certainly notice these overlaps and sometimes even contradictions. There remains a substantial amount of research to be done and experience to be gathered in order to arrive at a more consistent picture.

Initially, it was my intention to win contributors not only from academia and TSOs but also from wind turbine manufacturers, as these have valuable experience to share. However, with the exception of one wind turbine manufacturer, the design of this book was considered to be too 'academic' by those I approached. In my opinion, wind turbine manufacturers have been developing and introducing interesting solutions for the integration of wind power into power systems and should present such solutions in any possible future edition of this book.

I would like to thank all authors of the individual chapters for supporting this time-consuming project. I would also like to thank Kathryn Sharples, Simone Taylor, Emily Bone, Lucy Bryan, Rachael Catt and Claire Twine from Wiley for their continuous support and great patience, and Dörte Müller from Powerwording.com for her language editing, which has improved the book's general readability. I would also like to thank Professor Lennart Söder and the entire Department of Electrical Engineering at the Royal Institute of Technology, Stockholm, Sweden. Special thanks go also to Göran Andersson, now with the Swiss Federal Institute of Technology, Zurich, Switzerland, who was very open to the initial idea of holding workshops on these subjects. He also provided valuable comments on the workshops and this book.

I hope that the book proves to be a useful source of information and basis for discussion for readers with diverse backgrounds.

In connection with this publication, the editor will introduce a website (http://www.windpowerinpowersystems.info) with more information regarding this book, a discussion group and information on forthcoming workshops and other events.

Part A

Theoretical Background and Technical Regulations

2

Historical Development and Current Status of Wind Power

Thomas Ackermann

2.1 Introduction

The power of the wind has been utilised for at least 3000 years. Until the early twentieth century wind power was used to provide mechanical power to pump water or to grind grain. At the beginning of modern industrialisation, the use of the fluctuating wind energy resource was substituted by fossil fuel fired engines or the electrical grid, which provided a more consistent power source.

In the early 1970s, with the first oil price shock, interest in the power of the wind re-emerged. This time, however, the main focus was on wind power providing electrical energy instead of mechanical energy. This way, it became possible to provide a reliable and consistent power source by using other energy technologies – via the electrical grid – as a backup.

The first wind turbines for electricity generation had already been developed at the beginning of the twentieth century. The technology was improved step by step from the early 1970s. By the end of the 1990s, wind energy has re-emerged as one of the most important sustainable energy resources. During the last decade of the twentieth century, worldwide wind capacity doubled approximately every three years. The cost of electricity from wind power has fallen to about one sixth of the cost in the early 1980s. And the trend seems to continue. Some experts predict that the cumulative capacity will be growing worldwide by about 25 % per year until 2005 and costs will be dropping by an additional 20 to 40 % during the same time period (*Wind Power Monthly* 1999).

Wind Power in Power Systems   Edited by T. Ackermann
© 2005 John Wiley & Sons, Ltd   ISBN: 0-470-85508-8 (HB)

Table 2.1 Development of wind turbine size between 1985 and 2004 (Reproduced by permission of John Wiley & Sons, Ltd.)

Year	Capacity (kW)	Rotor diameter (m)
1985	50	15
1989	300	30
1992	500	37
1994	600	46
1998	1500	70
2003	3000–3600	90–104
2004	4500–5000	112–128

Wind energy technology itself also moved very fast in new dimensions. At the end of 1989 a 300 kW wind turbine with a 30-meter rotor diameter was state of the art. Only 10 years later, 2000 kW turbines with a rotor diameter of around 80 meters were available from many manufacturers. The first demonstration projects using 3 MW wind turbines with a rotor diameter of 90 meter were installed before the turn of the century. Now, 3 to 3.6 MW turbines are commercially available. By the time of writing (early 2004), 4–5 MW wind turbines are under development or have already been tested in demonstration projects (see also Table 2.1), and 6–7 MW turbines are expected to be built in the near future.

In Section 2.2 a more detailed historical overview of wind power development is provided; in Section 2.3 the current status is presented and in Section 2.4 a brief introduction to different wind turbine designs is given. The current status regarding network integration (e.g. penetration levels in different countries) will be presented in Chapter 3.

2.2 Historical Background

The following historical overview divides the utilisation of the natural resource wind into the generation of mechanical power and the production of electricity (The historical development of wind turbine technology is documented in many publications, for instance see Ancona, 1989; Gipe, 1995; Heymann, 1995; Hill, 1994; Johnson, 1985; Kealey, 1987; Koeppl, 1982; Putnam, 1948; Righter, 1996; Shepherd, 1990, 1994).

2.2.1 Mechanical power generation

The earliest windmills recorded were vertical axis mills. These windmills can be described as simple drag devices. They have been used in the Afghan highlands to grind grain since the seventh century BC.

The first details about horizontal axis windmills are found in historical documents from Persia, Tibet and China at about 1000 AD. This windmill type has a horizontal shaft and blades (or sails) revolving in the vertical plane. From Persia and the Middle East, the horizontal axis windmill spread across the Mediterranean countries and Central Europe. The first horizontal axis windmill appeared in England around 1150, in France in 1180, in Flanders in 1190, in Germany in 1222 and in Denmark in 1259. This fast development was most likely influenced by the Crusaders, taking the knowledge about windmills from Persia to many places in Europe.

In Europe, windmill performance was constantly improved between the twelfth and nineteenth centuries. By the end of the nineteenth century, the typical European windmill used a rotor of 25 meters in diameter, and the stocks reached up to 30 meters. Windmills were used not only for grinding grain but also for pumping water to drain lakes and marshes. By 1800 about 20 000 modern European windmills were in operation in France alone, and in the Netherlands 90 % of the power used in the industry was based on wind energy. Industrialisation then led to a gradual decline in windmills, but in 1904 wind energy still provided 11 % of the Dutch industrial energy and Germany had more than 18 000 installed units.

When the European windmills slowly started to disappear, windmills were introduced by settlers in North America. Small windmills for pumping water to livestock became very popular. These windmills, also known as American Windmills, operated fully self-regulated, which means they could be left unattended. The self-regulating mechanism pointed the rotor windward during high-speed winds. The European style windmills usually had to be turned out of the wind or the sailing blades had to be rolled up during extreme wind speeds, to avoid damage to the windmill. The popularity of windmills in the USA reached its peak between 1920 and 1930, with about 600 000 units installed. Various types of American Windmills are still used for agricultural purposes all over the world.

2.2.2 Electrical power generation

In 1891, the Dane Poul LaCour was the first to build a wind turbine that generated electricity. Danish engineers improved the technology during World Wars 1 and 2 and used the technology to overcome energy shortages. The wind turbines by the Danish company F. L. Smidth built in 1941–42 can be considered forerunners of modern wind turbine generators. The Smidth turbines were the first to use modern airfoils, based on the advancing knowledge of aerodynamics at this time. At the same time, the American Palmer Putnam built a giant wind turbine for the American company Morgan Smith Co., with a diameter of 53 meters. Not only was the size of this machine significantly different but also the design philosophy differed. The Danish philosophy was based on an upwind rotor with stall regulation, operating at slow speed. Putnam's design was based on a downwind rotor with variable pitch regulation. Putnam's turbine, however, was not very successful. It was dismantled in 1945. See Table 2.2 for an overview of important historical wind turbines.

After World War 2, Johannes Juul in Denmark developed the Danish design philosophy further. His turbine, installed in Gedser, Denmark, generated about 2.2 million kWh between 1956 and 1967. At the same time, the German Hütter developed a new approach. His wind turbine comprised two slender fibreglass blades mounted downwind of the tower on a teetering hub. Hütter's turbine became known for its high efficiency.

Table   2.2 Historical wind turbines

Turbine and country	Diameter (m)	Swept Area (m²)	Power (kW)	Specific Power (kW/m²)	Number of blades	Tower height (m)	Date in service
Poul LaCour, Denmark	23	408	18	0.04	4	—	1891
Smith-Putnam, USA	53	2231	1250	0.56	2	34	1941
F. L. Smidth, Denmark	17	237	50	0.21	3	24	1941
F. L. Smidth, Denmark	24	456	70	0.15	3	24	1942
Gedser, Denmark	24	452	200	0.44	3	25	1957
Hütter, Germany	34	908	100	0.11	2	22	1958

Source: Gipe, 1995, page 18 (Reproduced by permission of John Wiley & Sons, Ltd.).

Despite the early success of Juul's and Hütter's wind turbines, the interest in large-scale wind power generation declined after World War 2. Only small-scale wind turbines for remote-area power systems or for battery charging received some interest. With the oil crises at the beginning of the 1970s, the interest in wind power generation returned. As a result, financial support for research and development of wind energy became available. Countries such as Germany, the USA and Sweden used this money to develop large-scale wind turbine prototypes in the megawatt range. Many of these prototypes, however, did not perform very successfully most of the time (see Table 2.3) because of various technical problems (e.g. with the pitch mechanisms).

Table   2.3 Performance of the first large-scale demonstration wind turbines

Turbine and country	Diameter (m)	Swept area (m²)	Capacity (MW)	Operating hours	Generated GWh	Period
Mod-1, USA	60	2827	2	—	—	1979–83
Growian, Germany	100	7854	3	420	—	1981–87
Smith-Putnam, USA	53	2236	1.25	695	0.2	1941–45
WTS-4, USA	78	4778	4	7 200	16	1982–94
Nibe A, Denmark	40	1257	0.63	8 414	2	1979–93
WEG LS-1, GB	60	2827	3	8 441	6	1987–92
Mod-2, USA	91	6504	2.5	8 658	15	1982–88
Näsudden I, Sweden	75	4418	2	11 400	13	1983–88
Mod-OA, USA	38	1141	0.2	13 045	1	1977–82
Tjæreborg, Denmark	61	2922	2	14 175	10	1988–93
École, Canada	64	4000	3.6	19 000	12	1987–93
Mod-5B, USA	98	7466	3.2	20 561	27	1987–92
Maglarp WTS-3, Sweden	78	4778	3	26 159	34	1982–92
Nibe B, Denmark	40	1257	0.63	29 400	8	1980–93
Tvind, Denmark	54	2290	2	50 000	14	1978–93

Source: Gipe, 1995, page 104.

Nevertheless, owing to special government support schemes in certain countries (e.g. in Denmark) further development in the field of wind energy utilisation took place. The single most important scheme was the Public Utility Regulatory Policies Act (PURPA), passed by the US Congress in November 1978. With this Act, President Carter and the US Congress aimed at an increase of domestic energy conservation and efficiency, and thereby at decreasing the nation's dependence on foreign oil. PURPA, combined with special tax credits for renewable energy systems, led to the first wind energy boom in history. Along the mountain passes east of San Francisco and northeast of Los Angeles, huge wind farms were installed. The first of these wind farms consisted mainly of 50 kW wind turbines. Over the years, the typical wind turbine size increased to about 200 kW at the end of the 1980s. Most wind turbines were imported from Denmark, where companies had developed further Poul LaCour's and Johannes Juul's design philosophy of upwind wind turbines with stall regulation. At the end of the 1980s, about 15 000 wind turbines with a capacity of almost 1500 MW where installed in California (see also Chapter 12).

At this time, the financial support for wind energy slowed down in the USA but picked up in Europe and later in India. In the 1990s, the European support scheme was based mainly on fixed feed-in tariffs for renewable power generation. The Indian approach was mainly based on tax deduction for wind energy investments. These support schemes led to a fast increase in wind turbine installations in some European countries, particularly in Germany, as well as in India.

Parallel to the growing market size, technology developed further. By the end of the twentieth century, 20 years after the unsuccessful worldwide testing of megawatt wind turbines, the 1.5 to 2 MW wind turbines had become the technical state of the art.

2.3 Current Status of Wind Power Worldwide

The following section will provide a brief overview of the wind energy status around the world at the end of the twentieth century. Furthermore, it will present major wind energy support schemes. The overview is divided into grid-connected wind power generation and stand-alone systems.

2.3.1 Overview of grid-connected wind power generation

Wind energy was the fastest growing energy technology in the 1990s, in terms of percentage of yearly growth of installed capacity per technology source. The growth of wind energy, however, has not been evenly distributed around the world (see Table 2.4). By the end of 2003, around 74 % of the worldwide wind energy capacity was installed in Europe, a further 18 % in North America and 8 % in Asia and the Pacific.

2.3.2 Europe

Between the end of 1995 and the end of 2003, around 76 % of all new grid-connected wind turbines worldwide were installed in Europe (see Tables 2.4 and 2.5). The countries with the largest installed wind power capacity in Europe are Germany, Denmark and

Table 2.4 Operational wind power capacity worldwide

Region	Installed capacity (MW) by end of year						
	1995	1997	1999	2000	2001	2002	2003
Europe	2518	4766	9307	12972	17500	21319	28706
North America	1676	1611	2619	2695	4245	4708	6677
South and Central America	11	38	87	103	135	137	139
Asia and Pacific	626	1149	1403	1795	2330	2606	3034
Middle East and Africa	13	24	39	141	147	149	150

Sources: *Wind Power Monthly*; European Wind Energy Association.

Table 2.5 Operational wind power capacity in Europe

Country	Installed capacity (MW) at end of year	
	1995	2003
Germany	1136	14609
Denmark	619	3110
Spain	145	6202
Netherlands	236	912
UK	200	649
Sweden	67	399
Italy	25	904
Greece	28	375
Ireland	7	186
Portugal	13	299
Austria	3	415
Finland	7	51
France	7	239
Norway	4	101
Luxembourg	0	22
Belgium	0	68
Turkey	0	19
Czech Republic	7	10
Poland	1	57
Russia	5	7
Ukraine	1	57
Switzerland	0	5
Latvia	0	24
Hungary	0	3
Estonia	0	3
Cyprus	0	2
Slovakia	0	3
Romania	0	1
Total	2518	28706

Sources: *Wind Power Monthly*; European Wind Energy Association.

Spain. For a map of the wind power installations in Germany see also Plate 1; for a similar map for Denmark see Plate 2.

In these countries, the main driver of wind power development has been the so-called fixed feed-in tariffs for wind power. Such feed-in tariffs are defined by the governments as the power purchase price that local distribution or transmission companies have to pay for local renewable power generation that is fed into the network. Fixed feed-in tariffs reduce the financial risk for wind power investors as the power purchase price is basically fixed over at least 10 to 15 years.

In Germany, for instance, the Renewable Energy Sources Act (EEG) defines the purchase price (feed-in tariffs) for wind energy installation in 2004 as follows: 8.8 eurocents per kWh for the first five years and 5.9 eurocents per kWh for the following years. The German government currently works at changing the EEG and the power purchase price. The aim is to introduce incentives for offshore wind power development through higher power purchase prices. At the same time, onshore wind power is expected to be forced to become more competitive by decreasing power purchase prices over the next years. It is also important to mention that the EEG and similar laws in other countries require network companies to connect wind turbines or wind farms whenever technically feasible.

More and more European countries (e.g. England, the Netherlands and Sweden) are switching to an approach that is known as 'fixed quotas combined with green certificate trading'. This approach means that the government introduces fixed quotas for utilities regarding the amount of renewable energy per year the utilities have to sell via their networks. At the same time, producers of renewable energy receive a certificate for a certain amount of energy fed into the grid. The utilities have to buy these certificates to show that they have fulfilled their obligation.

Table 2.6 presents the average size of wind turbines installed in Germany for each year between 1988 and 2003. It shows that the average size per newly installed wind turbine increased from 67 kW in 1988 to approximately 1650 kW in 2003. That means that in 2003 the typical wind turbine capacity was approximately 25 times higher than 16 years earlier. Such a development is almost comparable to that of information technology. Heavy machinery industry, however, has never experienced such a rapid development. In other European countries, the introduction of multimegawatt turbines progressed at a slower pace because of the infrastructure required for the road transport and the building equipment (e.g. cranes). By 2003, however, multimegawatt wind turbines dominated the market almost all over Europe.

Finally, it must be mentioned that the first offshore projects have materialised in Europe (see Chapter 22 for more details).

2.3.3 North America

After the wind power boom in California during the mid-1980s, development slowed down significantly in North America. In the middle of the 1990s the dismantling of old wind farms sometimes exceeded the installations of new wind turbines, which led to a reduction in installed capacity.

In 1998 a second boom started in the USA. This time, wind project developers rushed to install projects before the federal Production Tax Credit (PTC) expired on 30 June

Table 2.6 Average size of yearly new installed wind capacity in Germany

Year	Average size (kW)
1988	66.9
1989	143.4
1990	164.3
1991	168.8
1992	178.6
1993	255.8
1994	370.6
1995	472.2
1996	530.5
1997	628.9
1998	785.6
1999	935.5
2000	1114
2001	1278
2002	1394
2003	~1650

Source: German Wind Energy Institute.

1999. The PTC added $0.016–$0.017 per kWh to wind power projects for the first 10 years of a wind plant's life. Between the middle of 1998 and 30 June 1999, more than 800 MW of new wind power generation were installed in the USA. That includes between 120 and 250 MW of 'repowering' development at several Californian wind farms. A similar development took place before the end of 2001, which added 1600 MW between mid-2001 and December 2001 as well as at the end of 2003, with an additional 1600 MW (see also Table 2.7). Early in 2004 the PTC was again on hold and wind power development in the USA slowed down. However, in September 2004 the PTC was renewed until the end of 2006. In addition to California and Texas, there are major projects in the states of Iowa, Minnesota, Oregon, Washington, Wyoming and Kansas (see also Table 2.8). The first large-scale wind farms have also been installed in Canada.

Table 2.7 Operational wind power capacity in North America

Country	Installed capacity (MW) by end of year	
	1995	2003
USA	1655	6350
Canada	21	327
Total	1676	6677

Sources: American Wind Power Association; *Wind Power Monthly*.

Table 2.8 Operational wind power capacity in the USA at end of 2003

State	Installed capacity (MW)
California	2042
Texas	1293
Minnesota	562
Iowa	471
Wyoming	284
Oregon	259
Washington	243
Colorado	223
New Mexico	206
Pennsylvania	129
Oklahoma	176
Kansas	113
North Dakota	66
West Virginia	66
Wisconsin	53
Illinois	50
New York	49
South Dakota	44
Hawaii	18
Nebraska	14
Vermont	6
Ohio	2
Tennessee	2
Alaska	1
Massachusetts	1
Michigan	1
Total	6350

Source: American Wind Energy Association.

The typical wind turbine size installed in North America at the end of the 1990s was between 500 and 1000 kW. In 1999, the first megawatt turbines were erected and, since 2001, many projects have used megawatt turbines. In comparison with Europe, however, the overall size of the wind farms is usually larger. In North America, typically, wind farms are larger than 50 MW, with some projects of up to 200 MW. In Europe, projects are usually in the range of 20 to 50 MW. The reason is the high population density in Central Europe and consequently the limited space. These limitations led to offshore developments in Europe. In North America, offshore projects are not a major topic.

In several states of the USA, the major driving force for further wind energy developments is the extension of the PTC as well as fixed quotas combined with green certificate trading, known in the USA as the Renewable Portfolio Standard (RPS). The certificates are called Renewable Energy Credits (RECs). Other drivers include financial incentives [e.g. offered by the California Energy Commission (CEC)] as well as green pricing programmes. Green pricing is a marketing programme offered by utilities to provide choices for electricity customers to purchase power from environmentally preferred sources. Customers thereby agree to pay higher tariffs for 'green electricity' and the utilities guarantee to produce the corresponding amount of electricity by using 'green energy sources' (e.g. wind energy).

2.3.4 South and Central America

Despite large wind energy resources in many regions of South and Central America, the development of wind energy has been very slow (Table 2.9) because of the lack of a sufficient wind energy policy as well as low electricity prices. Many wind projects in South America have been financially supported by international aid programmes. Argentina, however, introduced a new policy at the end of 1998 that offers financial support to wind energy generation, but with little success. In Brazil, some regional governments and utilities have started to offer higher feed-in tariffs for wind power. The typical size of existing wind turbines is around 300 kW. Larger wind turbines are difficult to install because of infrastructural limitations for larger equipment (e.g. cranes). Offshore wind projects are not planned, but further small to medium-size ($\leq$100 MW) projects are under development onshore, particularly in Brazil.

2.3.5 Asia and Pacific

India achieved an impressive growth in wind turbine installation in the middle of the 1990s, the 'Indian Boom'. In 1992/93, the Indian government started to offer special incentives for renewable energy investments (e.g. a minimum purchase rate was guaranteed, and a 100 % tax depreciation was allowed in the first year of the project). Furthermore, a 'power banking' system was introduced that allows electricity producers to 'bank' their power with the utility and avoid being cut off during times of load shedding. Power can be banked for up to one year. In addition, some Indian States have

Table 2.9 Operational wind power capacity in South and Central America at end of 2003

Country	Installed capacity (MW)
Costa Rica	71
Argentina	26
Brazil	22
Caribbean	13
Mexico	5
Chile	2
Total	139

Source: *Wind Power Monthly*.

Plate 1 Geographical distribution of wind power in Germany, indicating installed capacity (MW), as of January 2004. (Reproduced by permission of ISET, Kassel, Germany)

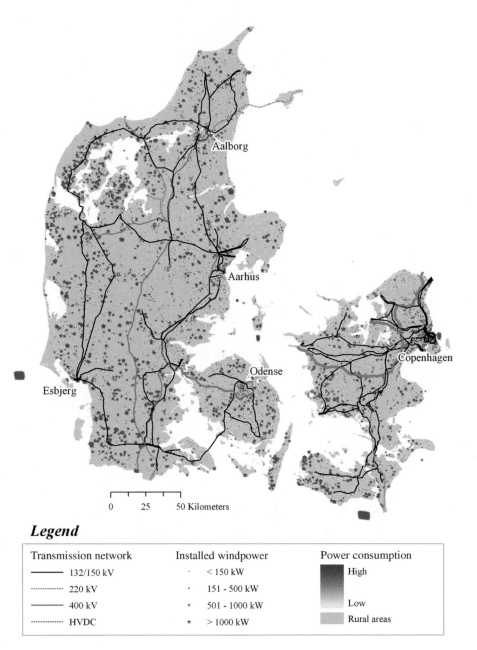

Legend

Transmission network	Installed windpower	Power consumption
——— 132/150 kV	· < 150 kW	High
·········· 220 kV	· 151 - 500 kW	
——— 400 kV	✳ 501 - 1000 kW	Low
·········· HVDC	★ > 1000 kW	Rural areas

Plate 2 Wind power capacity in relation to electricity consumption and the transmission grid in Denmark as of September 2003. (Reproduced by Permission of Bernd Möller, Aalborg University, Denmark)

This map was produced by Bernd Möller, Aalborg University, using a geographic information system (GIS). It has been composed from wind turbine and transmission line data sets and a distribution of electricity consumption. Rural areas are not included. The data base of wind turbines, their locations and technical properties originate from the reports of the transmission system operators as of September 2003, available from the Danish Energy Authority. The location and properties of transmission lines were derived from the online mapping system EnergyData, owned by the Danish Energy Authority. The distribution of electricity consumption was calculated from a building density grid available from the Area Information System of the Danish Ministry of the Environment, to which annual power demand for the year 2003 was related by building and consumer type.

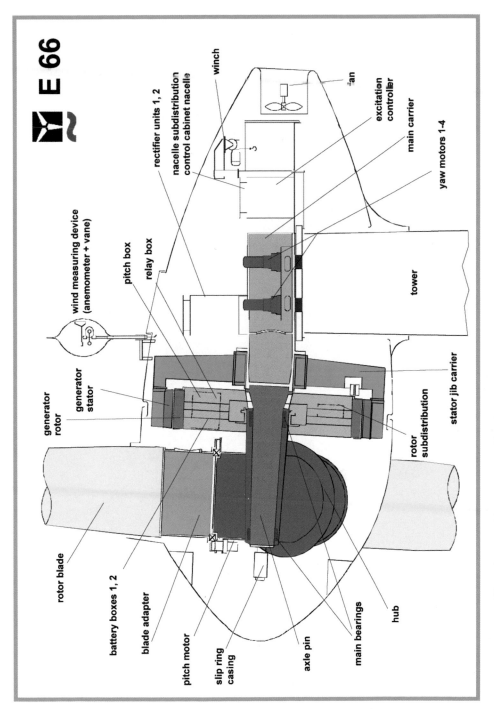

Plate 3   Nacelle Enercon E66 1.5 MW. (Reproduced by permission of Enercon, Germany)

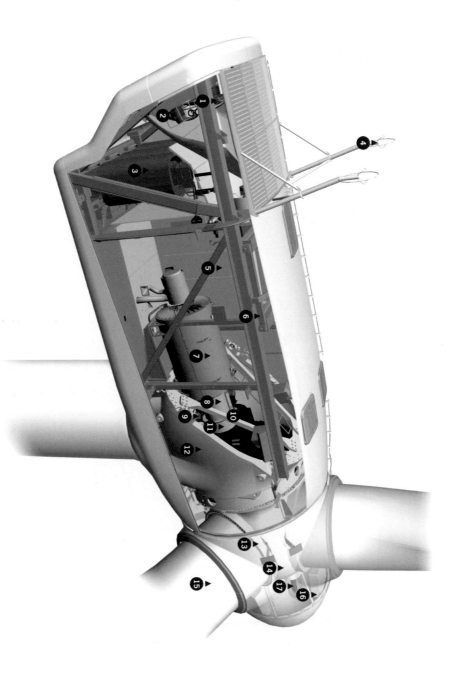

Plate 4 Nacelle Vestas V90 3 MW. Note: 1= oil cooler; 2 = generator cooler; 3 = transformer; 4 = ultrasonic wind sensors; 5 = VMP-Top controller with converter; 6 = service crane; 7 = generator; 8 = composite disc coupling; 9 = yaw gears; 10 = gearbox; 11 = parking brake; 12 = machine foundation; 13 = blade bearing; 14 = blade hub; 15 = blade; 16 = pitch cylinder; 17 = hub controller. (Reproduced by permission of Vestas Wind Systems A/S, Denmark)

introduced further incentives (e.g. investment subsidies). This policy led to a fast development of new installations between 1993 and 1997. Then the development slowed down as a result of uncertainties regarding the future of the incentives but picked up again in the new millennium after a more stable policy towards wind power was provided (for some aspects of wind power in India, see Chapter 15).

The wind energy development in China is predominately driven by international aid programmes, despite some government programmes to promote wind energy (e.g. the 'Ride-the-Wind' programme of the State Planning Commission). In Japan, the development has been dominated by demonstration projects testing different wind turbine technologies. At the end of the 1990s the first commercial wind energy projects started operation on the islands of Hokkaido as well as Okinawa. Interest in wind power is constantly growing in Japan. Also, at the end of the 1990s, the first wind energy projects materialised in New Zealand and Australia. The main driver for wind energy development in Australia is a green certification scheme.

In China and India, the typical wind turbine size is around 300–600 kW; however, some megawatt turbines have also been installed. In Australia, Japan and New Zealand, the 1–1.5 MW range is predominantly used (for installed capacity in countries in Asia and the Pacific, see Table 2.10).

2.3.6 Middle East and Africa

Wind energy development in Africa is very slow (see also Table 2.11). Most projects require financial support from international aid organisations, as there is only limited regional support. Projects are planned in Egypt, where the government agency for the New and Renewable Energy Authority (NREA) would like to build a 600 MW project near the city of Zafarana. Further projects are planned in Morocco as well as in Jordan (25 MW). The typical wind turbine size used in this region is around 300 kW, but there are plans to use 500–600 kW turbines in future projects.

Table   2.10 Operational wind power capacity in Asia and Pacific at end of 2003

Country	Installed capacity (MW)
India	1900
China	468
Japan	401
Australia	196
New Zealand	50
South Korea	8
Taiwan	8
Sri Lanka	3
Total	3034

Source: *Wind Power Monthly.*

Table 2.11 Operational wind power capacity
in Middle East and Africa at end of 2003

Country or Region	Installed capacity (MW)
Egypt	69
Morocco	54
Iran	11
Israel	8
Jordan	2
Rest of Africa	3
Total	150

Source: *Wind Power Monthly*.

2.3.7 Overview of stand-alone generation

Stand-alone systems are generally used to power remote houses or remote technical applications (e.g. for telecommunication systems). The wind turbines used for these purposes can vary from between a few watts and 50 kW. For village or rural electrification systems of up to 300 kW, wind turbines are used in combination with a diesel generator and sometimes a battery system. For more details about the current status of stand-alone systems, see Chapter 14.

2.3.8 Wind power economics

Over the past 10 years, the cost of manufacturing wind turbines has declined by about 20 % each time the number of manufactured wind turbines has doubled. Currently, the production of large-scale, grid-connected wind turbines doubles almost every three years. Similar cost reductions have been reported for photovoltaic solar and biomass systems, even though these technologies have slightly different doubling cycles. A similar cost reduction was achieved during the first years of oil exploitation about 100 years ago, but the cost reduction for electricity production between 1926 and 1970 in the USA, mainly from economies of scale, was higher. For this period, an average cost reduction of 25 % for every doubling of production has been reported (Shell, 1994).

The Danish Energy Agency (1996) predicts that a further cost reduction of 50 % can be achieved until 2020, and the EU Commission estimates in its *White Book* that energy costs from wind power will be reduced by at least 30 % between 1998 and 2010.[1] Other authors emphasise, though, that the potential for further cost reduction is not unlimited and is very difficult to estimate (Gipe, 1995).

A general comparison of electricity production costs is very difficult as production costs vary significantly between countries, because of the differing availability of

---

[1] The *White Book* is available at http://europa.eu.int/comm/off/white/index_en.htm.

resources, different tax structures and other reasons. In particular, the impact of wind speed on the economics of wind power must be stressed: a 10 % increase in wind speed, achieved at a better location for example, will in principal result in 30 % higher energy production at a wind farm (see also Chapter 3).

The competitive bidding processes for renewable power generation in England and Wales [*The Non-Fossil Fuel Obligation (NFFO)*] in the 1990s, however, provide a good comparison of power production prices for wind power and other generation technologies. The NFFO was based on a bidding process that invited potential project developers of renewable energy projects to bid for building new projects. The developers bid under different technology brands (e.g. wind or solar) for a feed-in tariff or for an amount of financial incentives to be paid for each kilowatt-hour fed into the grid by renewable energy systems. The best bidder(s) were awarded their bid feed-in tariff for a predefined period.

Owing to changes in regulations, only the price development of the last three bidding processes can be compared. They are summarised in Table 2.12. It shows that wind energy bidding prices decreased significantly; for example, between 1997 (NFFO4) and 1998 (NFFO5), the average decrease was 22 %. Surprisingly, the average price of all renewables for NFF05 is 2.71 British pence per kWh, with some projects as low as 2.34 pence per kWh, with the average power purchase price (PPP) on the England and Wales spot market, based on coal, gas and nuclear power generation, was 2.455 pence per kWh between April 1998 and April 1999.

The question arises, why would a project developer accept a lower-priced contract from NFFO if he or she could also sell the energy for a higher price via the spot market? The reason probably is that NFFO offered a 15-year fixed contract, which means a reduced financial risk, and additional costs for trading via the spot market make the trade of a small amount of energy unfeasible. Furthermore, as project developers have a period of five years to commission their plants, some developers have used cost predictions for their future projects based on large cost reductions during the following five years. In summary, wind power can be competitive in some countries, depending on the available wind speed, the prices of competing energy resources and the tax system.

Table   **2.12** Successful bidding prices in British pence per kilowatt-hour

	NFFO3 (1994)	NFFO4 (1997)	NFFO5 (1998)
Large wind	3.98–5.99	3.11–4.95	2.43–3.14
Small wind	—	—	3.40–4.60
Hydro	4.25–4.85	3.80–4.40	3.85–4.35
Landfill gas	3.29–4.00	2.80–3.20	2.59–2.85
Waste system	3.48–4.00	2.66–2.80	2.34–2.42
Biomass	4.90–5.62	5.49–5.79	—

Source: Office of Electricity Regulation, 1998 — There was no bidding within NFFO3 and NFFO4 for small wind projects.
Note: 1 ecu = 1 euro = 1.15 US dollar = 0.7 pounds sterling, as at January 1999; NFFO = Non-Fossil Fuel Obligation.

2.3.9 Environmental issues

Wind energy can be regarded as environmentally friendly; however, it is not emission-free. The production of the blades, the nacelle, the tower and so on, the exploration of the material and the transport of equipment leads to the consumption of energy resources. This means that emissions are produced as long as these energy resources are based on fossil fuel. Such emissions are known as indirect emissions. Table 2.13 provides an overview of the most important emissions related to electricity production based on different power generation technologies. The data comprise direct emissions and indirect emissions. The calculation is based on the average German energy mix and on typical German technology efficiency.

In addition, the noise and the visual impact of wind turbines are important considerations for public acceptance of wind energy technology, particularly if the wind turbines are located close to populated areas. The noise impact can be reduced through technical means (e.g. through use of variable speed or reduced rotational speed). The noise impact as well as the visual impact can also be reduced with an appropriate siting of wind turbines in the landscape (helpful guidelines as well as important examples for the appropriate siting of wind turbines can be found in Nielsen, 1996; in Pasqualetti, Gipe and Righter, 2002; Stanton, 1996).

Table   2.13 Comparison of energy amortisation time and emissions of various energy technologies

Technology	Energy payback time in month	Emissions per GWh			
		SO_2 (kg)	NO_x (kg)	CO_2 (t)	CO_2 and CO_2 equivalent for methane (t)
Coal-fired (pit)	1.0–1.1	630–1370	630–1560	830–920	1240
Nuclear	—	—	—	—	28–54
Gas (CCGT)	0.4	45–140	650–810	370–420	450
Hydro:					
large	5–6	18–21	34–40	7–8	5
mico	9–11	38–46	71–86	16–20	—
small	8–9	24–29	46–56	10–12	2
Windturbine:					
4.5 m/s	6–20	18–32	26–43	19–34	—
5.5 m/s	4–13	13–20	18–27	13–22	—
6.5 m/s	2–8	10–16	14–22	10–17	11

Note: All figures include direct and indirect emissions based on the average German energy mix, technology efficiency and lifetime. The last column also includes methane emissions, based on CO_2 equivalent. — Data not available in the source studies.

Sources: Payback time and SO_2, NO_x and CO_2 emissions: Kaltschmitt, Stelzer and Wieser, 1996. CO_2 and CO_2 equivalent for methane, Fritsch, Rausch and Simon, 1989; Lewin, 1993. For a summary of all studies in this field, see AWEA, 1992; for a similar Danish study, see Schleisner, 2000.

2.4 Status of Wind Turbine Technology

Wind energy conversion systems can be divided into those that depend on aerodynamic drag and those that depend on aerodynamic lift. The early Persian (or Chinese) vertical axis wind wheels utilised the drag principle. Drag devises, however, have a very low power coefficient, with a maximum of around 0.16 (Gasch and Twele, 2002).

Modern wind turbines are based predominately on aerodynamic lift. Lift devices use airfoils (blades) that interact with the incoming wind. The force resulting from the airfoil body intercepting the airflow consists not only of a drag force component in the direction of the flow but also of a force component that is perpendicular to the drag: the lift forces. The lift force is a multiple of the drag force and therefore the relevant driving power of the rotor. By definition, it is perpendicular to the direction of the air flow that is intercepted by the rotor blade and, via the leverage of the rotor, it causes the necessary driving torque (Snel, 1998).

Wind turbines using aerodynamic lift can be further divided according to the orientation of the spin axis into horizontal axis and vertical axis turbines. Vertical axis turbines, also known as Darrieus turbines after the French engineer who invented them in the 1920s, use vertical, often slightly curved, symmetrical airfoils. Darrieus turbines have the advantage that they operate independently of the wind direction and that the gearbox and generating machinery can be placed at ground level. High torque fluctuations with each revolution, no self-starting capability as well as limited options for speed regulation in high winds are, however, major disadvantages. Vertical axis turbines were developed and commercially produced in the 1970s until the end of the 1980s. The largest vertical axis wind turbine was installed in Canada, the Ecole C with 4200 kW. Since the end of the 1980s, however, the research and development of vertical axis wind turbines has almost stopped worldwide.

The horizontal axis, or propeller-type, approach currently dominates wind turbine applications. A horizontal axis wind turbine consists of a tower and a nacelle that is mounted on the top of the tower. The nacelle contains the generator, gearbox and the rotor. Different mechanisms exist to point the nacelle towards the wind direction or to move the nacelle out of the wind in the case of high wind speeds. On small turbines, the rotor and the nacelle are oriented into the wind with a tail vane. On large turbines, the nacelle with the rotor is electrically yawed into or out of the wind, in response to a signal from a wind vane.

Horizontal axis wind turbines typically use a different number of blades, depending on the purpose of the wind turbine. Two-bladed or three-bladed turbines are usually used for electricity power generation. Turbines with 20 or more blades are used for mechanical water pumping. The number of rotor blades is indirectly linked to the tip speed ratio, λ, which is the ratio of the blade tip speed and the wind speed:

$$\lambda = \frac{\omega R}{V}, \tag{2.1}$$

where ω is the frequency of rotation, R is the radius of the aerodynamic rotor and V is the wind speed.

Wind turbines with a high number of blades have a low tip speed ratio but a high starting torque. Wind turbines with only two or three blades have a high tip speed ratio but only a low starting torque. These turbines might need to be started if the wind speed reaches the operation range. A high tip speed ratio, however, allows the use of a smaller and therefore lighter gearbox to achieve the required high speed at the driving shaft of the power generator.

Currently, three-bladed wind turbines dominate the market for grid-connected, horizontal axis wind turbines. Three-bladed wind turbines have the advantage that the rotor moment of inertia is easier to understand and therefore often better to handle than the rotor moment of inertia of a two-bladed turbine (Thresher, Dodge and Darrell, 1998). Furthermore, three-bladed wind turbines are often attributed 'better' visual aesthetics and a lower noise level than two-bladed wind turbines. Both aspects are important considerations for wind turbine applications in highly populated areas (e.g. European coastal areas).

Two-bladed wind turbines have the advantage that the tower top weight is lighter and therefore the whole supporting structure can be built lighter, and the related costs are very likely to be lower (Gasch and Twele, 2002; Thresher, Dodge and Darrell, 1998). As visual aesthetics and noise levels are less important offshore, the lower costs might be attractive and lead to the development of two-bladed turbines for the offshore market.

2.4.1 Design approaches

Horizontal axis wind turbines can be designed in different ways. Thresher, Dodge and Darrell (1998) distinguish three main design philosophies.

- The first philosophy aims at withstanding high wind loads and is optimised for reliability and operates with a rather low tip speed ratio. The precursor of this approach is the Gedser wind turbine built in the 1950s in Denmark.
- The second design philosophy has the goal to be compliant and shed loads and is aimed at optimised performance. The approach is represented by the Hütter turbine, developed in the 1950s in Germany. It has a single blade and a very high tip speed ratio.
- Modern grid-connected wind turbines usually follow the third design philosophy, which aims at managing loads mechanically and/or electrically. This approach uses a lower tip speed ratio than the second design philosophy. Therefore, visual disturbance is lower than in the first design philosophy, as are material requirements, because of the fact that the structure does not need to withstand high wind loads. This means lower costs. Finally, this approach usually leads to a better power quality, because short-term wind speed variations (within seconds) are not directly translated into power output fluctuations.

Each of the design approaches leaves a lot of options regarding certain design details. The details of modern grid-connected wind turbines can vary significantly, even though they follow the same principal design philosophy. Plates 3 and 4 provide examples of wind turbines that follow the same philosophy (i.e. variable speed operation with a low tip speed ratio); however, the actual designs show significant differences.

The wind turbine designs developed over the past decade, in particular those developed within the third design philosophy, use to a great extent power electronics. Most of the power electronic solutions were developed in the late 1980s and early 1990s, based on important technical and economic developments in the area of power electronics.

Chapter 4 therefore presents an overview of current wind turbine designs, as well as a treatment of generator and power electronic solutions, in more detail. Possible future designs are also briefly discussed in Chapter 4 as well as in Chapter 22.

2.5 Conclusions

Wind energy has the potential to play an important role in future energy supply in many areas of the world. Within the past 12 years, wind turbine technology has reached a very reliable and sophisticated level. The growing worldwide market will lead to further improvements, such as larger wind turbines or new system applications (e.g. offshore wind farms). These improvements will lead to further cost reductions and in the medium term wind energy will be able to compete with conventional fossil fuel power generation technology. Further research, however, will be required in many areas, for example, regarding the network integration of a high penetration of wind energy.

Acknowledgements

The author would like to acknowledge valuable discussions with Per-Anders Löf, Irene Peinelt and Jochen Twele (Technical University Berlin).

References

[1] Ancona, D. F. (1989) 'Power Generation, Wind Ocean', in *Wilk's Encycopedia of Architecture: Design, Engineering and Construction, Volume 4*, John Wiley & Sons, Ltd/Inc., New York, pp. 32–39.

[2] AWEA (American Wind Energy Association) (1992) 'Energy and Emission Balance favours Wind', *Wind Energy Weekly* number 521, 9 November 1992.

[3] Danish Energy Agency (1996) *Energy 21: The Danish Government's Action Plan* April, Danish Energy Agency, Denmark.

[4] Fritsch, U., Rausch, L., Simon, K. H. (1989) *Umweltwirkungsanalyse von Energiesystemen, Gesamt-Emissions-Modell integrierter Systeme*, Hessisches Ministerium für Wirtschaft und Technology (Ed.), Wiesbaden.

[5] Gasch, R., Twele, J. (2002) *Wind Power Plants: Fundamentals, Design, Construction and Operation*, James and James, London, and Solarpraxis, Berlin.

[6] Gipe, P. (1995) *Wind Energy Comes of Age*, John Wiley & Sons, Ltd/Inc., New York.

[7] Heymann, M. (1995) *Die Geschichte der Windenergienutzung 1890–1990* (The History of Wind Energy Utilisation 1890–1990), Campus, Frankfurt are Main.

[8] Hills, R. L. (1994) *Power From Wind – A History of Windmill Technology*, Cambridge University Press, Cambridge.

[9] Johnson, G. L. (1985) *Wind Energy Systems*, Prentice Hall, New York.

[10] Kaltschmitt, M., Stelzer, T., Wiese, A. (1996) 'Ganzheitliche Bilanzierung am Beispiel einer Bereitstellung elektrischer Energie aus regenerativen Energien', *Zeitschrift für Energiewirtschaft*, **20**(2) 177–178.

[11] Kealey, E. J. (1987) *Harvesting the Air: Windmill Pioneers in Twelfth-century England*, University of California Press, Berkeley, CA.

[12] Koeppl, G. W. (1982) *Putnam's Power from the Wind*, 2nd edition, Van Norstrand Reinhold Company, New York (this book is an updated version of Putnam's original book, *Power from the Wind*, 1948).

[13] Lewin, B. (1993) *CO2-Emission von Energiesystemen zur Stromerzeugung unter Berücksichtigung der Energiewandlungsketten*, PhD thesis, Fachbereich 16, Bergbau und Geowissenschaften, Technical University Berlin, Berlin.

[14] Nielsen, F. B. (1996) *Wind Turbines and the Landscape: Architecture and Aesthetics*, prepared for the Development Programme for Renewable Energy of the Danish Energy Agency (originally appeared as *Vindmøller og Landskab: Arkitektur og Æstetik*).

[15] Office of Electricity Regulation (1998) *Fifth Renewable Order for England and Wales*, September 1998, Office of Electricity Regulation, UK.

[16] Pasqualetti, M., Gipe, P., Righter, R. (2002) *Wind Power in View – Energy Landscapes in a Crowded World*, Academic Press, San Diego, CA.

[17] Putnam, P. C. (1948) *Power From the Wind*, Van Nostrand, New York (reprinted 1974).

[18] Righter, R. (1996) *Wind Energy in America: A History*, University of Oklahoma Press, USA.

[19] Schleisner, L. (2000) 'Life Cycle Assessment of a Wind Farm and Related Externalities', *Renewable Energy* **20** 279–288.

[20] Shell (Shell International Petroleum Company) (1994) 'The Evolution of the World's Energy System 1860–2060', in *Conference Proceedings: Energy Technologies to Reduce CO_2-emissions in Europe: Prospects, Competition, Synergy*, International Energy Agency, Paris, pp. 85–113.

[21] Shepherd, D. G. (1990) *Historical Development of the Windmill*, DOE/NASA-5266-2, US Department of Energy, Washington, DC.

[22] Shepherd, D. G. (1994) 'Historical Development of the Windmill', in *Wind Turbine Technology*, SAME Press, New York.

[23] Snel, H. (1998) 'Review of the Present Status of Rotor Aerodynamics', *Wind Energy* **1**(S1) 46–69.

[24] Stanton, C. (1996) *The Landscape Impact and Visual Design of Windfarms*, School of Landscape Architecture, Edinburgh, Scotland.

[25] Thresher, Dodge, R. W., Darrell, M. (1998) 'Trends in the Evolution of Wind Turbine Generator Configurations and Systems', *Wind Energy* **1**(S1) 70–85.

[26] *Wind Power Monthly* (1999) **15**(5) p. 8.

3

Wind Power in Power Systems: An Introduction

Lennart Söder and Thomas Ackermann

3.1 Introduction

This chapter provides a basic introduction to the relevant engineering issues related to the integration of wind power into power systems. It also includes links to further, more detailed, reading in this book. In addition, the appendix to this chapter provides a new basic introduction to power system engineering for nonelectrical engineers.

3.2 Power System History

Soon after Thomas Alva Edison installed the first power systems in 1880 entrepreneurs realized the advantages of electricity, and the idea spread around the world. The first installations had one thing in common: the generation unit(s) were installed close to the load, as the low-voltage direct-current transmission led to high losses.

With the development of transformers, alternating current became the dominant technology, and it was possible to link power stations with loads situated further away. In 1920, each large load centre in Western Europe had its own power system (Hughes, 1993). With the introduction of higher transmission line voltages, the transport of power over larger distances became feasible, and soon the different power systems were interconnected. In the beginning, only stations in the same region were interconnected. Over the years, technology developed further and maximum possible transmission line voltage increased step by step.

In addition to the increasing interconnections between small power systems, an institutional and organizational structure in the electricity industry started to emerge. After the turn of the century, municipally owned companies started operation, often

Wind Power in Power Systems   Edited by T. Ackermann
© 2005 John Wiley & Sons, Ltd   ISBN: 0-470-85508-8 (HB)

side by side with private companies. Municipally owned companies operated mainly as cooperatives for the users of electricity. In many countries, municipally owned companies had been taking over private companies, partly because it was easier for those to obtain the capital investment required for building the electric power system.

A strong impetus for further electrification came from government and industry, as electricity was seen as an important step into a modern world. Governments also promoted the idea that the electricity sector (i.e. electricity generation, transmission and distribution) should be considered as natural monopolies. That meant that with increasing output levels average costs were expected to fall continuously.

The main driver of technological development was now to realize economies of scale by installing larger units. In the 1930s, the most cost-effective size of thermal power stations was about 60 MW. In the 1950s it was already 180 MW, and by the 1980s about 1000 MW. The location of the fuel resource (e.g. coal mines) or the most convenient transport connection (e.g. seaports) usually determined where these huge thermal power plants were built. As the availability of such locations was limited, large power stations were often built next to each other. Further economies of scale or simply administrative convenience were the reasons for this development. There was a similar development regarding hydropower units. The only difference was that more and more units were built along the same river. When nuclear power stations were introduced into power systems in the 1960s they soon followed a similar development pattern. At the end of the 1980s, a typical nuclear power station consisted of three to five blocks of 800 to 1000 MW each (Hunt and Shuttleworth, 1996).

Despite the fact that the Dane Poul la Cour developed the first electricity generating wind turbine in 1891, wind power played hardly any role in the development of electricity supply. Interestingly enough, in 1918 wind turbines were already supplying 3 % of Danish electricity demand. However, large steam turbines dominated the electricity generation industry worldwide because of their economic advantages.

3.3 Current Status of Wind Power in Power Systems

In most parts of the world wind energy supplies only a fraction of the total power demand, if there is any wind power production at all. In other regions, for example in Northern Germany, Denmark or on the Swedish Island of Gotland, wind energy supplies a significant amount of the total energy demand. In 2003, wind energy supplied around 4200 GWh of the total system demand of 13 353 GWh (energy penetration of 31.45 %) in the German province of Schleswig–Holstein (Ender, 2004)[1]. In the network area of the Danish system operator Eltra (Jutland and Funen), wind power supplied 3800 GWh of 20 800 GWh (18 %)[2], and, on the Swedish island of Gotland, wind power supplied 200 GWh of a total system demand of 900 GWh (a penetration of 22 %)[3]. For further details, see also Table 3.1.

In the future, many countries around the world are likely to experience similar penetration levels as wind power is increasingly considered not only a means to reduce

[1] For more details on the situation in Germany, see Chapter 11.
[2] For more details on the situation in Eltra's work, see also Chapter 10.
[3] For more details on Gotland, see also Chapter 13.

Table   3.1 Examples of wind power penetration levels, 2002

Country or region	Installed wind capacity (MW)	Total installed power capacity (MW)	Average annual penetration level[a] (%)	Peak penetration level[b] (%)
Western Denmark:[c]	2 315	7 018	~18	>100
Thy Mors[d]	~40	—[e]	>50	~300
Germany:	12 000	119 500	~5	n.a.
Schleswig Holstein[f]	1 800	—[g]	~28	>100
Papenburg[h]	611	—[g]	~55	>100
Spain:	5 050	53 300	~5	n.a.
Navarra[i]	550	Part of the Spanish System	~50	>100
Island systems:				
Swedish island of Gotland[j]	90	No Local Generation in normal state	~22	>100
Greak island of Crete[k]	70	640	~10	n.a.
Denham, Australia[l]	690	2 410 2.4	~50	~70

n.a. = Not available.
[a] Wind Energy production as share of system consumption.
[b] Level at high wind production and low energy demand, hence, if peak penetration level is > 100% excess energy is exported to other regions.
[c] For further details, see Chapter 10.
[d] Local distribution area in Denmark.
[e] Part of the Western Danish System.
[f] German costal province.
[g] Part of the German System.
[h] Local network area in Germany.
[i] Spanish province.
[j] The island of Gotland has a network connection to the Swedish mainland; see also Chapter 13.
[k] Crete has no connection to the mainland.
[l] Isolated wind–diesel (flywheel) system; see Chapter 14, in particular Figure 14.7, for details.

CO_2 emissions but also an interesting economic alternative in areas with appropriate wind speeds. The integration of high penetration levels of wind power into power systems that were originally designed around large-scale synchronous generators may require new approaches and solutions. As the examples of isolated wind–diesel systems show (see also Chapter 14) it is technically possible to develop power systems with a very high wind power peak penetration of up to 70 %. Whether such penetration levels are economically feasible for large interconnected systems remains to be seen. When integrating wind power, wind–diesel systems, however, have the advantage of being able to neglect existing large, and sometimes not very flexible, generation units.

In other words: the integration of high penetration levels of wind power (>30 %) into large existing interconnected power systems may require a step-by-step redesign of the existing power system and operation approaches. This is, however, more likely an economic than a technical issue (see also Chapter 18). For many power systems, the current challenge is not suddenly to incorporate very high penetration levels but to deal with a gradual increase in wind power. This chapter will focus mainly on the issues related to the incorporation of low to medium wind power energy penetration levels (<20–30 %).

3.4 Network Integration Issues for Wind Power

The examples mentioned in Section 3.1 (see Table 3.1) show that the integration of significant penetration levels of wind power not only is possible but also often does not require a major redesign of the existing power system. However, in this context it is important to point out that the areas with very high penetration levels in Denmark, Germany and Spain are part of very large national, or even multinational, strongly interconnected power systems.

From a technical perspective, power system engineers have to keep in mind that the main aim of a power system is to supply network costumers with electricity whenever the customers have a demand for it. Now, if wind power is introduced into the power system the main aim of the power system must still be fulfilled. The challenge wind power introduces into power system design and operation is related to the fluctuating nature of wind as well as to the comparatively new generator types (e.g. doubly-fed induction generators)[4] that are used in wind turbines but that are not commonly used in traditional power systems.

The basic challenge regarding the network integration of wind power consists therefore of the following two aspects:

- How to keep an acceptable voltage level for all consumers of the power system: customers should be able to continue to use the same type of appliances that they are used to.
- How to keep the power balance of the system: that is, how can wind power production and other generation units continuously meet consumers' needs?

In general, power system engineers have always worked with such challenges. When nuclear power generation was introduced, for instance, engineers faced the problem that nuclear power is a very inflexible generation source while load varies all the time. Hence many countries had to increase the flexibility of other power sources (e.g. by using hydropower together with nuclear power). In Sweden, for instance, the power system consists mainly of comparatively inflexible nuclear power generation, which is used for base load supply, and very flexible hydro generation, for load following (Kaijser, 1995). Other countries have developed other system solutions in order always to be able to fulfil customers' needs (e.g. in Japan pump storage systems are part of the power system to provide more flexibility to a power system with a very high penetration level of nuclear power).

The brief examples in this section show that the integration issues related to wind power are very much dependent on the power system (i.e. they depend on the specific case). However, the general methods that power system engineers have been applying for a long time can also be applied to the network integration of wind power. Some

---

[4] For more information on wind turbine generators, see Chapter 4.

methods may require modifications, though, in order to allow the design and operation of a power system that is able to meet customers' needs.

3.5 Basic Electrical Engineering

We will start with some power engineering basics. In a power system, there are voltages and currents. Since we have an alternating current (AC) system this means that they can be represented as:

$$u(t) = U_M \cos(\omega t),$$
$$i(t) = I_M \cos(\omega t - \varphi), \tag{3.1}$$

where

$u(t)$ = voltage as a function of time;
U_M = maximum voltage amplitude;
$\omega = 2\pi f$;
f = frequency, normally 50 or 60 Hz;
$i(t)$ = current as a function of time;
I_M = maximum current amplitude;
φ = phase shift between voltage and current.

The power is the product of the voltage and the current and can be calculated as:

$$p(t) = u(t)i(t) = U_M \cos(\omega t) I_M \cos(\omega t - \varphi)$$
$$= P[1 + \cos(2\omega t)] + Q \sin(2\omega t), \tag{3.2}$$

where

$P = \frac{U_M}{\sqrt{2}} \frac{I_M}{\sqrt{2}} \cos \varphi$ = active power;
$Q = \frac{U_M}{\sqrt{2}} \frac{I_M}{\sqrt{2}} \sin \varphi$ = reactive power;
$\cos \varphi$ = power factor.

Figure 3.1 shows the voltage, current and power as a function of time.

When power systems are studied, the most common method is to use complex notations of voltages, currents and power. This method simplifies the calculations compared with calculations with these variables expressed as a function of time [see Equations (3.3)–(3.4)]. The complex voltage and current are:

$$U = |U|e^{j \arg(U)},$$
$$I = |I|e^{j \arg(I)}, \tag{3.3}$$

where

U = complex voltage;
$|U| = \frac{U_M}{\sqrt{2}}$ = root mean square (RMS) phase voltage;
I = complex current;
$|I| = \frac{I_M}{\sqrt{2}}$ = RMS current.

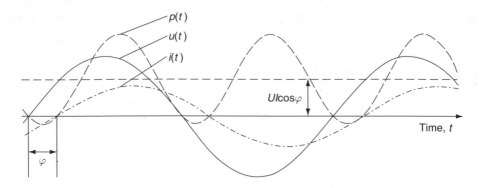

Figure   3.1 Voltage, current and power as a function of time, t [$u(t)$, $i(t)$ and $p(t)$, respectively]

The complex power, S, can be defined as:

$$S = |S|e^{j\arg(S)} = P + jQ = UI$$
$$= |U||I|\, e^{j[\arg(U)-\arg(I)]} = |U||I|\, e^{j\varphi} \tag{3.4}$$

In most power systems, the power is transmitted using three phases. The general goal is to arrive at a three-phase power that is as *symmetric* as possible. For a perfect symmetric system, the voltages (and currents) in the three phases will have the same amplitude, and the phase shift between the three phases, a–c, will be exactly 120°; that is:

$$U_a = |U|\, e^{j0°},$$
$$U_b = |U|\, e^{-j120°},$$
$$U_c = |U|\, e^{j120°},$$
$$I_a = |I|\, e^{-j\varphi°}, \tag{3.5}$$
$$I_b = |I|\, e^{-j(120+\varphi°)°},$$
$$I_c = |I|\, e^{j(120-\varphi°)°}.$$

The line-to-line voltage can be determined as

$$U_{ab} = U_a - U_b = |U_a|(1 - e^{-j120°}) = \sqrt{3}|U_a|e^{j30°}; \tag{3.6}$$

that is, the size of the line-to-line voltage is $\sqrt{3}$ times larger than the phase voltage. The convention concerning power system analysis is that a particular voltage level (e.g. 10 kV) corresponds to the RMS value of the line-to-line voltage. The benefit of symmetric systems is that the total three-phase power is constant and, as opposed to single-phase power, does not vary with time [see Equation (3.2)]. Another benefit is that the sum of the currents in the three phases is equal to zero (i.e. no current will flow in the return wire, if there is any).

Loads, transmission lines and transformers can be represented with impedances, Z, when analysing power systems. This is illustrated in Figure 3.2

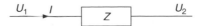

Figure   3.2 Voltages U_1 and U_2 either side of an impedance, Z, with current I

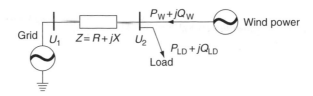

Figure   3.3 Grid connection of wind power. U_1, U_2 = voltage either side of impedance Z; P_w, P_{LD} = active power from wind power and of load; Q_w, Q_{LD} = reactive power of wind power and load

With voltages (RMS values of line-to-line voltages) on each side of the impedance (which may represents a transmission line, for instance), and the current (RMS value of the phase current), the voltage drop over the impedance can be calculated as:

$$U_1 - U_2 = \sqrt{3}\,I\,Z. \tag{3.7}$$

Figure 3.3 shows the basic problem of the grid connection of a wind farm. The grid can be seen as a voltage source, U_1, next to the impedance, Z. The impedance represents the impedance in all transmission lines, cables and transformers in the feeding grid. At the point of connection of the wind farm, there is also a local load. The short circuit power, S_k, in the wind power connection point can be calculated as

$$S_k = \frac{U_1^2}{Z^*}. \tag{3.8}$$

Changes in wind power production will cause changes in the current through the impedance Z. These current changes cause changes in the voltage U_2. If Z is large (in a weak grid, S_k is small) there is not as much room for wind power as there is in a situation where Z is small (in a strong grid, S_k is large).

In Figure 3.3, wind power production and load are represented as $P + jQ$. P here is the active power, and Q is the reactive power. The reactive power depends on the phase shift, φ, between the voltage and the current, such as:

$$Q = \frac{\sin \varphi}{\cos \varphi} P; \tag{3.9}$$

see Equation (3.2).

The voltage U_2 in the figure can be calculated as:

$$U_2 = \left\{ -\frac{2a_1 - U_1^2}{2} + \left[\left(\frac{2a_1 - U_1^2}{3} \right)^2 - (a_1^2 + a_2^2) \right]^{1/2} \right\}^{1/2}, \tag{3.10}$$

where

$$a_1 = -R(P_W - P_{LD}) - X(Q_W - Q_{LD}),$$

$$a_2 = -X(P_W - P_{LD}) + R(Q_W - Q_{LD}).$$

This equation shows that the reactive power production in the wind farm, Q_W, has an impact on the voltage U_2. The impact is dependent on the local load and on the feeding grid impedance. We will now look at the different possibilities of generating wind power:

An asynchronous generator consumes reactive power and the amount is not controllable. The reactive consumption is normally partly compensated with shunt capacitors, the so-called phase compensation (which produces reactive power and reduces φ). If there are several capacitors, the voltage can be controlled stepwise by changing the number of connected capacitors.

In synchronous generators (not with permanent magnets) and in converters it is possible to control the reactive generation or consumption. This makes it possible to control the voltage as shown in Equation (3.10).

3.6 Characteristics of Wind Power Generation

In the following section we will discuss in more detail wind as a power generation source, including its fluctuating character and the physical limitations for utilising this natural resource. After that, the typical characteristics of wind power generation will be analysed.

3.6.1 The wind

Air masses move because of the different thermal conditions of these masses. The motion of air masses can be a global phenomenon (i.e the jet stream) as well as a regional and local phenomenon. The regional phenomenon is determined by orographic conditions (e.g. the surface structure of the area) as well as by global phenomena.

Wind turbines utilise the wind energy close to the ground. The wind conditions in this area, known as boundary layer, are influenced by the energy transferred from the undisturbed high-energy stream of the geostrophic wind to the layers below as well as by regional conditions. Owing to the roughness of the ground, the local wind stream near the ground is turbulent.

The wind speed varies continuously as a function of time and height. The time scales of wind variations are presented in Figure 3.4 as a wind frequency spectrum. The turbulent peak is caused mainly by gusts in the subsecond to minute range. The diurnal peak depends on daily wind speed variations (e.g. land–sea breezes caused by different temperatures on land and sea) and the synoptic peak depends on changing weather patterns, which typically vary daily to weekly but include also seasonal cycles.

From a power system perspective, the turbulent peak may affect the power quality of wind power production. As discussed in Chapters 5, 6 and 16, the impact of turbulences on power quality depends very much on the turbine technology applied. Variable speed wind turbines, for instance, may absorb short-term power variations by the immediate storage of energy in the rotating masses of the wind turbine drive train. That means that the power output is smoother than for strongly grid-coupled turbines. Diurnal and

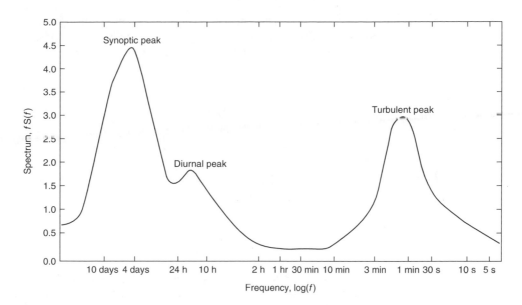

Figure **3.4** Wind spectrum based on work by van Hoven (Reproduced from T. Burton, D. Sharpe N. Jenkins and E. Bossanyi, 2001, *Wind Energy Hand book*, by permission of John Wiley & Sons, Ltd.)

synoptic peaks, however, may effect the long-term balancing of the power system as discussed mainly in Chapters 10 and 11. Wind speed forecasts play a significant role for the long-term balancing of the power system, see also Chapter 17.

Another important issue is the long-term variations of the wind resources. There are several studies regarding this issue (see Petersen *et al.*, 1998; Söder, 1997; 1999). Based on these studies, it has been estimated that the variation of the yearly mean power output from one 20-year period to the next has a standard deviation of 10 % or less (see also Chapter 9). Hence, over the lifetime of a wind turbine, the uncertainty of the resource wind is not large, which is an important factor for the economic evaluation of a wind turbine. In many locations of the world, the uncertainty regarding the availability of water of a longer time period (more than 1 year) for hydropower generation exceed that of wind power (Hills, 1994).

3.6.2 The physics

The power of an air mass that flows at speed V through an area A can be calculated as follows:

$$\text{power in wind} = \frac{1}{2}\rho A V^3 (\text{watts}), \tag{3.11}$$

where

ρ = air density (kg m^{-3});
V = wind speed (m s^{-1}).

The power in the wind is proportional to the air density ρ, the intercepting area A (e.g. the area of the wind turbine rotor) and the velocity V to the third power. The air density is a function of air pressure and air temperature, which both are functions of the height above see level:

$$\rho(z) = \frac{P_0}{RT}\exp\left(\frac{-gz}{RT}\right),\tag{3.12}$$

where

$\rho(z)$ = air density as a function of altitude (kg m^{-3});
P_0 = standard sea level atmospheric density $(1.225\,\text{kg m}^{-3})$;
R = specific gas constant for air $(287.05\,\text{J kg}^{-1}\text{K}^{-1})$;
g = gravity constant $(9.81\,\text{m s}^{-2})$;
T = temperature (K);
z = altitude above sea level (m).

The power in the wind is the total available energy per unit of time. The power in the wind is converted into the mechanical–rotational energy of the wind turbine rotor, which results in a reduced speed in the air mass. The power in the wind cannot be extracted completely by a wind turbine, as the air mass would be stopped completely in the intercepting rotor area. This would cause a 'congestion' of the cross-sectional area for the following air masses.

The theoretical optimum for utilising the power in the wind by reducing its velocity was first discovered by Betz, in 1926. According to Betz, the theoretical maximum power that can be extracted from the wind is

$$P_{\text{Betz}} = \frac{1}{2}\rho A V^3 C_{\text{P Betz}} = \frac{1}{2}\rho A V^3 \times 0.59.\tag{3.13}$$

Hence, even if power extraction without any losses were possible, only 59 % of the wind power could be utilised by a wind turbine (Gasch and Twele, 2002)

3.6.3 Wind power production

In the following subsections typical characteristics of wind power production are briefly discussed.

3.6.3.1 Power curve

As explained by Equation (3.11), the available energy in the wind varies with the cube of the wind speed. Hence a 10 % increase in wind speed will result in a 30 % increase in available energy.

The power curve of a wind turbine follows this relationship between cut-in wind speed (the speed at which the wind turbine starts to operate) and the rated capacity, approximately (see also Figure 3.5). The wind turbine usually reaches rated capacity at a wind speed of between $12-16\,\text{m s}^{-1}$, depending on the design of the individual wind turbine.

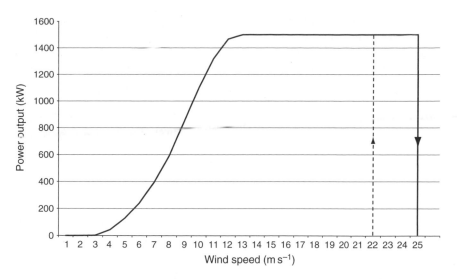

Figure 3.5 Typical power curve of a 1500 kW pitch regulated wind turbine with a cut-out speed of 25 m s^{-1} (the broken line shows the hysteresis effect)

At wind speeds higher than the rated wind speed, the maximum power production will be limited, or, in other words, some parts of the available energy in the wind will be 'spilled'. The power output regulation can be achieved with *pitch-control* (i.e. by feathering the blades in order to control the power) or with *stall control* (i.e. the aerodynamic design of the rotor blade will regulate the power of the wind turbine). Hence, a wind turbine produces maximum power within a certain wind interval that has its upper limit at the cut-out wind speed. The cut-out wind speed is the wind speed where the wind turbine stops production and turns out of the main wind direction. Typically, the cut-out wind speed is in the range of 20 to 25 m s^{-1}.

The power curve depends on the air pressure (i.e. the power curve varies depending on the height above sea level as well as on changes in the aerodynamic shape of the rotor blades, which can be caused by dirt or ice). The power curve of fixed-speed, stall-regulated wind turbines can also be influenced by the power system frequency (see also Chapter 15).

Finally, the power curve of a wind farm is not automatically made up of the scaled-up power curve of the turbines of this wind farm, owing to the shadowing or wake effect between the turbines. For instance, if wind turbines in the first row of turbines that directly face the main wind direction experience wind speeds of 15 m s^{-1}, the last row may 'get' only 10 m s^{-1}. Hence, the wind turbines in the first row will operate at rated capacity, 1500 kW for the turbine in Figure 3.5, whereas the last row will operate at less than rated capacity (e.g. 1100 kW for the same turbines).

3.6.3.2 Hysteresis and cut-out effect

If the wind speed exceeds the cut-out wind speed (i.e. 25 m s^{-1} for the wind turbine in Figure 3.5) the turbine shuts down and stops producing energy. This may happen during a storm, for instance. When the wind drops below cut-out wind speed, the turbines will not immediately start operating again. In fact, there may be a substantial delay, depending on the individual wind turbine technology (pitch, stall and variable speed) and the wind regime in which the turbine operates. The restart of a wind turbine, also referred to as the hysteresis loop (see the broken line in Figure 3.5), usually requires a drop in wind speed of 3 to 4 m s^{-1}.

For the power system, the production stop of a significant amount of wind power arising from wind speeds that exceed the cut-out wind speed may result in a comparatively sudden loss of significant amounts of wind power. In European power systems, where wind power is installed in small clusters distributed over a significant geographic area (for more details, see Chapters 10 and 11), this shut down of a significant amount of wind power as a result of the movement of a storm front usually is distributed over several hours. However, for power systems with large wind farms installed in a small geographic area, a storm front may lead to the loss of a significant amount of wind power in a shorter period of time (less than 1 hour). The hysteresis loop then determines when the wind turbines will switch on again after the storm has passed the wind farm.

To reduce the impact of a sudden shutdown of a large amount of wind power and to solve the issues related to the hysteresis effect, some wind turbine manufacturers offer wind turbines with power curves that, instead of an sudden cut out, reduce power production step by step with increasing wind speeds (see Figure 3.6). This certainly reduces the possible negative impacts that very high wind speeds can have on power system operation.

3.6.3.3 Impact of aggregation of wind power production

The aggregation of wind power provides an important positive effect on power system operation and power quality. Figure 3.7 shows the principal of aggregated wind power production. The positive effect of wind power aggregation on power system operation has two aspects:

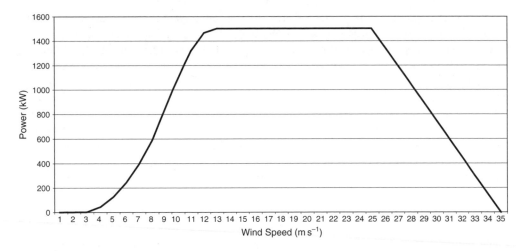

Figure   3.6 The power curve of a 1500 kW pitch-regulated wind turbine with smooth power reduction during very high wind speeds

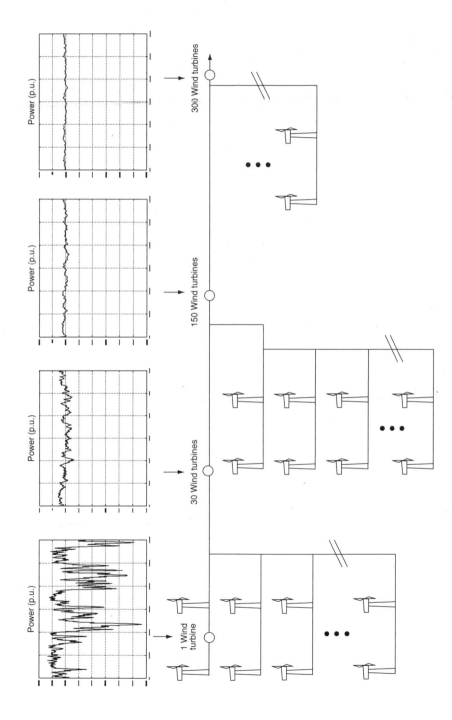

Figure 3.7 Impact of geographical distribution and additional wind turbines on aggregated power production. Time scale on graphics: seconds. Based on simulations by Pedro Rosas (Reproduced from P. Rosas, 2003, *Dynamic Influences of Wind Power on the Power System* (PLD thesis, Ørsted Institute and Technical University of Denmark), by permission of Pedro Rosas.)

- an increased number of wind turbines within a wind farm;
- the distribution of wind farms over a wider geographical area.

Increased number of wind turbines within a wind farm

An increased number of wind turbines reduces the impact of the turbulent peak (Figure 3.4) as gusts do not hit all the wind turbines at the same time. Under ideal conditions, the percentage variation of power output will drop as $n^{-1/2}$, where n is the number of wind generators. Hence, to achieve a significant smoothing effect, the number of wind turbines within a wind farm does not need to be very large.

Distribution of wind farms over a wider geographical area

A wider geographical distribution reduces the impact of the diurnal and synoptic peak significantly as changing weather patterns do not affect all wind turbines at the same time. If changing weather patterns move over a larger terrain, maximum up and down ramp rates are much smaller for aggregated power output from geographically dispersed wind farms than from a very large single wind farm. The maximum ramp rate of the aggregated power output from geographically dispersed wind farm clusters in the range of 10 to 20 MW with a total aggregated capacity of 1000 MW, for instance, can be as low as 6.6 MW per minute (see also Chapter 9), and single wind farms with an installed capacity of 200 MW have shown ramp rates of 20 MW per minute or more. The precise smoothing effect of the geographical distribution depends very much on local weather effects and on the total size of the geographical area (see also Chapter 8). It must be emphasised that a distribution of wind farms over a larger geographical area usually has a positive impact on power system operation.

3.6.3.4 Probability density function

As explained in Section 3.6.1, the power production of a wind power plant is related to the wind speed [see Equation (3.11)]. Since wind speed varies, power production varies, too. There are two exceptions, though. If the wind speed is below the cut-in wind speed or is higher than the cut-out wind speed then power production will be zero.

Figure 3.8 shows the structure of the probability density function (pdf), $f_P(x)$, of the total wind power from several wind power units during a specific period (e.g. one year).

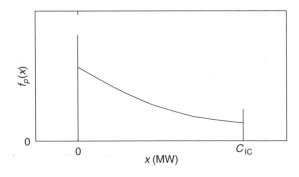

Figure   3.8 Probability density function for the available power production from several wind power units

The total installed capacity (IC) is assumed to be C_{IC}. There is one discrete probability of zero production, p_0, when the wind speed is below the cut-in wind speed for all wind turbines or when the wind turbines are shut down because of too high winds. There is also one discrete probability of installed capacity, p_{IC}, when the wind speed at all pitch-controlled units lies between the rated wind speed and cut-out wind speed, and when for all stall-controlled turbines the wind speed lies in the interval that corresponds to installed capacity. Between these two levels there is a continuous curve where for each possible production level there is a probability. There is a structural difference between wind power production and conventional power plants, such as thermal power or hydro-power. The difference is that these power plants have a much higher probability, p_{IC}, that the installed capacity is available and a much lower probability, p_0, of zero power production. For conventional power plants, there are normally only these two probabilities, which means that the continuous part of the pdf in Figure 3.8 will be equal to zero.

The values of p_0 and p_{IC} decrease with an increasing total amount of wind power. This is owing to the fact that if there is a larger amount of wind power, the turbines have to be spread out over a wider area. This implies that the probability of zero wind speed at all sites at the same time decreases. The probability of high (but not too high) wind speeds at all sites at the same time will also decrease.

The mean power production of all units can be calculated as

$$P_m = \int_0^{C_{IC}} x f_P(x) \mathrm{d}x. \tag{3.14}$$

3.6.3.5 Capacity factor

The ratio P_m/C_{IC} is called the *capacity factor* (CF) and can be calculated for individual units or for the total production of several units (for a more detailed discussion on the capacity factor, see Chapter 9). The capacity factor depends on the wind resources at the location and the type of wind turbine, but lies often in the range of about 0.25 (low wind speed locations) to 0.4 (high wind speed locations). The *utilisation time* in hours per year is defined as 8760 P_m/C_{IC}. This value lies, then, in the range of 2200–3500 hours per year. In general, if the utilisation time is high, the unit is most likely to be operating at rated capacity comparatively often.

The yearly energy production, W, can be calculated as

$$\begin{aligned} W = P_m \times 8760 &= C_{IC} \times 8760 \times \mathrm{CF} \\ &= C_{IC} \times t_{util} \end{aligned} \tag{3.15}$$

where

  CF = capacity factor;
  t_{util} = utilization time.

Occasionally, the utilisation time is interpreted as 'equivalent full load hours', and this is correct from an energy production perspective, but it is sometimes even misunderstood and assumed to be the operating time of a wind power plant.

Compared with base-load power plants, such as coal or nuclear power plants, the utilisation time of wind power plants is lower. This implies that in order to obtain the same energy production from a base-load power plant and a wind farm, the installed wind farm capacity must be significantly larger than the capacity of the base-load power plant.

3.7 Basic Integration Issues Related to Wind Power

The challenge that the integration of wind power poses can be illustrated using Figure 3.9. In this power system, there are industries and households that consume power P_D and a wind power station that delivers power P_W. The additional power, P_G, is produced at another location. The impedances Z_1–Z_3 represent the impedances in the transmission lines and transformers between the different components.

In an electric power system, such as the one illustrated in Figure 3.9, power cannot disappear. This means that there will always be a balance in this system, as

$$P_G = P_D + P_L - P_W, \tag{3.16}$$

where

P_G = additional required power production;
P_D = power consumption;
P_L = electrical losses in the impedances Z_1–Z_3;
P_W = wind power production.

Equation (3.16) is valid for any situation; it does not matter whether we look at a short time period (e.g. minutes) or a long period (e.g. a year). Equation (3.16) also implies that electricity cannot be stored within an electric power system. Hence any change in electricity demand (or wind power) must be simultaneously balanced by other generation sources within the power system.

3.7.1 Consumer requirements

As mentioned in Section 3.4, the main aim of a power system is to supply consumers with the required electricity at any given time at a reasonable cost. From the consumer's perspective, three main requirements can be defined:

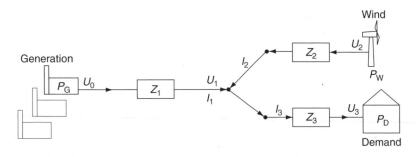

Figure   3.9 Illustrative power system

- Customer requirement 1 (CR1): the voltage level at the connection point has to stay within an acceptable range [see Equation (3.10)], as most customer appliances (e.g. lighting equipment, motors, computers, etc.) require a specific voltage range (e.g. 230 V ± 10 %) for reliable operation.
- Customer requirement (CR2): the power should be available at exactly the time the consumers need it in order to use their various appliances (i.e. when a customer switches on a certain device).
- Customer requirement 3 (CR3): the consumed power should be available at a reasonable cost (this may also include low external costs to reflect the low environmental impact of electricity production). CR1 and CR2 also concern the reliability of the power supply. Greater reliability leads to higher costs and hence a conflict arises between CR1 and CR2 and the demand for reasonable costs.

3.7.2 Requirements from wind farm operators

Similar to consumers, wind farm owners or operators have certain demands on the existing power system in order to be able to sell the wind power production:

- Wind power requirement 1 (WP1): similar to consumer appliances, wind farms require a certain voltage level at the connection point as wind turbines are usually designed to operate within a specific voltage range (e.g. nominal voltage ±10 %). For the consumers, the requirements are rather homogeneous, since all consumers use the same type of equipment. For wind farms the requirements can sometimes be softened, since wind farms can be designed to handle different quality levels.
- Wind power requirement 2 (WP2): in addition, wind farm owners want to be able to sell their power production to the grid when wind power production is possible (i.e. when the wind speed is sufficient), otherwise, the production has to be spilled, which means that the wind farm owner loses possible financial income.
- Wind power requirement 3 (WP3): WP1 and WP2 also concern the reliability of the power system at the connection point of the wind farm. As mentioned in Section 3.7.1, there is always a trade-off between costs (i.e. low system costs) and reliability. The higher the reliability the higher the costs.

3.7.3 The integration issues

The challenge that the integration of wind power poses is to meet CR1 and CR2 and WP1 and WP2 in an economically efficient way (i.e. CR3 and WP3), even in the case of significant wind power penetration in the system.[5] This challenge will now be discussed in more detail using the simplified power system in Figure 3.9.

---

[5] In this context, it is also important to consider that the amount of the losses, P_L in a power system depend on the size of the loads and the wind power production (see Chapter 9).

3.7.3.1 Customer requirement 1: voltage level at the connection point of the consumer

First we will assume that no wind power is installed in the power system shown in Figure 3.9 and that the voltage U_0 is kept constant by the P_G generators. Now, if the load P_D varies, the currents I_3 and consequently I_1 will vary. Hence, there will be a voltage drop [Equation (3.7)] over the corresponding impedances Z_3 and Z_1. If impedances Z_3 and Z_1 are large (e.g. in the case of long lines or a comparatively low voltage) voltage U_3 will vary substantially when P_D varies. Possible measures to avoid large voltage variations in U_3, are as follows.

- Measure (a): use a stronger grid (i.e. impedances Z_1 and Z_3 are small). This is possible by using higher voltages in the lines and transformers that are not too small.
- Measure (b): control the voltage U_3 by using controllable transformers close to U_3.
- Measure (c): control voltage U_1 by using controllable transformers and/or some voltage controlling equipment such as shunt capacitors and/or shunt reactors.

We will now assume that wind power, P_W, is connected to the power system in Figure 3.9. As the wind power production P_W will vary, current I_2 varies. Hence current I_1 will vary, which will cause a voltage drop in Z_1. Furthermore, U_1 will vary and possibly also the voltage close to the customer connection point, voltage U_3. The impact of wind power variations on the voltage variations in U_3 depend mainly on the size of impedance Z_1. On the one hand, if Z_1 is large, there will be a strong connection between wind power variations and voltage variations in U_3. On the other hand, if Z_1 is very small, voltage U_3 will be more or less independent of wind power variations. In reality, only consumers that are rather close to wind farms may be affected by variations in wind power production. To avoid problems for consumers close to wind turbines, Measures (a)–(c) can be used as well as the following measure:

- Measure (d): use local control of voltage U_2 at the wind farm. Depending on the wind turbine technology, the voltage control of U_2 can be performed by the wind farm (see also Chapters 12 and 19).

3.7.3.2 Wind power requirement 1: voltage level at the connection point of the wind farm

The voltage U_2 will also depend on P_W, P_D and the size of Z_1 and Z_2. The difference to the discussion in Sub-section 3.7.3.1, on CR1 is that the size of Z_2 is important instead of the size of Z_3. This is mainly of interest when wind power is located at a larger distance from the consumers. In this situation, Measures (a), (b) and (d) explained above are of a certain interest as well as the following:

- Measure (e) use a controllable transformer close to U_2. A controllable transformer is normally slower than application of Measure (d), though.

3.7.3.3 Customer requirement 2: power availability on demand

Again, we will first assume that there is no wind power connected to the power system shown Figure 3.9. Hence, when consumers increase their consumption, the power will be delivered directly from the conventional power plants [see Equation (3.16)].

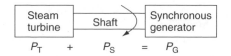

Figure   3.10 The Power balance in a conventional power plant

Conventional power plants generally use synchronous generators, which can be modelled as shown in Figure 3.10. The figure includes the generator and the steam or hydro turbine. The turbine rotates and drives the rotor of the generator. P_T is the power delivered by the turbine and P_S is the power delivered from kinetic energy stored in the rotating mass consisting of turbine, shaft and rotor. During normal operation P_S is 0. P_G is the electric power delivered to the power system.

If the load P_D increases, the power generation, P_G, will directly increase. However, the initial increase in power production is not due to an increase in the power production in the steam or hydro turbine. The increase of P_G will originate from the stored kinetic energy, P_S. Since the kinetic energy is used, the turbine–shaft–generator rotational system will slow down. Now, as the rotor speed of a synchronous generator is strongly coupled to the power system frequency, a decrease in rotor speed will result in a decrease in electric frequency. Therefore, a load increase will lead to a decrease in electric frequency.

In order to limit the decrease in power system frequency, some power plants are equipped with a so-called *primary control system* (see also Appendix A). This system measures the power system frequency and adjusts the power production of the power plant (i.e. P_T) when the frequency changes. The reaction time of primary control units depends on the power plants (e.g. how fast the production of steam can be increased). Usually, primary control units can increase their production by a few percent of rated capacity within 30 seconds to 1 minute. Secondary control (i.e. power plants with a slower response), will take over the capacity tasks of the primary control 10 to 30 minutes later and will thereby free up capacity that is used for primary control (see also Chapters 9 and 18).

The requirement of load balancing (i.e. so that consumers' demand for power can be met) means that:

- a power system must have sufficient primary and secondary control capacity available in order to be able to respond to changes in demand;
- these power plants must always have sufficient reserve margins to increase the power production to the level required for always meeting the system demand (i.e. the aggregated consumer demand).

If wind power is added to such a power system there will be an additional fluctuating source in the power system. With increasing wind power penetration, the requirements regarding power system balancing may increase (the details depend on the system design and load characteristics; see also Chapter 8). However, the primary and secondary control system will operate in exactly the same way as described above. Also, if wind power production decreases this has exactly the same effect on the system as an increase in demand [see Equation (3.16)].

The consequence is that there will be more variations that have to be balanced by primary and secondary control. Experience from Europe shows that even very high wind power penetration levels (up to 20 %) may not require additional primary control capacity as long as the installed wind power capacity is geographically distributed over a wide area (see also Chapter 8). The smoothing effect related to geographical distribution will result in low short-term variations in wind power production. However, owing to current limitations in wind speed forecast technologies, any mismatch between fore-casted wind power production and actual wind power production has eventually to be handled by the secondary control capacity (on wind power forecasts, see Chapter 17; on the experience in Denmark, see Chapter 10; for additional discussion, see Chapters 8, 9 and 18). The requirements for secondary control capacity are therefore significantly influenced by the wind power penetration level.

The additional system requirements for keeping the system balanced at all times depends very much on the individual system; that is, it depends on the load characteristics, the flexibility of the existing conventional power plants and the wind power penetration as well as the geographic distribution of the wind farms. The cost of meeting the increased requirements depends on the type of power plant (i.e. P_G), the size of interconnections to neighbouring systems and, of course, on the additional requirements.

3.7.3.4 Wind power requirement 2: network availability on demand

As already mentioned, wind farms want to sell into the power system whenever wind power production is possible. Depending on the power system design and the wind power penetration this may cause transmission congestions as well as stability issues (i.e. voltage stability). For a more detailed discussion of these issues we refer the reader to Chapter 20 as well as to Part 4 of this book, in particular to the introduc-tion of Part 4, Chapter 24.

3.7.3.5 Customer requirement 3: economical power supply

Again, we will first assume there is no wind power connected to the power system shown in Figure 3.9. In general, power system design must analyse the costs and benefits of a certain reliability level. No power system will be built with a 100 % reliability, as the costs will be in no relation to the benefits. Also, the cost of increasing reliability is very high if one already has a high reliability. There are two issues that have to be taken into account.

First, the power system must have a sufficient amount of power capacity, P_G, to meet maximum demand, $P_D + P_L$. We assume that we have a reliability of 99.9999 % [i.e. during 1 hour per year (the expected mean value) there is not enough installed capacity to meet the load]. If we want to increase reliability, we have to build a new plant that is used only during one hour per year. In this case it is sometimes considered economically more efficient to disconnect consumers (paid voluntary disconnections or forced dis-connections) than to build a new power plant that is used only 1 hour per year.

For the dimensioning of a power system, the so called $N-1$ criterion is usually applied. The $N-1$ criterion means that an outage in the largest power plant should not cause the disconnection of any consumer. An outage in a large plant causes a decrease in frequency and the activation of the primary control system, which increases power production in some other units in order to compensate this outage (see sub-section 3.7.3.3). Because of the $N - 1$ criterion, it is not acceptable to disconnect consumers [P_D remains the same in Equation (3.16)], which means system reserve margins must be at least of the same size as the largest unit in the system.

Second, there must be sufficient grid capacity to transmit the power from the generators to the consumers. The consumers are actually distributed in the grid and there has to be sufficient capacity to supply the maximum load of each individual consumer. However, no component in a power system has a reliability of 100 %. Therefore, redundancy within the power system is required (e.g. redundant transmission lines and backup power plants). For defining how large the redundancy should be the costs of keeping the redundancy have to be compared with the costs associated with disconnecting consumers as a result of insufficient redundancy.

The impact of wind power on system reliability is discussed in the following sub-section, on WR3.

3.7.3.6 Wind power requirement 3: power system reliability

The introduction of wind power modifies the economic trade-off between reliability costs and consumer costs for insufficient reliability. Some of the trade-offs apply only to wind farms (i.e. they affect income for wind farm owners, but not consumers) and some affect the whole system.

An important reliability issue is related to the capacity margin in a given power system (i.e. there must be enough capacity available in a power system to cover the peak load). If we assume a certain power system, there is always a probability that the available power plants are not sufficient to cover the peak load. If wind power is now added to a power system, reliability will increase as there is a certain probability that there will be a certain amount of wind power production available during a peak load situation, which will decrease the risk of capacity deficit. Adding more wind power capacity to a power system may also allow a decrease in the installed capacity of other power plants in the system without reducing the system reliability. This is treated in more detail in Chapter 9.

In addition depending on the power system design and the wind power penetration, power system reliability might be affected by the introduction of wind power because of the influence it has on stability in the case of a fault in the power system. For a more detailed discussion of these issues we refer the reader to Part D of this book, in particular to the introduction of Part D, Chapter 24, but also to Chapters 10, 11 and 20.

It should also be considered that, in contrast to systems with only varying load, active power balancing in a system with both wind power and load varying may require more balancing equipment to keep a certain system reliability level, since the total variation may increase. However, the cost–benefit analysis should consider that, for instance, the largest possible decrease in wind power (which requires an increased production from other power plants) can coincide with high wind power production. In such a situation, other power plants may have previously reduced their power production as a result of increasing wind power production. Hence, these (conventional) power plants may be able to increase production if wind power production decreases. But also in this case the trade-off between the consumer benefit of high reliability and system costs for back-up and reserves have to be taken into account. An important issue here is, for instance, how fast aggregated wind power production can decrease during times when aggregated load levels typically increase very fast.

If we consider now the reliability issues related to the installation of a certain wind farm, we have to discuss the dimensioning of the transmission system between the wind farm and the rest of the grid. We can start with the assumption that good wind resources are located in remote areas, at a larger distance from the rest of the power system (i.e. the cost of Z_2-lines will be rather high), hence a back-up transmission system between the wind farm and the main grid will be rather expensive compared with the economic benefits a back-up transmission system may provide (for a more discussion on back-up transmission systems for offshore wind farms, see Chapter 22) Hence the lack of a redundant transmission line may have a negative impact on the technical availability of wind farms. In this case, the compromise must consider the costs of a redundant transmission system and the lost income for wind farm owners during times when the transmission system is interrupted.

Furthermore, the required power quality level (U_2) at the connection point of the wind farm might be of importance. If a very stable voltage level is required, the grid, represented by Z_2, has to incorporate some voltage regulating equipment. An alternative is that the wind turbines that are used have a reduced sensitivity to voltage variations and/or have a voltage control capability.

It must also be noted that the current variations from the wind farms will affect the voltage drop over Z_1. The most extreme cases occur during times with maximum wind power production (P_W) and minimum consumption (P_D), and zero wind power production (P_W) and maximum consumption (P_D). This implies that U_1 will show increased variation with increasing installed wind power, which may require additional voltage control equipment. However, it might be important to consider the probability of those extreme cases. The probability of full production in all wind farms at the same time that the load is at its minimum, for instance, might be low, so that additional voltage control equipment will not be required. The question is, then, for how severe situations the voltage control equipment should be designed. It is not economically relevant for all points within a network to keep a voltage within certain limits 100 % of the time.

3.8 Conclusions

This chapter has provided an overview of the challenges regarding the integration of wind power into power systems. The overall aim of power system operation, independent of wind power penetration levels, is to supply an acceptable voltage to consumers and continuously to balance production and consumption. Furthermore, the power system should have an acceptable reliability level, independent of wind power penetration. Therefore this chapter has discussed the impact of wind power on voltage control and overall system balance.

The chapter has included basic electric power system theory and the basics of how wind power production behaves in order to arrive at a better understanding of integration issues. The appendix gives a mechanical equivalent to a power system with wind power in order to illustrate how a power system is operated.

Appendix: A Mechanical Equivalent to Power System Operation with Wind Power

For engineers that are not very familiar with power system analysis it may be sometimes rather difficult to understand the relevant problems related to the interconnection of wind power into a power system. Therefore, the following mechanical equivalent to a power system was developed. The mechanical equivalent consists of a bike with multiple seats (see Figure 3.11), referred to as 'long bike' throughout the following text (this is a shortened and modified version of Söder, 2002).

When taking the image of the long bike, the active power balance refers to the operation of the bike at a constant speed. The goal to keep a constant voltage (i.e. to keep the reactive power balance) corresponds to the challenge to keep the bike in balance (i.e. to kept the bike on the road without it falling to one side).

Introduction

Assume a very long bike according to Figure 3.11. The bike is made of a nonrigid material, and the chains between the different chain rings are slightly elastic. When the bike stands still, all pedals will be at exactly the same position (e.g. at the bottom position). We can now assume that the bike runs on a flat, straight road and the aim is to keep the bike at a constant speed and in an upright position. The rolling resistance and air resistance are neglected.

Figure 3.11 A long bike (Reproduced by permission of the Royal Institute of Technology, Sweden.)

Active power balance

On the bike, there are some cyclists (pedalling cyclists are generators, and braking cyclists are loads) who pedal continuously, but different cyclists apply different force. Since all chain rings are connected via a chain, the chain rings will rotate with the same speed. A so-called 'bike rpm' (system frequency) related to the speed of the bike is obtained. Some of the braking cyclists try to stop their pedals from rotating (motor loads). It is now important to note that the chain rings connected to these pedals also have the same bike rpm. As air resistance and rolling resistance are not considered, the bike will roll at a constant speed if the total force of the biking cyclists (total generation) is exactly the same as the total braking power of all braking cyclists (total load). Otherwise, the speed of the bike will change (the frequency in the power system will change).

Synchronous machines

Some cyclists, so called 'synchronous cyclists', have their pedals connected directly to the chain rings. This is the most common way of connecting pedals with chain rings for a 'grown-up' cyclists (large power plants), but for cyclists who bike only when it is windy (wind power plants), this connection is very rare. This implies that all synchronous cyclists will bike at the same speed. It must be noted, though, that the chains between the cyclists are slightly elastic. This means that there will be an angle difference between the different cyclists; that is, the pedals will not be at the bottom position at exactly the same time [the voltage angles, $\arg(U)$, in different points in the power system are different]. It can be noted that some cyclists (who could be hydropower stations) prefer to bike slower. They have to have a gear (several machine poles) between pedals and sprocket. It becomes obvious that a sudden change of the force when pedalling causes oscillations (power system angle dynamics) since there is some inertia in the sprocket connected to the pedals, and the chains to neighbouring chain rings are elastic.

Asynchronous machines

Some cyclists, usually children, that cannot bike as well as others (generators with low power production, such as wind power or distributed generation) prefer to use a softer connection between pedals and sprocket. Instead, they use a belt, which implies that they, depending on the belt slip, have to bike slightly faster than the bike rpm. These cyclists are therefore called 'asynchronous cyclists'. There are also some asynchronous cyclists who use the belt connection to the sprocket and toe clips. They try to stop the bike (asynchronous motor loads).

Power electronic interfaces

Another technique is used by 'variable cyclists' (power plants with an electronic interface between the generator and the grid). These cyclists use a stepless transmission (e.g. a wind power plant with a double-feed asynchronous generator) between pedals and a sprocket instead of a gearbox. This way, the speed of the pedals is more independent of the bike rpm.

Frequency control

Some of the (normally synchronous) cyclists continuously look at the speed of the bike (power plants used for frequency control). If the speed of the bike decreases, because of a nonreliable cyclist (forced outage in a power plant) or too many braking cyclists (the load on the system increases), these 'speed controllers' adjust their stroke on the pedals in order to adjust the speed. They always have to keep a margin (power plants cannot run at maximum installed capacity) and cannot continuously bike at their maximum capacity. If the speed of the bike changes too often or too much, they require other cyclists to change their pedalling efforts. On some bikes (in deregulated power systems) a special person, who is not biking and who is in charge of the speed control (independent system operator), finds out who is prepared to make these changes least expensively (cheapest bid on balancing market) and asks this cyclist to change the pedal stroke.

Wind power

Some cyclists only bike when it is windy. When the wind speed is low, these cyclists will pedal with less force. But still the bike rpm is about the same. The consequence is that the bike runs a bit slower and the bike speed controllers increase their force on the pedals. This is from the bike's perspective exactly what happens when some cyclists increase their braking force. When it is windy some other cyclists can bike less. These cyclists can drink some water, to build up energy for the time without wind (storage of water in hydro reservoirs). Remember that it is still the speed controller who makes the bike run at the same speed (keeps the balance between production and consumption).

Reactive power balance

A problem on the bike is that the seats of some cyclists and braking people are not positioned at the midpoint of the bike. The consequence is that the bike will fall to the side (a so-called voltage collapse, which occurs when there are no reactive power sources to keep an acceptable voltage) if these forces are not balanced. Figure 3.12 shows what such a four-person bike might look like from above.

P_G, the biking cyclists, try to accelerate the bike, whereas P_D, the braking cyclists, try to decrease the speed. Q_L are forces that try to overturn the bike to the left (producers of reactive power), whereas Q_R try to overturn it to the right (consumers of reactive power). The aim is to keep the bike in an exact vertical position (voltages = 1 per unit = nominal voltage), but a slight angle difference from vertical is still acceptable (a slight voltage deviation).

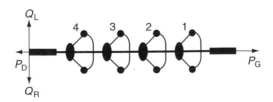

Figure   3.12 Forces on a long bike: a four-person bike

Asynchronous machines

The asynchronous cyclists, the ones with the belt, always sit a bit to the right of the bike midpoint. This implies that this causes a force on the bike (they consume reactive power) that overturns it to the right if this force is not compensated. It does not matter if they bike (asynchronous generators) or if they brake (asynchronous motors).

Capacitors

There are some special high-capacity cyclists, who never bike or brake. They just sit on a seat, located to the left-hand side of the midpoint of the bike (capacitors producing reactive power). The best location is to have them close to asynchronous cyclists in order to compensate them directly (it is common to have capacitors close to wind power plants with asynchronous generators). If the high-capacity people are located at another part of the bike, extra forces will be induced on the bike frame (if there is no local compensation of the reactive power consumption in wind power asynchronous generators, the reactive power has to be produced at other locations and transported to the locations where it is used, which will increase losses in the system, for instance).

Synchronous Machines

The synchronous cyclists have seats that could move from the right-hand side of the midpoint (reactive power consumption) to the left-hand side (reactive power production). By studying the balance of the bike they move from left to right in order to keep the bike in an upright position (to control reactive power to keep an acceptable voltage). It can be noted that the possibility of moving from left to right is limited by the force on the pedals. When they bike a lot they cannot move so much from left to right (the total current limits reactive power production or consumption at high active power production).

Power electronic interfaces

The variable cyclists use the stepless transmission. This type of system often lets the person sit slightly to the right or to the left of the midpoint (with the possibility of controlling the voltage by modifying the reactive power production or consumption). A common solution is to let the biker sit on the midpoint of the bike when the cyclist bikes as much as possible (power factor = 1 at installed capacity in wind power plants with variable speed).

References

[1] Burton, T., Sharpe, D., Jenkins, N., Bossanyi, E. (2001) *Wind Energy Handbook*, John Wiley & Sons, Ltd/ Inc., Chichester.
[2] Gasch, R., Twele, J. (2002) *Wind Power Plants*, James & James, London.
[3] Ender, C. (2004) 'Windenergienutzung in der Bundesrepublik Deutschland – Stand 31.12.2003' (Wind Energy Use in Germany – Status 31.12.2003), *DEWI Magazin* Number 24 (February 2004) pp. 6–18.

[4] Kaijser, A. (1995) 'Controlling the Grid: The Development of High-tension Power Lines in the Nordic Countries', in *Nordic Energy Systems – Historical Perspectives and Current Issues*. Science History Publication, Catson, MA, USA, pp. 31–54.

[5] Hills, R. L. (1994) *Power from Wind – A History of Windmill Technology*, Cambridge University Press, Cambridge.

[6] Hughes, T. P. (1993) *Networks of Power*, The John Hopkins University Press, Baltimore, MD.

[7] Hunt, S., Shuttleworth, G. (1996) *Competition and Choice in Electricity*, John Wiley & Sons, Ltd/Inc., Chichester.

[8] Petersen, E. L., Mortensen, N. G., Landberg, L., Højstrup, J., Frank, H. P. (1998) 'Wind power Meteorology, Part II: Siting and Models', *Wind Energy*; **1**(2) 55–72.

[9] Rosas, P. (2003) *Dynamic Influences of Wind Power on the Power System*, PhD thesis, Ørsted Institute and Technical University of Denmark, March 2003.

[10] Söder, L. (1997) 'Vindkraftens effektvärde' (Capacity Credit of Wind Power; in Swedish); Elforsk Rapport 97:27, Stockholm, Sweden, December 1997.

[11] Söder, L. (1999) 'Wind Energy Impact on the Energy Reliability of a Hydro-thermal Power System in a Deregulated Market', paper presented at the 13th Power Systems Computation Conference, 28 June to 2 July 1999, Trondheim, Norway.

[12] Söder, L. (2002) 'Explaining Power System Operation to Nonengineers', *Power Engineering Review, IEEE* **22**(4) 25–27.

4

Generators and Power Electronics for Wind Turbines

Anca D. Hansen

4.1 Introduction

Today, the wind turbines on the market mix and match a variety of innovative concepts with proven technologies both for generators and for power electronics. This chapter presents from an electrical point of view the current status of generators and power electronics in wind turbine concepts. It describes classical and new concepts of generators and power electronics based on technical aspects and market trends.

4.2 State-of-the-art Technologies

This section will describe the current status regarding generators and power electronics for wind turbines. In order to provide a complete picture, we will first briefly describe the common power control topologies of wind turbines.

4.2.1 Overview of wind turbine topologies

Wind turbines can operate either with a fixed speed or a variable speed.

4.2.1.1 Fixed-speed wind turbines

In the early 1990s the standard installed wind turbines operated at fixed speed. That means that regardless of the wind speed, the wind turbine's rotor speed is fixed and determined by the frequency of the supply grid, the gear ratio and the generator design.

Wind Power in Power Systems   Edited by T. Ackermann
© 2005 John Wiley & Sons, Ltd   ISBN: 0-470-85508-8 (HB)

It is characteristic of fixed-speed wind turbines that they are equipped with an induction generator (squirrel cage or wound rotor) that is directly connected to the grid, with a soft-starter and a capacitor bank for reducing reactive power compensation. They are designed to achieve maximum efficiency at one particular wind speed. In order to increase power production, the generator of some fixed-speed wind turbines has two winding sets: one is used at low wind speeds (typically 8 poles) and the other at medium and high wind speeds (typically 4–6 poles).

The fixed-speed wind turbine has the advantage of being simple, robust and reliable and well-proven. And the cost of its electrical parts is low. Its disadvantages are an uncontrollable reactive power consumption, mechanical stress and limited power quality control. Owing to its fixed-speed operation, all fluctuations in the wind speed are further transmitted as fluctuations in the mechanical torque and then as fluctuations in the electrical power on the grid. In the case of weak grids, the power fluctuations can also lead to large voltage fluctuations, which, in turn, will result in significant line losses (Larsson, 2000).

4.2.1.2 Variable-speed wind turbines

During the past few years the variable-speed wind turbine has become the dominant type among the installed wind turbines.

Variable-speed wind turbines are designed to achieve maximum aerodynamic efficiency over a wide range of wind speeds. With a variable-speed operation it has become possible continuously to adapt (accelerate or decelerate) the rotational speed ω of the wind turbine to the wind speed v. This way, the tip speed ratio λ is kept constant at a predefined value that corresponds to the maximum power coefficient.[1] Contrary to a fixed-speed system, a variable-speed system keeps the generator torque fairly constant and the variations in wind are absorbed by changes in the generator speed.

The electrical system of a variable-speed wind turbine is more complicated than that of a fixed-speed wind turbine. It is typically equipped with an induction or synchronous generator and connected to the grid through a power converter. The power converter controls the generator speed; that is, the power fluctuations caused by wind variations are absorbed mainly by changes in the rotor generator speed and consequently in the wind turbine rotor speed.

The advantages of variable-speed wind turbines are an increased energy capture, improved power quality and reduced mechanical stress on the wind turbine. The disadvantages are losses in power electronics, the use of more components and the increased cost of equipment because of the power electronics.

The introduction of variable-speed wind-turbine types increases the number of applicable generator types and also introduces several degrees of freedom in the combination of generator type and power converter type.

---

[1] Tip speed ratio, λ, is equal to $\omega R/v$, where R is the radius of the rotor.

4.2.2 Overview of power control concepts

All wind turbines are designed with some sort of power control. There are different ways to control aerodynamic forces on the turbine rotor and thus to limit the power in very high winds in order to avoid damage to the wind turbine.

The simplest, most robust and cheapest control method is the *stall control* (passive control), where the blades are bolted onto the hub at a fixed angle. The design of rotor aerodynamics causes the rotor to stall (lose power) when the wind speed exceeds a certain level. Thus, the aerodynamic power on the blades is limited. Such slow aerodynamic power regulation causes less power fluctuations than a fast-pitch power regulation. Some drawbacks of the method are lower efficiency at low wind speeds, no assisted startup and variations in the maximum steady-state power due to variations in air density and grid frequencies (for an example, see also Chapter 15).

Another type of control is *pitch control* (active control), where the blades can be turned out or into the wind as the power output becomes too high or too low, respectively. Generally, the advantages of this type of control are good power control, assisted startup and emergency stop. From an electrical point of view, good power control means that at high wind speeds the mean value of the power output is kept close to the rated power of the generator. Some disadvantages are the extra complexity arising from the pitch mechanism and the higher power fluctuations at high wind speeds. The instantaneous power will, because of gusts and the limited speed of the pitch mechanism, fluctuate around the rated mean value of the power.

The third possible control strategy is the *active stall control*. As the name indicates, the stall of the blade is actively controlled by pitching the blades. At low wind speeds the blades are pitched similar to a pitch-controlled wind turbine, in order to achieve maximum efficiency. At high wind speeds the blades go into a deeper stall by being pitched slightly into the direction opposite to that of a pitch-controlled turbine. The active stall wind turbine achieves a smoother limited power, without high power fluctuations as in the case of pitch-controlled wind turbines. This control type has the advantage of being able to compensate variations in air density. The combination with the pitch mechanism makes it easier to carry out emergency stops and to start up the wind turbine.

4.2.3 State-of-the-art generators

In the following, the most commonly applied wind turbine configurations are classified both by their ability to control speed and by the type of power control they use. Applying speed control as the criterion, there are four different dominating types of wind turbines, as illustrated in Figure 4.1.

Wind turbine configurations can be further classified with respect to the type of power (blade) control: stall, pitch, active stall. Table 4.1 indicates the different types of wind turbine configurations, taking both criteria (speed control and power control) into account. Each combination of these two criteria receives a label; for example, Type A0 denotes the fixed-speed stall-controlled wind turbine. The grey zones in Table 4.1 indicate the combinations that are not used in the wind turbine industry today (e.g. Type B0).

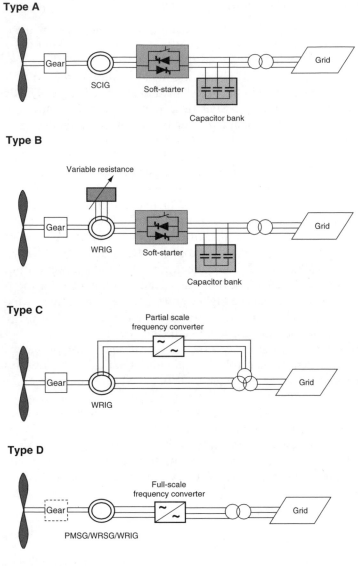

Figure   4.1 Typical wind turbine configurations. *Note*: SCIG = squirrel cage induction generator; WRIG = wound rotor induction generator; PMSG = permanent magnet synchronous generator; WRSG = wound rotor synchronous generator. The broken line around the gearbox in the Type D configuration indicates that there may or may not be a gearbox

In this chapter, we will look mainly at the standard wind turbine types, depicted in Figure 4.1 and Table 4.1. Other alternative, slightly different, wind turbine designs will not be discussed. Therefore, only the typical wind turbine configurations and their advantages as well as disadvantages will be presented in the following discussion.

Table   4.1 Wind turbine concepts

Speed control		Power control		
		Stall	Pitch	Active stall
Fixed speed	Type A	Type A0	Type A1	Type A2
Variable speed	Type B	Type B0	Type B1	Type B2
	Type C	Type C0	Type C1	Type C2
	Type D	Type D0	Type D1	Type D2

Note: The grey zones indicate combinations that are not in use in the wind turbine industry today.

4.2.3.1 Type A: fixed speed

This configuration denotes the fixed-speed wind turbine with an asynchronous squirrel cage induction generator (SCIG) directly connected to the grid via a transformer (see Figure 4.1). Since the SCIG always draws reactive power from the grid, this configuration uses a capacitor bank for reactive power compensation. A smoother grid connection is achieved by using a soft-starter.

Regardless of the power control principle in a fixed-speed wind turbine, the wind fluctuations are converted into mechanical fluctuations and consequently into electrical power fluctuations. In the case of a weak grid, these can yield voltage fluctuations at the point of connection. Because of these voltage fluctuations, the fixed-speed wind turbine draws varying amounts of reactive power from the utility grid (unless there is a capacitor bank), which increases both the voltage fluctuations and the line losses. Thus the main drawbacks of this concept are that it does not support any speed control, it requires a stiff grid and its mechanical construction must be able to tolerate high mechanical stress.

All three versions (Type A0, Type A1 and Type A2) of the fixed-speed wind turbine Type A are used in the wind turbine industry, and they can be characterised as follows.

Type A0: stall control
This is the conventional concept applied by many Danish wind turbine manufacturers during the 1980s and 1990s (i.e. an upwind stall-regulated three-bladed wind turbine concept). It has been very popular because of its relatively low price, its simplicity and its robustness. Stall-controlled wind turbines cannot carry out assisted startups, which implies that the power of the turbine cannot be controlled during the connection sequence.

Type A1: pitch control
These have also been present on the market. The main advantages of a Type A1 turbine are that it facilitates power controllability, controlled startup and emergency stopping. Its major drawback is that, at high wind speeds, even small variations in wind speed result in large variations in output power. The pitch mechanism is not fast enough to avoid such power fluctuations. By pitching the blade, slow variations in the wind can be compensated, but this is not possible in the case of gusts.

Type A2: active stall control

These have recently become popular. This configuration basically maintains all the power quality characteristics of the stall-regulated system. The improvements lie in a better utilisation of the overall system, as a result the use of active stall control. The flexible coupling of the blades to the hub also facilitates emergency stopping and startups. One drawback is the higher price arising from the pitching mechanism and its controller.

As illustrated in Figure 4.1 and Table 4.1, the variable speed concept is used by all three configurations, Type B, Type C and Type D. Owing to power limitation considerations, the variable speed concept is used in practice today only together with a fast-pitch mechanism. Variable speed stall or variable speed active stall-controlled wind turbines are not included here as potentially they lack the capability for a fast reduction of power. If the wind turbine is running at maximum speed and there is a strong gust, the aerodynamic torque can get critically high and may result in a runaway situation. Therefore, as illustrated in Table 4.1, Type B0, Type B2, Type C0, Type C2, Type D0 and Type D2 are not used in today's wind turbine industry.

4.2.3.2 Type B: limited variable speed

This configuration corresponds to the limited variable speed wind turbine with variable generator rotor resistance, known as OptiSlip®.[2] It uses a wound rotor induction generator (WRIG) and has been used by the Danish manufacturer Vestas since the mid-1990s. The generator is directly connected to the grid. A capacitor bank performs the reactive power compensation. A smoother grid connection is achieved by using a soft-starter. The unique feature of this concept is that it has a variable additional rotor resistance, which can be changed by an optically controlled converter mounted on the rotor shaft. Thus, the total rotor resistance is controllable. This optical coupling eliminates the need for costly slip rings that need brushes and maintenance. The rotor resistance can be changed and thus controls the slip. This way, the power output in the system is controlled. The range of the dynamic speed control depends on the size of the variable rotor resistance. Typically, the speed range is 0–10 % above synchronous speed. The energy coming from the external power conversion unit is dumped as heat loss.

Wallace and Oliver (1998) describe an alternative concept using passive components instead of a power electronic converter. This concept achieves a 10 % slip, but it does not support a controllable slip.

4.2.3.3 Type C: variable speed with partial scale frequency converter

This configuration, known as the doubly fed induction generator (DFIG) concept, corresponds to the limited variable speed wind turbine with a wound rotor induction generator (WRIG) and partial scale frequency converter (rated at approximately 30 % of nominal generator power) on the rotor circuit (Plate 4, in Chapter 2 shows the nacelle of a Type C turbine). The partial scale frequency converter performs the reactive power

---

[2] OptiSlip is a registered trademark of Vestas Wind Systems A/S.

compensation and the smoother grid connection. It has a wider range of dynamic speed control compared with the OptiSlip®, depending on the size of the frequency converter. Typically, the speed range comprises synchronous speed −40 % to +30 %. The smaller frequency converter makes this concept attractive from an economical point of view. Its main drawbacks are the use of slip rings and protection in the case of grid faults.

4.2.3.4 Type D: variable speed with full-scale frequency converter

This configuration corresponds to the full variable speed wind turbine, with the generator connected to the grid through a full-scale frequency converter. The frequency converter performs the reactive power compensation and the smoother grid connection. The generator can be excited electrically [wound rotor synchronous generator (WRSG) or WRIG) or by a permanent magnet [permanent magnet synchronous generator (PMSG)].

Some full variable-speed wind turbine systems have no gearbox (see the dotted gearbox in Figure 4.1). In these cases, a direct driven multipole generator with a large diameter is used, see Plate 3, in Chapter 2 for instance. The wind turbine companies Enercon, Made and Lagerwey are examples of manufacturers using this configuration.

4.2.4 State-of-the-art power electronics

The variable-speed wind turbine concept requires a power electronic system that is capable of adjusting the generator frequency and voltage to the grid. Before presenting the current status regarding power electronics, it is important to understand why it is attractive to use power electronics in future wind turbines: Table 4.2 illustrates the

Table 4.2 Advantages and disadvantages of using power electronics in wind turbine systems

Power electronics properties	Advantages	Disadvantages
Controllable frequency (important for the wind turbine)	▫ Energy optimal operation ▫ Soft drive train ▫ Load control ▫ Gearless option ▫ Reduced noise	Extra costs Additional losses
Power plant characteristics (important for the grid)	▫ Controllable active and reactive power ▫ Local reactive power source ▫ Improved network (voltage) stability ▫ Improved power quality ◊ reduced flicker level ◊ filtered out low harmonics ◊ limited short circuit power	High harmonics

implications of using power electronics in wind turbines both for the wind turbine itself and for the grid to which the wind turbine is connected.

Power electronics have two strong features:

• Controllable frequency: power electronics make it possible actually to apply the variable-speed concept, and it is therefore important from a wind turbine point of view. This feature results in the following direct benefits to wind turbines: (1) optimal energy operation; (2) reduced loads on the gear and drive train, as wind speed variations are absorbed by rotor speed changes; (3) load control, as life-consuming loads can be avoided; (4) a practical solution for gearless wind turbines, as the power converter acts as an electrical gearbox; and (5) reduced noise emission at low wind speeds. Regarding the wind turbine, the disadvantages of power electronics are the power losses and the increased costs for the additional equipment.
• Power plant characteristics: power electronics provide the possibility for wind farms to become active elements in the power system (Sørensen *et al.*, 2000). Regarding the grid, this property results in several advantages: (1) the active or reactive power flow of a wind farm is controllable; (2) the power converter in a wind farm can be used as a local reactive power source (e.g. in the case of weak grids); (3) the wind farm has a positive influence on network stability; and (4) power converters improve the wind farm's power quality by reducing the flicker level as they filter out the low harmonics and limit the short-circuit power. As far as the grid is concerned, power electronics have the disadvantage of generating high harmonic currents on the grid.

Power electronics include devices such as soft-starters (and capacitor banks), rectifiers, inverters and frequency converters. There is a whole variety of different design philosophies for rectifiers, inverters and frequency converters (Novotny and Lipo, 1996).

The basic elements of power converters are diodes (uncontrollable valves) and electronic switches (controllable valves), such as conventional or switchable thyristors and transistors. Diodes conduct current in one direction and will block current in the reverse direction. Electronic switches allow the selection of the exact moment when the diodes start conducting the current (Mohan, Undeland and Robbins, 1989). A conventional thyristor can be switched on by its gate and will block only when there is a zero crossing of the current (i.e. when the direction of the current is reversing), whereas switchable thyristors and transistors can freely use the gate to interrupt the current. The most widely known switchable thyristors and transistors are gate turn-off (GTO) thyristors, integrated gate commutated thyristors (IGCTs), bipolar junction transistors (BJTs), metal oxide semiconductor field effect transistors (MOSFETs) and insulated gate bipolar transistors (IGBTs). Table 4.3 compares characteristics and ratings of five of these switches. The values for voltage, current and output power are maximum output ratings. The switching frequency defines the operational frequency range.

Conventional thyristors can control active power, while switchable thyristors and transistors can control both active and reactive power (for details, see Mohan, Undeland and Robbins, 1989).

Today, variable-speed wind turbine generator systems can use many different types of converters. They can be characterised as either grid-commutated or self-commutated converters (Heier, 1998). The common type of grid-commutated converter is a thyristor.

It is cheap and reliable, but it consumes reactive power and produces current harmonics that are difficult to filter out. Typical self-commutated converters consist of either GTO thyristors or transistors. Self-commutated converters are interesting because they have high switching frequencies. Harmonics can be filtered out more easily and thus their disturbances to the network can be reduced to low levels. Today the most common transistor is the IGBT. As illustrated in Table 4.3, the typical switching frequency of an IGBT lies in the range of 2 to 20 kHz. In contrast, GTO converters cannot reach switching frequencies higher than about 1 kHz; therefore, they are not an option for the future.

Self-commutated converters are either voltage source converters (VSCs) or current source converters (CSCs; see Figure 4.2). They can control both the frequency and the voltage.

Table 4.3 Switches: maximum ratings and characteristics

	Switch type				
	GTO	IGCT	BJT	MOSFET	IGBT
Voltage[a] (V)	6000	6000	1700	1000	6000
Current[a] (A)	4000	2000	1000	28	1200
Switching frequency[b] (kHz)	0.2–1	1–3	0.5–5	5–100	2–20
Drive requirements	High	Low	Medium	Low	Low

[a] Maximum output rating.
[b] Operational range.
Note: GTO = gate turn-off thyristor; IGCT = integrated gate commutated thyristor;
BJT = bipolar junction transistor; MOSFET = metal oxide semiconductor field effect transistor;
IGBT = insulated gate bipolar transistor.
Source: L. H. Hansen *et al.*, 2001.

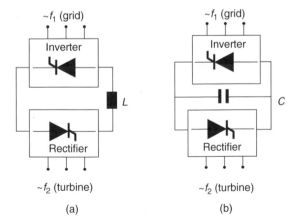

Figure 4.2 Types of self-commutated power converters for wind turbines: (a) a current source converter and (b) a voltage source converter. Reproduced from S. Heier, 1998, *Grid Integration of Wind Energy Conversion Systems*, by permission of John Wiley & Sons Ltd

VSCs and CSCs supply a relatively well-defined switched voltage waveform and a current waveform, respectively, at the terminals of the generator and the grid. In the case of a VSC, the voltage in the energy storage (the DC bus) is kept constant by a large capacitor. In a CSC, it is just the opposite; the current in the energy storage (the DC bus) is kept constant by a large inductor. It has to be stressed that voltage source conversion and current source conversion are different concepts. They can be implemented in several ways: six-step, pulse amplitude modulated (PAM) or pulse width modulated (PWM). By using the PWM technique, the low-frequency harmonics are eliminated and the frequency of the first higher-order harmonics lie at about the switching frequency of the inverter or rectifier.

4.2.5 State-of-the-art market penetration

Table 4.4 contains a list of the world's top-10 wind turbine suppliers for the year 2002 (BTM Consults Aps., 2003), with respect to installed power. The table includes the two largest (i.e. latest) wind turbines produced by each of the top-10 manufacturers. The applied configuration, control concept, generator type, generator voltage and generator or rotor speed range for each wind turbine have been evaluated by using data publicly available on the Internet or based on email correspondence with the manufacturers.

The Danish company Vestas Wind Systems A/S is the largest manufacturer of wind turbines in the world, followed by the German manufacturer Enercon. Danish NEG Micon and Spanish Gamesa are in third and fourth positions, respectively.

All top-10 manufacturers produce wind turbines in the megawatt range. For the time being, the most attractive concept seems to be the variable-speed wind turbine with pitch control. Of the top-10 suppliers, only the Danish manufacturer Bonus consistently uses only the active-stall fixed-speed concept. All the other manufacturers produce at least one of their two largest wind turbines based on the variable-speed concept. The most commonly used generator type is the induction generator (WRIG and SCIG). Only two manufacturers, Enercon and Made, use synchronous generators. All top-10 manufacturers use a step-up transformer for coupling the generator to the grid. Only one, Enercon, offers a gearless variable-speed wind turbine.

Comparing Table 4.4 with the analysis made by L. H. Hansen *et al.* (2001), there is an obvious trend towards the configuration using a DFIG (Type C1) with variable speed and variable pitch control. We wanted to illustrate this trend by looking at specific market shares and have therefore carried out detailed market research regarding the market penetration of the different wind turbine concepts from 1998 to 2002. The analysis is based on the suppliers' market data provided by BTM Consults Aps. and on evaluating the concept of each individual wind turbine type sold by the top-10 suppliers over the five years considered. This information was gathered from the Internet. The investigation processed information on a total of approximately 90 wind turbine types from 13 different manufacturers that were among the top-10 suppliers of wind turbines between 1998 and 2002: Vestas (Denmark), Gamesa (Spain), Enercon (Germany), NEG Micon (Denmark), Bonus (Denmark), Nordex (Germany and Denmark), GE-Wind/Enron (USA), Ecotechnia (Spain), Suzlon (India), Dewind (Germany), Repower (Germany), Mitsubishi (Japan) and Made (Spain). Table 4.5 presents

Table 4.4 Concepts applied by the top-10 wind turbine manufacturers (with respect to installed power in 2002); including the two largest (i.e. latest) wind turbines from each manufacturer

Turbine, by manufacturer	Concept[a]	Power and speed control features	Comments
Vestas, Denmark:			
V80, 2.0 MW	Type C1	Pitch Limited variable speed	WRIG (DFIG concept) Generator voltage: 690 V Generator speed range: 905–1915 rpm Rotor speed range: 9–19 rpm
V80, 1.8 MW	Type B1	Pitch Limited variable speed	WRIG Generator voltage: 690 V Generator speed range: 1800–1980 rpm Rotor speed range: 15.3–16.8 rpm
Enercon, Germany:			
E112, 4.5 MW	Type D1	Pitch Full variable speed	Multiple WRIG Generator voltage: 440 V Generator and rotor speed range: 8–13 rpm
E66, 2 MW	Type D1	Pitch Full variable speed	Multiple WRIG Generator voltage: 440 V Generator and rotor speed range: 10–22 rpm
NEG Micon, Denmark:			
NM80, 2.75 MW	Type C1	Pitch Limited variable speed	WRIG (DFIG concept) Generator stator/rotor voltage: 960 V/690 V Generator speed range: 756–1103 rpm Rotor speed range: 12–17.5 rpm
NM72, 2 MW	Type A2	Active stall Fixed speed	SCIG Generator voltage: 960 V Two generator speeds: 1002.4 rpm and 1503.6 rpm Two rotor speeds: 12 rpm and 18 rpm
Gamesa, Spain:			
G83, 2.0 MW	Type C1	Pitch Limited variable speed	WRIG Generator voltage: 690 V Generator speed range: 900–1900 rpm Rotor speed range: 9–19 rpm
G80, 1.8 MW	Type B1	Pitch Limited variable speed	WRIG (OptiSlip® concept) Generator voltage: 690 V Generator speed range: 1818–1944 rpm Rotor speed range: 15.1–16.1 rpm
GE Wind, USA:			
GE 104, 3.2 MW	Type C1	Pitch Limited variable speed	WRIG (DFIG concept) Generator stator/rotor voltage: 3.3 kV/690 V Rotor speed range: 7.5–13.5 rpm Generator speed range: 1000–1800 rpm
GE 77, 1.5 MW	Type C1	Pitch Limited variable speed	WRIG Generator voltage: 690 V Rotor speed range: 10.1–20.4 rpm Generator speed range: 1000–2000 rpm
Bonus, Denmark:			
Bonus 82, 2.3 MW	Type A2	Active stall Fixed speed	SCIG Generator voltage: 690 V Two generator speeds: 1000 rpm and 1500 rpm Two rotor speeds: 11 rpm and 17 rpm
Bonus 76, 2 MW	Type A2	Active stall Fixed speed	SCIG Generator voltage: 690 V Two generator speeds: 1000 rpm and 1500 rpm Two rotor speeds: 11 rpm and 17 rpm

(continued overleaf)

Table 4.4 (*continued*)

Turbine, by manufacturer	Concept[a]	Power and speed control features	Comments
Nordex, Germany:			
N80, 2.5 MW	Type C1	Pitch Limited variable speed	WRIG (DFIG concept) Generator voltage: 660 V Generator speed range: 700–1300 rpm Rotor speed range: 10.9–19.1 rpm
S77, 1.5 MW	Type C1	Pitch Limited variable speed	WRIG (DFIG concept) Generator voltage: 690 V Generator speed range: 1000–1800 rpm Rotor speed range: 9.9–17.3 rpm
Made, Spain:			
Made AE-90, 2 MW	Type D1	Pitch Full variable speed	WRSG Generator voltage: 1000 V Generator speed range: 747–1495 rpm Rotor speed range: 7.4–14.8 rpm
Made AE-61, 1.32 MW	Type A0	Stall Fixed speed	SCIG Generator voltage: 690 V Two generator speeds: 1010 rpm and 1519 rpm Two rotor speeds: 12.5 rpm and 18.8 rpm
Repower, Germany:			
MM 82, 2 MW	Type C1	Pitch Limited variable speed	WRIG (DFIG concept) Generator voltage: 690 V Generator speed range: 900–1800 rpm Rotor speed range: 10–20 rpm
MD 77, 1.5 MW	Type C1	Pitch Limited variable speed	WRIG (DFIG concept) Generator voltage: 690 V Generator speed range: 1000–1800 rpm Rotor speed range: 9.6–17.3 rpm
Ecotecnia, Spain:			
Ecotecnia 74, 1.67 MW	Type C1	Pitch Limited variable speed	WRIG (DFIG concept) Generator voltage: 690 V Generator speed range: 1000–1950 rpm Rotor speed range: 10–19 rpm
Ecotecnia 62, 1.25 MW	Type A0	Stall Fixed speed	SCIG Generator voltage: 690 V Two generator speeds: 1012 rpm and 1518 rpm Two rotor speeds: 12.4 rpm and 18.6 rpm

[a] For definitions of concept types, see Table 4.1 and Section 4.2.3 in text.

the results, which show that the capacity installed by these 13 suppliers in each year corresponds to more than 90 % of the total capacity installed during these years. This makes this analysis highly credible. These data cover approximately 76 % of the accumulated worldwide capacity installed at the end of 2002. They are therefore very comprehensive and illustrative.

Table 4.5 presents a detailed overview of the market share of each wind turbine concept from 1998 to 2002. It becomes obvious that the market share of the fixed-speed wind turbine concept (Type A) decreased during this period, with the DFIG concept (Type C) becoming

Table 4.5 World market share of wind turbine concepts between 1998 and 2002

Concept[a]	Share (%)				
	1998	1999	2000	2001	2002
Type A	39.6	40.8	39.0	31.1	27.8
Type B	17 8	17.1	17.2	15.4	5.1
Type C	26.5	28.1	28.2	36.3	46.8
Type D	16.1	14.0	15.6	17.2	20.3
Total installed power[b] (MW)	2349	3788	4381	7058	7248
Percentage share of the world market (supply)[b]	92.4	90.1	94.7	97.6	97.5

[a] For definitions of concept types, see Table 4.1 and Section 4.2.3 in text.
[b] For the 13 suppliers studied (Vestas, Gamesa, Enercon, NEG Micon, Bonus, Nordex, GE Wind/ Enron, Ecotechnia, Suzlon, Dewind, Repower, Mitsubishi and Made).
Source: Internet survey.

the dominant concept. The full variable-speed concept (Type D) slightly increased its market share, thus becoming the third most important concept in 2001 and 2002.

The market position of the OptiSlip concept (Type B) was almost constant during the first three years of the period analysed, but during the last two years it decreased, ceding its market share to the Type C concept. From the trend depicted in Table 4.5, Type B is likely to lose its market altogether. This is mainly because of the variable speed range of Type B, which is much more limited than is the variable speed range of Type C.

Table 4.5 also provides information on the total capacity installed annually by the 13 suppliers. It shows that the total installed capacity in 2002 was three times that in 1998, primarily because of the increased rated power of the wind turbines.

4.3 Generator Concepts

Basically, a wind turbine can be equipped with any type of three-phase generator. Today, the demand for grid-compatible electric current can be met by connecting frequency converters, even if the generator supplies alternating current (AC) of variable frequency or direct current (DC). Several generic types of generators may be used in wind turbines:

- Asynchronous (induction) generator:
  - squirrel cage induction generator (SCIG);
  - wound rotor induction generator (WRIG):
    - OptiSlip induction generator (OSIG),
    - Doubly-fed induction generator (DFIG).
- Synchronous generator:
  - wound rotor generator (WRSG);
  - permanent magnet generator (PMSG).

- Other types of potential interest:
  - high-voltage generator (HVG);
  - switch reluctance generator (SRG);
  - transverse flux generator (TFG).

In this section, we will summarise the essential properties of these generic generator types. For a detailed analysis of generator types, see the standard literature on this field (Heier, 1998; Krause, Wasynczuk and Sudhoff, 2002).

4.3.1 Asynchronous (induction) generator

The most common generator used in wind turbines is the induction generator. It has several advantages, such as robustness and mechanical simplicity and, as it is produced in large series, it also has a low price. The major disadvantage is that the stator needs a reactive magnetising current. The asynchronous generator does not contain permanent magnets and is not separately excited. Therefore, it has to receive its exciting current from another source and consumes reactive power. The reactive power may be supplied by the grid or by a power electronic system. The generator's magnetic field is established only if it is connected to the grid.

In the case of AC excitation, the created magnetic field rotates at a speed determined jointly by the number of poles in the winding and the frequency of the current, the synchronous speed. Thus, if the rotor rotates at a speed that exceeds the synchronous speed, an electric field is induced between the rotor and the rotating stator field by a relative motion (slip), which causes a current in the rotor windings. The interaction of the associated magnetic field of the rotor with the stator field results in the torque acting on the rotor.

The rotor of an induction generator can be designed as a so-called short-circuit rotor (squirrel cage rotor) or as a wound rotor.

4.3.1.1 Squirrel cage induction generator

So far, the SCIG has been the prevalent choice because of its mechanical simplicity, high efficiency and low maintenance requirements (for a list of literature related to SCIGs, see L. H. Hansen *et al.*, 2001).

As illustrated in Figure 4.1, the SCIG of the configuration Type A is directly grid—coupled. The SCIG speed changes by only a few percent because of the generator slip caused by changes in wind speed. Therefore, this generator is used for constant-speed wind turbines (Type A). The generator and the wind turbine rotor are coupled through a gearbox, as the optimal rotor and generator speed ranges are different.

Wind turbines based on a SCIG are typically equipped with a soft-starter mechanism and an installation for reactive power compensation, as SCIGs consume reactive power. SCIGs have a steep torque speed characteristic and therefore fluctuations in wind power are transmitted directly to the grid. These transients are especially critical during the grid connection of the wind turbine, where the in-rush current can be up to 7–8 times the rated current. In a weak grid, this high in-rush current can cause severe voltage disturbances. Therefore, the connection of the SCIG to the grid should be made gradually in order to limit the in-rush current.

During normal operation and direct connection to a stiff AC grid, the SCIG is very robust and stable. The slip varies and increases with increasing load. The major problem is that, because of the magnetising current supplied from the grid to the stator winding, the full load power factor is relatively low. This has to be put in relation to the fact that most power distribution utilities penalise industrial customers that load with low power factors. Clearly, generation at a low power factor cannot be permitted here either. Too low a power factor is compensated by connecting capacitors in parallel to the generator.

In SCIGs there is a unique relation between active power, reactive power, terminal voltage and rotor speed. This means that in high winds the wind turbine can produce more active power only if the generator draws more reactive power. For a SCIG, the amount of consumed reactive power is uncontrollable because it varies with wind conditions. Without any electrical components to supply the reactive power, the reactive power for the generator must be taken directly from the grid. Reactive power supplied by the grid causes additional transmission losses and in certain situations, can make the grid unstable. Capacitor banks or modern power electronic converters can be used to reduce the reactive power consumption. The main disadvantage is that the electrical transients occur during switching-in.

In the case of a fault, SCIGs without any reactive power compensation system can lead to voltage instability on the grid (Van Custem and Vournas, 1998). The wind turbine rotor may speed up (slip increases), for instance, when a fault occurs, owing to the imbalance between the electrical and mechanical torque. Thus, when the fault is cleared, SCIGs draw a large amount of reactive power from the grid, which leads to a further decrease in voltage.

SCIGs can be used both in fixed-speed wind turbines (Type A) and in full variable-speed wind turbines (Type D). In the latter case, the variable frequency power of the machine is converted to fixed-frequency power by using a bidirectional full-load back-to-back power converter.

4.3.1.2 Wound rotor induction generator

In the case of a WRIG, the electrical characteristics of the rotor can be controlled from the outside, and thereby a rotor voltage can be impressed. The windings of the wound rotor can be externally connected through slip rings and brushes or by means of power electronic equipment, which may or may not require slip rings and brushes. By using power electronics, the power can be extracted or impressed to the rotor circuit and the generator can be magnetised from either the stator circuit or the rotor circuit. It is thus also possible to recover slip energy from the rotor circuit and feed it into the output of the stator. The disadvantage of the WRIG is that it is more expensive and not as robust as the SCIG. The wind turbine industry uses most commonly the following WRIG configurations: (1) the OptiSlip® induction generator (OSIG), used in the Type B concept and (2) the doubly-fed induction generator (DFIG) concept, used in the Type C configuration (see Figure 4.1).

OptiSlip induction generator
The OptiSlip® feature was introduced by the Danish manufacturer Vestas in order to minimise the load on the wind turbine during gusts. The OptiSlip® feature allows the

generator to have a variable slip (narrow range) and to choose the optimum slip, resulting in smaller fluctuations in the drive train torque and in the power output. The variable slip is a very simple, reliable and cost-effective way to achieve load reductions compared with more complex solutions such as full variable-speed wind turbines using full-scale converters.

OSIGs are WRIGs with a variable external rotor resistance attached to the rotor windings (see Figure 4.1). The slip of the generator is changed by modifying the total rotor resistance by means of a converter, mounted on the rotor shaft. The converter is optically controlled, which means that no slip rings are necessary. The stator of the generator is connected directly to the grid.

The advantages of this generator concept are a simple circuit topology, no need for slip rings and an improved operating speed range compared with the SCIG. To a certain extend, this concept can reduce the mechanical loads and power fluctuations caused by gusts. However, it still requires a reactive power compensation system. The disadvantages are: (1) the speed range is typically limited to 0–10 %, as it is dependent on the size of the variable rotor resistance; (2) only poor control of active and reactive power is achieved; and (3) the slip power is dissipated in the variable resistance as losses.

Doubly-fed induction generator

As depicted in Table 4.5, the concept of the DFIG is an interesting option with a growing market. The DFIG consists of a WRIG with the stator windings directly connected to the constant-frequency three-phase grid and with the rotor windings mounted to a bidirectional back-to-back IGBT voltage source converter.

The term 'doubly fed' refers to the fact that the voltage on the stator is applied from the grid and the voltage on the rotor is induced by the power converter. This system allows a variable-speed operation over a large, but restricted, range. The converter compensates the difference between the mechanical and electrical frequency by injecting a rotor current with a variable frequency. Both during normal operation and faults the behaviour of the generator is thus governed by the power converter and its controllers.

The power converter consists of two converters, the rotor-side converter and grid-side converter, which are controlled independently of each other. It is beyond the scope of this chapter to go into detail regarding the control of the converters (for more detail, see Leonhard, 1980; Mohan, Undeland and Robbins, 1989; Pena, Clare and Asher, 1996). The main idea is that the rotor-side converter controls the active and reactive power by controlling the rotor current components, while the line-side converter controls the DC-link voltage and ensures a converter operation at unity power factor (i.e. zero reactive power).

Depending on the operating condition of the drive, power is fed into or out of the rotor: in an oversynchronous situation, it flows from the rotor via the converter to the grid, whereas it flows in the opposite direction in a subsynchronous situation. In both cases – subsynchronous and oversynchronous – the stator feeds energy into the grid.

The DFIG has several advantages. It has the ability to control reactive power and to decouple active and reactive power control by independently controlling the rotor excitation current. The DFIG has not necessarily to be magnetised from the power grid, it can be magnetised from the rotor circuit, too. It is also capable of generating reactive power that can be delivered to the stator by the grid-side converter. However, the grid-side converter normally operates at unity power factor and is not involved in the reactive power exchange between the turbine and the grid. In the case of a weak grid, where the voltage may fluctuate, the DFIG may be ordered to produce or absorb an amount of reactive power to or from the grid, with the purpose of voltage control.

The size of the converter is not related to the total generator power but to the selected speed range and hence to the slip power. Thus the cost of the converter increases when the speed range around the synchronous speed becomes wider. The selection of the speed range is therefore based on the economic optimisation of investment costs and on increased efficiency. A drawback of the DFIG is the inevitable need for slip rings.

4.3.2 The synchronous generator

The synchronous generator is much more expensive and mechanically more complicated than an induction generator of a similar size. However, it has one clear advantage compared with the induction generator, namely, that it does not need a reactive magnetising current.

The magnetic field in the synchronous generator can be created by using permanent magnets or with a conventional field winding. If the synchronous generator has a suitable number of poles (a multipole WRSG or a multipole PMSG), it can be used for direct-drive applications without any gearbox.

As a synchronous machine, it is probably most suited for full power control as it is connected to the grid through a power electronic converter. The converter has two primary goals: (1) to act as an energy buffer for the power fluctuations caused by an inherently gusting wind energy and for the transients coming from the net side, and (2) to control the magnetisation and to avoid problems by remaining synchronous with the grid frequency. Applying such a generator allows a variable-speed operation of wind turbines.

Two classical types of synchronous generators have often been used in the wind turbine industry: (1) the wound rotor synchronous generator (WRSG) and (2) the permanent magnet synchronous generator (PMSG).

4.3.2.1 Wound rotor synchronous generator

The WRSG is the workhorse of the electrical power industry. Both the steady-state performance and the fault performance have been well-documented in a multitude of research papers over the years, (see L. H. Hansen *et al.*, 2001).

The stator windings of WRSGs are connected directly to the grid and hence the rotational speed is strictly fixed by the frequency of the supply grid. The rotor winding is excited with direct current using slip rings and brushes or with a brushless exciter with a rotating rectifier. Unlike the induction generator, the synchronous generator does not need any further reactive power compensation system. The rotor winding, through which direct current flows, generates the exciter field, which rotates with synchronous speed. The speed of the synchronous generator is determined by the frequency of the rotating field and by the number of pole pairs of the rotor.

The wind turbine manufacturers Enercon and Lagerwey use the wind turbine concept Type D (see Figure 4.1) with a multipole (low-speed) WRSG and no gearbox. It has the advantage that it does not need a gearbox. But the price that has to be paid for such a gearless design is a large and heavy generator and a full-scale power converter that has to handle the full power of the system. The wind turbine manufacturer Made also applies the wind turbine concept Type D, but with a four-pole (high-speed) WRSG and a gearbox (see Table 4.4).

4.3.2.2 Permanent magnet synchronous generator

Many research articles have suggested the application of PMSGs in wind turbines because of their property of self-excitation, which allows an operation at a high power factor and a high efficiency, (see Alatalo, 1996).

In the permanent magnet (PM) machine, the efficiency is higher than in the induction machine, as the excitation is provided without any energy supply. However, the materials used for producing permanent magnets are expensive, and they are difficult to work during manufacturing. Additionally, the use of PM excitation requires the use of a full-scale power converter in order to adjust the voltage and frequency of generation to the voltage and the frequency of transmission, respectively. This is an added expense. However, the benefit is that power can be generated at any speed so as to fit the current conditions. The stator of PMSGs is wound, and the rotor is provided with a permanent magnet pole system and may have salient poles or may be cylindrical. Salient poles are more common in slow-speed machines and may be the most useful version for an application for wind generators. Typical low-speed synchronous machines are of the salient-pole type and the type with many poles.

There are different topologies of PM machines presented in the literature. The most common types are the radial flux machine, the axial flux machine and the transversal flux machine. A detailed description of all these types is given in Alatalo (1996).

The synchronous nature of the PMSG may cause problems during startup, synchronisation and voltage regulation. It does not readily provide a constant voltage (Mitcham and Grum, 1998). The synchronous operation causes also a very stiff performance in the case of an external short circuit, and if the wind speed is unsteady. Another disadvantage of PMSGs is that the magnetic materials are sensitive to temperature; for instance, the magnet can lose its magnetic qualities at high temperatures, during a fault, for example. Therefore, the rotor temperature of a PMSG must be supervised and a cooling system is required.

Examples of wind turbine manufacturers that use configuration Type D with PMSGs are Lagerwey, WinWind and Multibrid.

4.3.3 Other types of generators

In the following, we will briefly present other types of generators that are possible future candidates for the wind turbine industry.

4.3.3.1 Highvoltage generator

Most commonly, wind turbine generators are operated at 690 V (see Table 4.4) and they therefore require a transformer in the nacelle or at the bottom of the tower. The main motivation for increasing the voltage of the generator is to reduce the current and thereby reduce the losses and the amount of heat that has to be dissipated. This can lead to a reduction in the size of the generator and to a higher efficiency of the wind turbine, especially at higher loads. If the voltage of the machine matches the grid voltage, a grid connection is possible without a transformer.

HVGs are manufactured both as synchronous generators and as asynchronous generators. HVGs are a potentially interesting alternative for large wind turbines exceeding 3 MW. The major disadvantages are the high cost of the total system, the uncertainty regarding its long-term performance and the safety requirements, which are more complex than those for low-voltage machines. The price of the HVG, the power electronics and the auxiliary equipment, such as switchgears, increases substantially with the size of the generator. The price could decrease in the future if the number of wind turbines with HVGs increases significantly.

So far, only very few prototype wind turbines that have been designed by utilities or large manufacturers of electric equipment have applied HVGs: Tjæreborg, with 2 MW, applies an induction generator with an output voltage of 10 kV, and Growian, with 3 MW, uses a DFIG with an output voltage of 6.3 kV. Different companies have initiated, with varying success, research projects on HVG wind turbines, during the past few years. Lagerwey, for example, has started the series production of its LW72 2 MW turbine that has a synchronous generator with an output voltage of 4 kV. Contrary to what was expected, the Windformer/ABB 3 MW concept has not been successful. However, at present there are not many commercially available wind turbines with HVGs. Instead of using HVGs, the trend has so far been towards moving the step-up transformer up into the nacelle.

4.3.3.2 The switched reluctance generator

The SRG machine has shown over the years that it has a robust and simple mechanical structure, a high efficiency, reduced costs and that it provides the opportunity to eliminate the gearbox (Kazmierkowski, Krishnan and Blaabjerg, 2002). It is interesting for aerospace applications because of its ability to continue operating at reduced output in the presence of faults in the generator itself. A survey of the positive and negative properties of SRGs is included in L. H. Hansen *et al.* (2001). The literature on SRGs related to wind turbines is not substantial, and much research remains to be done before the SRG will be adapted to wind turbine applications.

The SRG is a synchronous generator with a doubly salient construction, with salient poles on both the stator and the rotor. Excitation of the magnetic field is provided by the stator current in the same way as it is provided for the induction generator. The SRG is considered inferior to the PMSG machine because of its lower power density. The SRG requires a full-scale power converter in order to operate as a grid-connected generator. Moreover, the SRG has a lower efficiency than a PMSG and a lower power factor than asynchronous generators (Kazmierkowski, Krishnan and Blaabjerg, 2002).

4.3.3.3 Transverse flux generator

The TFG machine topology is fairly new, but seems to be interesting. However, more research is required before the TFG machine will be adapted so that it can be used as a wind power generator.

The transverse flux (TF) principle may be applied to a range of machine types. It could be used both in permanent magnet and in reluctance machines, for instance. The machine will inherently behave as the generic type that is applied, but will have characteristics that are influenced by the TF design. The high ratio of torque per kilogram of active material seems to be very attractive (Dubois, Polinder and Fereira, 2000).

The nature of its operation is equal to that of a synchronous machine, and it will function in principle in a way that is similar to any other PM machines. It can comprise a very large number of poles, which may make it suitable for direct gearless applications. However, the TFG has a relatively large leakage inductance. In the reluctance version, this may cause the power factor to become very low during normal operation, and the short-circuit current is insufficient to trip normal protection. There are similar problems regarding the PM version, but because of the PM, they will not be as severe.

A disadvantage of the TFG is the large number of individual parts that it requires and that a lamination technology has to be used. With the advance of powder technology, this situation may improve.

4.4 Power Electronic Concepts

Power electronics is a rapidly developing technology. Components can handle higher current and voltage ratings, the power losses decrease and the devices become more reliable. The devices are also very easy to control with a megascale power amplification. The price/power ratio is still decreasing, and power converters are becoming more and more attractive as a means of improving the performance of wind turbines. In this section we will present the power converter topologies that are most commonly used in wind turbine applications, including their advantages and disadvantages.

4.4.1 Soft-starter

The soft-starter is a simple and cheap power electrical component used in fixed-speed wind turbines during their connection to the grid (see Figure 4.1, Types A and B). The soft-starter's function is to reduce the in-rush current, thereby limiting the disturbances to the grid. Without a soft-starter, the in-rush current can be up to 7–8 times the rated current, which can cause severe voltage disturbances on the grid.

The soft-starter contains two thyristors as commutation devices in each phase. They are connected antiparallel for each phase. The smooth connection of the generator to the grid, during a predefined number of grid periods, is achieved by adjusting the firing angle (α) of the thyristors. The relationship between the firing angle (α) and the resulting amplification of the soft-starter is highly nonlinear and is additionally a function of the power factor of the connected element. After the in-rush, the thyristors are bypassed in order to reduce the losses of the overall system.

4.4.2 Capacitor bank

The capacitor bank is used in fixed-speed or limited variable-speed wind turbines (see Figure 4.1, Types A and B). It is an electrical component that supplies reactive power to the induction generator. Thus the reactive power absorbed by the generator from the grid is minimised.

The generators of wind turbines can have a full load dynamic compensation, where a certain number of capacitors are connected or disconnected continuously, depending on the average reactive power demand of the generator over a predefined period of time.

The capacitor banks are usually mounted at the bottom of the tower or to the nacelle (i.e. at the top of the wind turbine). They may be heavy loaded and damaged in the case of overvoltages on the grid and thereby may increase the maintenance cost of the system.

4.4.3 Rectifiers and inverters

A traditional frequency converter, also called an adjustable speed drive, consists of:

- a rectifier (as AC-to-DC conversion unit) to converts alternating current into direct current, while the energy flows into the DC system;
- energy storage (capacitors);
- an inverter (DC-to-AC with controllable frequency and voltage) to convert direct current into alternating current, while the energy flows to the AC side.

Diodes can be used only in rectification mode, whereas electronic switches can be used in the rectifying as well as in the inverting mode.

The most common rectifier solution is the diode rectifier, because of its simplicity, its low cost and low losses. It is nonlinear in nature and, consequently, it generates harmonic currents (Kazmierkowski, Krishnan and Blaabjerg, 2002). Another drawback is that it allows only a unidirectional power flow; it cannot control the generator voltage or current. Therefore, it can be used only with a generator that can control the voltage and with an inverter (e.g. an IGBT) that can control the current.

The thyristor (grid-commutated) based inverter solution is a cheap inverter, with low losses and, as its name indicates, it needs to be connected to the grid to be able to operate. Unfortunately, it consumes reactive power and produces large harmonics (Heier, 1998). The increasing demands on power quality make thyristor inverters less attractive than self-commutated inverters, such as GTO inverters and IGBTs. The advantage of a GTO inverter is that it can handle more power than the IGBT, but this feature will be less important in the future, because of the fast development of IGBTs. The disadvantage of GTOs is that the control circuit of the GTO valve is more complicated.

The generator and the rectifier must be selected as a combination (i.e. a complete solution), while the inverter can be selected almost independently of the generator and the rectifier. A diode rectifier or a thyristor rectifier can be used together only with a synchronous generator, as it does not require a reactive magnetising current. As opposed to this, GTO and IGBT rectifiers have to be used together with variable-speed induction generators, because they are able to control the reactive power. However, even though IGBTs are a very attractive choice, they have the disadvantages of a high price and high losses. The synchronous generator with a diode rectifier, for example, has a much lower total cost than the equivalent induction generator with an IGBT inverter or rectifier (Carlson, Hylander and Thorborg, 1996).

There are different ways to combine a rectifier and an inverter into a frequency converter. There are five applicable technologies for adjustable speed: back-to-back, multilevel, tandem, matrix and resonant. An evaluation of each power converter topology is included in L. H. Hansen *et al.* (2001).

4.4.4 Frequency converters

During recent years different converter topologies have been investigated as to whether they can be applied in wind turbines:

- back-to-back converters;
- multilevel converters;
- tandem converters;
- matrix converters;
- resonant converters.

A presentation of each of these converter topologies and their working principles, including their advantages and disadvantages, is beyond the scope of this chapter. For a detailed description, see L. H. Hansen *et al.* (2001). It is evident that the back-to-back converter is highly relevant to wind turbines today. It constitutes the state of the art and may therefore be used for benchmarking the other converter topologies. The analysis in L. H. Hansen *et al.* shows that the matrix and multilevel converter are the most serious competitors to the back-to-back converter and thus are recommended for further studies.

This section focuses mainly on the back-to-back converter, as it is today the most widely used three-phase frequency converter. However, we will also include some brief comments regarding multilevel and matrix converters.

The back-to-back converter is a bidirectional power converter consisting of two conventional pulse-width modulated (PWM) VSC converters. The topology is shown in Figure 4.3. The DC link voltage is boosted to a level higher than the amplitude of the grid line-to-line voltage in order to achieve full control of the grid current. The presence

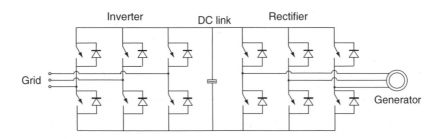

Figure **4.3** Structure of the back-to-back frequency converter

of the boost inductance reduces the demands on the input harmonic filter and offers some protection for the converter against abnormal conditions on the grid.

The capacitor between the inverter and rectifier makes it possible to decouple the control of the two inverters, allowing the compensation of asymmetry on both the generator side and the grid side, without affecting the other side of the converter. The power flow at the grid-side converter is controlled to keep the DC link voltage constant, and the control of the generator-side converter is set to suit the magnetisation demand and the desired rotor speed. The control of the back-to-back converter in wind turbine applications is described in several papers (e.g. Bogalecka, 1993; Pena, Clare and Asher, 1996).

Kim and Sul (1993) mention that the presence of the DC link capacitor in a back-to-back converter reduces the overall lifetime and efficiency of the system compared with a converter without a DC link capacitor, such as the matrix converter (Schuster, 1998). However, the protection of the matrix converter in a fault situation is not as good as that of the back-to-back converter. Another aspect of the back-to-back converter is the high switching losses compared with the switching losses of the matrix converter (Wheeler and grant, 1993). The disadvantages of the matrix converter compared with the back-to-back converter are higher conduction losses and the limitation of the output voltage converter (Wheeler and Grant, 1993).

In comparison with converters with constant DC link voltage and only two output levels, the output harmonic content of the matrix converter is lower because of the fact that the output voltage of the matrix converter is composed of three voltage levels. However, considering the harmonic performance, the multilevel converter excels by being the converter with the lowest demands on the input filters and therefore with the best spectra on both the grid side and the generator side (Rodriguez et al., 1999).

To summarise the findings on the presented converters, it is concluded that the back-to-back converter, the matrix converter and the multilevel converter are the ones recommended for further studies in different generator topologies.

4.5 Power Electronic Solutions in Wind Farms

Today, and in the future, wind turbines will be sited in large concentrations with hundreds of megawatts of power capacity. Wind farms of this size will often be connected directly to the transmission grid and will, sooner or later, replace conventional power plants. This means that the wind turbines will be required to have power plant characteristics (A. D. Hansen, 2001; Sørensen et al., 2000), namely, to be able to behave as active controllable components in the power system. Such large wind farms will be expected to meet very high technical demands, such as to perform frequency and voltage control, to regulate active and reactive power and to provide quick responses during transient and dynamic situations in the power system. The traditional wind turbines, where the active power is controlled by a simple pitching of the blades or by using a dumping device or by disconnecting the wind turbines, do not have such control capabilities and cannot contribute to power system stability as will be required. Storage technologies may be an option, too, but at the present such technologies are rather expensive. Also, they are not a satisfactory solution in the case of large wind farms, because of voltage stability issues. Power electronic technology will therefore becomes more and more attractive for large wind farms that will have to fulfil future high demands.

Presently, there are significant research activities to develop the electrical control layout of such wind farms with different types of power electronic converters in order for them to be able to comply with the high requirements and to be as cheap as possible to install. Many control topologies are being investigated and some are already being implemented in practice.

Depending on how the power electronic devices are used inside a wind farm, there are several different topology options, each with its particular advantages and disadvantages. The topology can include the following.

- A completely decentralised control structure with an internal AC network connected to the main grid, with each turbine in the wind farm having its own frequency converter and its own control system, has the advantage that each wind turbine can operate at its optimum level with respect to its local wind conditions. A practical implementation of this structure is the 160 MW Horns Rev offshore wind farm, based on wind turbines with doubly-fed induction generators of Type C (see Figure 4.1).
- A partly centralised, partly decentralised control structure where the power converter is 'split up' and the output of each turbine is locally rectified and fed into a DC network, with the whole farm still connected through a central inverter, has been proposed by Dahlgren et al. (2000), who suggest use of a multipole high voltage synchronous generator with permanent magnets. However, this solution has not yet been implemented in practice. This configuration provides all the features of the variable-speed concept, since each wind turbine can be controlled independently. The generators could be SCIGs as well, if a VSC were used as rectifier.
- A completely centralised control structure has a central power electronic converter connected at the wind farm's connection point. The turbines either could have SCIGs or could have WRSGs. The advantage of such structure is that the internal behaviour of the wind turbines is separated from the grid behaviour, and thus the wind farm is robust to possible failures of the grid. The disadvantage of this concept is that all wind turbines are rotating with the same average angular speed and not at an individual optimal speed, thereby giving up some of the features of the variable-speed concept, for each individual wind turbine.

    An option of this concept is the centralised reactive power compensation topology with an advanced static VAR compensation (ASVC) unit. Reactive power compensation units are widely used in power systems in order to provide the reactive power balance and to improve voltage stability. ASVCs are inverters based on self-commutated switches (i.e. with full, continuous control of the reactive power). They have the advantage that, in the case of a voltage decrease (e.g. during a grid fault) their available maximum reactive power decreases more slowly compared with the static VAR compensation (SVC) units. In Rejsby Hede, Denmark, test equipment with an 8 Mvar ASVC unit has been incorporated into a 24 MW wind farm with 40 stall-regulated wind turbines. Equipment and test results are described in Stöber et al. (1998).

Another option is the use of a high-voltage DC (HVDC) link as power transmission. Such an installation is running on Gotland, Sweden (for more details, see Chapter 13). All wind turbines are connected to the same power converter, and the entire wind farm is connected to the public supply grid through another power converter. These two converters are connected to each other through a long, HVDC link cable.

The application of power electronic technology in large wind farms seems thus to be very promising. It plays a key role in complying with the high requirements that utility companies impose on wind farms. This technology therefore needs substantial additional research and development.

4.6 Conclusions

This chapter has provided a brief and comprehensive survey of the generator and power electronic concepts used by the modern wind turbine industry. A short introduction, presenting the basic wind turbine topologies and control strategies, was followed by the state of the art of wind turbines, from an electrical point of view. Old and new potentially promising concepts of generators and power electronics based on technical aspects and market trends were presented.

It is obvious that the introduction of variable-speed options in wind turbines increases the number of applicable generator types and further introduces several degrees of freedom in the combination of generator type and power converter type.

A very significant trend for wind turbines is that large wind farms will have to behave as integral parts of the electrical power system and develop power plant characteristics. Power electronic devices are promising technical solutions to provide wind power installations with power system control capabilities and to improve their effect on power system stability.

References

[1] Alatalo, M. (1996) *Permanent Magnet Machines with Air Gap Windings and Integrated Teeth Windings*, technical report 288, Chalmers University of Technology, Sweden.

[2] Bogalecka, E. (1993) 'Power Control of a Doubly Fed Induction Generator without Speed or Position Sensor', in *EPE 5th European Conference on Power Electronics and Application*, Volume 8, pp. 224–228.

[3] BTM Consults Aps. (2003) *International Wind Energy Department – Word Market Update 2002*, forecast 2003–2007, www.btm.du.

[4] Carlson, O., Hylander, J., Thorborg, K. (1996) 'Survey of Variable Speed Operation of Wind Turbines', in *1996 European Union Wind Energy Conference, Sweden*, pp. 406–409.

[5] Dahlgren, M., Frank, H., Leijon, M., Owman, F., Walfridsson, L. (2000) Wind power goes large scale, *ABB Review* 3 31–37.

[6] Dubois, M. R., Polinder, H., Fereira, J. A. (2000) 'Comparison of Generator Topologies for Direct-drive Wind Turbines', in *IEEE Nordic Workshop on Power and Industrial Electronics*, IEEE, New York, pp. 22–26.

[7] Hansen, A. D., Sørensen, P., Janosi, L., Bech, J. (2001) 'Wind Farm Modelling for Power Quality', in *Proceedings of IECON'01, Denver*.

[8] Hansen, L. H., Helle L., Blaabjerg F., Ritchie E., Munk-Nielsen S., Bindner, H., Sørensen, P., Bak-Jensen, B. (2001) *Conceptual Survey of Generators and Power Electronics for Wind Turbines*, Risø-R-1205(EN), Risø National Laboratory, Denmark.

[9] Heier, S. (1998) *Grid Integration of Wind Energy Conversion Systems*, John Wiley & Sons Ltd, Chichester, UK, and Kassel University, Germany.

[10] Kazmierkowski, M. P., Krishnan, R., Blaabjerg, F. (2002) *Control in Power Electronics – Selected problems*, Academic Press, London and New York.

[11] Kim, J. S., Sul, S. K. (1993) 'New Control Scheme for AC–DC–AC converter without DC link Electrolytic Capacitor', in *IEEE, PESC'93*, IEEE, New York, pp. 300–306.

[12] Krause, P. C., Wasynczuk, O., Sudhoff, S. D. (2002) *Analysis of Electric Machinery and Drive Systems*, John Wiley & Sons Inc., New York.

[13] Larsson, Å. (2000) *The Power Quality of Wind Turbines*, PhD dissertation, Chalmers University of Technology, Göteborg, Sweden.

[14] Leonhard, W. (1980) *Control of Electrical Drives*, Springer, Stuttgart.

[15] Mitcham, A. J., Grum, N. (1998) 'An Integrated LP Shaft Generator for the more Electric Aircraft', in *IEE Coloquium on All Electric Aircraft*, Institute of Electrical Engineers, London, pp. 8/1–8/9.

[16] Mohan N., Undeland, T. M., Robbins, W. P. (1989) *Power Electronics: Converters, Applications and Design*, Clarendon Press, Oxford, UK.

[17] Novotny, D. V., Lipo, T. A. (1996) *Vector Control and Dynamics of AC Drives*, Clarendon Press, Oxford.

[18] Pena, R., Clare, J. C., Asher, G. M. (1996) 'Doubly Fed Induction Generator using Back-to-back PWM Converters and its Application to Variable Speed Wind-energy Generation', *IEE Proceedings on Electronic Power Application* **143**(3) 231–241.

[19] Rodriguez, J. Moran, L., Gonzalez, A., Silva, C. (1999) 'High Voltage Multilevel Converter with Regeneration Capability', in *IEEE, PESC's 99, Volume 2*, IEEE, New York, pp. 1077–1082.

[20] Schuster, A. (1998) 'A Matrix Converter without Reactive Clamp Elements for an Induction Motor Drive System', in *IEEE PESC's, Volume 1*, IEEE, New York, pp. 714–720.

[21] Sørensen, P., Bak-Jensen, B., Kristian, J., Hansen, A. D., Janosi, L., Bech, J. (2000) 'Power Plant Characteristics of Wind Farms', in *Wind Power for the 21st Century; Proceedings of the International Conference Kassel, Germany*.

[22] Stöber, R., Jenkins, N., Pedersen, J. K., Søbrink, K. H., Helgesen Pedersen, K. O. (1998) *Power Quality Improvements of Wind Farms*, Eltra, Fredericia, Denmark.

[23] Van Custem, T., Vournas, C. (1998) *Voltage Stability of Electric Power Systems*, Kluwer Academic Publishers, Boston, MA.

[24] Wallace, A. K., Oliver, J. A. (1998) 'Variable-speed Generation Controlled by Passive Elements', paper presented at the International Conference in Electric Machines, Turkey.

[25] Wheeler, P. W., Grant, D. A. (1993) 'A Low Loss Matrix Converter for AC Variable-speed Drives', in *EPE 5th European Conference on Power Electronics and Application, Volume 5*, EPE, pp. 27–32.

5

Power Quality Standards for Wind Turbines

John Olav Tande

5.1 Introduction

Injection of wind power into an electric grid affects the voltage quality. As the voltage quality must be within certain limits to comply with utility requirements, the effect should be assessed prior to installation. To assess the effect, knowledge about the electrical characteristics of the wind turbines is needed or else the result could easily be an inappropriate design of the grid connection. The electrical characteristics of wind turbines are manufacturer specific but not site specific. This means that by having the actual parameter values for a specific wind turbine the expected impact of the wind turbine type on voltage quality when deployed at a specific site, possibly as a group of wind turbines, can be calculated.

Seeing the need for consistent and replicable documentation of the power quality characteristics of wind turbines, the International Electrotechnical Commission (IEC) started work to facilitate this in 1996. As a result, IEC 61400-21 (IEC, 2001) was developed and, today, most large wind turbine manufacturers provide power quality characteristic data accordingly.

IEC 61400-21 describes procedures for determining the power quality characteristics of wind turbines. These procedures are not presented in this chapter, but rather the application of the characteristics is explained. First, a brief description of the power quality characteristics defined in IEC 61400-21 is given, and thereafter

Wind Power in Power Systems   Edited by T. Ackermann
© 2005 John Wiley & Sons, Ltd   ISBN: 0-470-85508-8 (HB)

their application to determine the impact of wind turbines on voltage quality is explained through a case study considering a $5 \times 750\,\mathrm{kW}$ wind farm on a $22\,\mathrm{kV}$ distribution feeder. The case study further demonstrates that possible voltage quality problems due to installation of a wind turbine may in many cases be overcome simply by selecting an appropriate wind turbine type and/or by adjusting wind turbine control parameters.

Until the development of IEC 61400-21 there were no standard procedures for determining the power quality characteristics of a wind turbine, and simplified rules were often applied for dimensioning the grid connection of wind turbines (e.g. requiring a minimum short-circuit ratio of 25 or that the wind farm should not cause a voltage increment of more than 1 %). This approach has proved generally to ensure acceptable voltage quality; however, it has been costly by imposing grid reinforcements not needed and has greatly limited the development of wind farms in distribution grids. Actually, an application of the '1 % rule' would allow only one $550\,\mathrm{kW}$ wind turbine on the case study grid. However, a detailed assessment that applies the power quality characteristics of the wind turbines agrees with measurements that the $5 \times 750\,\mathrm{kW}$ wind farm can be accepted. Hence, simplified rules should be used with care, and it is preferable to apply a systematic approach as outlined in this chapter. Certainly, if the wind farm is large or the grid is very weak, additional analyses may be required to assess the impact on power system stability and operation.

5.2 Power Quality Characteristics of Wind Turbines

This section gives a brief description of the power quality characteristics of wind turbines.

According to IEC 61400-21, the following parameters are relevant for characterising the power quality of a wind turbine:

- rated data (P_n, Q_n, S_n, U_n and I_n);
- maximum permitted power, $\mathrm{P_{mc}}$ (10-minute average);
- maximum measured power, P_{60} (60-second average) and $P_{0.2}$ (0.2-second average);
- reactive power, Q, as 10-minute average values as a function of active power;
- flicker coefficient $c(\psi_k, v_a)$ for continuous operation as a function of the network impedance phase angle ψ_k and annual average wind speed v_a;
- maximum number of specified switching operations of the wind turbine within a 10-minute period, N_{10}, and a 2-hour period, N_{120};
- flicker step factor, $k_f(\psi_k)$, and voltage change factor, $k_u(\psi_k)$, for specified switching operations of the wind turbine as a function of the network impedance phase angle, ψ_k;
- maximum harmonic currents, I_h, during continuous operation given as 10-minute average data for each harmonic up to the 50th.

In the following sections, the above parameters are briefly described, providing information to understand their relevance and indicating variations depending on the wind turbine type.

5.2.1 Rated data

The rated data of a wind turbine, P_n, Q_n, S_n and I_n are defined as follows:

- Rated power, P_n, is the maximum continuous electric output power which a wind turbine is designed to achieve under normal operating conditions.
- Rated reactive power, Q_n, is the reactive power from the wind turbine while operating at rated power and nominal voltage and frequency.
- Rated apparent power, S_n, is the apparent power from the wind turbine while operating at rated power and nominal voltage and frequency.
- Rated current, I_n, is the current from the wind turbine while operating at rated power and nominal voltage and frequency.

5.2.2 Maximum permitted power

The 10-minute average output power of a wind turbine may, depending on the wind turbine design, exceed its rated value. Thus, the maximum permitted power parameter, P_{mc}, serves to provide a clear definition of the maximum 10-minute average power that can be expected from the wind turbine.

Wind turbines with active control of output power (i.e. by blade pitching and/or speed control) typically provide $P_{mc} = P_n$.

Wind turbines with passive control of output power (i.e. fixed-speed, stall-controlled wind turbines) are commonly set up with P_{mc} some 20 % higher than P_n.

5.2.3 Maximum measured power

The maximum measured power, P_{60}, measured as a 60-second average value, and $P_{0.2}$, measured as 0.2-second average value, serves two purposes. First, P_{60} and $P_{0.2}$ should be considered in conjunction with relay protection settings; second, they may be of particular relevance for the operation of wind turbines on isolated grids.

A variable-speed wind turbine may typically provide $P_{0.2} = P_{60} = P_n$, whereas for fixed-speed wind turbines, stall or pitch controlled, $P_{0.2}$ will commonly be larger than P_n.

5.2.4 Reactive power

The reactive power of the wind turbine is to be specified in a table as 10-minute average values as a function of the 10-minute average output power for 0.10 %, ..., 90 %, 100 % of the rated power. Also, the reactive power at P_{mc}, P_{60} and $P_{0.2}$ has to be specified.

Wind turbines with an induction generator connected directly to the grid consume reactive power as a function of the output active power. The consumption is typically compensated by capacitors that may be connected in steps (see also Chapters 4 and 19).

Wind turbines employing modern frequency converters are commonly capable of controlling the reactive power to zero or possibly of supplying or consuming reactive power according to needs, although this is limited by the size of the converter.

5.2.5 Flicker coefficient

The power fluctuations from wind turbines during continuous operation cause corresponding voltage fluctuations on the grid. The amplitude of the voltage fluctuations will depend not only on the grid strength relative to the amplitude of the power fluctuations but also on the network impedance phase angle and the power factor of the wind turbine.

Voltage fluctuations may cause annoying changes in the luminance from lamps. The impression of this is denoted 'flicker' and may be measured by using a flickermeter as described in IEC 61000-4-15 (IEC, 1997). The flickermeter takes voltage as input and gives the flicker severity as output. As can be seen from the normalised response of the flickermeter in Figure 5.1, even a quite small voltage fluctuation can be annoying if it persists at certain frequencies.

The flicker coefficient is a normalised measure of the maximum flicker emission (99th percentile) from a wind turbine during continuous operation:

$$c(\psi_k, v_a) = P_{st} \frac{S_k}{S_n}, \tag{5.1}$$

where

P_{st} is the flicker emission from the wind turbine,
S_n is the rated apparent power of the wind turbine,
S_k is the short-circuit apparent power of the grid.

The flicker coefficient has to be given as the 99th percentile for specified values of the network impedance phase angle (30°, 50°, 70° and 85°) and annual average wind speed (6 m/s, 7.5 m/s, 8.5 m/s and 10 m/s).

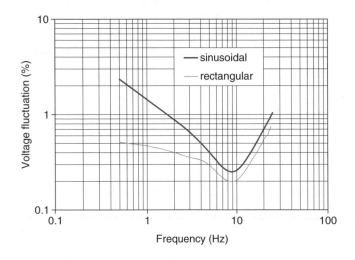

Figure   5.1 Normalised flickermeter response for voltage fluctuations (peak-to-peak)

Variable-speed wind turbines are commonly expected to yield fairly low flicker coefficients, whereas values for fixed-speed wind turbines may range from average (stall-controlled) to high (pitch-controlled); see also Figure 16.6 (page 353) and Figure 16.7 (page 354).

5.2.6 Maximum number of wind turbine switching operations

The following cases of switching operations are relevant as these may cause significant voltage variations:

- wind turbine startup at cut-in wind speed;
- wind turbine startup at rated wind speed;
- the worst case of switching between generators (applicable only to wind turbines with more than one generator or a generator with multiple windings).

The acceptance of switching operations depends not only on their impact on the grid voltage but also on how often these may occur. Hence, the maximum number of the above-specified switching operations within a 10-minute period, N_{10}, and a 2-hour period, N_{120}, should be stated. N_{10} and N_{120} may be governed by modern wind turbine control system settings.

5.2.7 Flicker step factor

The flicker step factor is a normalised measure of the flicker emission due to a single switching operation of a wind turbine:

$$k_f(\psi_k) = \frac{1}{130} \frac{S_k}{S_n} P_{st} T_p^{0.31},$$
(5.2)

where

T_p is the duration of the voltage variation due to the switching operation;
P_{st} is the flicker emission from the wind turbine;
S_n is the rated apparent power of the wind turbine;
S_k is the short-circuit apparent power of the grid.

The flicker step factor has to be given for specified values of the network impedance phase angle (30°, 50°, 70° and 85°) and for the specified types of switching operations (see Section 5.2.6).

Variable-speed wind turbines are commonly expected to yield fairly low flicker step factors, whereas values for fixed-speed wind turbines may range from average (pitch-controlled) to high (stall-controlled).

5.2.8 Voltage change factor

The voltage change factor is a normalised measure of the voltage change caused by a single switching operation of a wind turbine:

$$k_u(\psi_k) = \sqrt{3} \frac{U_{\max} - U_{\min}}{U_n} \frac{S_k}{S_n},$$
(5.3)

where

> $U_{\min}$ and $U_{\max}$ are the minimum and maximum voltage [root mean square (RMS) phase-to-neutral] due to the switching;
> U_n is the nominal phase-to-phase voltage;
> S_n is the rated apparent power of the wind turbine;
> S_k is the short-circuit apparent power of the grid.

The voltage change factor has to be given for specified values of the network impedance phase angle (30°, 50°, 70° and 85°) and for the specified cases of switching operations (see Section 5.2.6).

The voltage change factor k_u is similar to the in-rush current factor, k_i being the ratio between the maximum inrush current and the rated current, though k_u is a function of the network impedance phase angle. The highest value of k_u will be numerically close to k_i.

Variable-speed wind turbines are commonly expected to yield fairly low voltage change factors, whereas values for fixed-speed wind turbines may range from average (pitch-controlled) to high (stall-controlled).

5.2.9 Harmonic currents

The emission of harmonic currents during the continuous operation of a wind turbine with a power electronic converter has to be stated. The individual harmonic currents will be given as 10-minute average data for each harmonic order up the 50th at the output power giving the maximum individual harmonic current, and, further, the maximum total harmonic current distortion also has to be stated.

Harmonic emissions have been reported from a few installations of wind turbines with induction generators but without power electronic converters. There is, however, no agreed procedure for measurements of harmonic emissions from induction machines, and there is no known instance of customer annoyance or damage to equipment as a result of harmonic emissions from such wind turbines. IEC 61400-21 therefore does not require measurements of harmonic emissions from such wind turbines.

5.2.10 Summary power quality characteristics for various wind turbine types

Typical power quality characteristics for various wind turbine types are summarised in Table 5.1. The actual parameter values are specific to the wind turbine type and variations are to be expected depending on the detailed solutions applied by the manufacturer. Hence, accurate assessment requires manufacturer-specific data to be collected.

Table 5.1 Typical power quality characteristics for various wind turbine types

Quantity	Wind turbine type[a]					
	A0	A1	A2	B	C	D
Maximum permitted power, P_{mc}	$P_{mc} > P_n$	$P_{mc} = P_n$	$P_{mc} = P_n$	$P_{mc} = P_n$	$P_{mc} = P_n$	$P_{mc} = P_n$
Maximum measured power 60 s average, P_{60}	$P_{60} > P_n$	$P_{60} = P_n$	$P_{60} = P_n$	$P_{60} = P_n$	$P_{60} = P_n$	$P_{60} = P_n$
Maximum measured power 0.2 s average, $P_{0.2}$	$P_{0.2} > P_n$	$P_{0.2} > P_n$	$P_{0.2} > P_n$	$P_{0.2} = P_n$	$P_{0.2} = P_n$	$P_{0.2} = P_n$
Reactive power, Q^b	$f(P)$	$f(P)$	$f(P)$	$f(P)$	0	0
Flicker coefficient, $c(\psi_k, v_a)$	Average	High	Average	Low	Low	Low
Maximum number of switching operations in a 10 min period, N_{10}	CPS	CPS	CPS	CPS	CPS	CPS
Maximum number of switching operations in a 2 h period, N_{120}	CPS	CPS	CPS	CPS	CPS	CPS
Flicker step factor, $k_f(\psi_k)$	High	Average	Average	Low	Low	Low
Flicker change factor, $k_u(\psi_k)$	High	Average	Average	Low	Low	Low
Maximum harmonic current, $I_h{}^c$	—	—	—	—	Low	Low

[a] For more information turbine types A0, A1, A2, B, C and D, see Section 4.2.3.
[b] Regarding wind turbine types A and B, use of capacitors that are connected in steps or power electronics may lead to enhanced control of reactive power. Regarding wind turbine types C and D, assuming use of modern frequency converters, the reactive power may be fully controlled within the rating of the converter.
[c] This is relevant only for wind turbine types C and D. The indicated low emission of harmonic currents is based on the assumption that modern frequency converters are used.
— Not applicable
Note: CPS = control parameter setting.

5.3 Impact on Voltage Quality

This section gives a brief description of the impact that wind turbines may have on the voltage quality. The application of wind turbine power quality characteristics to determine the impact of wind turbines on voltage quality is explained through a case study considering a 5×750 kW wind farm on a 22 kV distribution feeder. The case study further demonstrates that possible voltage quality problems resulting from a wind turbine installation in many cases may be overcome simply by selecting an appropriate wind turbine type and/or by adjusting wind turbine control parameters.

5.3.1 General

Ideally, the voltage should form a perfect sinusoidal curve with constant frequency and amplitude. However, in any real-life power system, grid-connected appliances will cause the voltage to deviate from the ideal. Basically, any mismatch between generation and demand causes voltage frequency deviation, whereas line losses cause deviations in the voltage amplitude. Large interconnected systems are normally associated with smaller frequency deviations than island systems. This is because there will be relatively less demand variations in large interconnected systems, and the spinning generating capacity will be greater. Deviations in the voltage amplitude depend on the relative strength of the grid. Commonly, power systems have strong grids for the transmission of power that keeps the voltage amplitude within a narrow band, whereas distribution grids are weaker and are associated with greater deviations in the voltage amplitude.

EN 50160 (EN, 1995) states the supply voltage characteristics that can be expected at customer inlets at a low or medium voltage level during normal network operating conditions. In this section, only those characteristics that might be influenced by the normal operation of wind turbines are considered. Other characteristics, such as supply interruptions, temporary or transient over-voltages and voltage unbalance, are not assumed to be influenced by the normal operation of wind turbines and are not dealt with in this section. The characteristics stated by EN 50160 are for European countries. For other countries, different values may apply, though the principles are still relevant.

5.3.2 Case study specifications

The example network with a 5×750 kW wind farm given in Figure 5.2 is used as an illustration. The wind turbines are of a conventional design, operating at fixed speed and using stall for power limitation at high wind speeds (wind turbine type A0). Each wind turbine is equipped with power electronics that limits the in-rush current to the induction generator during startup and capacitors that are switched to maintain the power

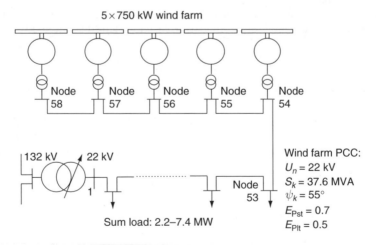

Figure   5.2 Example network with a 5×750 kW wind farm. *Note*: PCC = point of common coupling; U_n = nominal phase-to-phase voltage; S_k = short-circuit apparent power of the grid; ψ_k = network impedance phase angle; E_{Pst} = short-term flicker emission limit; E_{Ptt} = long-term flicker emission limit

Table   5.2 Power quality data for example wind turbine

Quantity	Value
Rated power, P_n (kW)	750
Rated reactive power, Q_n (kvar)	0
Nominal phase-to-phase voltage, U_n (kV)	0.69
Maximum permitted power, P_{mc}	$1.2\,P_n$
Flicker coefficient, $c(\psi_k = 55°, v_a = 8.2\,\text{m/s})$	10.9
Maximum number of switching operations in a 10 min period, N_{10}	1
Maximum number of switching operation in a 2 h period, N_{120}	12
Flicker step factor, $k_f(\psi_k = 55°)$	1.2
Voltage change factor, $k_u(\psi_k = 55°)$	1.5

Note: ψ_k = network impedance phase angle; v_a = annual average wind speed.

factor close to unity during operation. Further detailed power quality characteristics of each wind turbine, specified according to IEC 61400-21, are given in Table 5.2, though, for simplicity, only the data relevant for the example network conditions are listed.

The wind farm is connected to a 22 kV distribution feeder with short-circuit apparent power at the point of common coupling (PCC) about 10 times the installed wind power capacity. Hence, the grid appears relatively weak and the operation of the wind farm may be expected significantly to impact the voltage quality at customers connected to the 22 kV feeder. The 22 kV feeder is connected via a constant rail voltage transformer with a dead-band of ±1 % to a strong 132 kV transmission line. Given the strength of the 132 kV connection point, it is assumed that the 5 × 750 kW wind farm will not significantly influence the operation at this high voltage level. Consequently, the scope of analysis for this case study is limited to the 22 kV distribution feeder.

5.3.3 Slow voltage variations

Load-flow analyses may be carried out to assess the slow voltage variations (i.e. variations in the voltage amplitude expressed as 10 min average values). In general, all possible load cases should be included in the assessment of slow voltage variations. For the example system, however, given its simplicity and that the wind turbines are operated at a power factor close to unity, it is sufficient to consider the two load cases specified below that give minimum and maximum voltage amplitudes, respectively:

- maximum consumer loads at feeder and zero wind power production;
- minimum consumer loads at feeder and maximum continuous wind power production.

Figure 5.3 shows the results for the two load cases. Node 1 denotes the medium-voltage (MV) node at the high-voltage (HV) transformer. The 5 × 750 kW wind turbines are connected at nodes 54–58 (see Figure 5.2), whereas the other nodes connect consumers. The voltage shown for low voltage at minimum load and maximum wind power refers to the point immediately after the low-voltage (LV) transformer, whereas the

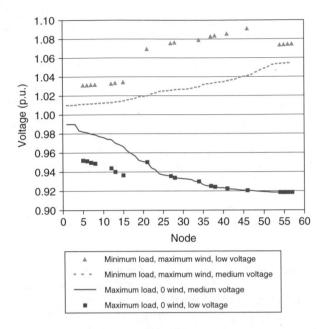

Figure **5.3** Result of load-flow analysis

voltage shown for low voltage at maximum load and no wind power is at the far end of the LV line. The difference between the per-unit (p.u.) voltages at medium and low voltage are a result of the voltage drop at the LV lines and the assumed tap-changer position of the LV transformers.

According to EN 50160, the slow voltage variations measured as 10 min averages at customer inlets should lie within $\pm 10\,\%$ of U_n during 95 % of a week. In addition, for low voltage only, the slow voltage variations should always be within $-15\,\%$ and $+10\,\%$ of U_n. Hence, our example system lies within these limits. However, slow voltage variations may be a constraint on the further expansion of the wind farm as an expansion would cause the maximum voltage to increase. However, this constraint may easily be overcome by adjusting the power factor of the wind turbines, for instance. A modest reduction of the power factor from unity to 0.98 (inductive) decreases the maximum voltage by 1.5 % and makes room for more wind power. Actually, the wind farm may expand to a total of eight 750 kW wind turbines operated at a power factor of 0.98 before slow voltage variations again become a constraint on further expansion.

Certainly, possible uncertainties in estimates of minimum and maximum load levels may require safety margins. This does not, however, change the suggestion that a possible slow voltage variation constraint may be counteracted by adjusting the power factor of the wind turbines. It can be argued that a reduced power factor causes increased network losses. This implies that a regulation of the power factor should be used with care, and that alternative options should be assessed. Examples of alternative options are grid reinforcement by the installation of new lines and a voltage-dependent reduction of the wind farm power (see e.g. Tande, 2000). It is further relevant to consider voltage-dependent power factor control that probably could substantially reduce the increase of network losses. It is possible, with clever control and allowing overcompensation at high load and low wind conditions, to achieve a net reduction in grid losses. To assess this, repeated load flow analyses need to be conducted, taking properly into account the expected distribution of consumption and wind power production during the year [e.g. by probabilistic load flow, as in Hatziargyriou, Karakatsanis and Papadopoulos (1993) and in Tande and Jørgensen (1996)].

5.3.4 Flicker

Flicker and/or rapid voltage changes are commonly due to rapid changes in the load or to switching operations in the system. According to EN 50160, a rapid voltage change should generally be less than 5 % of U_n, though a change of up to 10 % of U_n may occur several times a day under certain circumstances.

The flicker severity can be given as a short-term value, P_{st}, measured over a period of 10 minutes, or as a long-term value, P_{lt}, corresponding to a period of 2 hours, calculated from a sequence of P_{st} values:

$$P_{lt} = \left[\sum_{i=1}^{12} \left(\frac{P_{st,i}^3}{12} \right) \right]^{1/3}. \tag{5.4}$$

According to EN 50160, the long-term flicker severity has to be less than or equal to 1 during 95 % of a week. It is noted that reaction to flicker is subjective, so that in some cases people may be annoyed by $P_{lt} = 1$, for instance, whereas in other cases higher values can be accepted.

To ensure $P_{lt} \leq 1$ at the customer inlets, each source of flicker connected to the network can be allowed only a limited contribution; for example, in the example network, $E_{Pst} = 0.7$ and $E_{Plt} = 0.5$ at the PCC of the wind farm, where E_{Pst} and E_{Plt} are the short-term and long-term flicker emission limits, respectively. In other networks, different values may be found by using IEC 61000-3-7 (IEC, 1996b) as a guide.

Wind turbines emit flicker as a result of switching operations, such as startups, and as a result of rapid fluctuations in the output power during continuous operation.

Following the recommendations in IEC 61400-21, the flicker emission from a single wind turbine or wind farm may be assessed. Procedures are given for assessing flicker emission due to switching operations and due to continuous operation.

5.3.4.1 Switching operations

The procedure for assessing flicker emission due to switching operations assumes that each wind turbine is characterised by a flicker step factor, $k_f(\psi_k)$, which is a normalised measure of the flicker emission due to a single worst-case switching operation. The worst-case switching operation is commonly a startup, although IEC 61400-21 also requires the assessment of switching operations between generators (e.g. to obtain two-speed operation), if applicable to the wind turbine in question. Further, the procedure assumes that, for each wind turbine, information is given on the maximum number of starts, N_{10} and N_{120}, that can be expected within a 10-minute and 2-hour period, respectively. Based on these characteristics, the flicker emission due to worst-case switching of wind turbines can be calculated (for a deduction of the equations, see IEC, 2001; Tande, 2002):

$$P_{st} = \frac{18}{S_k} \left\{ \sum_{i=1}^{N_{wt}} N_{10} \left[k_{f,i}(\psi_k) S_{n,i} \right]^{3.2} \right\}^{0.31}, \tag{5.5}$$

$$P_{lt} = \frac{8}{S_k} \left\{ \sum_{i=1}^{N_{wt}} N_{120} \left[k_{f,i}(\psi_k) S_{n,i} \right]^{3.2} \right\}^{0.31}, \tag{5.6}$$

where N_{wt} is the total number of wind turbines.

For the example specifications, $P_{st} = 0.71$ and $P_{lt} = 0.68$ (i.e. exceeding the assumed example limits). Hence, the flicker emission due to startups may be a constraint on the operation of the example wind farm. This constraint may, however, be overcome quite easily by using another type of wind turbine with a smaller flicker step factor, k_f [e.g. a pitch-regulated or a (semi)variable-speed type]. Another alternative is to ensure that only a reduced number of wind turbines are allowed to start within the same 10-minute and 2-hour period. This involves altering the control system settings of the wind turbine to a smaller value for N_{120} and introducing a wind farm control system that allows only a reduced number of wind turbines to start within the same 10-minute period, in effect altering N_{wt} in Equation (5.5).

5.3.4.2 Continuous operation

The procedure for assessing flicker emission due to continuous operation assumes that each wind turbine is characterised by a flicker coefficient, $c(\psi_k, v_a)$, which is a normalised measure of the maximum expected flicker emission during continuous operation of the wind turbine. To arrive at the flicker emission from a single wind turbine, the flicker coefficient with the relevant ψ_k and v_a is simply multiplied by S_n/S_k, whereas the emission from a wind farm can be found from the following equation (for a deduction of this equation, see IEC, 2001; Tande, 2002):

$$P_{st} = P_{lt} = \frac{1}{S_k} \left\{ \sum_{i=1}^{N_{wt}} \left[c_i(\psi_k, v_a) S_{n,i} \right]^2 \right\}^{0.5}. \tag{5.7}$$

$P_{st} = P_{lt}$ in Equation (5.7) because it is probable that conditions during the short-term period persist over the long-term period. Further, Equation (5.7) assumes that the maximum power levels between wind turbines are uncorrelated. At special conditions, however, wind turbines in a wind farm may 'synchronise', causing power fluctuations to coincide. Equation (5.7) would then underestimate the flicker emission. This was assessed in (Tande, Relakis and Alejandro, 2000), and it was concluded that, for common conditions, Equation (5.7) will provide a good estimate (see also Figure 5.4).

For the example specifications, $P_{st} = P_{lt} = 0.49$, which lies just within the assumed example limit of $E_{Plt} = 0.5$. If the wind farm was expanded with more wind turbines of

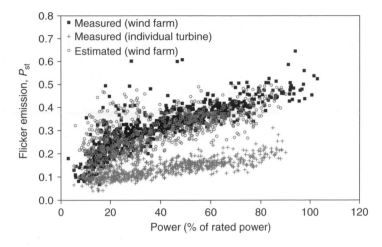

Figure 5.4 Measured and estimated flicker emission from a $5 \times 750\,\text{kW}$ wind farm. The estimated flicker emission is calculated as the measured emission from one wind turbine within the farm multiplied by the square root of the number of wind turbines in the farm. The good agreement between the estimated and measured flicker emission verifies Equation (5.7)

the same type, the flicker emission would further increase above the acceptable limit. Hence, for the example system, flicker emission due to continuous operation may be a constraint for a further expansion of the wind farm. To overcome this constraint, the straightforward approach is to select a different wind turbine type that has a smaller value of c. First of all, $c = 10.9$ is rather high for a fixed-speed, stall-regulated wind turbine, so this value should not be assumed to be typical for this type of wind turbine. Certainly, a (semi)variable-speed wind turbine would yield a much lower value of c, whereas a fixed-speed, pitch-regulated type could actually yield a higher value of c.

5.3.5 Voltage dips

A voltage dip is defined in EN 50160 as a sudden reduction of the voltage to a value between 1 % and 90 % of U_n followed by a voltage recovery after a short period of time, conventionally 1 ms to 1 min. The expected number of voltage dips during a year may vary from a few dozen to a thousand. According to EN 50160, dips with depths of between 10 % and 15 % of U_n are commonly due to the switching of loads, whereas larger dips may be caused by faults.

The startup of a wind turbine may cause a sudden reduction of the voltage followed by a voltage recovery after a few seconds. Assuming that each wind turbine is characterised by a voltage change factor, $k_u(\psi_k)$, the sudden voltage reduction may be assessed (deducted directly for definition):

$$d = 100\, k_u(\psi_k) \frac{S_n}{S_k}. \tag{5.8}$$

As it is unlikely that several wind turbines in a wind farm start up at the exact same time, Equation (5.8) is not a function of the number of wind turbines as are the quantities in Equations (5.5)–(5.7).

For the example wind turbines, $d = 3.0\,\%$. This sudden voltage reduction is in most cases acceptable, especially considering that this would imply a voltage of less than 0.90 p.u. (i.e. a voltage dip) only for the case when startups coincide with a high load on the network. Further, as Equation (5.8) is not a function of the number of wind turbines in the farm, it can be concluded that, for the example system, voltage dips are not a constraint on further expansion of the wind farm.

5.3.6 Harmonic voltage

Nonlinear loads (i.e. loads that draw a current that is not a linear function of the voltage) distort the voltage waveform and may in severe cases cause an overheating of neutral conductors and electrical distribution transformers, the malfunctioning of electronic equipment and the distortion of communication systems. Examples of nonlinear loads are power electronic converters, arc furnaces, arc welders, fluorescent lamps and mercury lamps. The distorted waveform may be expressed as a sum of sinusoids with various frequencies and amplitudes, by application of the Fourier transform. The sinusoids with frequencies equal to an integer multiple of the fundamental frequency are denoted harmonics, whereas the others are denoted interharmonics.

Harmonic voltages, U_h, where h denotes the harmonic order (i.e. an integer multiple of 50 Hz) can be evaluated individually by their relative amplitude:

$$u_h = \frac{U_h}{U_n}. \tag{5.9}$$

According to EN 50160, the 10-minute mean RMS values of each u_h have to be less than the limits given in Figure 5.5 during 95 % of a week. Further, the total harmonic distortion (THD) of the voltage, calculated according to Equation (5.10), has to be $\leq 8\%$:

$$\mathrm{THD} = \left[\sum_{h=2}^{40}(u_h)^2\right]^{1/2}. \tag{5.10}$$

With regard to higher-order harmonics, EN 50160 does not specify any limits but states that higher-order harmonics are usually small, though largely unpredictable. In the same manner, EN 50160 does not indicate any levels for interharmonics. These are under consideration, though.

To ensure that the harmonic voltages are kept within acceptable limits, each source of harmonic current can be allowed only a limited contribution. The limits depend on network specifications, such as harmonic impedance and voltage level, and can be found using IEC 61000-3-6 (IEC, 1996a) as a guide.

A wind turbine with an induction generator directly connected to the grid without an intervening power electronic converter (i.e. wind turbine Types is A and B; see Section

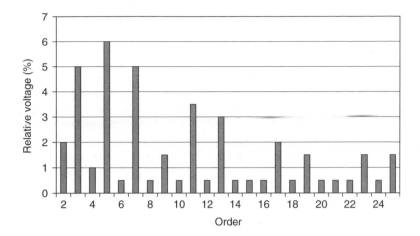

Figure 5.5 Values of individual harmonic voltages as a percentage of the nominal phase-to-phase voltage, U_n, according to EN 50160 (EN, 1995); the third harmonic is commonly significantly lower at medium voltage

4.2.3) is not expected to distort the voltage waveform. Power electronics applied for soft-start may give a short-term burst of higher-order harmonic currents, though the duration and magnitude of these are, in general, so small that they can be accepted without any further assessment. So, for the example system with fixed-speed wind turbines, emission limits for harmonics are not a constraint. However, if we considered variable-speed wind turbines using power electronic converters, their emission of harmonic currents should be assessed against given or calculated limits.

Thyristor-based converters are expected to emit harmonic currents that may influence the harmonic voltages. EN 50160 includes limits for this. Such converters are, however, rarely used in modern wind turbines. Their converters are, rather, transistor-based and operate at switching frequencies above 3 kHz. As a consequence, the impact of such new wind turbines on the voltage waveform is commonly negligible and probably not a constraint for wind power development.

In general, the connection of electric equipment does change the harmonic impedance of the network; for example, e.g. capacitor banks, which are often installed as part of wind farms consisting of fixed-speed wind turbines, may shift the resonance frequency of the harmonic impedance. Hence, possible harmonic sources already present in the network may, then, given unfortunate conditions, cause unacceptable harmonic voltages. Consequently, for networks with significant harmonic sources, the connection of new appliances such as wind farms with capacitor banks should be carefully designed in order to avoid an ill-conditioned modification of the harmonic impedance.

5.4 Discussion

Prior to the development of IEC 61400-21 there were no standard procedures for determining the power quality characteristics of a wind turbine. Basically, the manufacturer provided only rated data, giving little support for assessing the impact of wind turbines on voltage quality. Therefore, utilities often applied simplified rules (e.g. requiring the voltage increment due to a wind turbine installation to be less than 1 %). A large number of wind turbines have been connected and successfully operated applying such simplified rules. Therefore, it is relevant to discuss if the more detailed approach suggested in this chapter will lead to any real improvement in securing voltage quality and still allow a fair amount of wind power to be connected at a minimum cost.

A simplified assessment would typically consider only the rated data of the wind turbine installation (e.g. for the case study wind farm the active power, P, is 3750 kW, and the reactive power, Q, is 0 kvar). Then, the feeder impedance between the wind farm and an assumed stiff point would be applied to calculate the voltage increment due to the wind farm. For the example system, the resistance, R, is 8.83 Ω and the reactance, X, is 9.17 Ω, giving a voltage increment $\Delta U \approx 6.8\%$ by applying Equation (5.11):

$$\Delta U \approx \frac{RP + XQ}{U_n^2}. \tag{5.11}$$

Assuming that only a 1 % voltage increment is permitted, which is common for such simplified assessments, the wind farm cannot be connected according to this 1 % criterion. Hence, the simplified assessment would conclude that the grid must be reinforced or the wind farm be limited. In fact, leaving the grid as it is, the application of the 1 % rule would allow only one 550 kW wind turbine.

This is indeed very much in contrast to the results obtained by the detailed assessment made in Section 5.3, suggesting that the wind farm can be connected to the existing grid, and it is also supported by measurements on a similar real-life system (SINTEF TR A5330; see Tande, 2001). Hence, it can be concluded that the simplified assessment unnecessarily limits the acceptable wind farm capacity. A simplified assessment is therefore not recommended for dimensioning the grid connection of a wind farm.

5.5 Conclusions

Procedures for determining the power quality characteristics of wind turbines are specified in IEC 61400-21. The need for such consistent quantification of the power quality characteristics of wind turbines was verified in Section 5.2. In Section 5.3, the application of these characteristics for assessing the impact of wind turbines on voltage quality was explained.

A 5×750 kW wind farm on a 22 kV distribution feeder was used as an illustration. The detailed analysis suggested that the wind farm capacity can be operated at the grid without causing an unacceptable voltage quality. For comparison, a simplified design criterion was considered, assuming that the wind farm is allowed to cause a voltage increment of only 1 %. According to this criterion, only a very limited wind power capacity would be allowed. Measurements, however, confirm the suggestion of the detailed analysis, and it is concluded that a simplified design criterion such as the '1 % rule' should not be used for dimensioning the grid connection of a wind farm. Rather, the following analyses are suggested, which are made possible through the development of IEC 61400-21:

- Load-flow analyses should be conducted in order to assess whether slow voltage variations remain within acceptable limits (Section 5.3.3).
- Maximum flicker emission from the wind turbines should be estimated and compared with given or calculated limits (Section 5.3.4).
- Possible voltage dips due to wind turbine startup should be assessed (Section 5.3.5).
- Maximum harmonic currents from the wind turbines (if relevant) should be estimated and compared with given or calculated limits (Section 5.3.6).

It should be emphasised that the above may have to be supported by additional analyses, for instance, regarding the system stability in the case of large wind farms or weak grids, and regarding the impact on grid frequency in systems where wind power covers a high fraction of the load (e.g. in isolated systems).

References

[1] EN (1995), *Voltage Characteristics of Electricity Supplied by Public Distribution Systems*, EN 50160, www.cenelec.org.

[2] Hatziargyriou, N. D., Karakatsanis, T. S., Papadopoulos, M. (1993) 'Probabilistic Load Flow in Distribution Systems Containing Dispersed Wind Power Generation', *IEEE Transactions on Power Systems*, **8** (1), 159–165.

[3] IEC (International Electrotechnical Commission) (1996a) *EMC, Part 3: Limits. Section 6: Assessment of Emission Limits for Distorting Loads in MV and HV Power Systems*, IEC 61000-3-6, Basic EMC publication (technical report), www.iec.ch.

[4] IEC (International Electrotechnical Commission) (1996b) *EMC, Part 3: Limits. Section 7: Assessment of Emission Limits for Fluctuating Loads in MV and HV Power Systems*, IEC 61000-3-7, Basic EMC publication (technical report), www.iec.ch.

[5] IEC (International Electrotechnical Commission) (1997) *EMC, Part 4: Testing and Measurement Techniques. Section 15: Flickermeter – Functional and Design Specifications*, IEC 61000-4-15, IEC. www.iec.ch.

[6] IEC (International Electrotechnical Commission) (2001) *Measurement and Assessment of Power Quality Characteristics of Grid Connected Wind Turbines*, IEC 61400-21, IEC. www.iec.ch.

[7] Tande, J. O. G. (2000) 'Exploitation of Wind-Energy Resources in Proximity to Weak Electric Grids', *Applied Energy* **65** 395–401.

[8] Tande, J. O. G. (2001) 'Wind Power in Distribution Grids – Impact on Voltage Conditions', SINTEF Energy Research TR A5330 (in Norwegian).

[9] Tande, J. O. G. (2002) 'Applying Power Quality Characteristics of Wind Turbines for Assessing Impact on Voltage Quality', *Wind Energy*, **5**, 37–52.

[10] Tande, J. O. G., Jørgensen, P. (1996) 'Wind Turbines' Impact on Voltage Quality', in Ed. E. Sesto, *Proceedings of the EWEA Special Topic Conference: Integration of Wind Power Plants in the Environment and Electric Systems, Rome, Italy, 7–9 October 1996*, ISES, Rome, paper 3.4, pp. 3–8.

[11] Tande, J. O. G., Relakis, G., Alejandro, O. A. M. (2000) 'Synchronisation of Wind Turbines', in *Proceedings of the EWEA Special Topic Conference: Wind Power for the 21st Century, 25–27 September 2000, Kassel, Germany*, WIP-Renewasle Energies, Germany, pp. 152–155.

6

Power Quality Measurements

Fritz Santjer

6.1 Introduction

The term 'power quality' in relation to a wind turbine describes the electrical performance of the turbine's electricity generating system. It reflects the generation of grid
interferences and thus the influence of a wind turbine on the power and voltage quality
of the grid. The main influences of wind turbines on the grid concerning power quality
are voltage changes and fluctuations, especially at the local level, and harmonics for
wind turbines with electronic inverter systems. Other parameters are reactive power,
flicker, power peaks and in-rush currents.

  The grid interferences have different causes, which are mostly turbine-specific. Average power production, turbulence intensity and wind shear (i.e. the difference of wind
speed between the top and the bottom of the rotor) refer to causes that are determined
by meteorological and geographical conditions. However, the technical performance of
the wind turbine may also have an influence on grid interferences. This performance of
the wind turbine is determined not only by the characteristics of the electrical components, such as generators, transformers and so on, but also by the aerodynamic and
mechanical behaviour of rotor and drive train.

  Wind turbines and their power quality will be certified on the basis of measurements
according to national or international guidelines. These certifications are an important
basis for utilities to evaluate the grid connection of wind turbines and wind farms.

Wind Power in Power Systems   Edited by T. Ackermann
© 2005 John Wiley & Sons, Ltd   ISBN: 0-470-85508-8 (HB)

6.2 Requirements for Power Quality Measurements

6.2.1 Guidelines

The following guidelines provide rules and requirements for the measurement of the power quality of wind turbines:

- International Electrotechnical Commission (IEC) guideline IEC 61400-21: 'Wind Turbine Generator Systems, Part 21: Measurement and Assessment of Power Quality Characteristics of Grid Connected Wind Turbines' (IEC, 2001);
- MEASNET guideline: 'Power Quality Measurement Procedure of Wind Turbines' (MEASNET, 2000);
- The German Fördergesellschaft Windenergie (FGW) guideline: 'Technische Richtlinie für Windenergieanlagen, Teil 3: Bestimmung der Elektrischen Eigenschaften' (Technical guidelines for wind turbines, part 3: determination of the electrical characteristics; FGW, 2002.

IEC 61400-21 is one of the most important guidelines for power quality measurements of wind turbines. The first edition of this guideline was published at the end of 2001. It specifies which characteristics that are relevant to the power quality of wind turbines have to be measured (e.g. flicker, harmonics, power, switching operations) and which measurement methods should be used. As a result of a power quality measurement according to IEC 61400-21 the measuring institute will issue a data sheet, which comprises all the relevant electrical characteristics of the measured wind turbine. All the measurement results that are contained in the data sheet are given in a normalised form, independent of the specific site and of the specific network of the measured wind turbine. Thus the measured data can be used for nearly all sites and networks, and it is necessary only to measure the electrical characteristics of one wind turbine of each type. Often, the prototype will be measured. All the other wind turbines of this type should have the same electrical characteristics, as long as they are identical to the measured wind turbine. Measurements and investigations on several wind turbines of the same type, but at different sites, have shown that the electrical characteristics, measured according to IEC 61400-21 at the one site, are similar or identical to the electrical characteristics measured at another site. Thus it is not necessary to measure each individual wind turbine, but only one wind turbine of each type.

The data sheet with the electrical characteristics of the wind turbine provides the basis for the utility's assessment regarding a grid connection of a wind turbine or of a wind farm. In order to decide whether to connect a wind farm project at a specific site to its grid, the utility will also require site-specific data (e.g. the number of wind turbines and the annual average wind speed), and data of the specific network (e.g. the short-circuit power and the impedance angle of the grid). In addition to the measurement requirements, IEC 61400-21 also gives recommendations on how a utility should perform such an assessment. Unless there are specific guidelines – such as that of the Verband der Elektrizitätswirtschaft (VDEW, 1998) or the Danish utilities Research Association (DEFU, 1998) – the utility has the option to follow the recommendations of IEC 61400-21.

MEASNET is a measuring network of wind energy institutes, which has the aim to harmonise measuring procedures and recommendations in order to achieve comparability and mutual recognition of the measurement results by its member institutes. Regular round robin tests (mutual checks) ensure the high quality of the measurement results of the member institutes. The 'Power Quality Measurement Procedure' of MEASNET (2000) coincides with the requirements of IEC 61400-21. However, the MEASNET guideline requires more extended measurements regarding harmonic currents.

In Germany, the FGW technical guideline (FGW, 2002) is used for measuring the power quality of wind turbines. The guideline was introduced at the beginning of the 1990s and since then has proved to be a useful instrument for obtaining objective criteria for the grid connection of wind turbines. It has been continuously improved and updated and incorporates new research results and new turbine concepts. In principle, the methods and the procedures given in this guideline are similar to those in IEC 61400-21. There are, however, some differences between the methods of the two guidelines, so that the data measured according to IEC 61400-21 are not comparable to the data measured according to the german guideline. The specific differences between the two guidelines are explained elsewhere (Santjer, 2000). A data sheet, the so-called 'Extract of the Test Report', summarises the results of a power quality measurement according to the technical guideline (FGW, 2002). This data sheet contains all the relevant power quality characteristics of the measured turbine and, together with the VDEW (1998) guideline, which includes the requirements and limits to be observed, it is used by German utilities for the assessment of the grid connection.

6.2.2 Specification

According to the above-mentioned guidelines, the power quality measurements are performed for harmonics, flicker and transients, during normal and switching operations, separately, as well as for power factor, reactive power consumption and power peaks.

Although measurement and evaluation methods of the IEC guideline and the MEASNET guideline are similar to the regulations of the German technical guideline, results may differ considerably. This is partly because of the different averaging periods used but also because of the differences in the evaluation methods, which in some cases have a strong influence on the results. An overview of the requirements of the guidelines is given in Table 6.1.

Power peaks

A power peak is the maximum active power output of the wind turbine over a specified averaging time during continuous operation (without start or stop of the turbine). Power peaks of the turbine are determined over three different time intervals: instantaneous, 1 minute and 10 minutes. The power peaks have to be measured. Only under the IEC guideline is the 10-minute power peak determined based on manufacturers' information.

Table 6.1 Requirements of the guidelines

Requirement	IEC 61400-21	MEASNET	German technical guideline
Instantaneous power peaks	Average value over 0.2 s	Identical to IEC 61400-21	Average value over 8 to 16 line periods
1-minute power peak	Included	Included	Included
10-minute power peak	Maximum permitted power, based on manufacturer's information	Identical to IEC 61400-21	Measurement at 10-minute intervals
Reactive power or power factor	Reactive power as 10 minute average value	Identical to IEC 61400-21	Power factor as 1-minute average value
Flicker in normal operation	10-minute intervals	Identical to IEC 61400-21	1-minute intervals
Integer harmonic currents	Up to 50th order, 10-minute average values	Up to 50th order, grouping, 10-minute average values	Up to 40th order, 8 line period value
Interharmonic currents	No	Up to 2 kHz, grouping, 10-minute average values	Up to 2 kHz, 8 line period value
Current distortion in the higher frequency range	No	Frequency range 2–9 kHz, grouping, 10-minute average values, bandwidth 200 Hz	Frequency range 2–9 kHz, 8 line period value, bandwidth 200 Hz
Flicker and voltage change during switching operations of the wind turbine	For switching operations: • cut-in at cut-in wind speed • cut-in at nominal wind speed • switching between generator stages Separate evaluation of flicker and voltage change	Identical to IEC 61400-21	For switching operations: • cut-in at cut-in wind speed • cut-in at nominal wind speed • cut-off at nominal wind speed • switching between generator stages Same evaluation of flicker and voltage change

Reactive power and power factor

The reactive power is measured as a 10-minute average value (IEC guideline) or as a 1-minute average value (German guideline) over the whole power range of the wind turbine. The reactive power is given for each 10 % step of the active power from 0 % to 100 % of rated power. Instead of reactive power, the German guideline requires the measurement of the power factor.

Harmonics

According to the guidelines, harmonic measurements are not required for fixed-speed wind turbines (Type A), where the induction generator is directly connected to the grid.[1] Harmonic measurements are required only for variable-speed turbines with electronic power converters (Types C and D).

Wind turbines with electronic power converters produce harmonic currents. The three guidelines include different requirements concerning harmonic measurements. The IEC guideline only requires the measurement of integer harmonic currents up to the 50th order. In general, the power converters of wind turbines are pulse-width modulated (PWM) inverters, though, which have clock frequencies in the range of 2–3 kHz and produce mainly interharmonic currents. Thus the requirements of the IEC guideline do not really reflect the harmonic emission of the wind turbine.

The MEASNET guideline and also the German guideline require measurements of interharmonic currents of up to 2 kHz and of current distortions in the higher frequency range of between 2–9 kHz. That means that these two guidelines reflect the harmonic emission of wind turbines. Even though both guidelines require measurements in the same frequency range, the results of such measurements are not comparable. The MEASNET guideline requires measuring intervals of 10 minutes. The relevant value is the maximum 10-minute value of each frequency. Integer harmonic currents and interharmonic currents are grouped according to IEC 61000-4-7 Ed.2.0 (IEC, 2002). In this context, grouping means that all interharmonic currents between two neighbouring integer harmonics are combined into a single interharmonic current, for example. Current distortions in the higher frequency range are grouped into bandwidths of 200 Hz.

The German guideline requires measuring intervals of 8 line periods. For each frequency, a 99 % value of the harmonic (or interharmonic) current is determined (i.e. the value that is not exceeded in 99 % of all cases). These 99 % values are relevant. Grouping for interharmonics or integer harmonics is not used. Thus, interharmonics are given in steps of 6.25 Hz. Similar to the MEASNET guideline, the current distortions in the higher frequency range are grouped into bandwidths of 200 Hz.

In practice, the main difference between the German and the MEASNET guideline lies in the different averaging intervals. In general, the harmonic emission of wind turbines with power converters is not steady. The harmonics, interharmonics and higher frequency distortions behave stochastically. There is an averaging effect, which leads to lower values in the measurements according to the MEASNET guideline. Owing to the very short measurement intervals of the German guideline, the corresponding values are higher.

---

[1] For definitions of turbine Types A to D see Section 4.2.3.

One of the main problems of the harmonic measurements at wind turbines is the influence of the already existing harmonic voltages in the grid. The voltage waveshape of the grid is, of course, not sinusoidal. There are always already harmonic voltages in the grid, such as integer harmonics of 5th and 7th order, which affect the measurements. In many cases, the wind turbine behaves like a consumer of such harmonics. It makes no sense to evaluate such harmonics, because they originate from the grid. The problem is to identify and separate the harmonic emission originating from the turbine and that originating from the grid. A first indicator may be the phase angle of the harmonic and the time or power dependency. We also have to know whether there are wind turbines with power converters in the vicinity of the turbine that is measured. The harmonic emission of the neighbouring turbines can influence the harmonic measurement, particularly as the harmonic emission of the neighbouring turbines may also be correlated to the wind speed.

Flicker

The term 'flicker' means the flickering of light caused by fluctuations of the mains voltage, which can cause distortions or inconvenience to people as well as other electrical consumers. Flicker is defined as the fluctuation of voltage in a frequency range of up to 35 Hz. Flicker evaluation is based on IEC 61000-3-7 (IEC, 1996). The basis for the evaluation is the curve given in Figure 6.1, with a threshold value for the short-term flicker disturbance factor, P_{st}, of $P_{st} = 1$. This curve shows the level where flicker is a visible disturbing factor to people. The most sensitive frequency is 8.8 Hz.

IEC 61000-4-15 (IEC, 2003) specifies a flickermeter that can be used to measure flicker directly. In general, flicker is the result of the flicker that is already present in the grid and of the emissions to be measured. A direct measurement will require an undisturbed constant-impedance power supply. However, this is not feasible for wind turbines because of their size. The flicker measurement is therefore based on measurements of three instantaneous phase voltages and currents, which are followed by an analytical determination of P_{st} for different grid impedance angles. The measured

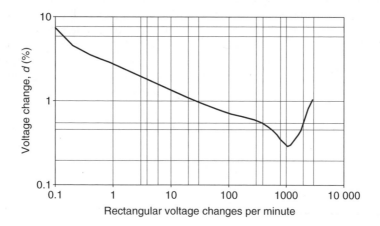

Figure   6.1 Curve of the regular rectangular voltage charge (as a percentage of the nominal value) against the number of changes per minute, for a short-term flicker disturbance, P_{st}, of 1

voltage and current time series of the wind turbines are the input to a simple grid model (see Figure 6.2). For different grid impedances, especially grid impedance angles, time series of voltage fluctuations are calculated for this fictitious grid. The simulated time series of instantaneous voltage fluctuations are the input to the standard voltage flicker algorithm in order to generate the flicker emission values P_{st}. The algorithm is described in IEC 61000-4-15. Such P_{st} values are calculated for a larger number of measured time series over the whole power range of the turbine. Weighted with standard wind-speed distributions, a 99 percentile of the P_{st} values is calculated, which ensures that for 99 % of the time, the flicker of the wind turbine will be within this 99 percentile. Last, a flicker coefficient is calculated from this 99 percentile of the P_{st} values. The flicker coefficient gives a normalised, dimensionless measure of the flicker, independent of the network situation and thus independent of the selected short-circuit apparent power of the fictitious grid. It determines the ratio between short-circuit power and generator-rated apparent power, which is necessary to achieve a long-term flicker level, P_{lt}, of 1. The flicker coefficient c is defined as:

$$c(\Psi_k, V_a) = P_{lt}\frac{S_k}{S_n} \tag{6.1}$$

where

$c(\Psi_k, V_a)$ is the flicker coefficient, dependent on the grid impedance angle Ψ_k and on the annual average wind speed V_a.
S_k is the short-circuit power of the grid at the point of common coupling (PCC).
S_n is the apparent power of the wind turbine at rated power .
P_{lt} is the long-term flicker emission.

This flicker coefficient can be used by utilities to calculate the flicker emission of a wind turbine or a wind farm at a specific site.

The procedures of IEC 61400-21 (IEC, 2001) and the German guideline are similar. The only difference is that the German guideline requires 1-minute time intervals and thus 1-minute P_{st} values, and the IEC guideline requires 10-minute intervals. This leads to different results, with the flicker coefficients of the German guideline in general slightly exceeding those of the IEC guideline.

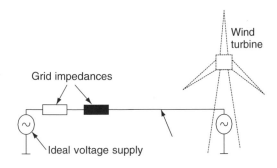

Figure 6.2 Simulation model for flicker measurements

Switching operations

Switching operations can cause voltage fluctuations due to in-rush currents and can cause flicker. Therefore, the German guideline and the IEC guideline as well as the MEASNET guideline require measurements concerning flicker and voltage fluctuations during switching operations. Owing to the already existing flicker and voltage fluctuations in the grid, the measurement procedure is similar to the procedure for flicker measurements during normal operation. This means that the measured time series during switching operations is inserted into a grid model with specific grid impedances (see Figure 6.2). The model produces the time series of voltages, from which the voltage fluctuations are calculated. The normalised voltage fluctuations lead to a voltage change factor, which is a dimensionless value of the voltage fluctuations during switching operations. In addition, the standard flicker algorithm is used to calculate a flicker step factor from the voltage fluctuations that the grid model generates. This flicker step factor is the basis for the calculation of the flicker impression that the switching operations of a wind turbine or farm at a specific site will cause.

The German guideline requires calculation methods that are similar to those of the IEC guideline. In the German guideline, though, the flicker impression and the voltage fluctuation are combined into a single grid-related switching factor. This factor is very easy for the utilities to apply. It is, however, less accurate than the procedure of the IEC and MEASNET guidelines.

The following switching operations have to be measured:

- cut-in at cut-in wind speed;
- cut-in at rated wind speed;
- switching operations between generator stages;
- service cut-out at rated power (German guideline only).

6.2.3 Future aspects

In some countries, such as Germany and Denmark, there is a high penetration of electricity generated from wind energy. Grid operators will have to take precautions in order to ensure a safe and stable operation of the grid, taking into account the expected increase in production from wind energy. Therefore, some utilities have established new additional requirements (EON, 2003; VDN, 2004; Eltra, 2000; NGC, 2004 for a discussion of the new grid requirements, see Chapter 7).

Until now, the philosophy regarding wind farms connected to the grid has been to switch off the farm as soon as there is a fault on the grid. The new guidelines are a radical change to this philosophy. They require wind farms to remain connected in the case of grid faults (e.g. short-term voltage drops). The wind farms are expected to support the grid. Therefore, wind farms have to control reactive power over a wide range, deliver reactive power in the case of voltage drops and remain connected during short-term voltage drops. And they must be able to operate over a wide frequency range.

In order to ensure that wind turbines and wind farms will fulfil these new requirements, in Germany, an additional measurement guideline has been developed (FGW, 2003). This guideline includes methods to check the compatibility of the turbine or farm with the requirements of EON (2003) and VDN (2004). The guideline includes the following measurements and checks:

- range of reactive power (capacitive as well as inductive);
- power gradient after grid losses;
- operation at underfrequency and overfrequency;
- power reduction through set-point signal;
- protection system concerning frequency and voltage changes;
- behaviour of the wind turbine in the case of short voltage drops (ride-through capability).

6.3 Power Quality Characteristics of Wind Turbines and Wind Farms

The grid interferences of wind turbines or wind farms have different causes, which are mostly turbine-specific. The relevant parameters are listed in Table 6.2. Average power production, turbulence intensity and wind shear refer to causes that are determined by meteorological and geographical conditions. All the other causes are attributable to the technical performance of the wind turbine. This performance is determined not only by the characteristics of the electrical components, such as generators, transformers and so on, but also by the aerodynamic and mechanical behaviour of rotor and drive train. The turbine type (i.e. variable versus fixed speed stall versus pitched controlled) is of major importance to the power quality characteristics of wind turbines and wind farms.

6.3.1 Power peaks

Variable-speed wind turbines (Types C and D) can control the power output of the inverter system by pitch control, thus smoothing power fluctuations as well as power peaks. Thus the power peaks lie within the range of the rated power. Instantaneous power peaks of fixed-speed wind turbines (Type A) often exceed rated power by 30 % or more, even in the case of pitch-controlled fixed-speed turbines. The pitch control is not

Table 6.2 Grid interferences caused by wind turbines and wind farms

Parameter	Cause
Voltage rise	Power production
Voltage fluctuations and flicker	Switching operations
	Tower shadow effect
	Blade pitching error
	Yaw error
	Wind shear
	Fluctuations of wind speed
Harmonics	Frequency inverter
	Thyristor controller
Reactive power consumption	Inductive components or generating systems (asynchronous generator)
Voltage peaks and drops	Switching operations

fast enough to control fast power peaks. Slower power peaks, such as 1-minute and 10-minute power peaks, can be reduced by the pitch control. These slower power peaks are therefore also close to rated power, also for pitch-controlled fixed-speed turbines (Type A1). The power peaks of a stall-controlled fixed-speed turbine (Type A0) depend on air pressure and air temperature, among other things. This has the effect that 1-minute and 10-minute power peaks often exceed rated power by about 10 % to 20 %.

The number of wind turbines on a wind farm is important for smoothing power peaks. In particular, fast power peaks at the individual wind turbines of a wind farm are in general uncorrelated and thus smoothed throughout the wind farm. IEC 61400-21 includes a formula for this smoothing effect, which in principal adds up the power peaks of the single wind turbines geometrically.

6.3.2 Reactive power

The reactive power demand of the asynchronous generator of fixed-speed wind turbines (Type A) is partly compensated by capacitor banks. Thus the power factor (i.e. the ratio of active power and apparent power) lies, in general, at about 0.96. Variable-speed wind turbines with PWM inverter systems (Types C and D) have an inverter to control the reactive power. Thus, these wind turbines have, in general, a power factor of 1.00. These turbines can control the reactive power over a wide range (inductive and capacitive). It is therefore possible to control the voltage and to keep it more stable at the grid connection point of the wind farm or wind turbine (see also Chapter 19).

6.3.3 Harmonics

Today's variable-speed turbines (Types C and D) are equipped with self-commutated inverter systems, which are mainly PWM inverters, using an insulated gate bipolar transistor (IGBT; see also Chapter 4). This type of inverter has the advantage that both the active power and the reactive power can be controlled. It has the disadvantage, though, that it produces harmonic currents. In general, the inverter generates harmonics in the range of some kilohertz. Therefore, filters are necessary to reduce the harmonics.

Two types of PWM inverters are used: those with a fixed clock frequency and those with a variable clock frequency. Figure 6.3 gives examples of the harmonic emission of wind turbines with such inverter systems. The main difference between each type is that the inverter with a fixed clock frequency [Figure 6.3(a)] produces single interharmonics in the range of the clock frequency and multiples of the clock frequency. Inverters with variable clock frequency [Figure 6.3(b)] have a wide band of interharmonics and integer harmonics. Resonances of the grid are excited by this wide band of interharmonics and integer harmonics. The result are interharmonic and harmonic currents, as Figure 6.3(b) illustrates, where the specific maximum occurs at the resonance frequency of the grid.

The measurement of the harmonic currents poses one of the biggest challenges to the measurement of power quality. Harmonic current measurements require great accuracy, even for high frequencies, because the measurements refer to interharmonics that are in the range of 0.1 % of the rated current for frequencies of up to 9 kHz (for MEASNET and the German guideline). Therefore, up to 9 kHz, current clamps need to have a linear ratio.

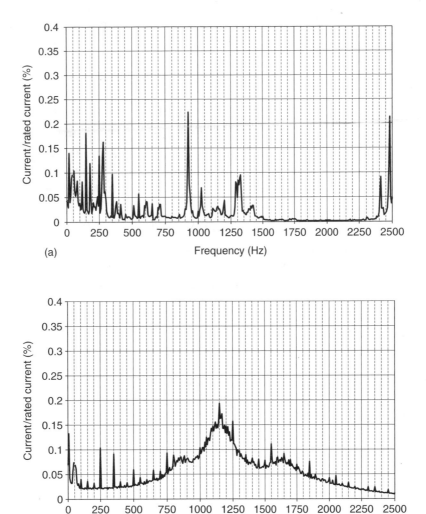

Figure 6.3 Harmonic current emission of a variable-speed wind turbine where the inverter has (a) a fixed clock frequency and (b) a variable clock frequency

Harmonic measurements at the low-voltage side of the wind turbine may give results that are different from measurements at the medium-voltage side. Investigations have shown that in some cases the transformer of the turbine has an influence on the harmonics (Santjer, 2003). The transformers of wind turbines are often connected at the medium-voltage side in delta operation and at the low-voltage side in star operation. Owing to this connection, a single phase at the medium-voltage side is influenced by two phases at the low-voltage side. This can lead to a smoothing effect that reduces some of the harmonics. In particular, the so-called 'zero current' may be eliminated. Thus harmonic measurements at the medium-voltage side of the transformer often produce lower values than at the low-voltage side, but, for practical reasons, measurements at the medium-voltage side are often more difficult than at the low-voltage side.

6.3.4 Flicker

Fluctuations of active and/or reactive power of wind turbines cause flicker. Active power fluctuations of wind turbines may be caused by the effect of the wake of the tower, by yaw errors, by wind shear, by wind turbulences or by fluctuations in the control system. The main reason for flicker in fixed-speed turbines (Type A) is the wake of the tower. Each time a rotor blade passes the tower, the power output of the turbine is reduced. This effect causes periodical power fluctuations with a frequency of about 1 Hz. Figure 6.4 gives an example of such power fluctuations in a fixed-speed wind turbine. Power fluctuations due to wind-speed fluctuations have lower frequencies and thus are less critical for flicker. In general, the flicker of fixed-speed turbines reaches its maximum at high wind speeds. Owing to smoothing effects, large wind turbines generally produce lower flicker than small wind turbines, in relation to their size. An example of the flicker behaviour of a fixed-speed turbine (Type A) is given in Figure 6.5.

For variable-speed turbines (Types C and D), fast power fluctuations are smoothed and the wake of the tower does not affect power output. Therefore, the flicker of variable-speed turbines (Types C and D) is in general lower than the flicker of fixed-speed turbines (Type A). Figure 6.6 gives an example of the power output of a variable-speed wind turbine.

In wind farms, power fluctuations are smoothed because of the fact that the power fluctuations of the single wind turbines are uncorrelated. The flicker of a wind farm is

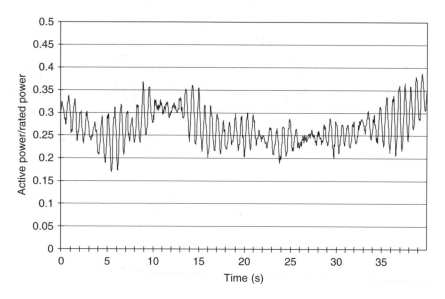

Figure   6.4 Active power output of a fixed-speed wind turbine (Type A); periodical fluctuations are due to the wake of the tower

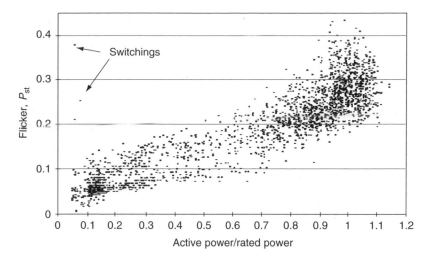

Figure   6.5 Flicker behaviour of a fixed-speed wind turbine as a function of active power output

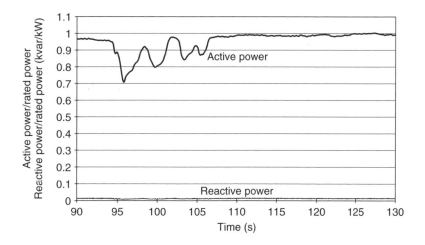

Figure   6.6 Power output of a variable-speed wind turbine

the geometrical sum of the flicker of all single turbines in the farm. This means that for a wind farm consisting of n single turbines of the same type, the flicker of the farm is $\sqrt{n}$ times the flicker of a single wind turbine.

6.3.5 Switching operations

Voltage changes during switching operations are due to in-rush currents and the respective changes in the active and reactive power of a wind turbine. For fixed-speed turbines, a soft-starter limits the in-rush current of the asynchronous generator. The soft-starter is generally based on thyristor technology and limits the highest root mean square (RMS) value of the in-rush current to a level below twice the rated current of the generator. Figure 6.7(a) gives an example of the cut-in of a fixed-speed turbine (Type A). Before cut-in, the rotational speed of the rotor and that of the asynchronous generator increase, driven by the wind. Once the synchronous speed is reached, the generator will be connected to the grid. The soft-starter is in operation for about 1 or 2 seconds and limits in-rush current. During this period, the generator needs reactive power to become magnetised. A few seconds after the generator is connected, the capacitors are switched on to minimise the demand of reactive power. The fast power changes during the switching operation cause flicker. The (large) changes in active and reactive power cause voltage fluctuations.

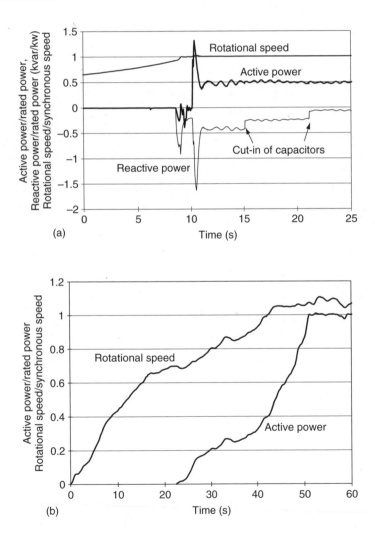

Figure   6.7 Cut-in of (a) a fixed-speed wind turbine (Type A) and (b) a variable-speed wind turbine (Types C and D)

In general, variable-speed turbines (Types C and D) do not show such in-rush currents. Figure 6.7(b) gives an example of a cut-in. Before the turbine generates power, the rotational speed of the rotor increases, driven by the wind. If the rotational speed is high enough the power production starts without high in-rush currents. The power production increases to the power level that is determined by the actual wind speed. In the example of Figure 6.7(b) the power reaches rated power as a result of the actual wind speed, which is higher than the nominal wind speed. The smoothing behaviour of the variable-speed turbines (Types C and D) during switching operations generally leads to relatively low voltage changes and low flicker impression.

In wind farms, usually there are only one or a few wind turbines that start or stop at the same time. Only on very rare occasions will all wind turbines in a wind farm stop at the same time. For the calculation of the voltage change due to switching operations, it is therefore sufficient to consider one turbine or a small number of turbines.

6.4 Assessment Concerning the Grid Connection

IEC 61400-21 includes recommendations for the assessment of the grid connection of wind turbines and wind farms regarding power quality. The calculation methods are based on the equations given in Table 6.3. The influence of the turbine or farm on the voltage quality of the grid depends not only on the power quality of the turbine but also on how stiff or weak the grid is at the point where the farm or turbine is connected

Table 6.3 Calculation methods for grid connection

Item	Calculation method		Limiting value
	IEC 61400-21	German guideline	
Steady-state voltage change	Load flow calculation, or $d = \frac{S_{60}}{S_k}\lvert\cos(\psi_k + \varphi)\rvert$	As for IEC 61400-21	National guidelines: 2 %, ..., 10 %
Flicker	$P_{lt} = c(\psi_k, v_a)\frac{S_n}{S_k}$	$P_{lt} = c_\psi \frac{S_n}{S_k}$	IEC 61000-3-7: $P_{lt} \leq 0.25$ Germany: $P_{lt} \leq 0.46$
Harmonics	Not applicable	Not applicable	National guideline or IEC 61000-3-6
Switching operations	$d = k_u(\psi_k)\frac{S_n}{S_k}$ $P_{lt} = 8N_{120}^{0.31}k_f(\psi_k)\frac{S_n}{S_k}$	$d = k_{i\psi}(\psi_k)\frac{S_n}{S_k}$	$d \leq 2\%, ..., 10\%$ $P_{lt} \leq 0.25$ (IEC 61000-3-7)

Note: $d =$ steady-state voltage change as a percentage of the nominal voltage; $S_{60} = 1$ minute maximum apparent power; $S_k =$ short-circuit power; $\psi_k =$ network impedance phase angle; $\varphi =$ phase difference; $P_{lt} =$ long-term flicker disturbance factor; $c(\psi_k, V_a) =$ flicker coefficient for continuous operation as a function of ψ_k and the annual average wind speed, V_a; $c_\psi =$ flicker coefficient (according to FGW, 2002); $S_n =$ rated apparent power of the turbine; $k_u(\psi_k) =$ voltage change factor as a function of ψ_k; $k_f =$ flicker step factor; $k_{i\psi} =$ grid-dependent switching current factor; $N_{120} =$ maximum number of one type of switching within a 120 minute period.

[i.e. the point of common coupling (PCC)]. Short-circuit power and grid impedance angle are the parameters that describe the strength of the grid (an example of the assessment of the grid connection of a wind farm is given in EC, 2001).

On the local level, the steady-state voltage change often is one of the limiting factors for a grid connection. For an accurate assessment it is therefore important to carry out a complex load flow calculation of the grid including the wind farm or wind turbine. The fact that wind turbines can control reactive power with the inverter system, for instance, can be used to control and stabilise voltage and thus to minimise voltage changes (see also Chapter 19).

The harmonic emission may sometimes pose a problem for the grid connection. Germany has strict limiting values for harmonic emissions, especially for the frequency range above 2 kHz (VDEW, 1998) but also for interharmonics of up to 2 kHz. Denmark has strict limiting values for interharmonic currents of up to 2.5 kHz (DEFU, 1998). International guidelines generally include integer harmonics of up to 50th or 40th order only, but the inverter systems of variable-speed wind turbines mainly produce inter-harmonics or harmonic emissions in excess of 2 kHz. Thus integer harmonic limits are often not a problem for wind turbines. One difficulty for the assessment of the harmonic emission is that the international or national guidelines only provide limits for harmonic voltages. That means that the impact of harmonic current emissions of wind turbines or wind farms on the voltage at the PCC has to be calculated. For this, the grid impedances for frequencies above 50 Hz have to be known, because at this level grid resonances have a large influence. This makes the calculation rather difficult. Only some national guide-lines, such as the German or Danish guidelines, include limits of harmonic currents, which simplifies the assessment. They are often very strict, though.

Flicker is a minor problem for the grid connection of today's wind turbines. In the mid-1990s, when most of the turbines were small fixed-speed turbines (Type A), flicker sometimes was a limiting issue. Today's wind turbines, especially variable-speed tur-bines (Types C and D), have an improved flicker behaviour.

6.5 Conclusions

The power quality of wind turbines is described by their electrical characteristics. International (IEC 61400-21 and MEASNET) and national guidelines set the require-ments for power quality measurements. Power peaks, harmonic emission, reactive power, flicker and the electrical behaviour in the case of switching operations are measured according to these guidelines. Variable-speed wind turbines (Types C and D) have a smoothed power output compared with fixed-speed wind turbines (Type A) and are able to control active and reactive power fast. Variable-speed turbines have the disadvantage of producing harmonic emission, though. On the local level, the steady-state voltage change is often the limiting factor for a grid connection.

There are additional aspects regarding the grid connection of wind turbines and wind farms, such as the support of the grid in the case of grid faults (voltage drops) as well as the control of active and reactive power. These aspects are of increasing importance and are already included in some of the national guidelines. A new measurement guideline in Germany includes the requirement to check if the wind turbines comply with these new, additional, aspects.

References

[1] DEFU (Darish Utilities Research Association) (1998) *Connection of Wind Turbines to Low and Medium Voltage Networks*, October 1998, Committee Report 111-E, DEFU, DK-2800 Lyngby.

[2] EC (2001) (European Commission) *Wind Turbine Grid Connection and Interaction*, Deutsches Windenergie-Institut GmbH (DEWI), Tech-wise A/S, DM Energy; available from DEWI, http://www.dewi.de.

[3] Eltra (2000) *Tilslutningsbetingelser for vindmoelleparker tilsluttet transmissionsnettet*, 17 April 2000, http://www.eltra.dk/media/1030_12368.pdf.

[4] EON (2003) *Netzanschlussregeln, Hoch und Höchstspannung*, E.ON Netz GmbH, http://www.eon-nctz.com/Ressources/downloads/ENE_NAR_HS_01082003.pdf   or,   http://www.eon-netz.com/Resources/downloads/enenarhseng1.pdf.

[5] FGW (Fördergesellschaft Windenergie e.V.) (2002) *Technische Richtlinie für Windenergieanlagen, Teil 3: Bestimmung der Elektrischen Eigenschaften*, http//www.wind-fgw.de.

[6] FGW (Fördergesellschaft Windenergie e.V.) (2003) *Technische Richtlinie für Windenergieanlagen, Teil 4: Bestimmung der Netzanschlussgrößen*, http//www.wind-fgw.de.

[7] IEC (International Electrotechnical Commission) (1996) *EMC, Part 3: Limits. Section 7: Assessment of Emission Limits for Fluctuating Loads in MV and HV Power Systems*, IEC 61000-3-7 (10.1996), Basic EMC publication (technical report), http://www.iec.ch.

[8] IEC (International Electrotechnical Commission) (2002) *Electromagnetic Compatibility EMC, Parts 4–7: Testing and Measurement Techniques: General Guide on Harmonics and Interharmonics Measurements and Instrumentation, for Power Supply Systems and Equipment Connected there to IEC 61000-3-6: (10.1996), EMC. Part 3: Limits. Section 6: Assessment of Emission Limits for Distorting Loads in MV and HV Power Systems*, IEC 61000-4-7 Ed.2.0 (8.2002), basic EMC publication (technical report), http://www.iec.ch.

[9] IEC (International Electrotechnical Commission) (2003) *Electromagnetic Compatibility (EMC), Part 4: Testing and Measuring Techniques. Section 15: Flickermeter – Functional and Design Specifications*, IEC 61000-4-15 (02.2003), http://www.iec.ch.

[10] IEC (International Electrotechnical Commission) (2001) *Wind Turbine Generator Systems, Part 21: Measurement and Assessment of Power Quality Characteristics of Grid Connected Wind Turbines*, IEC 61400-21 (12.2001), 1st edn, http://www.iec.ch.

[11] MEASNET (2000) *Power Quality Measurement Procedure of Wind Turbines: Measuring Network of Wind Energy Institutes (MEASNET)*, Version 2, November 2000; copy available from Deutsches Windenergie Institut GmbH, http://www.dewi.de.

[12] NGC (National Grid Company plc) (2004) *The Grid Code*, http://www.nationalgridinfo.co.uk/grid_code/mn_current.html.

[13] Santjer, F. (2000) *Power Quality of Wind Turbines in Germany and in Denmark/Sweden*, Deutsches Windenergie-Institut GmbH (DEWI), *DEWI-Magazine* number 16, www.dewi.de.

[14] Santjer, F. (2003) *Influence of Transformers on Harmonics*, Fritz Santjer, Rainer Klosse, Deutsches Windenergie-Institut GmbH (DEWI), EWEC 2003, Madrid.

[15] VDEW (Verband des Elektrizitätswirtschaft e.V.) (1998) *Eigenerzeugungsanlagen am Mittelspannungsnetz. Richtlinie für Anschluss und Parallelbetrieb von Eigenerzeugungsanlagen am Mittelspannungsnetz*, 2nd edn, VWEW-Verlag, Frankfurt.

[16] VDN (Verband der Netzbetreiber e.v.) (2004) *EEG-Erzeugungsanlagen am Hoch-und Höchstspannungsnetz*, http://www.vdn-berlin.de/global/downloads/Publikationen/Fachberichte/RL_EEG_HH_2004-08.pdf.

7

Technical Regulations for the Interconnection of Wind Farms to the Power System

Julija Matevosyan, Thomas Ackermann and Sigrid M. Bolik

7.1 Introduction

This chapter provides a brief discussion and analysis of the current status of interconnection regulations for wind power in Europe. The chapter starts with a short overview of the relevant technical regulation issues, which includes a brief description of the relevant interconnection regulations considered in this chapter. This is followed by a detailed comparison of the different interconnection regulations. The discussion also includes the capabilities of wind turbines to comply with these requirements. In addition, a new wind farm control systems is briefly explained, which was developed particularly to comply with network interconnection requirements. Finally, issues related to international interconnection practice are briefly discussed.

7.2 Overview of Technical Regulations

Technical standards that are adopted by the industry often originate from standards developed by the Institute of Electrical and Electronic Engineers (IEEE) or from the International Electrotechnical Commission (IEC). These standards are, however, voluntary unless a specific organisation or legislative ruling requires the adoption of these standards. Hence, there is a large number of additional national or regional standards,

Wind Power in Power Systems   Edited by T. Ackermann
© 2005 John Wiley & Sons, Ltd   ISBN: 0-470-85508-8 (HB)

requirements, guidelines, recommendations or instructions for the interconnection for wind turbines or wind farms worldwide.

At the end of the 1980s, distribution network companies in Europe had to deal with small wind turbines and wind farms that wanted to be connected to the distribution network. At this time, the IEEE Standard 1001 (IEEE, 1988) *IEEE Guide for Interfacing Dispersed Storage and Generation Facilities with Electric Utility Systems* was the only IEEE guide in place that partly covered the connection of generation facilities to distribution networks. The standard included the basic issues of power quality, equipment protection and safety. The standard expired and, therefore, in 1998, the IEEE Working Group SCC21 P1547 started to work on a general recommendation for the interconnection of distributed generation, the *IEEE Standard for Interconnecting Distributed Resources with Electric Power Systems* (IEEE, 2003). Four years later, in September 2002, the working group finally agreed on a new standard.

Back in the late 1980s, however, distribution network companies in Europe started to develop their own interconnection rules or standards. In the beginning, each network company that faced an increasing amount of interconnection requests for wind farms developed its own rules. During the 1990s, these interconnection rules where harmonised on a national level (e.g. in Germany or Denmark). This harmonisation process often involved national network associations as well as national wind energy associations, which represented the interests of wind farm developers and owners. In Europe, government organisations are hardly involved in the definition of interconnection guidelines. In other regions of the world, this can be different. In Texas and California, for instance, electricity regulation authorities were involved in defining the interconnection guidelines for distributed generation (DG) [see for instance the *Interconnection Guidelines for Distributed Generation in Texas* (PUCT, 1999) and the *Distributed Generation Interconnection Rules* (California Energy Commission, 2000) in California].

In Europe, the harmonisation of national interconnection rules was not the end of the development of such rules. National interconnection rules were continuously reformulated because of the increasing wind power penetration and the rapid development of wind turbine technology (i.e. wind turbine ratings increased rapidly, from around 200 kW in the early 1990s to 3–4 MW turbines in early 2004). In addition, wind energy technology introduced new technologies such as doubly fed induction generators (DFIGs). Until then, generation technologies that used DFIGs with a rating of up to 3 MW and combined a large number of these within one power station (i.e. wind farms) were unheard of in the power industry.

Not only was the increased size of the wind turbines new but too was the increasing size of the wind farms, which resulted in interconnection requests at the transmission level. Hence, interconnection rules for wind farms to be connected to the transmission level were required.

Unfortunately, the continuously changing network rules and the re-regulation of the power market make a comparison or evaluation of the already very complex interconnection rules very difficult and there is only very limited literature in this area (see Bolik *et al.*, 2003; CIGRE, 1998; Jauch, Sørensen and Bak-Jensen, 2004).

A comparison might be useful because:

- Interconnection rules are often the source of controversies between wind farm developers and network operators (see for instance Jörß, 2002; Jörß *et al.*, 2003), therefore a comparison and analysis would contribute to reduce such controversies (in some countries, governments have decided to solve such controversies by setting up clearing committees that help solve controversies between network companies and wind farm developers).
- It would provide a better understanding of the relevant issues for those countries, regions or utilities that are still in the process of developing interconnection rules for wind farms. This would help to harmonise interconnection rules worldwide.
- To comply with new connection requirements is a challenge for wind turbine manufacturers. New hardware and control strategies have to be developed. The comparison of connection requirements in different countries will give wind turbine manufacturers an overview of the existing rules.
- An understanding of the difference between the national rules will contribute to a harmonisation of interconnection rules in Europe and even beyond.

The following sections therefore provide a brief comparison and evaluation of the most important interconnection rules that are relevant for wind farms. Owing to the complexity of the regulations, only a few specific areas of the interconnection rules are compared.

In the remainder of this section will briefly introduce the regulations that are to be compared in Section 7.3. For that purpose, we will divide the interconnection rules according to their area of application:

- Wind farms or wind turbines connected to networks with voltage levels below 110 kV;
- Wind farms or wind turbines connected to networks with voltage levels above 110 kV, (for E.ON, voltage levels above 60 kV);
- All wind farms or wind turbines, irrespective of voltage level.

Additionally, it is important to mention that national, regional or utility interconnection rules in Europe often include references to other standards (e.g. power quality standards). These standards will not be analysed in this chapter, but the most common additional standards that interconnection rules refer to are:

- IEC 61400–21 (IEC, 2001): *Measurements and Assessment of Power Quality Characteristics of Grid Connected Wind Turbines* (see also Chapters 5 and 6);
- IEEE 519 (IEC, 1992), *IEEE Recommended Practices and Requirements for Harmonic Control in Electric Power Systems*, which applies to all static power converters used in the industry, stating voltage and current harmonic limits as well as voltage limits of irritation curves for utility practice and DG requirements.

7.2.1 Regulations for networks below 110 kV

In this section we provide a brief introduction to interconnection guidelines, recommendations and requirements relevant to distribution networks [low-voltage (LV) and medium-voltage (MV) level] that are discussed in this chapter. The relevant *guidelines* are: DEFU 111, developed in Denmark; AMP, developed in Sweden; as well as the German VDEW guidelines. In addition, we will also present new interconnection *requirements* for Denmark that are still under discussion at the time of writing. Finally, the engineering *recommendations* that are currently applied in the UK are considered.

The guidelines, recommendations and requirements are directed towards distribution network companies, wind turbine manufacturers and network operators as well as others who are interested in connecting wind farms to the LV or MV network. The objective of the technical standards is to establish guidelines, recommendations and requirements for wind turbines and networks in compliance with applicable standards for voltage quality and reliability of supply. The guidelines and requirements should be independent of the design approach used for the wind turbine and open enough to apply to synchronous or induction generators with or without inverters, for instance. They deal with the technical data needed to assess the impact of wind turbines on power quality and discuss the requirements to be met by networks to which wind turbines are to be connected (voltage quality at the customer side).

7.2.1.1 Denmark: DEFU 111

The Danish report, 'Connection of Wind Turbines to Low and Medium Voltage Networks', or DEFU *111*, was developed the by Research Institute of Danish Electric Utilities (DEFU, 1998) and works as an interconnection guideline in Denmark. The guidelines do not include network stability problems that can occur in regional or transmission networks with high penetration levels of wind turbines or wind farms or in the case of a low ratio between short-circuit power at the connection point and total installed wind power. By the time of writing, DEFU 111 was under review and a first draft of an updated version (DEFU, 2003) was under discussion, see Section 7.2.1.2. Therefore, the reviewed guidelines are likely to have been implemented by the time this book is published.[1]

7.2.1.2 Denmark: Eltra and Elkraft

The new requirements for distribution networks, which are likely to replace DEFU 111, have been designed by the two Danish transmission system operators (TSOs), Eltra (Western Denmark) and Elkraft System (Eastern Denmark). The new requirements are titled 'Wind Turbine Generators Connected to Networks with Voltage Levels below 100 kV (Draft)' (DEFU, 2003). The requirements apply, after 1 July 2004, to wind farms that are connected to networks with voltage levels lower than 100 kV. These requirements have the aim of ensuring that wind turbines have regulating and dynamic properties that are essential for maintaining a reliable power supply and voltage quality in the short and long term.

---

[1] Update: DEFU (2003) – now Eltra and Elkraft regulations – was approved 19 May 2004 and is now in force; see http://www.eltra.dk/media (15716, 1030)/Forskrift_for_vindm%F811er_maj_2004.pdf.

7.2.1.3 Sweden: AMP

The Swedish counterpart to DEFU 111 is known as AMP, which is an abbreviation for the Swedish title of a report that translates as 'Connection of Smaller Power Plants to Electrical Networks' (Svensk Energi, 2001). AMP follows in many parts DEFU 111. The guidelines do not include network stability problems that can occur in regional or transmission networks at high penetration levels of wind turbines or wind farms or in the case of a low ratio between the short-circuit power at the connection point and total installed wind power.

7.2.1.4 Germany: VDEW guidelines

The German guidelines, 'Generation in the Medium Voltage Network – Guidelines for the Connection and Operation of Generation Units in the Medium Voltage Network', were developed and published by the German Electricity Association, the Verband der Elektrizitätswirtscheft (VDEW, 1998).

The guideline requires power quality measurements for each wind turbine type, outlined in a publication from the Fördergesellschaft Windenergie (FGW, 2002). The detailed requirements for power quality measurements were first included in 1992 in order to develop measurement strategies for comparing wind turbines. The measurement guidelines are continuously revised by working groups consisting of wind turbine manufacturers, network operators and measurement institutes. In addition to the already published material, the groups currently work on the development of wind turbine models for researching grid connection requirements, and grid stability, among other things. For a more detailed discussion of requirements regarding power quality measurements in Germany, see Chapter 6.

The VDEW guideline does not include network stability problems that can occur in regional or transmission networks at high penetration levels of wind turbines and wind farms or in the case of a low ratio between short-circuit power at the connection point and total installed wind power.

7.2.1.5 United Kingdom: G59/1 and G75

In the UK, Engineering Recommendation G59/1 (ER G59 1990; see Electricity Association, 1990) is the main document covering the connection of DG, including wind power, for generators of up to 5 MW at 20 kV and below. The recommendation is supported by Engineering Technical Report 113 (ETR 113; see Electricity Association, 1995). The technical document provides guidance on the methods of meeting the requirements of G59/1. In addition, Engineering Recommendation G75 (ER G75 1996; see Electricity Association, 1996) covers the connection of generators above 5 MW at 20 kV voltage; see also ER G75 1996. It is important to mention that at the time of writing a change of the relevant regulations was under discussion.

7.2.2 Regulations for networks above 110 kV

In this section we provide a brief introduction to the interconnection requirements that apply to transmission networks. The relevant requirements were defined by the TSO in Western Denmark, Eltra, by E.ON Netz, one of five German TSOs, and by the

Electricity Supply Board National Grid (ESBNG) in Ireland.[2] Among others, most of these documents include requirements on fault tolerance, wind farm modelling and communication as well as possibilities of external wind farm control in order to ensure secure system operation during and after faults.

7.2.2.1 Denmark: Eltra

Eltra's '*Specifications for Connecting Wind Farms to the Transmission Network*' (Eltra, 2000) apply to wind farms that are connected to network voltage levels above 100 kV. These minimum requirements are set by the TSO, and the owners of wind farms have to comply with them in order to ensure proper operation of the power system regarding security of supply, reliability and power quality, in the short as well as in the long term (see also Chapter 8).

7.2.2.2 Germany: E.ON

The German TSO, E.ON Netz, has continuously updated its interconnection rules over the past years (see E.ON Netz, 2001a; E.ON Netz, 2001b; Santjer and Klosse, 2003). In this chapter, the 'Grid Code for High and Extra High Voltage' (E.ON Netz, 2003) from August 2003 is considered. It applies to wind farms connected to high-voltage (HV: 60 kV, 110 kV) and extra-high-voltage (EHV: 220 kV, 380 kV) networks. Similar to Eltra, the E.ON Netz rules aim at ensuring proper operation of the power system regarding security of supply, reliability and power quality, in the short as well as in the long term (for a discussion, see Chapter 11).

7.2.2.3 Ireland: ESB

The Irish Electricity Supply Board (ESB) is the utility company and consists of different business units. In February 2002, ESB NG elaborated a draft proposal of wind farm connection requirements (ESBNG, 2002a). This is mainly a clarification of how the existing grid code (ESBNG, 2002b) should be interpreted for the connection and operation of wind farms. Some requirements are also specially adapted to wind farms. A completed version of a 'grid code for wind farms' is expected to be available in mid-2004.[3]

7.2.3 Combined regulations

In this section we provide a brief introduction to the interconnection regulations that are relevant to wind farms or wind turbines, irrespective of the connection voltage level: the Swedish interconnection requirement of the TSO Svenska Kraftnät (SvK), the

---

[2] EirGrid Plc. will take over the operation of Ireland's electricity transmission system from ESBNG in the near future; see http://www.eirgrid.com/eirgridportal/Home Flash.aspx.
[3] Update: The Commission for Energy Regulation approved the proposed Wind Grid Code on 1 July 2004; see http://www.eirgrid.com/EirGridPortal/DesktopDefault.aspx?tabid=Wind&TreeLinkModID=1445&Tree LinkItemID=42.

Norwegian guideline from Sintef and the Scottish draft guideline proposed by the Scottish TSOs. These documents refer to wind farms of a certain installed capacity and, similar to the regulations discussed in Section 7.2.2, address the impact of wind farms on power system stability.

7.2.3.1 Sweden: Svenska Kraftnät

The 'Technical Instructions for the Design of New Production Installations to Ensure Secure Operation' (SvK, 2002), developed by the Swedish TSO Svenska Kraftnät (SvK), intends to establish prerequisites for secure operation of the power system. The requirements include: disturbance tolerance; voltage regulation; power regulation; shutdown and startup after exterior voltage loss; communication and controllability; tests and documentation. The requirements refer to different types of production installations, including wind turbines with a nominal power larger than 0.3 MW and even wind farms exceeding 100 MW. At the time of writing a change in regulations was under discussion.

7.2.3.2 Norway: Sintef

The Norwegian research institute Sintef has developed the 'Guidelines for the Connection of Wind Turbines to the Network' (Sintef, 2001). The report refers to wind turbines and wind farms that are to be connected to the distribution, regional or central network. Connection to island networks and/or LV distribution networks is not included in the report. The guidelines refer to the following topics: thermal limits, losses, slow voltage variations, flicker, voltage dips and harmonics. The report also discusses the protection of wind turbines – that is over voltage and undervoltage protection, short-circuit protection, protection against undesirable island operation and transient overvoltages. The guidelines are based on the recommendations in IEC 61400–21 (IEC, 2001), DEFU 111 (DEFU, 1998) and AMP (Svensk Energi, 2001).

7.2.3.3 Scotland: Scottish Power Transmission and Distribution and Scottish Hydro-Electric

The Scottish 'Guidance Note for the Connection of Large Wind Farms' (Scottish Hydro Electric, 2002) applies to all wind farms with a registered capacity of 5 MW and above. The guidance note refers to the Scottish grid code and deals mainly with control issues of a wind farm in order to sustain stable operation of the network. At the time of writing, the general regulations for Scotland, England and Wales were under discussion.

7.3 Comparison of Technical Interconnection Regulations

In the following, the interconnection regulations introduced in Section 7.2 will be compared. The comparison is divided as follows: active power control, frequency control, voltage control, tap changer, wind farm protection, modelling and communication requirements.

7.3.1 Active power control

In general, power production and consumption have to be in balance within a power system. Changes in power supply or demand can lead to a temporary imbalance in the system and affect operating conditions of power plants as well as affecting consumers.

In order to avoid long-term unbalanced conditions the power demand is predicted and power plants adjust their power production. The requirements regarding active power control of wind farms aim to ensure a stable frequency in the system (see also Section 7.3.2), to prevent overloading of transmission lines, to ensure compliance with power quality standards and to avoid large voltage steps and in-rush currents during startup and shutdown of wind turbines.

It should be noticed that the Scottish guidance note and ESBNG include requirements regarding maximum active power change during startup, shutdown and wind speed change in order to avoid impacts on system frequency (see Table 7.1). Eltra, Eltra and Elkraft and SvK state requirements regarding active power change in order to ensure sufficiently fast down regulation in case of necessity (e.g. overfrequencies). E.ON regulations define both maximum permissible active power change and minimum required active power reduction capability.

Table   7.1 Power control requirements

Requirement	Source
Active power:	
1 min average $\leq$ production limit $+5\,\%$ of maximum power of wind farm[a]	Eltra
1 min average $= \pm 5\,\%$ of rated power of the wind turbine from conditional set point (0–100 % of maximum power of wind farm)	Eltra and Elkraft
10 min average $\leq k \times$ registered capacity at any time $\leq$ registered capacity	DEFU 111, AMP, E.ON, ESBNG, VDEW
Active power change:	
Reduction to $<20\,\%$ of maximum power (by individual control of each wind turbine) when demanded: in 2 s (Eltra); in 5 sec (SvK)	Eltra, SvK
Power change from any operating point to a set point defined by E.ON	E.ON
Power reduction of a minimum of 10 % of registered capacity per minute	
Power increase $\leq 10\,\%$ of registered capacity per minute	
Adjustable in the range of 10–100 % of rated power per minute	Eltra and Elkraft
In any 15-minute period, active power change is limited to:	ESBNG
5 % rated power of wind farm per min ($P_{WF} < 100\,\text{MW}$)	
4 % rated power of wind farm per min ($P_{WF} < 200\,\text{MW}$)	
2 % rated power of wind farm per min ($P_{WF} > 200\,\text{MW}$)	
Specific reduction must be possible; reduction order comes from system operator	SvK
Active power change is limited to:	Scotland
60 MW per hour, 10 MW over 10 min, 3 MW over 1 min (for $P_{WF} < 15\,\text{M W}$);	
4× registerd capacity per hour, registered capacity/1.5 over 10 min, registered capacity/5 over 1 min (for $15\,\text{MW} < P_{WF} < 150\,\text{MW}$);	

Table 7.1 (*continued*)

Requirement	Source
600 MW per hour, 100 MW over 10 min, 30 MW over 1 min ($P_{WF} > 150$ MW) (may be exceeded at $f \neq 50$ Hz if farm provides frequency control)	
Startup: Wind farm shall contain a signal clarifying the cause of preceding wind farm shutdown. This signal should be a part of the logic managing startup of wind turbines for operation	Eltra
Has to comply with requirements regarding active power change	Scotland, E.ON
Has to comply with requirements regarding voltage quality	DEFU 111, AMP, Sintef, VDEW, Scotland, Eltra, E.ON
Shutdown: High wind speed must not cause simultaneous stop of all wind turbines	Eltra, SvK
No more than 2 % of registered capacity may be tripped. Phased reduction of output over 30 min period	Scotland
Has to comply with requirements regarding active power change	Scotland
Has to comply with requirement regarding voltage quality	DEFU 111, AMP, VDEW

[a] The production limit is an external signal deduced from the local values of, for example, frequency and/or voltage.

Note: P_{WF} = rated power of wind farm.

Sources: on DEFU 111, see Subsection 7.2.1.1; on Eltra and Elkraft, see Subsection 7.2.1.2; on AMP, see Subsection 7.2.1.3; on VDEW, see Subsection 7.2.1.4; on G59/1 and G75, see Subsection 7.2.1.5 (not mentioned in this table); on Eltra, see Subsection 7.2.2.1; on E.ON, see Subsection 7.2.2.2; on ESBNG, see Subsection 7.2.2.3; on SvK, see Subsection 7.2.3.1; on Sintef, see Subsection 7.2.3.2; on Scotland, see Subsection 7.2.3.3.

7.3.2 Frequency control

In the power system, the frequency is an indicator of the balance or imbalance between production and consumption. For normal power system operation, the frequency should be close to its nominal value. In European countries, the frequency usually lies between 50 ± 0.1 Hz and very seldom outside the range of 49–50.3 Hz.

In the case of an imbalance between production and consumption, primary and secondary control is used to return to a balanced system. If, for instance, consumption is larger than production, the rotational energy stored in large synchronous machines is utilised to keep the balance between production and consumption and, as a result, the rotational speed of the generators decrease. This results in a decrease of the system frequency. In a power system, there are some units that have frequency-sensitive equipment. These units are called primary control units. The primary control units will increase their generation until the balance between production and consumption is restored and frequency has stabilised. The time span for this control is 1–30 s.

In order to restore the frequency to its nominal value and release used primary reserves, the secondary control is employed with a time span of 10–15 min. The secondary control thus results in a slower increase or decrease of generation. In some countries, automatic generation control is used; in other countries the secondary control is accomplished manually by request from the system operator.

At normal operation, the power output of a wind farm can vary up to 15 % of installed capacity within 15 minutes. This could lead to additional imbalances between production and consumption in the system. Considerably larger variations of power production may occur during and after extreme wind conditions.

As wind turbines use other generation technologies than conventional power plants, they have a limited capability of participating in primary frequency control in the same way conventional generators do. However ESBNG, for instance, requires wind farms to include primary frequency control capabilities of 3–5 % (as required for thermal power plants) into the control of wind farm power output. ESBNG and some other regulations also require wind farms to be able to participate in secondary frequency control. During overfrequencies, this can be achieved by shutting down of some turbines in the wind farm or by pitch control. Since wind cannot be controlled, power production at normal frequency would be intentionally kept lower than possible in order for the wind farm to be able to provide secondary control at underfrequencies (see also Section 7.4). Figures 7.1 and 7.2 illustrate the requirements regarding frequency control of wind turbines in different countries. In the case of large frequency transients after system faults, Eltra's regulation requires the wind farm to contribute to frequency control (i.e. secondary control).

7.3.3 Voltage control

Utility and customer equipment is designed to operate at a certain voltage rating. Voltage regulators and the control of reactive power at the generators and consumption connection points is used in order to keep the voltage within the required limits and avoid voltage stability problems (see also Chapter 20). Wind turbines also have to contribute to voltage regulation in the system; the requirements either refer to a certain voltage range that has to be maintained at the point of connection of a wind turbine or wind farm, or to a certain reactive power compensation that has to be provided.

7.3.3.1 Reactive power compensation

Required reactive power compensation is defined in terms of power factor range. Figure 7.3 shows the requirements regarding reactive power compensation in distribution networks, and Figure 7.4 shows these requirements for transmission networks.

VDEW guidelines recommend a power factor of 1 but leave the final decision to the network company that handles the network integration.

In the Swedish regulations (SvK), the demand for reactive power compensation is expressed in terms of a permissible voltage range. According to these regulations, large (> 100 MW) and medium-size (20–50 MW) wind farms have to be able to maintain automatic regulation of reactive power, with voltage as the reference value. The reference value has to be adjustable within at least ±10 % of nominal operating voltage. The same requirement is stated in the Sintef regulations regarding all wind farms connected to the voltage level > 35 kV.

DEFU, VDEW and AMP also define the maximum permissible voltage increase from a wind turbine at the point of common connection (PCC), which is 1 % for DEFU, 2 % for VDEW and 2.5 % for AMP.

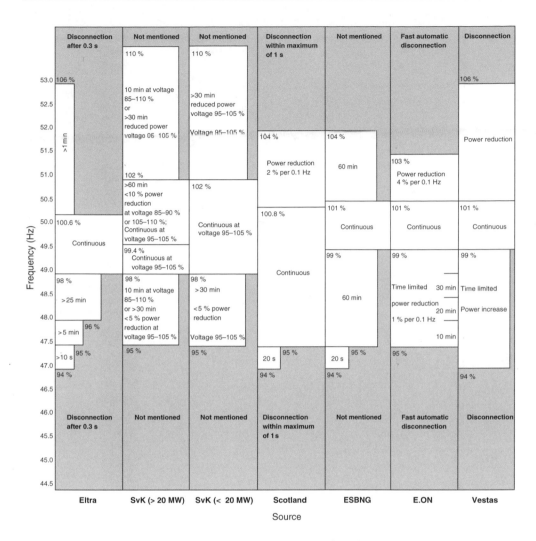

Figure **7.1** Overview of frequency control requirements: Eltra, SvK, Scotland, ESBNG, E.ON and Vestas
Sources: see Table 7.1.

7.3.3.2 Voltage quality

Voltage quality assessment of the wind farm is based on the following concepts:

- rapid voltage changes: single rapid change of voltage root mean square (RMS) value, where voltage change is of certain duration (such as during switching operations in the wind farm);
- voltage flicker: low-frequency voltage disturbances;
- harmonics: periodic voltage or current disturbances with frequencies $n \times 50\,\text{Hz}$, where n is an integer.

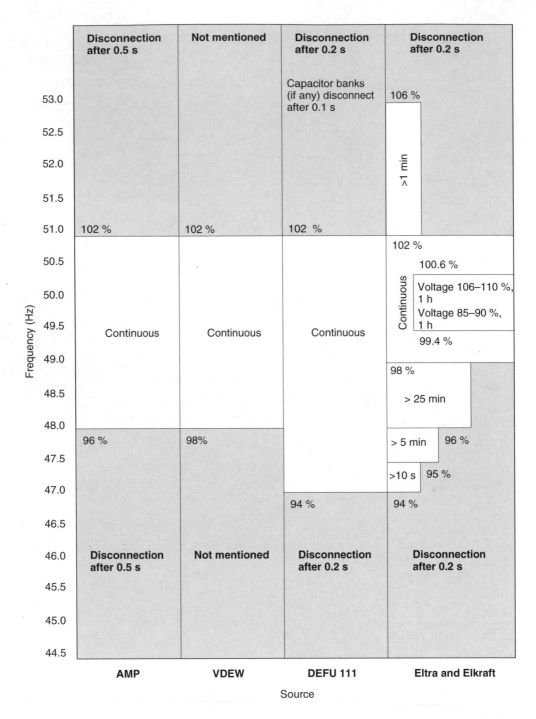

Figure   7.2 Overview of frequency control requirements: AMP, VDEW, DEFU 111, and Eltra and Elkraft
Sources: see Table 7.1.

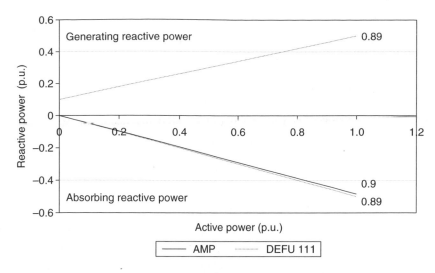

Figure 7.3 Requirements regarding reactive power (per unit of rated active power)
Sources: see Table 7.1.

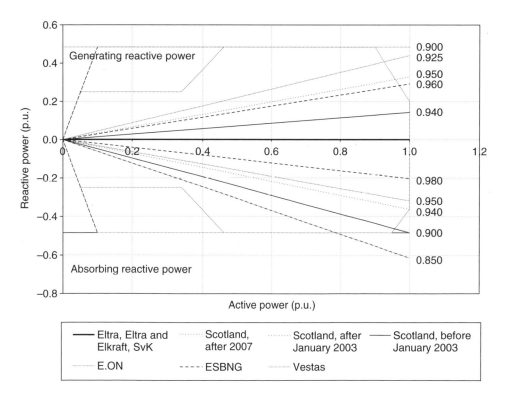

Figure 7.4 Requirements regarding reactive power and technical capability of a Vestas V80 2 MW turbine (per unit of rated active power)
Sources: see Table 7.1.

Voltage variations and harmonics can damage or shorten the lifetime of the utility and customer equipment. Voltage flicker causes visible variations of light intensity in bulb lamps. The requirements concerning voltage quality are listed in Table 7.2.

7.3.4 Tap changers

Tap-changing transformers are used to maintain predetermined voltage levels. This is achieved by alternating the transformer-winding ratio. The E.ON and Scottish Power Scottish Hydro Electric regulations include special requirements regarding the tap-changers.

The E.ON regulations recommend equipping the wind farm with a tap-changing grid transformer in order to be able to vary the transformer ratio.

Scottish Power and Scottish Hydro Electric states that wind farms of 100 MW and above have to have manual-control tap-changing transformers to allow the grid control to dispatch the desired reactive power output. Wind farms between 5 MW and 100 MW may use this method if they have their own transformer, or may use other methods of controlling reactive power agreed with Scottish Power at the application stage.

ESBNG requires that every wind farm that is connected to the network have an on-load tap-changer. The tap step should not alter the voltage ratio at the HV terminals by more than:

- 2.5 % on a 110 kV system;
- 1.6 % on 220 kV to 400 kV systems.

7.3.5 Wind farm protection

The dynamic behaviour of wind turbines during and after different disturbances and the transient stability of the power system is discussed in detail in Part D of this book. Recommendations for the connection of wind farms to distribution networks usually include the disconnection of wind farms in the case of a fault in the network (DEFU 111, AMP). However, this does not apply to large wind farms. If the fault occurs in the system, the immediate disconnection of large wind farms would put additional stress on the already troubled system.

After severe disturbances, it may happen that several transmission lines are disconnected and part of the network may be isolated (or 'islanded') and there may be an imbalance between production and consumption in this part of the network. As a rule, wind farms are not required to disconnect, as long as certain voltage and frequency limits are not exceeded (see also Section 7.4). Danish regulations additionally require wind farms to take part in frequency control (secondary control) in island conditions.

High short-circuit currents, undervoltages and overvoltages during and after the fault can also damage wind turbines and associated equipment. The relay protection system of the farm should therefore be designed to pursue two goals:

- to comply with requirements for normal network operation and support the network during and after the fault;
- to secure wind farms against damage from impacts originating from faults in the network.

Table   7.2 Overview of voltage quality

Requirement	Source
Rapid voltage changes (as a percentage of nominal voltage at the connection point):	
General limit $\leq 2\,\%$	VDEW
General limit $< 3\,\%$	Eltra, Scotland, Eltra and Elkraft (50–60 kV)
General limit $< 4\,\%$	Sintef, Eltra and Elkraft (10–20 kV)
Until a frequency of 10 per hour, $< 2.5\,\%$	Eltra
Until a frequency 100 per hour, $< 1.5\,\%$	Eltra
$k_f(\psi_k) \leq 0.04 R_k$ or $k_i \leq 0.04 R_k$	DEFU 111
Voltage flicker:	
$P_{st} < 0.30$	Eltra
$P_{st} < 0.7$	Sintef
$P_{st} < 0.35$	AMP
$P_{lt} < 0.25$	Eltra, DEFU 111, AMP
$P_{lt} \leq 0.46$	VDEW
$P_{lt} \leq 0.5$ (at main substation, if customers are also connected there)	DEFU 111
$P_{lt} \leq 0.5$	Eltra and Elkraft (10–20 kV), Sintef
$P_{lt} \leq 0.35$	Eltra and Elkraft (50–60 kV)
$k_f(\psi_k) < 0.031 R_k/N^{1/3.2}$ or $k_i < 0.031 R_k/N^{1/3.2}$ (during switchings)	DEFU 111
$C_c(\psi_k) < 0.25 R_k$ (during normal operation)	
$k_f(\psi_k) < P_{lt}/(8 N^{1/3}) S_k/(S_{park} S_{rG}^2)^{1/3}$ (during switchings)	Eltra and Elkraft
$C_c(\psi_k) < P_{lt} S_k/(S_{park} S_{rG})^{1/2}$ (during normal operation)	
Has to comply with Engineering Recommendation G5/4	Scotland
Has to comply with IEC/TR3 61000-3-7 standard (IEC, 1996b)	ESBNG
Has to comply with VDEW guideline 'Grundsätze für die Beurteilung von Netzrückwirkungen'	E.ON
Harmonics:	
$D_n < 1\,\%$ (of fundamental voltage) for $1 < n < 51$ THD $< 1.5\,\%$	Eltra
D_n (n is odd) $< 0.3\text{--}5\,\%$ (of fundamental current, depending on n)	DEFU 111, Sintef
D_n (n is even and interharmonic) $< 25\,\% D_n$ (n is odd) THD $< 5\,\%$	
D_n (n is odd) $< 0.8\text{--}3\,\%$ (of fundamental voltage, depending on n)	Eltra and Elkraft
D_n (n is even) $< 0.2\text{--}2\,\%$ (of fundamental voltage, depending on n)	
D_n (n is odd, $n < 26$) $< 0.01\text{--}0.115$ (in A/MVA for 10 kV, depending on n)	VDEW
D_n (n is even) and D_n (n is odd, $n > 26$) $< 0.06/n$ (for 10 kV)	

(continued overleaf)

Table   7.2 (*continued*)

Requirement	Source
D_n (*n* is odd, $n < 26$) < 0.005–0.058 (in A/MVA for 20 kV, depending on *n*)	
D_n (*n* is even) and D_n (*n* is odd, $n > 26$) $< 0.03/n$ (for 20 kV)	
D_n (*n* is odd) $< 4\%$ (depending on *n*)	AMP
D_n (*n* is even) $< 1\%$	
D_n *n* is interharmonic $< 0.3\%$	
THD $< 6\%$	
Should comply with Engineering Recommendation G5/4	Scotland
Has to comply with IEC/TR3 61000-3-6 (IEC, 1996a)	ESBNG

Note: $k_f(\psi_k)$ = the flicker emission factor during the switching operations at a certain network impedance phase angle ψ_k; $c_c(\psi_k)$ = flicker emission factor during normal operation; k_i = the in-rush current factor; R_k = the short-circuit ratio at the point of connection (ratio between short circuit power and maximum apparent power of the turbine); S_k = short-circuit capacity at the connection point; S_{park} = nominal apparent power of the wind farm; S_{rG} = nominal apparent power of a wind turbine; N = the maximum number of switching operations; P_{lt} = the weighted average flicker emission during 2 hours; P_{st} = the weighted average of the flicker contribution during 10 minutes; THD = the total harmonic distortion; D_n = the harmonic interference of each individual harmonic, *n*. THD, the individual harmonic interference and flicker contributions P_{lt} and P_{st} are defined in IEC/TR 61000-3-7 (IEC, 1996b) (DEFU and AMP also include the definitions).
Sources: see note to Table 7.1.

Table 7.3 compares the requirements regarding fault tolerance, undervoltage and overvoltage protection and requirements during islanding. Although wind farm protection regarding overfrequency and underfrequency, overvoltage and undervoltage, for instance, is not treated separately in some regulations it is assumed that wind farm protection systems comply with the requirements discussed in the preceding sections.

EBSN states that the protection system of a wind farm has to comply with the same requirements that are defined in the grid code (ESBNG, 2002b) for conventional generators. That means that internal faults should be disconnected within

- 120 ms for a 110 kV system;
- 100 ms for a 220 kV system;
- 80 ms for a 400 kV system.

Furthermore, power plants are required to be equipped with underfrequency and overfrequency protection, undervoltage and overvoltage protection, differential protection of the generator transformer, and backup protection (including generator overcurrent protection, voltage-controlled generator overcurrent protection or generator distance protection). The settings for the protection are not discussed in detail in ESBNG (2002b).

Table 7.3 Overview of protection issues

Requirement	Source
Fault tolerance: 3-phase faults on a random line or transformer with definitive disconnection without any attempt at reclosing	Eltra
2-phase fault on a random line with unsuccessful reclosing.	
3-phase fault in transmission network during 100 ms	Eltra and Eltkraft
2-phase faults and 2-phase to ground faults for 100 ms followed by another fault in 300–500 ms with a duration of 100 ms	

WT has to have enough capacity to fulfil these requirements at least in the case of:
   two 2-phase or 3-phase short circuits during 2 min
   six 2-phase or 3-phase short circuits with a 5 min interval	

During asymmetrical faults in the system, the WF has to be disconnected faster than the fault is cleared:
   If the connection point is within 50 m of the WT transformer, the cable between transformer and connection point has to be equipped with its own current-measuring fault protection in the connection point. If the length of the cable is substantial compared with the outer network, this protection has to be directional. Settings of the protection have to be coordinated with the protection of the respective network.
   If there is a risk of islanding, the fault protection that measures neutral point voltage in the network has to be installed at the connection point. The recommended setting for the disconnection is 5 seconds.	

Overcurrent protection:
This should take into account selectivity and provide maximum protection and at the same time avoid disconnection during normal operation. The following should be taken into account:
   maximum in-rush current (see Table 7.2 on voltage quality)
   maximum current at continuous operation (see Table 7.1 on power control)
	Sintef, DEFU 111, AMP

Tolerance of over frequency: See Figures 7.1 and 7.2	All
Tolerance of undervoltages (during and after the fault); see also Figure 7.5: $U = 0\%$, WF disconnection after > 0.25 s	SvK ($P_{WF} > 100$ MW)
$U < 90\%$, WF disconnection after > 0.75 s	SvK
$U = 25\%$, WF disconnection after > 0.25 s	SvK ($P_{WF} < 100$ MW)
$U = 0\%$, WF disconnection after > 0.1 s	Eltra
$U < 60\%, \ldots, 80\%$, WF disconnection after $2, \ldots, 20$ s	
$U = 20\%$, WF disconnection after > 0.7 s	E.ON
$U < 80\%$, WF disconnection after > 3 s	

(*continued overleaf*)

Table 7.3 (*continued*)

Requirement	Source
$0\% \leq U < 90\%$, WF (installed after January 2004) disconnection after $> 0.14\,\mathrm{s}$	Scotland
$15\% \leq U < 90\%$, WF (installed before January 2004) disconnection after $> 0.14\,\mathrm{s}$	
$U = 25\%$, WF disconnection after $> 0.1\,\mathrm{s}$	Eltra and Elkraft
$U < 75\%$, WF disconnection after $> 0.75\,\mathrm{s}$	
$U < 90\%$, WF disconnection after $> 10\,\mathrm{s}$	
$50\% \leq U < 90\%$, WF disconnection after $> 0.6\,\mathrm{s}$	ESBNG
Tolerance of undervoltages (general):	
$U < 90\%$, WF disconnection after $500\,\mathrm{ms}$	Scotland
$U < 80\%$, WF disconnection after $3\text{–}5\,\mathrm{s}$	E.ON
$U < 94\%$, WF disconnection after $60\,\mathrm{s}$	
$U < 80\%$, WF disconnection after $0.2\,\mathrm{s}$	
$U < 85\%, \ldots, 95\%$, WF disconnection after $60\,\mathrm{s}$[a]	DEFU 111
$U < 90\%$, WF disconnection after $60\,\mathrm{s}$	Sintef
$U < 90\%$, WF disconnection after $10, \ldots, 60\,\mathrm{s}$	Eltra and Eltkraft
$U < 70\%$, WF disconnection	VDEW
Tolerance of overvoltages (general):	
WF should not trip within normal operating voltage range, including the 15-minute overvoltage (defined in grid code)	Scotland
For overvoltages (not specified) a protection time of 30–60 seconds	Eltra
$U > 106\%$, WF disconnection after $60\,\mathrm{s}$	Eltra and Elkraft
$U > 110\%$, WF disconnection after $0.2\,\mathrm{s}$	
$U > 110\%$, WF disconnection after $60\,\mathrm{s}$	AMP, Sintef
$U > 120\%$, WF disconnection after $0.2\,\mathrm{s}$	AMP, Sintef
$U > 95\%, \ldots, 110\%$, WF disconnection after $50\,\mathrm{s}$[a]	DEFU 111
$U > 115\%$, WF disconnection	VDEW
Tolerance of overvoltages (island conditions):	
$U > 115\%$ (at $275\,\mathrm{kV}$) or $U > 120\%$ (at $132\,\mathrm{kV}$), WF disconnection after $250\,\mathrm{ms}$	Scotland
$U > 120\%$, $< 100\,\mathrm{ms}$, voltage reduction	Eltra
$U > 120\%$, disconnection after maximum of $100\,\mathrm{ms}$ (if equipped with capacitor banks)	DEFU 111
$U > 120\%$, disconnection after maximum of $200\,\mathrm{ms}$ (if not equipped with capacitor banks)	
Power control:	
Reduction to $\leq 20\%$ of maximum power in $2\,\mathrm{s}$ (by individual control of each WT), after fault clearance	Eltra
Motor mode protection:	
Disconnection after $5\,\mathrm{s}$ of operation in motor mode	AMP

[a] Settings for undervoltage and overvoltage protection required by DEFU 111 (DEFU, 1998) must be adjustable within a given range.

Note: WT = wind turbine; WF = wind farm; P_{WF} = rated power of wind farm; U = voltage.

Sources: See Table 7.1.

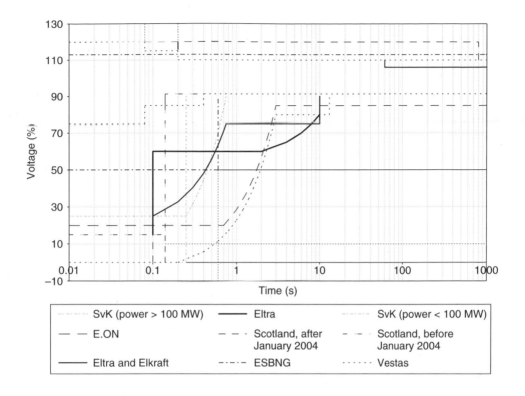

Figure **7.5** Requirements for tolerance of undervoltages during and after a fault in the system as well as technical capability of a standard Vestas V80 2 MW turbine and modified Vestas V80 2 MW turbine
Sources: See Table 7.1.

7.3.6 Modelling information and verification

The interaction between wind farm and power system during faults in the power system is usually verified through simulations. In order to make such simulations possible, wind farm owners have to provide the system operator with the necessary models.

In order to verify the wind farm model and the response of the farm to faults in the power system, registration equipment has to be installed (see Table 7.4).

7.3.7 Communication and external control

In most regulations, the wind farm owner is required to provide the signals necessary for the operation of the power system. Unlike the other aspects of regulations discussed above, the requirements on communication are quite similar in all considered documents. Table 7.5 summarises the requirements.

Table 7.4 Requirements regarding modelling information

Requirement	Source
Modelling information (dynamic parameters):	
The models have to be well documented and agree with tests on corresponding WT prototypes.	Eltra, Scotland
If the WF consists of several WT types, models for each individual WT type must be presented.	Eltra
Technical data of WF have to be documented.	SvK, E.ON
Modelling information (fault contribution):	
A detailed calculation of the fault infeed has to be provided to specify 3-phase short-circuit infeed from the system and how the value has been determined.	Scotland
Fault recorder (recorded variables):	
Voltage, active and reactive power, frequency and current in the WF connection point; voltage, active and reactive power, rotating speed for a single WT of each type within a WF.	Eltra
3-phase currents, 3-phase voltages and wind speeds.	Scotland

Note: WT = wind turbine; WF = wind farm.
Sources: See Table 7.1.

7.3.8 Discussion of interconnection regulations

The brief overview of the interconnection regulations presented above shows that the regulations vary considerably and that it is often difficult to find a general technical justification for the different technical regulations that are currently in use worldwide. This applies particularly to power quality regulations such as flicker and harmonic limits.

Many of the differences in the technical regulations are caused by different wind power penetration levels in the national power systems and different power system robustness. For instance, countries with a rather weak power system, such as Scotland or Ireland, have to consider the impact of wind power on network stability issues, which means that they require fault ride-through capabilities for wind turbines already at a lower wind power penetration level compared with countries that have very robust systems.

Discussions with wind turbine manufacturers show that they would prefer a greater harmonisation whenever possible, but they generally also say that they are able to comply with the different technical regulations. They are, however, much more concerned about the continuous changes in technical standards worldwide, the very short notice given for updates and changes and the little influence distributed resource manufacturers have on these aspects. Large wind turbine manufacturers, for instance, employ 4–5 or even more experts to keep track of the ongoing changes in technical regulations and to document the technical capabilities of their wind turbines. Smaller wind turbine manufacturers, which cannot employ as many experts in this area, often stay out of certain national markets because they cannot follow the changes in regulations and the corresponding required technical documentation and validation of the technical capabilities of their product.

Table 7.5 Requirements regarding communication and external control

Requirement	Source
Variables required from wind farm:	
Voltage	SvK, Eltra and Elkraft, Eltra, ESBNG, Scotland, E.ON
Active power	SvK, Eltra, Scotland, AMP, Eltra and Elkraft; ESBNG, E.ON
Reactive power	SvK, Eltra, Scotland, AMP, Eltra and Elkraft, ESBNG, E.ON
Operating status	SvK, Eltra, Scotland, Eltra and Elkraft, ESBNG
Wind speed	Eltra, Scotland, ESBNG
Wind direction	Scotland, ESBNG
Ambient temperature and pressure	ESBNG
Generator transformer tap position	ESBNG, E.ON
Regulation capability	SvK
Frequency control status (enabled or disabled)	Scotland
Abnormalities resulting in WF tripping or startup in 15 min	
External control possibilities[a]:	
WF > 20 MW, manual, local or remote control within 15 mins after the fault to allow: disconnection from the network, connection to the network and regulation of active and reactive power output	SvK
It must be possible to connect or disconnect the wind turbine generator externally	Eltra and Elkraft, Eltra, E.ON

[a] Some of the requirements on external control possibilities have already been mentioned in the preceding subsections in text.
Sources: See Table 7.1.

In addition, new network interconnection regulations increase the costs of wind turbines. According to wind turbine manufacturers, allowing wind turbines with doubly fed induction generators to 'ride through a fault', as defined in many European regulations for wind turbines, increases the total costs of a wind turbine by up to 5 %.

Hence, it can be summarised that interconnection regulations should be harmonised in areas that have little impact on the overall costs of wind turbines (i.e. power quality regulations). In other areas, interconnection regulations should take into account the specific power system robustness, the penetration level and/or the generation technology. Therefore, interconnection standards of different countries may also vary in future. It is important that national regulations should aim at an overall economically efficient solution; that is, costly technical requirements such as a 'fault ride-through' capability for wind turbines should be included only if they are technically required for reliable and stable power system operation.

7.4 Technical Solutions for New Interconnection Rules

In the following example, the control system of the recently installed Horns Rev off-shore wind farm is briefly presented. The wind farm is the first wind farm that had to comply with the requirements outlined by Eltra (2000), the TSO in Western Denmark. At the time of writing, the control system was still being tested, therefore, there were no details available of practical experience.

The offshore wind farm Horns Rev is located approximately 15 km into the North Sea. The wind farm has an installed power of 160 MW and comprises 80 wind turbines laid out in a square pattern. The turbines are arranged in 10 lines with 8 turbines each. Two lines constitute a cluster of 32 MW, with the turbines connected in series. Each cluster is connected to the offshore transformer substation where the 34/165 kV transformer is located (Christiansen and Kristoffersen, 2003).

From an electrical perspective, the wind farm had to comply with new specifications and requirements for connecting large-scale wind farms to the transmission network. As discussed before, the TSO (Eltra) had designed requirements for power control, frequency, voltage, protection, communication, verification and tests (Eltra, 2000). According to those requirements, the wind farm has to be able to contribute to control tasks on the same level as conventional power plants, constrained only by the limitations imposed at any time by the existing wind conditions. For example, during periods with reduced transmission capacity in the grid (e.g. as a result of service or replacement of components in the main grid) the wind farm may be required to operate at reduced power levels with all turbines running. The wind farm must also be able to participate in regional balance control (secondary control).

The general control principle of the farm has to take into account that the control range of the farm depends on the actual wind speed. Furthermore, as the wind speed cannot be controlled, the power output of the farm can only be downregulated. For instance, if the wind speed is around 11 m/s, the power output from the farm can be regulated to any value between 0 MW and approximatly 125 MW. In the following, some of the key elements of the overall control strategy are presented:

7.4.1 Absolute power constraint

This control approach limits the total power output of the wind farm to a predefined setpoint (see Figure 7.6).

7.4.2 Balance control

The balance control approach allows a reduction in the power production of the overall wind farm at a predefined rate and later an increase in the overall power output, also at a predefined ramp rate.

7.4.3 Power rate limitation control approach

This approach limits the *increase* in power production to a predefined setpoint (e.g. maximum increase in power production 2 MW per minute, see Figure 7.7). It is

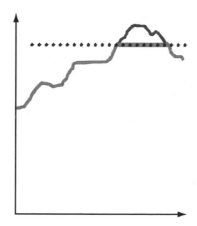

Figure 7.6 Absolute power control (Reproduced by permission of Vestas Wind Systems A/S, Denmark.)

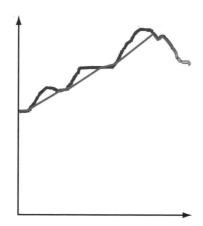

Figure 7.7 Power rate limitation (Reproduced by permission of Vestas Wind Systems A/S, Denmark.)

important to emphasise that this approach cannot limit the speed of power *reduction*, as the decrease in wind cannot be controlled. In some cases, however, this can be achieved when combined with the *delta control* approach.

7.4.4 Delta control

This reduces the amount of total power production of the wind farm by a predefined setpoint (e.g. 50 MW). Hence, if delta control is now combined with a balance control approach, the production of the farm can be briefly increased and decreased according to the power system requirements (see Figure 7.8). A wind farm equipped with such a control approach can be used to supply automatic secondary control to a power system.

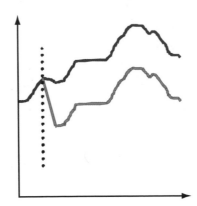

Figure   **7.8** Delta control (Reproduced by permission of Vestas Wind Systems A/S, Denmark.)

7.5 Interconnection Practice

While interconnection standards usually define the minimum requirements to be fulfilled by the wind turbines to be connected, they usually say little or nothing about the approach used for interconnection analysis (e.g. how much capacity can be connected at a given point, and how long this analysis may take).

A number of empirical studies (e.g. Alderfer Eldridge and Starrs, 2000; DOE, 2000; Jörß *et al.*, 2003) include a variety of case studies showing that wind farm project developers often have long, fairly expensive, discussions with network companies about the maximum capacity that can be connected at a proposed connection point. Typically, network companies do not publish network data, hence it is difficult independently to verify the capacity limitations defined by network companies. If there is a considerable lobby for a wind farm project, developers in some countries use public media to put pressure on network companies to reconsider initial capacity limitations. In situations with less public support, network companies may even consider not to react to any network interconnection application. In Sweden, for instance, a local distribution company did not react for 15 months to an interconnection application. After continuous complaints to the network company, the network company finally rejected the application. At the same time, however, the company informed the applicant in a personal discussion that the interconnection is technically possible but that the network company did not have any economic interest in an interconnection. Interestingly enough, the network company directly suggested that the applicant should sue the network company, because the network company meant that a legal clarification of such cases would be required, because the Swedish regulator did not provide the relevant clarification.

A possible solution to avoid such conflicts is a clear outline of the method to be used for defining the maximum interconnection capacity at a given network point as well as a definition of the maximum time for the network company to perform the relevant studies. Ideally, the method as well as the maximum time of a response to an interconnection application should be defined by a neutral authority, for instance a regulator.

The regulator in Texas, for instance, has clearly defined methods of how to analyse different interconnection requests. Figure 7.9, for instance, describes the methods for an interconnection request for distributed resources on the customer side of the meter. The regulations force the network companies to answer to any interconnection application within 6 weeks. If the application requires a network integration study, an additional 6 weeks can be added to this deadline (PUCT, 1999).

In addition, the regulator should consider setting up an independent body to help to clear up network interconnection disputes. Such a body could be approached by

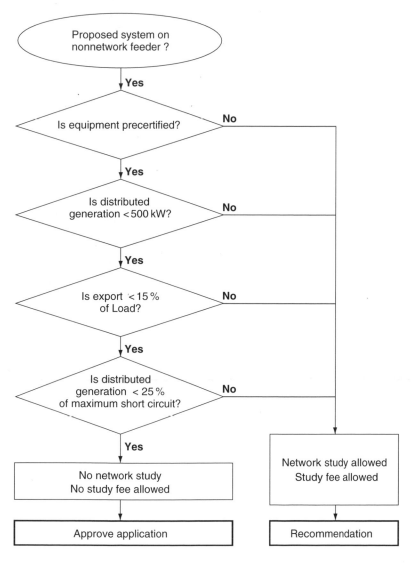

Figure 7.9 Method for interconnection requests for distributed resources on the customer side of the meter as outlined in regulations in Texas (PUCT, 1999)

network companies as well as by project developers. In Germany, for instance, a clearing office for network interconnection issues related to wind power has been operating for many years now. For many years, the office was part of the University of Aachen. It was recently relocated to the German Ministry for Environment, which is the ministry in charge of renewable energy.

7.6 Conclusions

This chapter has presented a comparison of the existing regulations for the interconnection of wind farms to the power system. Most of the analysed documents are still under revision and will probably undergo some changes in the future. The comparison revealed that the regulations differ significantly between countries. This depends on the properties of each power system as well as the experience, knowledge and policy of the TSOs.

In general, new interconnection regulations for wind turbines or wind farms tend to add the following requirements:

- to maintain operation of the turbine during a fault on the grid, know as 'fault ride-through' capability;
- to operate the wind turbine in the range of 47–52 Hz (for European networks);
- to control the active power during frequency variations (active power control);
- to limit the power increase to a certain rate (power ramp rate control);
- to supply or consume reactive power depending on power system requirements (reactive power control);
- to support voltage control by adjusting the reactive power, based on grid measurements (voltage control).

In general, the wind energy industry is able to comply with the increased requirements outlined in new interconnection standards. However, in some cases this can increase the total costs of a wind turbine or wind farm significantly. Hence, interconnection regulations should increase requirements only if needed for secure operation of the power system. In other words, countries with a very low wind power penetration do not necessarily need to adopt similar interconnection requirements as countries that have significant wind power penetration.

References

[1] Alderfer, B. R., Eldridge, M. M., Starrs, T. J. (2000) *Market Connections: Case Studies of Interconnection Barriers and their Impact on Distributed Power Projects*, National Renewable Energy Laboratory, Golden, USA, May 2000, available at http://www.nrel.gov/docs/fy00osti/28053.pdf.

[2] Bolik, S. M., Birk, J., Andresen, B., Nielsen, J. G. (2003) *Vestas Handles Grid Requirements: Advanced Control Strategy for Wind Turbines*, European Wind Energy Conference, Madrid, Spain, 2003.

[3] California Energy Commission (2000) 'Distributed Generation Interconnection Rules', Committee Report (Draft), Sacramento, CA, USA, September 2000.

[4] Christiansen, P., Kristoffersen, J. R. (2003) 'The Wind Farm Main Controller and the Remote Control System of the Horns Rev Offshore Wind Farm', in *Proceedings of the Fourth International Workshop on*

Large-Scale Integration of Wind Power and Transmission Networks for Offshore Wind Farms, October 2003, Eds Matevosyan, T. Ackermann, Royal Institute of Technology, Stockholm, Sweden.

[5] CIGRE [Conseil International des Grands Réseaux Électriques (International Council on Large Electric Systems)] (1998) 'Impact of Increasing Contribution of Dispersed Generation on the Power System', CIGRE Study Committee No. 37 (WG 37–23), Final Report, Paris, September 1998.

[6] DEFU (Danish Utilities Research Association) (1998) 'Connection of Wind Turbines to Low and Medium Voltage Networks', Committee report 111-E, October 1998, 2nd edn, published by DEFU, Lyngby, Denmark.

[7] DEFU (Danish Utilities Research Association) (2003) 'Vindmøller tilsluttet net med spændinger under 100 kV', Teknisk forskrift for vindmøllers egenskaber og regulering, 15 draft version, 17 June 2003, DEFU, Lyngby, Denmark.

[8] DOE (Department of Energy), (2000) 'Making Connections – Case Studies of Interconnection Barriers and their Impact on Distributed Power Projects', US Department of Energy, Washington, DC, USA, May 2000.

[9] Electricity Association (1990) 'Recommendations for the Connection of Embedded Generating Plant to the Regional Electricity Companies Distribution System', Engineering Recommendation G59/1 (ER G59 1990), Electricity Association, London.

[10] Electricity Association (1995) 'Notes of Guidance for the Protection of Embedded Generation Plant to 5 MW for Operation in Parallel with the Public Electricity Suppliers Distribution Systems', Engineering Technical Report No 113 (ETR 113), Electricity Association, London.

[11] Electricity Association (1996) 'Recommendations for the Connection of Embedded Generating Plant to the Public Electricity Supplies' Distribution Systems above 20 kV or with Outputs over 5 MW', Engineering Recommendation G75 (ER G75 1996), Electricity Association, London.

[12] Eltra (2000) 'Specifications for Connecting Wind Farms to the Transmission Networks', 2nd edn, Eltra, Fredericia, Denmark.

[13] E.ON (E.ON Netz) (2001a) 'Netzanschlussregeln-allgemein. Technische und organisatorische Regeln für den Netzanschluss innerhalb der Regelzone der E.ON Netz GmbH im Bereich der ehemaligen PreussenElektra Netz GmbH Stand: 1. December 2001', E.ON Netz GmbH, Bayreuth, Germany.

[14] E.ON (E.ON Netz) (2001b) 'Ergänzende Netzanschlussregeln für Windkraftanlagen: Zusätzliche technische und organisatorische Regeln für den Netzanschluss von Windkraftanlagen innerhalb der E.ON Netz GmbH', as of 1. December 2001; E.ON Netz GmbH, Bayreuth, Germany.

[15] E.ON (E.ON Netz) (2003) 'Netzanschlussregeln, Hoch- und Höchstspannung' (Grid code for high and extra high voltage), as of 1 August 2003; E.ON Netz GmbH, Bayreuth, Germany, available at http://www.eon-netz.com/.

[16] ESBNG (Electricity Supply Board National Grid) (2002a) 'Wind Farm Connection Requirements', draft Version 1.0, February 2002, ESBNG, Ireland.

[17] ESBNG (Electricity Supply Board National Grid) (2002b) 'Grid Code', Version 1.1, October 2002, ESBNG, Ireland

[18] FGW (Fördergesellschaft Windenergie e.V.) (2002) 'Technische Richtlinien für Windenergieanlagen. Teil 3: Bestimmung der Elektrischen Eigenschaften', 15 revision, as of 1 September 2002; FGW, Kiel, Germany, http://www.wind-fgw.de/.

[19] FGW (Fördergesellschaft Windenergie e.V.) (2003) 'Technische Richtlinien für Windenergieanlagen. Teil 4: Bestimmung der Netzanschlussgrössen', as of 1 September 2003; FGW, Kiel, Germany, http://www.wind-fgw.de/.

[20] IEC (International Electrotechnical Commission) (1996a) 'Electromagnetic Compatibility (EMC), Part 3: Limits, Section 6. Assessment of Emission Limits for Distorting Loads in MV and HV Power Systems – Basic EMC Publication', IEC/TR 61000-3-6, IEC, Geneva, Switzerland.

[21] IEC (International Electrotechnical Commission) (1996b) 'Electromagnetic Compatibility (EMC), Part 3: Limits, Section 7. Assessment of Emission Limits for Fluctuating Loads in MV and HV Power Systems – Basic EMC Publication', IEC/TR 61000-3-7, IEC, Geneva, Switzerland.

[22] IEC (International Electrotechnical Commission) (2001) 'Measurement and Assessment of Power Quality Characteristics of Grid Connected Wind Turbines', IEC 61400–21 IEC, Geneva, Switzerland.

[23] IEEE (Institute of Electrical and Electronic Engineers) (1988) 'IEEE Guide for Interfacing Dispersed Storage and Generation Facilities with Electric Utility Systems', IEEE Standard 1001, IEEE, New York.

[24] IEEE (Institute of Electrical and Electronic Engineers) (1992) 'IEEE Recommended Practices and Requirements for Harmonic Control in Electric Power Systems', IEEE Standard 519, IEEE, New York.

[25] IEEE (Institute of Electrical and Electronic Engineers) (2003) 'IEEE Standard for Interconnecting Distributed Resources with Electric Power Systems', IEEE Standard 1547, IEEE, New York.

[26] Jauch, C., Sørensen, P., Bak-Jensen, B. (2004) 'International Review of Grid Connected Requirements for Wind Turbines', in *Proceedings: Nordic Wind Power Conference, Chalmers University of Technology, Sweden, March 2004*.

[27] Jörß, W. (Project Coordinator) (2002) 'DECENT – Decentralised Generation Technologies, Potentials, Success Factors and Impacts in the Liberalised EU Energy Markets', Summary Report, May 2002, compiled by the Institute for Future Studies and Technology Assessment, Berlin, Germany.

[28] Jörß, W., Joergensen, B. H., Löffler, P., Morthorst, P. E., Uyterlinde, M., Sambeek, E. V., Wehnert, T. (2003) 'Decentralised Power Generation in the Liberalised EU Power Energy Markets', Results of DECENT Research Project, Springer, Heidelberg, Germany.

[29] PUCT (Public Utility Commission of Texas) (1999) 'Interconnection Guidelines for Distributed Generation in Texas', published by the Public Utility Commission of Texas, Office of Policy Development, as part of Project 20363: Investigation into Distributed Resources in Texas; PUCT, Austin, TX, USA.

[30] Santjer, F., Klosse, R. (2003) 'Die neuen ergänzenden Netzanschlussregeln von E.ON Netz GmbH – New Supplementary Regulations for Grid Connection by E.ON Netz GmbH', *DEWI Magazin* **22** (February 2003). 28–24 [German Wind Energy Institute (DEWI), Wilhelmeshaven, Germany].

[31] Scottish Hydro Electric (2002) 'Guidance Note for the Connection of Wind Farms', Issue 2.1.4, December 2002.

[32] Sintef (2001) 'Retningslinjer for nettilkobling av vindkraftverk (revidert utgave)', Sintef Energiforskning AS, Trondheim, Norway.

[33] Svensk Energi (2001) *AMP – Anslutning av mindre produktionsanläggningar till elnätet*. Svensk Energi, Stockholm, Sweden.

[34] Svk (2002) 'Affärsverket Svenska Kraftnäts föreskrifter om driftsäkerhetsteknisk utformning av produktionsanläggningar', Svenska Kraftnät, Stockholm, Sweden.

[35] VDEW (Verband Deutscher Elektrizitätswerke) (1998) Eigenerzeugungsanlagen am Mittelspannungsnetz, Richtlinie für Anschluss und Parallelbetrieb von Eigenerzeugungsanlagen am Mittelspannungsnetz, 2nd edn, VDEW Frankfurt, Germany.

8

Power System Requirements for Wind Power

Hannele Holttinen and Ritva Hirvonen

8.1 Introduction

The power system requirements for wind power depend mainly on the power system configuration, the installed wind power capacity and how the wind power production varies. Wind resources vary on every time scale: seconds, minutes, hours, days, months and years. On all these time scales, the varying wind resources affect the power system. An analysis of this impact will be based on the geographical area that is of interest. The relevant wind power production to analyse is that of larger areas, such as synchronously operated power systems, comprising several countries or states.

The integration of wind power into regional power systems is studied mainly on a theoretical level, as wind power penetration is still rather limited. Even though the average annual wind power penetration in some island systems (e.g. Crete in Greece; see also Chapter 14) or countries (e.g. Denmark; see also Chapters 3, 10 and 11) is already high, on average wind power generation represents only 1–2 % of the total power generation in the Scandinavian power system (Nordel) or the Central European system [Union for the Coordination of Transmission of Electricity UCTE]. And the penetration levels in the USA [North American Electricity Reliability Council (NERC) regions] are even lower. Most examples in this chapter come from Central and Northern Europe, as there is already some experience with large-scale integration of wind power, and there are far-reaching targets for wind power. In Central Europe, power production is based mostly on thermal production, whereas in the Nordic countries thermal production is mixed with a large share of hydro power.

We will refer to the energy penetration when we use the term wind power penetration in the system. The energy penetration is the energy produced by wind power (annually)

Wind Power in Power Systems   Edited by T. Ackermann
© 2005 John Wiley & Sons, Ltd   ISBN: 0-470-85508-8 (HB)

as a percentage of the gross electricity consumption. Low penetration means that less than 5 % of gross demand is covered by wind power production, high penetration is more than 10 %.

First, we will describe the power system and large-scale wind power production. We will then look at the effects of wind power production on power system operation as well as present results from studies in order to quantify these effects.

8.2 Operation of the Power System

Electric power systems include power plants, consumers of electric energy and transmission and distribution networks connecting the production and consumption sites. This interconnected system experiences a continuous change in demand and the challenge is to maintain at all times a balance between production and consumption of electric energy. In addition, faults and disturbances should be cleared with the minimum effect possible on the delivery of electric energy.

Power systems comprise a wide variety of generating plant types, which have different capital and operating costs. When operating a power system, the total amount of electricity that is provided has to correspond, at each instant, to a varying load from the electricity consumers. To achieve this in a cost-effective way, the power plants are usually scheduled according to marginal operation costs, also known as merit order. Units with low marginal operation costs will operate almost all the time (base load demand), and the power plants with higher marginal operation costs will be scheduled for additional operation during times with higher demand. Wind power plants as well as other variable sources, such as solar and tidal sources, have very low operating costs. They are usually assumed to be 0, therefore these power plants are at the top of the merit order. That means that their power is used whenever it is available. The electricity markets operate in a similar way, at least in theory. The price the producers bid to the market is slightly higher than their marginal cost, because it is cost-effective for the producers to operate as long as they get a price higher than their marginal costs. Once the market is cleared, the power plants that operate at the lowest bids come first.

If the electricity system fails the consequences are far-reaching and costly. Therefore, power system reliability has to be kept at a very high level. Security of supply has to be maintained both short-term and long-term. This means maintaining both flexibility and reserves that are necessary to keep the system operating under a range of conditions, also in peak load situations. These conditions include power plant outages as well as predictable or uncertain variations in demand and in primary generation resources, including wind.

The power system has to operate properly also in liberalised electricity markets. Usually, an independent system operator (ISO) is the system responsible grid company that takes care of the whole system operation, using active and reactive power reserves to maintain system reliability, voltage and frequency.

Reliability consists of system security and adequacy. The system security defines the ability of the system to withstand disturbances. The system adequacy describes the amount of production and transmission capacity in varying load situations.

8.2.1 System reliability

The planning of the power system is usually carried out according to mutually agreed principles. These principles include that the system has to withstand any single fault (e.g. the disconnection of a power plant, transmission line, substation busbar or power transformer) without major interruptions to power delivery. The consequences of faults for the power system depend on the power transmission (i.e. the production and consumption of electric energy at a given moment), on the topology of the system and on the type of the fault. The most severe fault that a power system can withstand and that will not lead to inadmissible consequences is called the dimensioning fault. The dimensioning fault varies according to the operational state of the system. Usually, it is the disconnection of the largest production unit or the busbar fault at a substation residing along an important transmission route.

Limits for power transfers are defined in predefined production and loading situations using power system analysis software, where the equipment (lines, substations, power plants and loads) is modelled together with connections and levels of production and load. In the simulations, the dimensioning fault(s) may not lead to situations where synchronous operation is lost, or there may be voltage collapses, load shedding, large deviations in voltage or frequency, overloads or undamped oscillations. The normal operational state of the power system is a power transfer state, where the system can withstand a dimensioning fault without the resulting disturbance spreading further than allowed. Within a normal operational area that consists of normal power transfer states, the faulty equipment can be disconnected in case of a fault. Disturbances are not allowed to spread to a larger area or cause a blackout of the system.

The system responsible ISO provides disturbance management that prevents faults from spreading and restores the system to normal operational state as soon as possible after the fault. The security of the power system is maintained by planning and operating the system in a way that minimises disturbances caused by faults. In order to manage disturbances, the system responsible grid operator keeps power transfers within the allowed limits and ensures that the system has enough reserves in power plants and in the transmission grid.

System adequacy is associated with static conditions of the system. It refers to the existence of sufficient electric energy production within the system to meet the load demand or constraints within the transmission and distribution system. The adequacy of the system is usually studied either by a simple generation–load model or by an extended bulk transmission system model consisting of generation, transmission, distribution and load. In a simple generation–load model, the total system production is examined to define its adequacy to meet the total system load. The estimation of the required production needs includes the system load demand and the maintenance needs of production units. The criteria that are used for the adequacy evaluation include the loss of load expectation (LOLE), the loss of load probability (LOLP) and the loss of energy expectation (LOEE), for instance.

The LOLP approach combines the applicable system capacity outage probability with the system load characteristics in order to arrive at the expected probability of loss of load. LOLE defines the number of days or hours per year with a probability of loss of load. LOEE defines the same values for energy (Billinton and Allan, 1988).

8.2.2 Frequency control

The power system that is operated synchronously has the same frequency. The frequency of a power system can be considered a measure of the balance or imbalance between production and consumption in the power system. With nominal frequency (e.g. in Europe 50 Hz, in the USA 60 Hz), production and consumption, including losses in transmission and distribution, are in balance. If the frequency is below 50 Hz, the consumption of electric energy is higher than production. If the frequency is above 50 Hz, the consumption of electric energy is lower than production. The better the balance between production and consumption, the less the frequency deviates from its nominal value. In the Nordic power system, for instance, the frequency is allowed to vary between 49.9 Hz and 50.1 Hz. Figure 8.1(a) shows an example of frequency variations during one day, and Figure 8.1(b) presents frequency variations during one week.

The primary frequency control in power plants is used to keep the frequency of the system within the allowed limits. The primary control is activated automatically if the frequency fluctuates. It is supposed to be fully activated when the maximum allowable frequency deviation (e.g. in the Nordic Power System, ±0.1 Hz) is reached. Figure 8.2 shows an example of the actual load in the system during 3 hours compared with the hourly load forecast, including forecast errors and short-term load deviations in the system.

If there is a sudden disturbance in balance between production and consumption in the power system, such as the loss of a power plant or a large load, primary reserves (also called disturbance reserves or instantaneous reserves) are used to deal with this problem. The primary reserve consists of active and reactive power supplied to the system. Figure 8.3 shows the activation of reserves and frequency of the system as a function of time, for a situation where a large power plant is disconnected from the power system. The activation time divides the reserves into primary reserve, secondary reserve (also called fast reserve) and long-term reserve (also called slow reserve or tertiary reserve), as shown in Figure 8.3.

The primary reserve is the production capacity that is automatically activated within 30 seconds from a sudden change in frequency. It consists of active and reactive power in power plants, on the one hand, and loads that can be shed in the industry, on the other hand. Usually, the amount of reserve in a system is defined according to the largest power plant of the system that can be lost in a single fault.

The secondary reserve is active or reactive power capacity activated in 10 to 15 minutes after the frequency has deviated from the nominal frequency. It replaces the primary reserve and it will be in operation until long-term reserves replace it, as shown in Figure 8.3. The secondary reserve consists mostly of rapidly starting gas turbine power plants, hydro (pump) storage plants and load shedding. Every country in an interconnected power system should have a secondary reserve. It corresponds to the amount of disconnected power during the dimensioning fault (usually loss of the largest power unit) in the country involved. In order to provide sufficient secondary power reserve, system operators may take load-forecast errors into account. In this case, the total amount of the secondary reserve may reach a value corresponding to about 1.5 times the largest power unit.

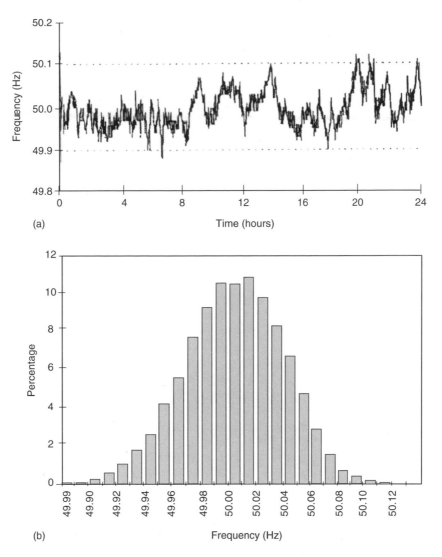

Figure 8.1 Examples of: (a) frequency variations in the system during one day; (b) frequency distribution in the system during one week. Reproduced, by permission, from R. Hirvonen, 2000, 'Material for course S-18.113 Sähköenergiajärjestelmät', Power Systems Laboratory, Helsinki University of Technology

8.2.3 Voltage management

The voltage level in the transmission system is kept at a technical and economical optimum by adjusting the reactive power supplied or consumed. Power plants and special equipment, (e.g. capacitors and reactors) control the reactive power. The voltage ratio of different voltage levels can be adjusted by tap-changers in power transformers. This requires a reactive power flow between different voltage levels.

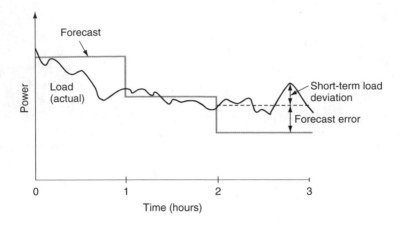

Figure 8.2 Example of actual load in the system during 3 hours compared to forecasted load Reproduced, by permission, from H. Holttinen, 2003, *Hourly Wind Power Variations and Their Impact on the Nordic Power System* (licentiate thesis), Helsinki University of Technology

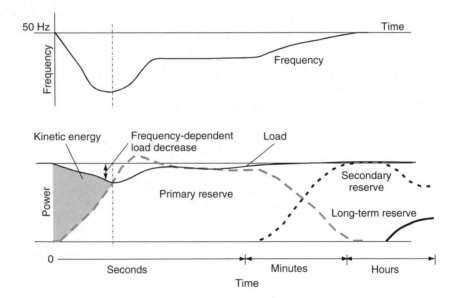

Figure 8.3 Activation of power reserves and frequency of power system as a function of time, for a situation where a large power plant is disconnected from the power system. Reproduced, by permission, from R. Hirvonen, 2000, 'Material for course S-18.113 Sähköenergiajärjestelmät', Power Systems Laboratory, Helsinki University of Technology

In order to manage the voltage level during disturbances, reactive reserves in power plants are allocated to the system. These reserves are used mainly as primary reserves in order to guarantee that the voltage level of the power system remains stable during disturbances.

Voltage level management has the aim to prevent undervoltages and overvoltages in the power system and to minimise grid losses. Voltage level management also guarantees that customer connection points have the voltages that were agreed by contracts.

8.3 Wind Power Production and the Power System

Wind energy is characterised by large variations in production (see Figure 8.4). If we look at the power system, we are interested in the wind power production of larger areas. Large-scale geographical spreading of wind power will reduce variability and increase predictability and there will be fewer instances of near zero or peak output.

For power systems, the relevant information on wind power production is the probability distribution, the range and seasonal or diurnal patterns of production, as well as the magnitude and the frequency of the variations (ramp rates).

8.3.1 Production patterns of wind power

Wind power production is highly dependent on the wind resources at the site. Therefore the average production, the distribution of the production as well as seasonal and diurnal variations can look very different at different sites and areas of the world. For most sites on land, the average power as the percentage of the nominal capacity (the capacity factor, CF), is between 20–40 %. This can be expressed as full load hours of 1800–3500 h/a. Full load hours are the annual production divided by the nominal capacity. Offshore wind power production, or some extremely good sites on land, can reach up to 4000–5000 full load hours (CF = 45−60 %).

We can compare that with other forms of power generation. Combined heat and power production (CHP) has full load hours in the range of between 4000–5000 h/a, nuclear power 7000–8000 h/a, and coal-fired power plants 5000–6000 h/a. However, full

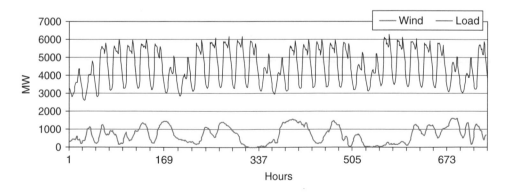

Figure 8.4 Example of large-scale wind power production: Denmark (both Zeeland and Jutland) in January 2000. Average wind power production in January was 687 MW and in 2000 the yearly average was 485 MW (approximately 24 % of the installed capacity). Reproduced, by permission, from H. Holttinen, 2003, *Hourly Wind Power Variations and Their Impact on the Nordic Power System* (licentiate thesis), Helsinki University of Technology

load hours are used only to compare different power plants. They do not tell us how many hours the power plant is actually in operation. Wind turbines, which operate most of the time at less than half of the nominal capacity, will typically produce power 6000–8000 h/a (70–90 % of the time).

The geographical spreading of the production evens out the variations of the total production from an area. There will be substantially less calm periods, as the wind will blow almost always somewhere in the area that the power system covers. However, maximum production levels will not reach the installed nominal capacity, as the wind will not have the same strength at all sites simultaneously. And out of hundreds or thousands of wind turbines not all will be technically available at each and every moment. The duration curve of dispersed wind power production shown in Figure 8.5 illustrates that: the production from one wind turbine is zero for 10–20 % of the time, and at nominal capacity 1–5 % of the time, whereas the production from large-scale wind power production is, in this example, rarely below 5 % or above 75 % of capacity.

Even for large-scale, geographically dispersed wind power production, the production range will still be large compared with other production forms. The maximum production will be three or even four times the average production, depending on the area (Giebel, 2000; Holttinen, 2003).

The available wind resources will vary from year to year. Wind power production during one year lies between ±15 % of the average long-term yearly production (Ensslin, Hoppe-Kilpper and Rohrig, 2000; Giebel, 2001). However, the year-to-year variation in the production from hydro power can be even larger.

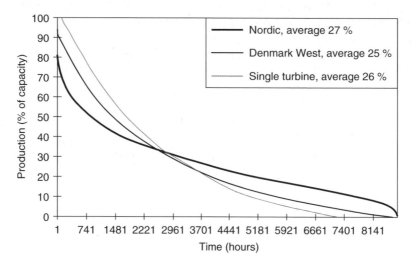

Figure 8.5 Duration curves: increased resources and geographical spreading lead to a flattened duration curve for wind power production. Data in this example are hourly data for 2000, with wind power production from turbines throughout the four Nordic countries (Denmark, Finland, Norway and Sweden) being compared with one of the wind turbines and one of the areas (Denmark West). Reproduced, by permission, from H. Holttinen, 2003, *Hourly Wind Power Variations and Their Impact on the Nordic Power System* (licentiate thesis), Helsinki University of Technology

There is often a distinct yearly (seasonal) and daily (diurnal) pattern in wind power production. In Central and Northern Europe, there is more production in winter than in summer (Figure 8.6).

Wind is driven by weather fronts. It may also follow a daily pattern caused by the sun. Depending on what is prevalent in the region, there is either a strong or hardly any diurnal pattern in the production. There are many sites where the wind often starts to blow in the morning and calms down in the evening (Ensslin, Hoppe-Kilpper and Rohrig, 2000; Hurley and Watson, 2002). In Northern Europe, this is most pronounced during the summer (see Figure 8.7). Diurnal variation can also be due to local phenomena. An example would be the mountain ranges in California, causing morning and evening peaks, when wind blows from the desert to the sea and in the opposite direction, respectively.

8.3.2 Variations of production and the smoothing effect

The wind speed varies on all time scales, and this has different effects on the power system. Wind gusts cause variations in the range of seconds or minutes. The changing weather patterns can be seen from the hourly time series of wind power production. This

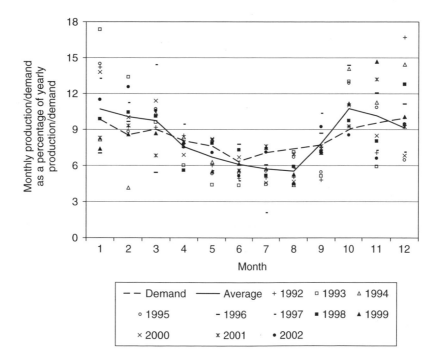

Figure 8.6 Seasonal variations of wind power production. *Note*: data are for Finland for the years 1992 to 2002. Average monthly production for 1992–2002 is shown (solid line) together with the electric consumption in 2002 (dotted line). Reproduced, by permission, from H. Holttinen, 2003, *Hourly Wind Power Variations and Their Impact on the Nordic Power System* (licentiate thesis), Helsinki University of Technology

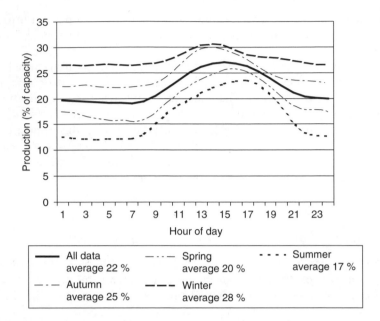

Figure 8.7 Diurnal variations in wind power production in Denmark: diurnal variations of wind power production are larger during the summer, in Northern Europe (Reproduced by permission of ISET, Kassel, Germany.)

time scale also illustrates the diurnal cycle. Seasonal cycles and annual variations, however, are important for long-term adequacy studies. For system planning, it is important to look at extreme variations of large-scale wind power production, together with the probability of such variations.

The larger the area, the longer the periods of time over which the smoothing effect extends. Figure 8.8 shows the decreasing correlation of the variations for different time scales (Ernst, 1999).[1] The correlation is here calculated for the differences between consecutive production values (ΔP). For the time series of production values, the correlation does not decrease as rapidly as shown here. Within one wind farm, gusts (seconds) will not effect all wind turbines at exactly the same moment. However, the hourly wind power production will follow approximately the same ups and downs.

---

[1] Cross-correlation r_{xy} is a measure of how well two time series, x and y, follow each other:

$$r_{xy} = \frac{1}{n\sigma_x\sigma_y}\sum_{i=1}^{n}(x_i - \mu_x)(y_i - \mu_y),$$

where μ_x and μ_y are averages, n is the number of points in the time series and σ_x and σ_y are the standard deviations.

It is near the maximum value, 1, if the 'ups and downs' of production occur simultaneously; it is near the minimum value, -1, if there is a tendency of decreasing production at one site and increasing production at the other site; and it is close to zero if the two are uncorrelated, and the ups and downs of production at two sites do not follow each other.

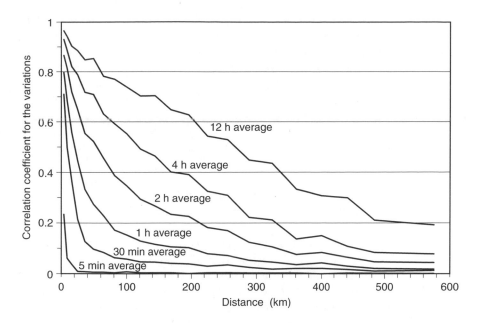

Figure 8.8 Variations will smooth out faster when the time scale is small. Correlation of variations for different time scales: Reproduced, by permission from B. Ernst, Y.-H. Wan, B. Kirby, 2000, 'Short-term Power Fluctuation of Wind Turbines: Analyzing Data from the German 250 MW Measurement Program from the Ancillary Services Viewpoint', *Proceedings of German Wind Energy Conference DEWEK 2000 7–8 June 2000, Wilhelmshaven, Germany*, Deutsches Windenergre Institute, Germany, pp. 125–128

In a larger area covering several hundreds of square kilometres, the weather fronts causing high winds will not pass simultaneously over the entire regions. However, high-wind and low-wind months will coincide for the whole area.

How large is the smoothing effect? It becomes more noticeable if there is a larger number of turbines spread over a larger area. The smoothing effect of a specified area has an upper limit. There will be a saturation in the amount of variation; that is, where an increase in the number of turbines will not decrease the (relative) variations in the total wind power production of the area. Beyond that point, the smoothing effect can be increased only if the area covered becomes larger. And there is a limit to that effect, too. The examples we use are from comparatively uniform areas. If wind power production is spread over areas with different weather patterns (coasts, mountains and desert) the smoothing effect will probably be stronger.

The smoothing effect is illustrated by the statistical parameters of the production (P) and fluctuation (ΔP) time series; that is by the maximum variations of production (extreme ramp rates), the probability distribution of the variations and the standard deviation (σ).

The second-to-second variations will be smoothed out already for one wind turbine. The inertia of the large rotating blades of a variable-speed wind turbine smoothes out very fast gusts. Second-to-second variations will be absorbed in the varying speed of the

rotor of a variable-speed wind turbine. The extreme ramp rates that were recorded for a
103 MW wind farm are: 4–7 % of capacity in a second, 10–14 % of capacity in a minute
and 50–60 % of capacity in an hour (Parsons, Wan and Kirby, 2001). However, system
operation is concerned with an area that is much larger than the area in this example.
For a larger area, with geographically dispersed wind farms, the second and minute
variations will not be significant, and the hourly variations will be considerably less than
50–60 % of capacity.

The largest hourly variations are about ±30 % of capacity when the area is in the
order of 200 × 200 km², such as West or East Denmark, about ±20 % of capacity when
the area is in the order of 400 × 400 km², such as Germany, Denmark, Finland or Iowa,
USA, and about ±10 % in larger areas covering several countries such as the Nordic
countries (Holttinen, 2003; ISET, 2002; Milligan and Factor, 2000). These are extreme
values. Most of the time the hourly variations will be within ±5 % of installed capacity
(Figure 8.9).

If the geographic dispersion of wind power increases, the standard deviation for
hourly time series decreases, which means that the variability in the time series is
reduced. The standard deviation of hourly time series decreases to 50–80 % of the single
site value (Focken, Lange and Waldl, 2001; Holttinen, 2003). The standard deviation of
the time series of fluctuations ΔP will decrease even faster, from about 10 % of capacity
for a single turbine to less than 3 % for an area such as Denmark or Finland, and to less
than 2 % for the four Nordic countries (Holttinen, 2003; Milborrow, 2001).

According to the Institut für Energieversorgungstechnik (ISET, 2002), in Germany,
maximum variation for 4 hours ahead is 50 % of capacity, and for 12 hours ahead is

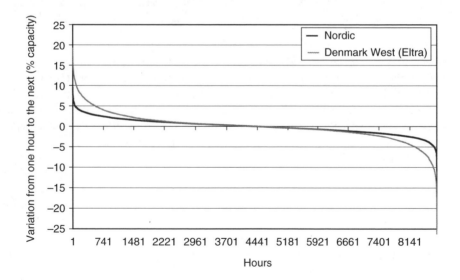

Figure 8.9 Variation of wind power production from one hour to the next, 2000: duration curve
of variations, as a percentage of installed capacity, for Denmark (Jutland) and for the theoretical
Nordic wind power production assuming equal production in each of the four countries. Reproduced, by permission, from H. Holttinen, 2003, *Hourly Wind Power Variations and Their Impact
on the Nordic Power System* (licentiate thesis), Helsinki University of Technology

85 % of capacity. If we take larger areas, such as Northern Europe, there is a ± 30 % variation in production 12 hours ahead only about once a year (Giebel, 2000). For longer time scales (i.e. 4–12 h variations), prediction tools give valuable information on the foreseeable variations of wind power production.

Diurnal variations in output can help to indicate at what time of the day significant changes in output are most likely to occur. The probability of significant variations is also a function of the output level. Significant variations are most likely to occur when wind farms operate at between 25–75 % of capacity. This output level corresponds to the steep part of the power curve when changes in wind speed produce the largest changes in power output of the turbines (Poore and Randall, 2001).

There are also means to reduce the variations of wind power production. Staggered starts and stops from full power as well as reduced (positive) ramp rates can reduce the most extreme fluctuations, in magnitude and frequency, over short time scales. However, this is at the expense of production. Therefore, frequent use of these options should be weighed against other measures (in other production units), regarding their cost effectiveness.

8.3.3 Predictability of wind power production

Wind power prediction plays an important part in the system integration of large-scale wind power. If the share of installed wind power is substantial, information regarding the online production and predictions of 1 to 24 h ahead are necessary. Day-ahead predictions are required in order to schedule conventional units. The starting up and shutting down of slow-starting units has to be planned in an optimised way in order to keep the units running at the highest efficiency possible and to save fuel and thus operational costs of the power plants. In liberalised electricity markets, this is dealt with at the day-ahead spot market. Predictions of 1–2 h ahead help to keep up the optimal amount of regulating capacity at the system operators' disposal.

Predictability is most important both at times of high wind power production and for a time horizon of up to 6 h ahead, which gives enough time to react to varying production. An estimate of the uncertainty, especially the worst-case error, is important information.

Forecast tools for wind power production are still under development and they will be improving (see also Chapter 17). The predictions of the power production 8 h ahead or more rely almost entirely on meteorological forecasts for local wind speeds. In northern European latitudes, for example, the variations of wind power production correspond to weather systems passing the area, causing high winds, which then calm down again. The wind speed forecasts of the Numerical Weather Prediction models contribute the largest error component. So far, an accuracy of $\pm 2 - 3\,\text{m/s}$ (level error) and $\pm 3 - 4\,\text{h}$ (phase error) has been sufficient for wind speed forecasts. However, the power system requires a more precise knowledge of the wind power production (see also Chapter 10).

For larger areas, the prediction error decreases. For East and West Denmark, for example, inclusion of East Denmark adds 100 km, or 50 % more, to West Denmark's area in the direction in which most weather systems pass. The errors of day-ahead predictions would cancel out each other to some extent for about a third of the time, when production is overpredicted in the West and underpredicted in the East, or vice versa (Holttinen, 2004).

8.4 Effects of Wind Energy on the Power System

The impact of wind power on the power system depends on the size and inherent flexibility of the power system. It is also related to the penetration level of wind power in the power system.

When studying the impact of wind power on power systems, we refer to an area that is larger than only one wind farm. According to the impact that is analysed, we have to look at the power system area that is relevant. For voltage management, only areas near wind power plants should be taken into account. Even though there should be enough reactive power reserve in the system during disturbances, the reserve should mainly be managed locally. For intrahour variations that affect frequency control for load following, we should look at the area of the synchronously operated system. Direct-current (DC) links connecting synchronously operated areas can also be automised to be used for primary power control (see Chapter 10). However, their power reserve capacity is usually allocated only as emergency power supply. For the day-ahead hourly production, a relevant area would be the electricity market. The Nordic power market, for instances, includes countries that are situated in different synchronous systems. Large interconnected areas lead to substantial benefits, unless there are bottlenecks in transmission (see Chapter 20).

If we analyse the incremental effects that a varying wind power production has on the power system, we have to study the power system as a whole. The power system serves all production units and loads. The system has only to balance the net imbalances.

Power system studies require representative wind power data. If the data from too few sites are upscaled the power fluctuations will be upscaled too. If large-scale wind power production with steadier wind resources (e.g. offshore or large wind turbines with high towers) is incorporated into the system, measurements from land or with too low masts will, in turn, overestimate the variations. In addition, most studies will require several years of data.

Figure 8.10 shows the impact that wind energy has on the system. These impacts can be categorised as follows:

- short-term: by balancing the system at the operational time scale (minutes to hours);
- long-term: by providing enough power during peak load situations.

These issues will be discussed in more detail in the following sections. For long-term trends affecting the integration of wind power into future power systems, see Section 8.4.4.

8.4.1 Short-term effects on reserves

The additional requirements and costs of balancing the system on the operational time scale (from several minutes to several hours) are primarily due to the fluctuations in power output generated from wind. A part of the fluctuations is predictable for 2 h to 40 h ahead. The variable production pattern of wind power changes the scheduling of the other production plants and the use of the transmission capacity between regions. This will cause losses or benefits to the system as a result of the incorporation of wind power. Part of the fluctuation, however, is not predicted or is wrongly predicted. This corresponds to the amount that reserves have to take care of.

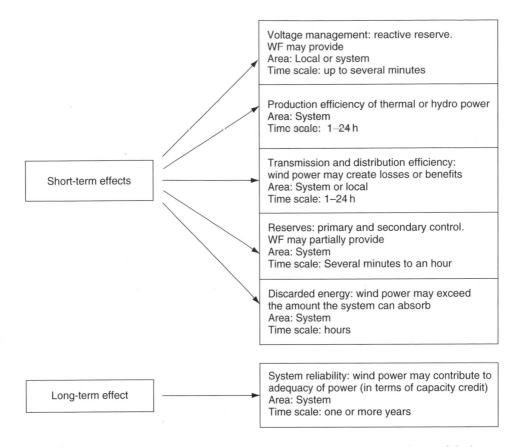

Figure   8.10 Power system impacts of wind power, causing integration costs. Some of the impacts can be beneficial to the system, and wind power can provide value, not only generate cost. Based on H. Holttinen, 2003, *Hourly Wind Power Variations and Their Impact on the Nordic Power System* (licentiate thesis), Helsinki University of Technology

The impact on reserves has to be studied on the basis of a control area. It is not necessary to compensate every change in the output of an individual wind farm by a change in another generating unit. The overall system reliability should remain the same, before and after the incorporation of wind power. The data used for wind power fluctuations are critical to the analysis. It is important not to upscale the fluctuations when wind power production in the system is upscaled. Any wind power production time series that is simulated or based on meteorological data should therefore follow the statistical characteristics that were presented in Section 8.3 (Holttinen, 2003; Milborrow, 2001).

The system needs power reserves for disturbances and for load following. Disturbance reserves are usually dimensioned according to the largest unit outage. As wind power consists of small units, there is no need to increase the amount of disturbance reserve (even large offshore wind farms still tend to be smaller than large condense plants). Hourly and less-than-hourly variations of wind power affect the reserves that are used for frequency control (load following), if the penetration of wind power is large enough to increase the total variations in the system.

Prediction tools for wind power production play an important role in integration. The system operator has to increase the amount of reserves in the system because, in addition to load swings, it has to be prepared to compensate unpredicted variations in production. The accuracy of the wind forecasts can contribute to risk reduction. An accurate forecast allows the system operator to count on wind capacity, thus reducing costs without jeopardising system reliability.

The requirement of extra reserves is quantified by looking at the variations of wind power production, hourly and intrahour, together with load variations and prediction errors. The extra reserve requirement of wind power, and the costs associated with it, can be estimated either by system models or by analytical methods using time series of wind power production together with system variables. Wind power production is not straightforward to model in the existing dispatch models, because of the uncertainty of forecast errors involved on several time scales, for instance (Dragoon and Milligan, 2003). Below, we will briefly describe analytical methods with statistical measures.

The effect of the variations can be statistically estimated using standard deviation. What the system sees is net load (load minus wind power production). If load and wind power production are uncorrelated, the net load variation is a simple root mean square (RMS) combination of the load and wind power variation:

$$(\sigma_{\text{total}})^2 = (\sigma_{\text{load}})^2 + (\sigma_{\text{wind}})^2, \tag{8.1}$$

where $\sigma_{\text{total}}, \sigma_{\text{load}}$ and σ_{wind} are the standard deviations of the load, net load and wind power production time series, respectively.

The larger the area in question and the larger the inherent load fluctuation in the system the larger the amount of wind power that can be incorporated into the system without increasing variations. The reserve requirement can be expressed as three times the standard deviation (3σ covers 99 % of the variations of a Gaussian distribution). The incremental increase from combining load variations with wind variations is 3 times ($\sigma_{\text{total}} - \sigma_{\text{load}}$). More elaborate methods allocating extra reserve requirements for wind power can be used, especially with nonzero correlations and any number of individual loads and/or resources (Hudson, Kirby and Wan, 2001; Kirby and Hirst, 2000).

On the time scale of seconds and minutes (primary control) the estimates for increased reserve requirements have resulted in a very small impact (Ernst, 1999; Smith *et al.*, 2004). This is because of the smoothing effect of very short variations of wind power production; as they are not correlated, they cancel out each another, when the area is large enough.

For the time scale of 15 min to 1 h (secondary control) it should be taken into account that load variations are more predictable than wind power variations. For this, data for load and wind predictions are needed. Instead of using time series of load and wind power variations, the time series of prediction errors one hour ahead are used and standard deviations are calculated from these. The estimates for reserve requirements as a result of use of wind power have resulted in an increasing impact if penetration increases. For a 10 % penetration level, the extra reserve requirement is in the order of 2–8 % of the installed wind power capacity (Holttinen, 2003; Milborrow, 2001; Milligan, 2003).

Both the allocation and the use of reserves cause extra costs. Regulation is a capacity service and does not involve net energy, as the average of regulation time series is zero. In most cases, the increase in reserve requirements at low wind power penetration can be handled by the existing capacity. This means that only the increased use of dedicated reserves, or increased part-load plant requirement, will cause extra costs (energy part). After a threshold, the capacity cost of reserves also has to be calculated. This threshold depends on the design of each power system. Estimates of this threshold suggest for Europe a wind power (energy) penetration of between 5 % and 10 % (Holttinen, 2003; Milborrow, 2001; Persaud, Fox and Flynn, 2000).

Estimates regarding the increase in secondary load following reserves in the UK and US thermal systems suggest €2–3 MWh^{-1} for a penetration of 10 %, and €3–4 MWh^{-1} for higher penetration levels (ILEX, 2003; Dale $et\,al.$, 2004; Smith $et\,al.$, 2004).[2] The figures may be exaggerated because the geographical smoothing effect is difficult to incorporate into wind power time series. In California, the incremental regulation costs for existing wind power capacity is estimated to €0.1 MWh^{-1} for a wind energy penetration of about 2 % (Kirby $et\,al.$, 2003).

Also, the recently emerged electricity markets can be used to estimate the costs for hourly production and power regulation. An ideal market will result in the same cost effectiveness as the optimisation of the system in order to minimise costs. However, especially at an early stage of implementation of a regulating market, or as a result of market power, the market prices for regulation can differ from the real costs that the producers have.

In a market-based study, Hirst (2002) estimated the increase in regulation (at the second and minute time scale) that would be necessary to maintain system reliability at the same level, before and after the implementation of wind power. The result was that the regulation cost for a large wind farm would be between €0.04 MWh^{-1} and €0.2 MWh^{-1}. This result applies to systems where the cost of regulation is passed on to the individual generators and is not provided as a general service by the system operator.

In West Denmark, with a wind penetration of about 20 %, the cost for compensating forecast errors in the day-ahead market at the regulating market amounted to almost €3 MWh^{-1} (see also Chapter 10).

In the electricity market, the costs for increased regulation requirements will be passed on to the consumers, and the production capacity providing for extra regulation will benefit from that. Regulation power nearly always costs more than the bulk power available on the market. The reason is that it is used during short intervals only, and that is has to be kept on standby. Therefore, any power continuously produced by that capacity cannot be sold to the electricity spot market. The cost of reserves depends on what kind of production is used for regulation. Hydro power is the cheapest option, and gas turbines are a more expensive option. The cost of extra reserves is important when

---

[2] The currency exchange rate for the end of 2003 has been used: €1 = \$1.263; €1 = £0.705.

the system needs an increasing amount of reserves, because of changes in production or consumption, such as increased load. The costs of regulation may rise substantially and suddenly during a phase when the cheapest reserves have already been used and the more expensive new reserves have to be allocated.

The cost estimates for thermal systems include the price for new reserve capacity and assume a price for lower efficiency and part load operation. To integrate wind power fully into the system in an optimal way means using the characteristics and flexibility of all production units in a way that is optimal for the system. Also, a wider range of options in order to increase flexibility can be used. Some examples for existing technologies that could be used to absorb more variable energy sources are:

- increased transmission between the areas, countries or synchronous systems;
- demand-side management (DSM) and demand-side-bidding (DSB);
- storage (e.g. thermal storage with CHP regulating); electrical storage may become cost-effective in the future, but is still expensive today;
- making the electricity production of CHP units flexible by using alternatives for heat demand (heat pumps, electric heating, electric boilers);
- short-term flexibility in wind farms. When based on reducing the output of wind power, this means loss of production. The desired flexibility can be achieved more cost-efficiently by conventional generation, if it requires an extensive reduction of wind power output.

Even simple statistical independence makes different variable sources more valuable than simply 'more of the same', such as wind power and solar energy. It may also be beneficial to combine wind power with energy-limited plants where the maximum effect cannot be produced continuously because the availability of energy is limited. This is the case of hydro power and biomass systems. Power systems with large hydro power reservoirs have the option to use hydro power to smooth out the variability of wind power by shifting the time of energy delivery (Krau, Lafrance and Lafond, 2002; Tande and Vogstad, 1999; Vogstad et al., 2000). This is possible also for short response times, within the operating constraints of flow and ramp rates of hydro power (Söder, 1999).

8.4.2 Other short-term effects

Other effects that wind power has at the operational level of the power system include its impact on losses in power systems (generation and transmission or distribution) and on the amount of fuel used and on emissions [e.g. carbon dioxide (CO_2)]. There is already technology that allows wind farms to benefit power system operation, such as by providing voltage management and reactive reserve (in the case of Type D turbines that are connected to the network or, in a limited way, as in the case of Type C turbines) as well as primary power regulation (Kristoffersson Christiansen and Hedevang, 2002).[3] This issue of reliability is not discussed in detail here.

---

[3] For a description of turbines of Type A, B, C and D, see Section 4.2.3.

Wind power can either decrease or increase the transmission and distribution losses, depending on where it is situated in relation to the load. However, large-scale wind power can result in increased transmission between regions. That can lead to increased transmission losses or a larger number of bottlenecks in transmission (see also Chapter 20). For the UK, concentrating the wind power generation in the north would double the estimated extra transmission costs to €2 MWh^{-1} and €3 MWh^{-1} at a penetration level of between 20 % and 30 %. This would not be the case if production were more geographically dispersed (ILEX, 2003). At more modest penetration levels, transmission costs would decrease.

Large amounts of intermittent wind power production can cause losses in conventional generation. The decreased efficiency of the system is caused by thermal or hydro plants operating below their optimum (startups, shutdowns, part load operation). The optimised unit commitment (i.e. planning the startups and shutdowns of slow-starting units) is complicated by the intermittent output from a wind resource. An accurate prediction of wind power production will help to solve this problem. However, even with accurate predictions, the large variations in wind power output can result in conventional power plants operating in a less efficient way. The effect on existing thermal and/or hydro units can be estimated by simulating the system on an hourly basis. At low penetration levels, the impact of wind power is negligible or small (Grubb, 1991; Söder, 1994), although costs for large prediction errors in a thermal system have been estimated to be about €1 MWh^{-1} (Smith et al., 2004).

If wind power production exceeds the amount that can be safely absorbed while still maintaining adequate reserves and dynamic control of the system, a part of the wind energy production may have to be curtailed. Energy is discarded only at substantial penetration levels. Whether such a measure is taken depends strongly on the operational strategy of the power system. The maximum production (installed capacity) of wind power is several times larger than the average power produced. This means that there are already some hours with nearly 100 % instant (power) penetration (wind power production equals demand during some hours), if about 20 % of yearly demand comes from wind power. There is experience from and studies on thermal systems that take in wind power production, but leave, even at high winds, the thermal plants running at partial load in order to provide regulation power. The results show that about 10 % (energy) penetration is the starting point where a curtailing of wind power may become necessary. When wind power production is about 20 % of yearly consumption, the amount of discarded energy will become substantial and about 10 % of the total wind power produced will be lost (GarradHassan, 2003; Giebel, 2001). For a small thermal island system (e.g. on Crete, Greece) discarded energy can reach significant levels already at a penetration of 10 % (Papazoglou, 2002).

For other areas, integration problems may arise during windy periods, if production in the area exceeds demand and also the transmission capacity to neighbouring systems. This can be especially pronounced during windy, cold periods when there is also a substantial share of local, prioritised CHP production, as is the case of Denmark (see also Chapter 10). When initially in West Denmark wind energy was discarded, this happened at penetration levels of 20 % rather than 10 %. With energy system models it has been estimated that by using the existing heat storage and boilers of CHP production units together with wind power, and assuming some flexible demand and electrical heating, a 50 % wind power penetration could be possible without discarding any energy (Lund and Münster, 2003).

Wind power is renewable energy, practically free from CO_2 emissions. CO_2 emissions from the manufacturing and construction are in the order of $10\,g\,CO_2\,kWh^{-1}$. If wind energy replaces generation that emits CO_2, CO_2 emissions from electricity production are reduced. The amount of CO_2 that will be abated depends on what production type and fuel is replaced at each hour of wind power generation. This will be the production form in use at each hour that has the highest marginal costs. Usually, this is the older coal-fired plants, resulting in a CO_2 abatement of about $800-900\,g\,CO_2\,kWh^{-1}$, often cited as the CO_2 abatement of wind energy. This is also true for larger amounts of wind power production, for countries that generate their electricity mainly from coal. In other countries, though, there may be a different effect if large amounts of wind power are added to the system. There may not be a sufficient number of old coal plants whose capacity can be replaced by wind power production throughout the year. During some hours of the year, wind power generation would replace other production forms, such as the production of gas-fired plants (CO_2 emissions of gas are $400-600\,g\,CO_2\,kWh^{-1}$), or even CO_2-free production (e.g. hydro, biomass or nuclear power). Instant (regulated) hydro production can be postponed and will replace condensing power at a later instant. Simulations of the Nordic system, for example, which is a mixed system of thermal and hydro production, result in a CO_2 reduction of $700\,g\,CO_2\,kWh^{-1}$ (Holttinen and Tuhkanen, 2004). This is the combined effect of wind power replacing other fuels.

8.4.3 Long-term effects on the adequacy of power capacity

The intermittent nature of wind energy poses challenges to utilities and system operators. These must be able to serve loads with a sufficiently low probability of failure. The economic, social and political costs of failing to provide adequate capacity to meet demand are so high that utilities have traditionally been reluctant to rely on intermittent resources for capacity.

Dimensioning the system for system adequacy usually involves estimations of the LOLP index. The risk at system level is the probability (LOLP) times the consequences of the event. For an electricity system, the consequences of a blackout are large, thus the risk is considered substantial even if the probability of the incident is small. The required reliability of the system is usually in the order of one larger blackout in 10–50 years.

What impact does wind power have on the adequacy of power production in the system – can wind power replace part of the (conventional) capacity in the system? To answer this question, it is critical we know wind power production during peak load situations. This also means that to assess the ability of wind power to replace conventional capacity (i.e. the capacity credits) it is important either to have representative data for several years (one year is not enough) or to make a variability assessment (Giebel, 2001; Milligan, 2000).

Some variable sources can be relied on to produce power at times of peak demand. Solar energy, for instance, follows air-conditioning loads, and wind energy reflects heating demand. If a diurnal pattern in wind power production coincides with the load (e.g. wind power production increases in the morning and decreases in the evening) this effect is beneficial. However, in most cases there is no correlation between load and the availability of this variable source. In Northern Europe, for example, even if the seasonal variations mean that more wind power is available in winter than in summer, there is not a strong correlation between the high loads in winter and high wind power production. In Denmark, the correlation is slightly positive (about 0.2), but there is usually less correlation during higher-load winter months than in the summer months.

In Northern Europe, the load is strongly correlated to outside temperature. The correlation between wind power production and temperature has an effect on the adequacy of power production, when determining the capacity value of wind power (see Figure 8.11). Looking at wind power production during the 10 highest peak load hours each year, one can see it ranges between 7–60 % of capacity (1999–2001 in the Nordic countries; Holttinen, 2003).

Nevertheless, variable sources can save thermal capacity. Since no generating plant is completely reliable, there is always a finite risk of not having enough capacity available. Variable sources may be available at the critical moment when demand is high and many other units fail. Fuel source diversity can also reduce risk.

It has been shown in several studies that if the capacity of a variable source is small (low system penetration) the capacity value equals that of a completely reliable plant generating the same average power at times when the system could be at risk. As the penetration increases, variable sources become progressively less valuable for saving thermal capacity (ILEX, 2003). The dispersion of wind power and a positive correlation between wind power and demand increase the value of wind power to the system. For very high penetration levels, the capacity credit tends towards a constant value – that is,

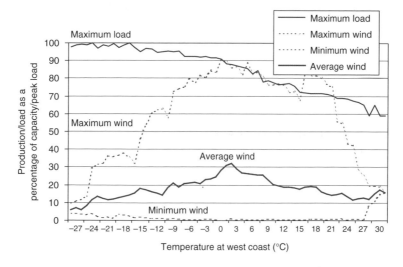

Figure 8.11 Correlation of temperature with wind power production and load in a cold climate (Finland), with geographically dispersed wind power. *Note*: there were 48 h (0.1 % of time) below −23 °C and 549 h (1.6 % of time) below −14 °C during the years 1999–2002. Reproduced, by permission, from H. Holttinen, 2003, *Hourly Wind Power Variations and Their Impact on the Nordic Power System* (licenciate thesis), Helsinki University of Technology

there is no increase in the capacity credit when increasing wind power capacity. This will be determined by the LOLP without wind energy and the probability of zero wind power (Giebel, 2001).

If there is a substantial amount of wind power in the system (greater than 5 % of peak load) an optimal system to accommodate wind power would contain more peaking and less base plants than a system without wind power. For hydro-dominated systems, where the system is energy restricted instead of capacity restricted, wind power can have a significant energy delivery value. As wind energy correlates only weakly with hydro power production, wind energy added to the system can have a considerably higher energy delivery value than the addition of more hydro power (Söder, 1999; see also Chapter 9).

8.4.4 Wind power in future power systems

Large-scale wind power still lies in the future for many countries. There are long-term trends that can influence the impact of wind power on the system. If there are large amounts of intermittent energy sources in the system, new capacity with lower investment costs (and higher fuel costs) will be favoured. The trend of increasing distributed generation from flexible gas turbines is beneficial for the integration of wind power, as is increasing load management. A greater system interconnection is highly beneficial as well: wind power spread all over Europe would be quite a reliable source. The use of electric vehicles will open new possibilities for variable and intermittent power production. Producing fuel for vehicles that are used only about 1000 hours per year will ease the flexibility needs in power systems.

The expected developments of wind power technology will affect the impact that wind power has on power systems. Very large wind farms (hundreds of megawatts) are one trend that can pose serious challenges to the integration of wind power, as they concentrate the capacity. As a result, the smoothing effect of variations by geographical spreading is lost. However, such large wind farms will also pave the way for other technologies that will help with integration. Increasingly sophisticated power electronics and computerised controls in wind farms, as well as an improved accuracy in wind forecasts, will lead to improvements in the predictability and controllability of wind power. Large wind energy power plants will mean that there are new requirements regarding the integration of wind power into the power system. Increasingly, wind farms will be required to remain connected to the system when there are faults in the system. They will be expected to withstand nearby faults without experiencing problems in power production during and after the faults. And they will be expected to provide reactive power support to the system during the fault.

8.5 Conclusions

Wind power will have an impact on power system reserves as well as on losses in generation and transmission or distribution. It will also contribute to a reduction in fuel usage and emissions.

Regarding the power system, the drawbacks of wind power are that wind power production is variable, difficult to predict and cannot be taken as given. However, integration of variable sources is much less complicated if they are connected to large power systems, which can take advantage of the natural diversity of variable sources. A large geographical spreading of wind power will reduce variability, increase predictability and decrease occasions of near-zero or peak output. The power system has flexible mechanisms to follow the varying load that cannot always be accurately predicted. As no production unit is 100 % reliable, part of the production can come from variable sources, with a similar risk level for the power system.

Power system size, generation capacity mix (inherent flexibility) and load variations have an effect on how intermittent production is assimilated into the system. If the proportion of intermittent power production is small and if wind power production is well dispersed over a large area and correlates with the load then wind power is easier to integrate into the system.

Short term mainly the variations in wind power production that affect power system operation. This refers to the allocation and use of extra reserves as well as cyclic losses of conventional power production units, and transmission or distribution network impacts. The area we have to look at for intrahour variations is the synchronously operated system. In a large system, the reserve requirements of different loads and wind power interact and partly compensate each other. The power system operation then needs only to balance the resulting net regulation. The variability introduced by wind power will not be significant until variations are of the same order as the variability of the random behaviour of electricity consumers. On the time scale of seconds and minutes (primary control), the estimates for increased reserve requirement have resulted in a very small impact. On the time scale of 15 min to 1 h, the estimated increase in reserve requirement is of the order of 2–10 % of installed wind power capacity, when the wind energy penetration level is 10 %.

Long-term, the expected wind power production at peak load hours has an impact on the power system adequacy. It is expressed as the capacity credit of wind power. For a low system penetration, the capacity credit equals that of a completely reliable plant generating the same average power at times when the system could be at risk. As the penetration increases, variable sources become progressively less valuable for saving thermal capacity.

There are no technical limits to the integration of wind power. However, as wind capacity increases, measures have to be taken to ensure that wind power variations do not reduce the reliability of power systems. There will be an increasing economic impact on the operation of a power system if wind power penetration exceeds 10 %.

Large-scale wind power still lies in the future for many countries, and there are long-term trends that can influence what impact wind power has on the system, such as the use of electricity for vehicles. At high penetration levels, an optimal system may require changes in the conventional capacity mix.

References

[1] Billinton, R. Allan, R. (1988) *Reliability Assessment of Large Electric Power Systems*, Kluwer, Boston, MA.
[2] Dale, L., Milborrow, D., Slark, R., Strbac, G., (2004) Total cost estimates for large-scale wind scenarios in UK. Energy Policy **32**(17) 1949–1956.

[3] Garrad Hassan (2003) 'Impacts of Increased Levels of Wind Penetration on the Electricity Systems of the Republic of Ireland and Northern Ireland: Final Report', a report commissioned by Commission for Energy Regulation in Republic of Ireland OFREG Northern Ireland; available at http://www.cer.ie/cerdocs/cer03024.pdf.

[4] Dragoon, K., Milligan, M. (2003) 'Assessing Wind Integration Costs with Dispatch Models: A Case Study', paper presented at AWEA Windpower 2003 Conference, May 2003, Austin, TX; available at http://www.nrel.gov/publications/.

[5] ILEX (2003) 'Quantifying the System Costs of Additional Renewables in 2020', a report commissioned by the UK Department of Trade and Industry: available at http://www2.dti.gov.uk/energy/developep/support.shtml 14.3.2003.

[6] Ensslin, C., Hoppe-Kilpper, M., Rohrig, K. (2000) 'Wind Power Integration in Power Plant Scheduling Schemes', European Wind Energy Conference (EWEC) Special Topic, Kassel, Germany. SEE [37]

[7] Ernst, B. (1999) 'Analysis of Wind Power Ancillary Services Characteristics with German 250 MW Wind Data', NREL Report TP-500-26969, available at http://www.nrel.gov/publications/.

[8] Ernst, B., Wan, Y.-H., Kirby, B. (2000) 'Short-term Power Fluctuation of Wind Turbines: Analyzing Data from the German 250 MW Measurement Program from the Ancillary Services Viewpoint', in *Proceedings of the German Wind Energy Conference DEWEK 2000, 7–8 June 2000, Wilhelmshaven, Germany*, Deutsches Windenergie Institut, pp. 125–128.

[9] Focken, U., Lange, M., Waldl, H.-P. (2001) 'Previento – A Wind Power Prediction System with an Innovative Upscaling Algorithm', in *Proceedings of European Wind Energy Conference, 2nd–6th July, 2001, Copenhagen*, WIP-Munich, Germany, pp. 826–829.

[10] Giebel, G. (2000) 'Equalizing Effects of the Wind Energy Production in Northern Europe Determined from Reanalysis Data', Risö-R-1182(EN), Roskilde, available at http://www.risoe.dk/rispubl/index.htm.

[11] Giebel, G., (2001) *On the Benefits of Distributed Generation of Wind Energy in Europe*, VDI Verlag, Düsseldorf, available at http://www.drgiebel.de/thesis.htm.

[12] Grubb, M. J. (1991) 'The Integration of Renewable Energy Sources', *Energy Policy* (September 1991) 670–689.

[13] Hirst, E. (2002) 'Integrating Wind Output with Bulk Power Operations and Wholesale Electricity Markets', *Wind Energy* **5** 19–36.

[14] Hirvonen, R. (2000) 'Material for Course S-18.113 Sähköenergiajärjestelmät', Power Systems Laboratory, Helsinki University of Technology, Helsinki (in Finnish).

[15] Holttinen, H. (2003) *Hourly Wind Power Variations and their Impact on the Nordic Power System Operation*, licenciate's thesis, Helsinki University of Technology, 2003; available at http://www.vtt.fi/renewables/windenergy/windinenergysystems.htm.

[16] Holttinen, H. (2004) 'Optimal Market for Wind Power', *Energy Policy* (in press).

[17] Holttinen, H., Tuhkanen, S. (2004) 'The Effect of Wind Power on CO_2 Abatement in the Nordic Countries', *Energy Policy* **32**(14) 1639–1652.

[18] Hudson, R., Kirby, B., Wan, Y. H. (2001) 'Regulation Requirements for Wind Generation Facilities', in *Proceedings of AWEA Windpower '01 Conference, 3–7 June 2001, Washington, DC*, American Wind Energy Association (AWEA).

[19] Hurley, T., Watson, R. (2002) 'An Assessment of the Expected Variability and Load Following Capability of a Large Penetration of Wind Power in Ireland', in *Proceedings of Global Wind Power Conference Paris, 2–5 April 2002.*

[20] ISET (Institut für Solare Energieversorgungstechnik) (2002) 'Wind Energy Report Germany', ISET, Kassel, Germany.

[21] Kirby, B., Hirst, E. (2000) 'Customer-specific Metrics for the Regulation and Load Following Ancillary Services', Oak Ridge National Laboratory, Oak Ridge, TN, USA.

[22] Kirby, B., Milligan, M., Hawkins, D., Makarov, Y., Jackson, K., Shui, H. (2003) 'California Renewable Portfolio Standard Renewable Generation Integration Cost Analysis, Phase I', available at http://cwec.ucdavis.edu/rpsintegration/.

[23] Krau, S., Lafrance, G., Lafond, L. (2002) 'Large Scale Wind Farm Integration: A Comparison with a Traditional Hydro Option', in *Proceedings of the Global Wind Power Conference, Paris, 2–5 April 2002.*

[24] Kristoffersen, J R., Christiansen, P., Hedevang, A. (2002) 'The Wind Farm Main Controller and the Remote Control System in the Horns Rev Offshore Wind Farm', in *Proceedings of the Global Wind Power Conference, Paris, 2–5 April 2002.*

[25] Lund, H., Münster, E. (2003) 'Management of Surplus Electricity Production from a Fluctuating Renewable Energy Source', *Applied Energy* **76** 65–74.

[26] Milborrow, D. (2001) 'Penalties for Intermittent Sources of Energy', submitted to prime minister for energy policy review, September 2001; available at http://www.pm.gov.uk/output/Page77.asp or directly at http://www.number10.gov.uk/output/Page3703.asp.

[27] Milligan, M. (2000) 'Modelling Utility-scale Wind Power Plants. Part 2: Capacity Credit', *Wind Energy*, *2000* **3** 167–206.

[28] Milligan, M. (2003) 'Wind Power Plants and System Operation in the Hourly Time Domain', paper presented at the American Wind Energy Association (AWEA) Windpower Conference, May 2003, Austin, TX; available at http://www.nrel.gov/publications/.

[29] Milligan, M., Factor, T. (2000) 'Optimizing the Geographic Distribution of Wind Plants in Iowa for Maximum Economic Benefit and Reliability', *Wind Engineering* **24**(4), 271–290.

[30] Papazoglou, T. P. (2002) 'Sustaining High Penetration of Wind Generation – The Case of Cretan Electric Power System', Blowing Network Meeting, Belfast, 22 November 2002; available at http://www.ee. qub.ac.uk/blowing/.

[31] Parsons, B., Wan, Y., Kirby, B. (2001) 'Wind Farm Power Fluctuations, Ancillary Services, and System Operating Impact Analysis Activities in the United States', in *Proceedings of European Wind Energy Conference, 2nd–6th July, 2001, Copenhagen*, WIP-Munich, Germany, pp. 1146–1149; also available as NREL Report CP-500-30547, available at http://www.nrel.gov/publications/.

[32] Persaud, S., Fox, B., Flynn, D. (2000) 'Modelling the Impact of Wind Power Fluctuations on the Load Following Capability of an Isolated Thermal Power System', *Wind Engineering* **24**(6) 399–415.

[33] Poore, R. Z., Randall, G. (2001) 'Characterizing and Predicting Ten Minute and Hourly Fluctuations in Wind Power Plant Output to Support Integrating Wind Energy into a Utility System', in *Proceedings of AWEA Windpower '01 Conference, 3–7, June 2001, Washington, DC*, American Wind Energy Association (AWEA).

[34] Smith, J. C., DeMeo, E. A., Parsons, B., Milligan, M. (2004) 'Wind Power Impacts on Electric Power System Operating Costs: Summary and Perspective on Work to Date', in *Proceedings of Global Wind Power Conference, April 2004, Chicago, USA*.

[35] Söder, L. (1994) 'Integration Study of Small Amounts of Wind Power in the Power System', KTH report TRITA-EES-9401, Royal Institute of Technology (KTH), Stockholm, Sweden.

[36] Söder, L. (1999) 'Wind Energy Impact on the Energy Reliability of a Hydro–Thermal Power System in a Deregulated Market', in *Proceedings of Power Systems Computation Conference, June 28–July 2, 1999, Trondheim, Norway*.

[37] Tande, J. O., Vogstad, K.-O. (1999) 'Operational Implications of Wind Power in a Hydro Based Power System', in *Proceedings of the European Wind Energy Conference, Nice, 1–4 March 1999*, James & James, UK, pp. 425–428.

[38] Vogstad, K.-O., Holttinen, H., Botterud, A., Tande J. O. (2000) 'System Benefits of Co-ordinating Wind Power and Hydro Power in a Deregulated Market', in *Proceedings of European Wind Energy Conference (EWEC) Special Topic, 25–27th September 2000, Kassel, Germany*.

9

The Value of Wind Power

Lennart Söder

9.1 Introduction

The aim of a power plant in a power system is to supply the load in an economical, reliable and environmentally acceptable way. Different power plants can fulfil these requirements in different ways. This chapter will describe the different requirements on a power plant and how these requirements can be met with wind power.

In Section 9.2, the different values attributable to a power plant will be described and in Section 9.3 these attributes will be discussed. Sections 9.2 and 9.3 will describe value as a decrease in total costs. In Sweden, the power market is deregulated, and in Section 9.4 an analysis is made of the value that is considered to decrease the costs for the market actor (for details of wind power in other markets, see e.g. Kirby *et al.*, 1997). In Section 9.5 some conclusions are presented.

9.2 The Value of a Power Plant

Different power plants have different characteristics concerning how they can be controlled in the power system. A common situation is that the value of a new power plant is considered to be the marginal value of the plant in an existing power system. The different types of value in relation to the new power plant are described in Sections 9.2.1–9.2.5.

9.2.1 Operating cost value

Operating cost value is the capability of the new power plant to decrease the operating costs in the existing power system. If a new power plant is added to the system, this power plant will supply energy to the system. This means that the energy production of

Wind Power in Power Systems   Edited by T. Ackermann
© 2005 John Wiley & Sons, Ltd   ISBN: 0-470-85508-8 (HB)

other power plants in the existing system will decrease. The consequence is that the operating costs in these plants decrease.

9.2.2 Capacity credit

Capacity credit refers to the capability of the new power plant to increase the reliability of the power system. In the existing power system, there is always a certain risk of a capacity deficit, the so-called loss of load probability (LOLP). This risk is normally very low. In the case of a capacity deficit, some load has to be disconnected. If a new power plant is added to this system, there is a certain chance that costumers do not have to be disconnected so often, since the installed capacity of the system increases. This implies that the reliability of the system increases, as a result of the new power plant.

9.2.3 Control value

Control value is a value related to the capability of the new power plant to follow demand. In a power system, there is a need for continuous production control, since total production always has to be equal to system load, including losses. Since load varies continuously, production also has to vary continuously. This value is different for different power plants. It can also become negative if the new power plant increases the need for control in the system.

9.2.4 Loss reduction value

Loss reduction value relates to the capability of the new power plant to reduce grid losses in the system. Power transmission always causes grid losses. If power is transmitted over long distances and/or at low voltages, the losses are relatively high. If a new power plant is located closer to the consumers, compared with existing power plants, the losses in the system decrease, since the amount of transmitted power decreases. This implies that the new power plant has extra value related to its capability of reducing these losses. However, a negative value means that the new power plant increases system losses.

9.2.5 Grid investment value

Grid investment value refers to the capability of the new power plant to decrease the need of grid investments in the power system. If a new power plant is located close to consumers exhibiting increasing demand the new power plant may reduce the need for new grid investments. This constitutes extra value in relation to this plant. Also, this value can be negative if the new power plant increases the need for grid investments.

9.3 The Value of Wind Power

As stated in Section 9.2, the value of a power plant can be divided into several subvalues. In this chapter, these subvalues will be described for wind power. Here, the system will be assumed to be an *ideal economic system* where the value is equal to the decrease of total costs. The numerical examples refer to the Swedish power system, with a comparatively high proportion of hydropower (50 %) but with strong connections to neighbouring power systems. The neighbouring systems use hydropower (Norway) and thermal power (Denmark, Finland, Germany and Poland). The value in the 'real world' (i.e. the existing market) is described in Section 9.4, where the Swedish deregulation framework is considered (for other markets, see Milligan *et al.*, 2002).

9.3.1 The operating cost value of wind power

The operating cost value considers the capability of wind power to decrease the operating costs of other power sources. The value means in reality that each kilowatt-hour that is produced by wind power will replace one kilowatt-hour that otherwise would have been produced from another power source. In the Swedish hydro system, hydro energy is (almost) never replaced by wind power production, since water that is stored in reservoirs during periods with high wind production can be used later for power production. There is one exception, though, during high inflows, when hydro production has to be very high in order to avoid spillage. If this coincides with very low load and very high winds it may happen that water has to be spilled because of the high wind power production. These situations are probably rather rare and in such cases excess power might also be sold to neighbouring systems.

The common situation is that wind power production replaces fuel-based generation in thermal power stations. The generation can be replaced directly (i.e. the production in thermal power stations is reduced if there is wind power production). An indirect way of replacing fuel-based generation in thermal power stations is to store water during good wind conditions and later use this 'stored hydro power' to replace fuel-based generation. In the thermal power part of the power system, the low-cost units are used first and those with higher operating costs are used later. This implies that wind power normally replaces fuels in the units with the highest operating costs, since one of the aims of a power system is to operate at the lowest cost possible.

9.3.2 The capacity credit of wind power

The capacity credit of wind power refers to the capability of wind power to increase the reliability of the power system. For defining the capacity credit we use the same method as for other types of power plants (e.g. see IEEE, 1978; the capacity credit of wind power is also included in Milligan, 2002). Figure 9.1 illustrates an example of weekly load. The available capacity is set to 3250 MW and the load is the real hourly load in the western part of Denmark during the week starting 2 January 2003. The available capacity is not the real one but has been chosen in order to illustrate the capacity credit of a certain amount of wind power. In this situation, there would have been a capacity deficit during 40 h of that week.

If, now, wind power is introduced into this system, the available capacity is increased (Figure 9.2). This figure includes the real wind power production during the same week illustrated in Figure 9.1. During this week, mean wind power production was 392 MW. On 31 December 2002, total installed wind power capacity was 1994 MW. During 2002,

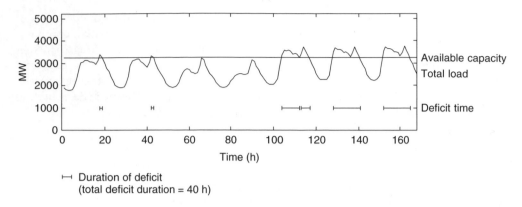

Figure   **9.1** Denmark, week starting 2 January 2003: capacity deficit without wind power

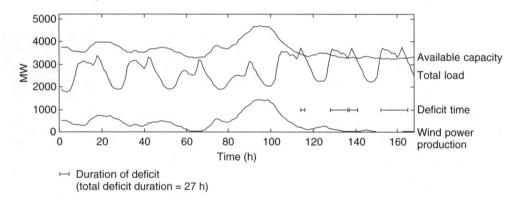

Figure   **9.2** Denmark, week starting 2 January 2003: capacity deficit with wind power

the mean wind power production in this area was 396 MW (i.e. the selected week is representative regarding the total energy). This amount of wind power resulted in a decrease in the number of hours with a capacity deficit, falling from 40 h to 27 h. This means that the reliability of the power system has increased as a result of the installed wind power. If the reliability was acceptable before the instalation of wind power, wind power production will enable the power system to meet a higher demand at the same reliability level.

Figure 9.3 shows that if the load increases by 300 MW during each hour (compared with the load shown in Figure 9.2) the number of hours with a capacity deficit increases to 40. This means that the capacity credit measured as the equivalent load-carrying capability is 300 MW. In this example, the capacity credit for an installed wind power capacity of 1994 MW is 300 MW, corresponding to 15 % of installed capacity. This is slightly lower than results from other studies, which indicate capacity credits in the range of 18–24 % of installed capacity (Munksgaard, Pedersen and Pedersen, 1995; van Wijk, 1990).

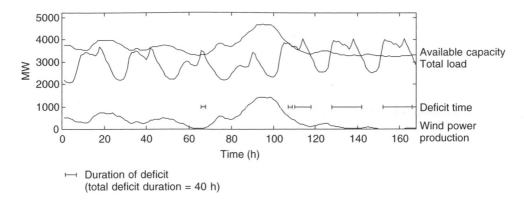

Figure 9.3 Denmark, week starting 2 January 2003: capacity deficit with wind power and with load +300 MW compared with situation illustrated in Figure 9.2

There remains the question of how wind power can have a capacity credit when there are situations with no wind? It has to be kept in mind that for any power source there is a risk that it is not available during peak loads. The method used here to define capacity credit is exactly the same as the method that is used to define the capacity credit for other sources (IEEE, 1978). The example above shows that the number of hours with capacity deficit decreases, but not to zero, when the amount of wind power increases. During hour 60, for instance, there is no wind, and it does not matter whether the amount of wind power increases during this hour, since there is no wind. But the figures also show that during peak hours there is sometimes wind, which means that wind power can decrease the risk of a capacity deficit. In the figures, one week is used for illustrative purposes, but the data of a longer period, probably over several years, have to be used to be able to draw general conclusions.

We can compare the capacity credit of wind power production with that of generation from a thermal plant. The mean weekly wind power production for 2002 was 396 MW; thus the yearly production was as follows:

$$\text{Power production in one year} = (396\,\text{MW}) \times (365 \times 24\,\text{h}) = 3469\,\text{GWh}.$$

Assume the thermal plant is available 92 % of the time and that it shuts down for 4 weeks for maintenance. In order to produce the same amount of energy as the wind power system (3469 GWh):

$$3469\,\text{GWh} = (\text{thermal power production}) \times \left(\frac{92}{100}\right) \times \left[\left(\frac{48}{52}\right) \times 8760\,\text{h}\right].$$

Therefore, rearranging, we obtain:

$$\text{thermal power production} = 3469\,\text{GWh} \times \left(\frac{100}{92}\right) \times \left[\left(\frac{52}{48}\right) \times \left(\frac{1}{8760\,\text{h}}\right)\right] = 0.466\,\text{GWh}$$

$$= 466\,\text{MW}.$$

The capacity credit for this thermal power plant is around 429 MW (92 % of 466 MW). This means, for this example, that the capacity credit of wind power production (300 MW) corresponds to approximately 70 % of the capacity credit for a thermal power plant with the same yearly energy production.

9.3.3 The control value of wind power

The control value of wind power considers the capability of wind power to participate in balancing production and consumption in the power system. Since electric power cannot be stored it is necessary to produce exactly as much power as is consumed, all the time. The balancing problem is handled differently depending on the time frame:

9.3.3.1 Primary control

Primary control considers the capability of the power system to keep a balance between production and consumption from second to second up to about a minute. Primary control can be summarised as follows: most of the generators in a power system are normally of a *synchronous* type. In these generators, there is a strong connection between the rotational speed of the axis and the electric frequency. When the rotational speed decreases, the electric frequency decreases, too. This means that if too little power is produced in a power system, the frequency decreases since the required extra power is taken from the rotational energy in the synchronous machines. When the rotational energy is used up, the rotational speed decreases, which also decreases the electric frequency. In some power plants (e.g. in Swedish hydro power plants) the frequency is measured and if it decreases the power production is increased until the frequency is stabilised. In this way, the short-term balance of the system is maintained.

Wind power plants do not contribute to primary control since they cannot keep any margins in their production units, which then could be used if frequency decreased. Wind power units normally produce at the maximum possible. That means that wind power generation increases the need of primary control reserve margins since additional variations are introduced into the system. There are two types of variations that are essential.

The first type is when the wind speed varies between cut-in wind speed and rated wind speed, v_R, as shown in Figure 9.4. Wind speed variations from second to second in different wind power plants can, on this time scale, be considered as independent variables. In Söder (1992), I provided an example with 3530 MW of wind power in a system with a total installed capacity of 15 589 MW (i.e. wind power corresponds to 23 % of installed capacity, and 15 % of yearly energy). This amount of wind power requires additional primary control reserve margins of 10 MW. The amount is so small because the load variations already require a rather large reserve (250 MW), and the 10 MW refers to the *increase* in these requirements.

The second type of variation refers to a situation where a storm front approaches a coast and may cause a sudden decrease in wind power production. During such a storm front, the wind speed may exceed the cut-out wind speed (25 m/s in Figure 9.4) in many wind turbines at the same time. However, the probability of that the amount of suddenly

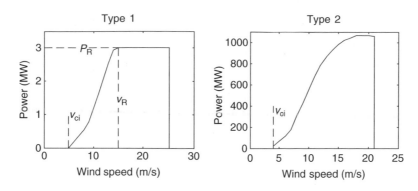

Figure 9.4 Wind speed power transfer function for a pitch-controlled wind power plant

stopped wind capacity reaches 1000 MW (equivalent to the production decrease if a large nuclear power station is stopped) is so low that the required amount of reserves will not have to be increased. The reason is that 1000 MW of wind power in reality means 500×2 MW, for instance, and the probability that these 500 units will be hit by the same storm front within 1 minute is extremely low. Table 9.1 shows examples of maximum wind gradients in two grids, E.ON Netz and Eltra.

The final conclusion is that the primary control value of wind power is negative, but the size of this value is rather low.

9.3.3.2 Secondary control

Secondary control refers to the capability of the power system to keep the balance between production and consumption during the time following primary control and up to one hour. A consequence of primary control is that the frequency is no longer 50 Hz and that reserve margins are used. This is handled by increasing the production in some units in the case of a frequency decrease. This means that some units have to be used as secondary reserve plants that can be started if the frequency is too low.

Wind power does not contribute to secondary control margins but instead requires extra secondary reserve margins, as it introduces additional variations in the period of up to one hour. That means that the secondary control value of wind power is negative. In Söder (1988), I provided an example with 3530 MW of installed wind power. The

Table 9.1 Comparison of maximum wind power gradients

	E.ON Netz		Eltra	
	For 4000 MW	Per 1000 MW	For 2300 MW	Per 1000 MW
Increasing	250 MW per 15 min	4.1 MW per 1 min	8 MW per 1 min	3.5 MW per 1 min
Decreasing	400 MW per 15 min	6.6 MW per 1 min	15 MW per 1 min	6.5 MW per 1 min

Source: compiled by T. Ackermann, based on data published by Eltra and E.ON Netz.

example uses the the same system as in the example of primary control (i.e. a system with a total installed capacity of 15 589 MW, with wind power corresponding to 23 % of installed capacity and 15 % of yearly energy). With this amount of wind power, secondary control reserves require an additional 230 MW, which is more significant than the increase in primary reserve requirements. The reason for the increase is, on the one hand, that wind speed can vary more within one hour than within 10 minutes, and, on the other hand, that wind speed variations at different sites are correlated to a larger extent.

9.3.3.3 Daily and weekly control

During a period of 24 hours, the control requirements are large, because power consumption is higher during days than during nights. Some wind power locations have higher wind speeds during the day than during the night. If this is the case, wind power contributes to the necessary daily control in the system. However, sometimes there may be higher winds during the night than during the daytime. If this is the case the daily control value is negative.

With increased daily (and weekly) variations in the power system, the other production units have to be controlled to a larger extent. Additional control in these units may decrease the efficiency of the controlled units. This issue is studied more thoroughly in Söder (1994). The simulation results show that, in Sweden, wind power plants of about 2–2.5 TWh per year do not affect the efficiency of the Swedish hydro system. At wind power levels of about 4–5 TWh per year, the installed amount of wind power has to be increased by about 1 % to compensate for the decreased efficiency in the hydro system. At wind power levels of about 6.5–7.5 TWh per year, the necessary compensation is probably about 1.2 %, but this figure has to be verified with more extended simulations. Larger amounts of wind power were not studied in this report. In that study, I found that the main reason for the decreased efficiency was that wind power variations caused larger variation in the hydro system, and, at very high production, the efficiency of hydro power plants (kWh per m^3 of water) in the studied hydro system decreased by several percent, compared with best efficiency operation point. However, correct conclusions can be drawn only if a large number of possible scenarios, including cases with decreased efficiencies (high load, low wind, high hydro) and increased efficiencies (high load, high wind, lowered hydro) are compared with the nonwind power system.

The consequence is that the daily (and weekly) control value of wind power is negative, but this value is not large, at least not for minor amounts of wind power.

9.3.3.4 Seasonal control

Seasonal control is needed in a power system where there are hydro reservoirs that can store water from one part of the year to another part. In this type of system, such as in Sweden and Norway, it is important to carry out this seasonal planning and scheduling in an economically efficient way. The uncertainty in the inflow forecasts has to be taken into account for this.

In Sweden, the load is higher in the winter than in the summer. Variations in wind power production follow the same pattern (i.e. production during the winter is higher than in the summer). This control value is therefore positive, in contrast to the other wind power control values. This value can be included in the operating cost value, unless the changes that wind power production causes to seasonal planning are not included in the method for determining this value.

9.3.3.5 Multiyear control

Multiyear control is needed in power systems with a large share of hydro power since the inflow is different during different years. This implies that water has to be stored between different years. However, the future inflow is uncertain, which means that it is not trivial to decide how much water should be stored, and having knowledge regarding this uncertainty is desirable. Especially, during dry years there is a certain risk that there will not be enough energy available to cover the load. At that point, the question is whether wind power is able to produce energy during years with low inflow. Here, it is important to look at the correlation between river inflow and wind energy production, as in Söder (1997, 1999b); in which I included an example that shows that the correlation between three different Swedish wind power sites and hydro inflow were 0.14, 0.15 and 0.44, respectively. This means that there is a slight tendency to lower winds during dry years compared with the multiyear mean value. This implies that wind power has a slightly negative impact on the requirement for multiyear control compared with thermal power stations, for example, which have the same potential during all years. But new wind power is doing slightly better than new hydro power, since the correlation between new hydro and old hydro power is much higher than the correlation between new wind power and old hydro power. The reason is that new hydro production will be located close to old hydro generation (high correlation in inflow), whereas new wind power generation will not necessarily be located in the same region as hydro production, and there is a limited correlation between wind in one region and inflow in another. Section 9.4.3 gives an example of the market value of multiyear variations in wind power.

9.3.4 The loss reduction value of wind power

The loss reduction value concerns the capability of wind power to reduce grid losses in the system. Wind power plants are often located rather far out in the distribution grid on a comparatively low voltage level. This is different from large power stations, which are located close to the transmission network. If the load close to the wind power plants is of the same size as the wind power plant, the power does not have to be transported far, which means that the losses are reduced as a result of wind power generation. In that case, the loss reduction value of wind power is positive. However, it could also happen that losses increase if there is no local load close to the wind power generation, which means that the power produced in the wind power plant has to be transported to the load centres on lines with comparatively high losses. If this is the case, the loss reduction value is negative.

An important issue here is that the *marginal losses* are changed. The losses on a power line can be calculated with use of the following equation:

$$P_{\mathrm{L}} = 3RI^2 = R\frac{P^2 + Q^2}{U^2} = R\left(\frac{1 + \tan^2 \varphi}{U^2}\right)P^2 = kP^2, \tag{9.1}$$

where

P_{L} = power losses on the line;
R = line resistance;
I = current on the line;
P = total transmitted active power on the line;
Q = total transmitted reactive power on the line;
U = voltage;
φ = phase shift between line voltage and current;

and

$$k = \frac{R(1 + \tan^2 \varphi)}{U^2}.$$

Assuming the approximation that U and φ are constant, the losses are proportional to the square of transmitted power. Assuming that a line has a transmitted power P_0, this results, according to Equation (9.1), in the losses (see also Figure 9.5):

$$P_{\mathrm{L0}} = kP_0^2. \tag{9.2}$$

For this transmission, the losses, as a percentage, can be calculated as follows:

$$P_{\mathrm{L0}}(\%) = 100\frac{P_{\mathrm{L0}}}{P_0} = 100kP_0. \tag{9.3}$$

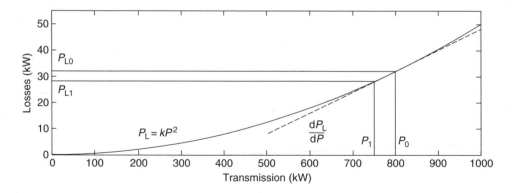

Figure 9.5 Impact of marginal losses

If a local power source now causes the transmission to decrease to P_1, losses decrease to P_{L1}. If the change in transmission is small, the new loss level can be calculated from a linearisation of the losses according to Figure 9.5:

$$P_{L1} \approx P_{L0} - \frac{dP_L}{dP}(P_0 - P_1) = P_{L0} - 2kP_0(P_0 - P_1). \tag{9.4}$$

The local production ΔP, equal to $P_0 - P_1$, thereby causes a reduction in the losses of ΔP_L, equal to $P_{L0} - P_{L1}$. The loss reduction as a percentage becomes

$$\Delta P_L(\%) = 100 \frac{\Delta P_L}{\Delta P} = 100 \frac{P_{L0} - P_{L1}}{P_0 - P_1}$$
$$\approx 100 \times 2kP_{L0} = 2P_{L0}(\%). \tag{9.5}$$

This means that the marginal losses, and thereby the marginal loss reduction, are twice as large as the mean losses. Whether wind power decreases or increases the losses in the system depends on the location of the wind power, the location of the load and the location of the power source that is replaced by the wind power. It is important to take into account the losses in the whole grid, and not simply a part of it.

In Söder (1999a) I studied an example of the loss reduction caused by 90 GWh of wind power on the Swedish island of Gotland. Table 9.2 summarises the results, comparing a wind farm with the alternatives where the same amount of yearly energy is produced at locations in Karlshamn and Gardikforsen. At both locations, the wind farms are connected directly to the transmission grid.

Table 9.2 shows that the wind power production on Gotland is the alternative that reduces the losses among the three studied alternatives the most. The difference compared with Karlshamn – 2.4 GWh (i.e. 4.6 GWh – 2.2 GWh) – is 2.7 % of the production of 90 GWh. The difference compared with Gardikforsen – 10.8 GWh [i.e. 4.6 GWh –(– 6.2 GWh)] – corresponds to 12 % of production. Even though wind power on Gotland increases the losses in the local grid on Gotland, this is compensated by decreased losses in the regional and transmission grids. The hydro power station at Gardikforsen is located in the northern part of Sweden, and the other sites are in the

Table 9.2 Loss reduction value of 90 GWh production at three different sites (Gotland, Karlshamn and Gardikforsen)

Grid	Loss (GWh)		
	Wind power, Gotland	Oil unit, Karlshamn	Hydro power, Gardikforsen
Local grid	−2.8	—	—
Regional grid	4.6	—	—
Transmission grid	2.8	2.2	−6.2
Total	4.6	2.2	−6.2

— Not applicable.

southern part. In Sweden, the dominating direction of power flow is from north to south. Therefore, Gardikforsen causes comparatively large losses.

9.3.5 The grid investment value of wind power

The grid investment value refers to the capability of wind power to decrease the need of grid investments in the power system. In order to be able to study this value it is first necessary to study how the grid is dimensioned without wind power and then to look at how the dimensioning changes if wind power is introduced. It is rather difficult to draw any general conclusions concerning this item. An important issue are the costs that are part of the required grid investments but that are also included in the cost of the wind farm. An example would be a medium-sized offshore wind farm with the costs for the transmission line from the wind farm to the transmission system being included in the total cost of the wind farm. The grid investment value is then probably comparatively small, if no other grid investments are needed.

I analysed this question in a more general way in Söder (1997), and an example of the market grid investment value of wind power is provided in Section 9.4.5.

9.4 The Market Value of Wind Power

In Sweden, new rules for the electricity market were introduced 1 January 1996. Some important changes included the following:

- All networks at all levels had to be opened;
- All network services had to be unbundled from generation and electricity sales;
- A transmission system operator (TSO) is responsible for the *short-term* power balance.

If a production source has a certain value, it is reasonable that the market be organised in a way that the owner of the production source is paid according to this value. This point will be discussed below in relation to the different values outlined in Section 9.3 and the deregulated market that has been introduced in Sweden, where wind power production can receive support from the state (e.g. through tax reductions and investment support). Currently, in Sweden there is also a system of certificates that supports owners of wind power plants. These issues will not be treated in detail, though.

9.4.1 The market operation cost value of wind power

In the Nordic market, power can be sold in three different ways:

- Sold bilaterally to consumers: if wind power is to be sold, this means in practice that someone has to handle the balance between wind power production and consumption; that is, the wind power control value has to be considered and a trader has to be involved.
- Sold bilaterally to traders: this means that the trader buys wind power and then sells this power, together with other production, directly to consumers and traders and/or

by putting bids on the exchanges. The prices at the exchange Nord Pool actually set the power price in the system.[1]

- Selling bids placed at the Nord Pool exchange or Elbas exchange: first, we have to mention the price level. The price level is set by bids at both these exchanges. The resulting prices are normally the prices defined by the units with the highest operating costs. This means that the selling companies are paid the avoided cost of the most expensive units. However, for wind power, the design of the exchange makes it difficult to bid. The bids at Nord Pool have to be put in 12–36 hours in advance of the physical delivery. This makes it almost impossible for wind power production to put in bids, since the forecasts normally are too inaccurate for this time horizon. At Elbas, the bids have to be put in at least 2 hours prior to physical delivery, therefore it is also of little interest for wind power production to participate. At Nord Pool, the prices are set at the price intersection of supply and demand. At Elbas, in contrast, the bids are accepted as they are put in (i.e. there is no price intersection). These exchanges will probably have to be changed if there is a future increase in the amount of wind power produced.

The conclusion is that, in the current deregulated market, it may be difficult for wind power to capture the income from the avoided costs of the most expensive units, as wind power generation has, in comparison with other power sources, rather limited options of actually trading in this market. From a physical point of view, the power system does not need 12–36 hours of forecasts. This structure reduces the value of wind power, or forces wind power generation to cooperate with competitors that are aware of the reduced value of wind power in the current structure.

9.4.2 The market capacity credit of wind power

Historically, capacity credit has not been an important issue in Sweden since the system LOLP has been very low. This was because the companies had to keep reserves for dry years. These reserves actually had become so large that there were huge margins for peak load situations. With deregulation, this has changed. Now the amount of reserves has decreased, resulting in an increased interest in capacity credit.

In Sweden, new regulations were introduced on 1 November 1999 in order to handle the risk of a capacity deficit in the system. Currently (2004), these regulations imply that the TSO Svenska Kraftnät can increase the regulating market price to €670 per MWh (about 20 times the normal price) in case there is a risk of a power deficit, and to €2200 per MWh if there is an actual power deficit. This method of handling the capacity problem means that wind power generation may receive very high prices during situations where there is a risk of a capacity deficit. The reason is that traders that expect customer demand to exceed the amount the traders have purchased have to pay €2200 per MWh for the deficit. This means that they are willing to pay up to €2200 per MWh

---

[1] See Nord Pool, http://www.nordpool.no/.

in order to avoid this cost. However, there are still problems related to the design of the exchange and the required forecasts as discussed in Section 9.4.1.

9.4.3 The market control value of wind power

In Sweden, primary and secondary control are handled through the so-called *balance responsibility*. The Electricity Act states that:

> An electricity retailer may only deliver electricity at points of connection where somebody is economically responsible for ensuring that the same amount of electricity that is taken out at the point of connection is supplied to the national power system (balance responsibility).

This implies that someone has to take on the balance responsibility for each point with a wind power plant – and the wind power owner has to pay for this. This means that problems related to primary and secondary control are included in the market structure. I am not aware of any detailed evaluation concerning whether these costs reflect the 'true costs' in the system. However, the TSO Svenska Kraftnät has to purchase any additionally required dispatch from the market, which means that, at least, the suppliers will be paid.

There may be problems regarding the competition in the so-called *regulating market*, which corresponds to secondary control. A power system normally does not have so much 'extra' reserve power, since the power system was dimensioned with 'sufficient' margins (i.e. very close to normal demand). The owners of such regulating resources tend to have little competition in cases when nearly all resources are needed. This means that they can bid much higher than marginal cost, since the buyer of the reserves, normally the TSO, has to buy, no matter what the price is. In a power system with larger amounts of wind power, the requirement and use of reserve power will increase. This means that, in a deregulated framework, there is a risk that profit maximisation in companies with a significant share of available regulating power can drive up the price for the regulation. This means that the price for regulating power that is needed to balance wind power can increase, which means that the 'integration cost' of wind power seems to be higher than the actual physical integration cost.

However, if competition is working, the daily control value will be included in the market, as the secondary control is used to handle the uncertainties. Again, the time frame required by the bidding process is a problem (see Section 9.4.1).

The seasonal control value is included in the market since the prices change over the year, corresponding mainly to a changed the hydro power situation.

9.4.3.1 Market multiyear control value

In this section, the impact of multiyear control on the value of wind power will be illustrated with use of the example of the Nordic market, with its significant amount of wind power. The prices in a hydro-dominated system will here be analysed in an approximate way. The aim is to estimate an economic value of the yearly energy production including the variations between different years.

As in every market, the prices on the power market decrease when there is a lot of power offered to the market. This implies that during wet years the prices in the Nordic market are significantly lower than during dry years. During a dry year, the prices are much higher as a large amount of thermal power stations have to be started up in order to compensate for the lack of water.

This means that the yearly mean price, P_{year}, can be approximated and described as a function of the inflow x (calculated as yearly energy production). In order to be able to analyse this from system statistics, the load changes between different years have to be taken into account. The price does not depend directly on the hydro inflow but rather on the amount of thermal power that is needed to supply the part of the load that is not covered by hydro power. Figure 9.6 shows the yearly mean power price in Sweden as a function of the required amount of thermal power (load – inflow). In the figure, the inflow is calculated as the sum of real hydro production and changes in reservoir contents. The x-axis does not show the total thermal production, since the inflow is used and not the hydropower production. The reason is as follows: the total reservoir capacity is comparatively large (120 TWh) and this means that the stored amount of water at the end of the year also influences the prices during the end of the year. If hydro power production is exactly the same during two years, but during one these years the content of the reservoir is rather small at the end of the year, the power price will be higher during this year. It simply has to be higher, otherwise the water would be saved for the next year.

A least square fit of the points in Figure 9.6 leads to:

$$P_{year} = a_1 + b_1 y, \tag{9.6}$$

where $a_1 = -0.0165$ and $b_1 = 0.000251$.

In the text below, it is easier to express the power price as a function of the inflow x. For a given load level, D, the inflow can be calculated as

$$x = D - y. \tag{9.7}$$

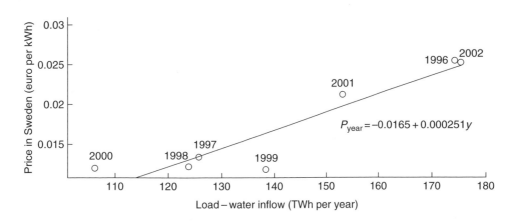

Figure 9.6 Yearly mean price as a function of total thermal production (for Sweden, Norway and Finland)

Equation (9.6) can now be rewritten as

$$P_{\text{year}} = a_1 + b_1(D - x) = (a_1 + b_1 D) - b_1 x = A + Bx, \tag{9.8}$$

where

$$
\begin{aligned}
A &= a_1 + b_1 D, \\
B &= -b_1.
\end{aligned}
\tag{9.9}
$$

Figure 9.7 shows the price as a function of the inflow, $P_{\text{year}}(x)$. The parameters in Equation (9.9) are then evaluated for a case with load figures from the year 2002 (i.e. $D = 353$ TWh).

The inflow varies between different years and can be approximated with a Gaussian distribution with a mean value m_x and a standard deviation σ_x as shown in Figure 9.7. Figure 9.7 shows the real case of Sweden, Norway and Finland with a normal inflow of $m_x = 195$ TWh. The applied standard deviation refers to the period 1985–2002, and the result is $\sigma_x = 19.3$ TWh. With the assumed linear relation between inflow and price, the expected price, $E(P_{\text{year}})$, and the standard deviation of the price, $\sigma(P_{\text{year}})$, can be calculated as follows:

$$
\begin{aligned}
E(P_{\text{year}}) &= E(A + Bx) = A + Bm_x, \\
\sigma(P_{\text{year}}) &= \sigma(A + Bx) = |B|\sigma_x.
\end{aligned}
\tag{9.10}
$$

The Gaussian distribution of the price is also shown in Figure 9.4. In this case, $E(P_{\text{year}})$ is €0.023 per kWh and the standard deviation is €0.0048 per kWh. The total value of the inflow V_x can be calculated as follows:

$$V_x = P_{\text{year}} x = Ax + Bx^2. \tag{9.11}$$

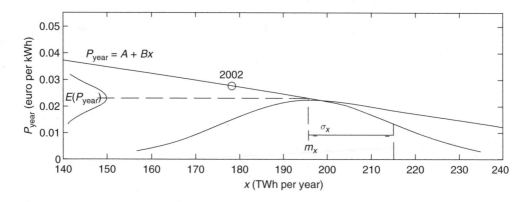

Figure 9.7 Yearly mean price as a function of inflow

The expected value can therefore be calculated as:

$$
\begin{aligned}
E(V_x) &= E(Ax + Bx^2) \\
&= Am_x + BE(x^2) \\
&= Am_x + B(\sigma_x^2 + m_x^2) \\
&= m_x(A + Bm_x) + B\sigma_x^2 < m_x E(P_{\text{year}}),
\end{aligned}
\qquad (9.12)
$$

since $B < 0$. The formula shows that the expected value of the inflow is lower than the mean inflow multiplied by the mean price. This is owing to the fact that high inflows give lower prices for large volumes, and this is not compensated by the higher prices that are paid during lower inflows.

9.4.3.2 The value of a new power source

We now assume that a new power source is installed and produces m_y TWh during a normal year. It is also assumed that the new power source supplies an increased load (or replaces old thermal plants) at the same yearly level as the mean production of the new source (i.e. m_y TWh). This implies that during a normal year, the price level will be the same, independent of the new source and the increased load (see Figure 9.8). This means that the new load and new power source will not affect the mean price level in the system. With this assumption, the price function in Equation (9.8) has to be rewritten as

$$
P'_{\text{year}} = A' + B(x + y). \qquad (9.13)
$$

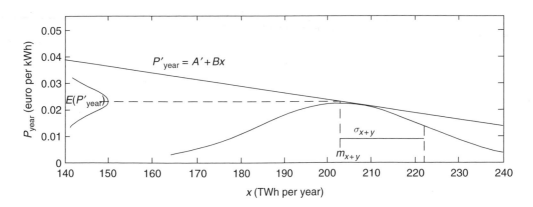

Figure   9.8 Yearly mean price as a function of inflow plus production in the new energy source with a new load

Only the A parameter is assumed to change, to A'. This implies that a decreased production from hydro and new energy source of 7 TWh, for example, is assumed to have the same effect on the price as reducing hydro production itself to a level of 7 TWh. With these assumptions, the mean price should still be the same; that is:

$$E(P'_{year}) = A' + B(x + y) = E(P_{year}),$$

which implies

$$A' = A - Bm_y. \tag{9.14}$$

We now assume that the new power source is renewable with a low marginal cost. This means that the yearly production is related to the amount of available primary energy: sunshine, water, waves or wind. It is assumed here that the new renewable power source cannot be stored from one year to the next. This implies that the value of existing hydro power x plus the new source y is given by:

$$V_{x+y} = A'(x + y) + B(x + y)^2. \tag{9.15}$$

The new renewable power source has a mean value of m_y, and a standard deviation of σ_y. There might also be a correlation between the yearly production of the new renewable source y and the existing hydro power x. The correlation coefficient is denoted ρ_{xy}. The expected value for hydro plus the new power source becomes:

$$
\begin{aligned}
E(V_{x+y}) =& E\left[A'(x + y) + B(x + y)^2\right] \\
=& A'(m_x + m_y) + B(\sigma_x^2 + m_x^2) + B(\sigma_y^2 + m_y^2 + 2\rho_{xy}\sigma_x\sigma_y + 2m_xm_y).
\end{aligned}
\tag{9.16}
$$

The expected incremental value of the new renewable source $E(V_y)$ can be estimated as the difference between the expected value of hydro plus new renewable source [see Equation (9.16)] y, and the expected value of only hydro power [see Equation (9.12)]. By applying Equation (9.14), this value can be written as

$$
\begin{aligned}
E(V_y) &= E(V_{x+y}) - E(V_x) \\
&= (A + Bm_x)m_y + B(\sigma_y^2 + \rho_{xy}\sigma_x\sigma_y).
\end{aligned}
\tag{9.17}
$$

We start by assuming the simple case where we have a constant new power source with $\sigma_y = 0$. The value for this source is

$$E[V_y(\sigma_y = 0)] = (A + Bm_x)m_y = E(P_{year})m_y, \tag{9.18}$$

that is, the produced energy is sold (as a mean value) to the mean price in the system.

The next example includes a variation between different years, but this variation is uncorrelated with the variation in the existing hydropower ($\rho_{xy} = 0$). Under these circumstances, the value is given by

$$E[V_y(\rho_{xy} = 0)] = E(P_{year})m_y + B\sigma_y^2, \tag{9.19}$$

which is equal to that of the hydro power in Equation (9.12). The lowered value ($B < 0$) depends on the fact that hydro together with the new power source, will vary more than only hydro itself, and larger variations lower the value [see Equation (9.12)].

With a positive correlation, $\rho_{xy} > 0$, between the new source and existing hydro power the value is even lower [see Equation (9.17)], where $B < 0$. This is owing to the fact that a positive correlation implies that the new power source produces more during wet years (with lower prices) than during dry years (with higher prices). If, however, the correlation is negative the value is higher.

9.4.3.3 Numerical examples

Here, the method presented above will be applied to three examples: new thermal power plants, new hydropower and new wind power. In all the three cases it is assumed that the expected yearly production is 7 TWh and that the load also increases by 7 TWh per year. The system studied is the Nordic system (i.e. Sweden, Norway and Finland the see data in Figure 9.7). This means that during a hydrologically normal year, the expected power price is $E(P_{year}) = A + Bm_x = €0.0230$ per kWh.

It is assumed here that a new thermal power plant will produce approximately the same amount of power per year, independent of hydropower inflow. This means that the operating cost of the plant is assumed to be rather low, because the amount of energy production per year is assumed to be the same during comparatively wet years, when the power price is lower. With these assumptions, the expected value of this power plant becomes, from Equation (9.18),

$$V(\text{thermal}) = E(P_{year})m_y.$$

The value is

$$\frac{V(\text{thermal})}{m_y} = E(P_{year}) = €0.0230 \text{ per kWh}.$$

Now, assume that new hydropower will be installed in Sweden. We assume that the variations of new hydro power in Sweden will follow the same pattern as old hydro power. This means that the standard deviation of 7 TWh of hydropower will be 0.895 TWh based on real inflows for the period 1985–2002. The inflows are not exactly the same for the Swedish and the Nordic system. The correlation coefficient is 0.890. With these assumptions, the expected value of hydropower becomes, from Equation (9.17),

$$V(\text{hydro}) = E(P_{year})m_y + B(\sigma_y^2 + \rho_{xy}\sigma_x\sigma_y)$$
$$= 0.023 \times 7 - 0.000251(0.895^2 + 0.890 \times 19.3 \times 0.895) = 0.1573.$$

The value is

$$\frac{V(\text{hydro})}{m_y} = €0.0225 \text{ per kWh}.$$

That is, the value of new hydro power is around 2.5% lower than the value of thermal power since more of the power production will be available during years with lower prices.

The amount of wind power also varies between different years. For Sweden, available hydropower and available wind power are correlated, as years with plenty of rain are often also years with higher winds. The correlation depends on where the wind power generation is located. In Sweden, hydropower is generated in the northern part of the country. If the wind turbines are located a long distance from the hydro power stations, the correlation is relatively low. In our example, the wind power is assumed to be spread out over different locations. The standard deviation for total wind power is, for this case, 0.7519 TWh (i.e. slightly lower than for hydro power). For the period 1961–1989, the correlation between installed wind power and Swedish hydro power is 0.4022. It is assumed here that the correlation between Swedish wind power and Nordic hydro power is $0.4022 \times 0.890 = 0.358$. With these assumptions, the expected value of wind power, from Equation (9.17), is

$$V(\text{wind}) = E(P_{\text{year}})m_y + B(\sigma_y^2 + \rho_{xy}\sigma_x\sigma_y)$$
$$= 0.023 \times 7 - 0.000251(0.7519^2 + 0.358 \times 19.3 \times 0.7519) = 0.1599.$$

The value is:

$$\frac{V(\text{wind})}{m_y} = €0.0228 \text{ per kWh}.$$

That is, the value of new wind power is around 1% lower than the value of thermal power, but it is 2% higher than the value for hydro power. This is owing to the fact that wind power has a slightly negative correlation to the power price, as during years with higher wind speeds prices are often lower.

9.4.4 The market loss reduction value of wind power

In Swedish deregulation, the grid companies levy tariff charges that include an energy part, which is related to the losses. We will now take the example used in Section 9.3.4, that is of 90 GWh of wind power on the Swedish island of Gotland, from the power market view (for more details, see Söder, 1999a).

In Sweden, the owners of wind power plants do not have to pay for increased losses in the local grid, Gotland in this case. However, concerning the supplying grid, a local grid owner has to pay the owners of wind power plants for the reduced tariff charges. Since less power is transmitted from the mainland, these charges, which include costs for losses, are reduced. In Table 9.3, the impact on the wind power owner is summarised.

On the one hand, can be concluded that, in Sweden, the local grid owner is not paid for the extra losses that wind power generation causes in the local grid. On the other hand, the regional grid owner does not pay for the actual loss reduction caused by wind power, since the tariffs are based on the mean cost of losses in the whole regional grid and not on marginal losses of the specific costumer. The result is that wind power

Table 9.3 Market loss reduction value for 90 GWh wind power in Gotland

Grid	True losses (GWh)	Market treatment for wind power owner	
		loss (GWh)	comment
Local grid	−2.8	0	Grid-owner pays
Regional grid	4.6	} } 3.7 }	Regional grid tariffs include transmission grid losses, and these tariffs are based on mean, not marginal, losses
Transmission grid	2.8		
Total	4.6	3.7	—

owners are paid less than the actual loss reduction. The ones that benefit from the system are other consumers in the regional grid, but the consumers in the local grid will have to pay higher tariffs because of the stipulation that wind power owners do not pay for increased local losses.

9.4.5 The market grid investment value of wind power

In Swedish deregulation, the grid companies are paid with tariff charges. It is important to ask whether these charges will be changed with the introduction of wind power. We will use a numerical example to illustrate this issue (for more details regarding this example, see Söder, 1999a). In this example we analyse how the power charges in the tariff for the supply of the Swedish island of Gotland could be reduced, depending on wind power generation on Gotland. The Gotland grid company pays this charge to the regional grid company on the mainland. The tariff of the regional grid includes the charges that the regional grid has to pay to the national transmission grid, owned by the TSO Svenska Kraftnät.

The basis for these calculations is therefore the tariff of the regional grid company, Vattenfall Regionnät AB, concerning southern Sweden. The aim of this example is mainly to illustrate how the calculations can be performed. The data are from 1997. Table 9.4 shows the tariff for the mainland node where the island of Gotland is connected. In Table 9.4:

- yearly power is defined as the mean value of the two highest monthly demand values;
- high load power is defined as the mean value of the two highest monthly demand values during high load. High load is during weekdays between 06:00 and 22:00 hours from January to March and November to December.

In relation to these two power levels, a subscription has to be signed in advance. If real 'yearly power' is higher than the subscribed value, an additional rate of 100 % becomes

Table 9.4 Subscription-based charges at the connection point of Gotland

	Charge (€/kW)
Yearly power charge	3.4
High load power charge	8

payable for the excess power (i.e. a total of €6.9 per kW). If real 'high load power' exceeds the subscribed value, an additional 50 % must be paid for the excess power (i.e. €12 per kW).

In addition, a monthly demand value can be calculated, which is the highest hourly mean value during the month.

As just mentioned, the subscribed values have to be defined in advance. The real power levels vary, of course, between different years. In the following, we will describe how the optimal subscribed values are estimated and give an example of how this level changes when wind power is considered.

Before a year starts the subscription levels for both the 'yearly power' and the 'high load power' have to be determined. We assume that both weeks with the dimensioning load have a weekly variation according to Figure 9.9. Figure 9.9 shows only one week but it is assumed that the other week in the other month has the same variation, but a different weekly mean level.

The load curve we use refers to a week from Monday 01:00 to Sunday 24:00 hours. The load is upscaled, and peak load during week 1 is 130 MW and during week 2 reaches 123.5 MW (i.e. it is 5 % lower). The following calculations are based on the assumption that these two weeks contain the dimensioning hours for the two dimensioning months. The peak load during both weeks is therefore on Thursday between 8:00 and 9:00 hours. These two hours fall into the high load period. Therefore, the dimensioning hours for the 'yearly power' and for the 'high load power' coincide. The mean value for these two occasions is 126.75 MW [i.e. $0.5 \times (130 + 123.5)$].

Statistics for several years have not been available for this study. Instead, it has been assumed that the mean value of the two highest monthly values varies by ±4 %

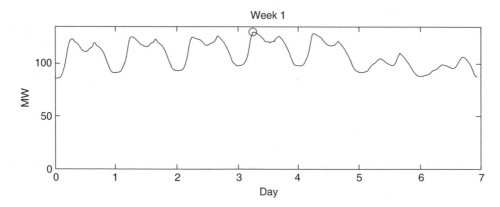

Figure 9.9 Hourly consumption during one week

(4 % standard deviation). The standard deviation implies that the level of 126.75 MW varies by ±4 % between different years. The aim is now to select the subscription levels in order to obtain a minimal expected power charge. If the subscribed level is too high then too much will always be paid. However, selection of a level that is too low will be too expensive, as there will be a need to pay extra for excess capacity. The optimal level can be obtained in the following way.

Assume that the real measured power, which defines the charge, varies between different years with a mean value m and a standard deviation σ. Assume also that the subscription level has been selected to be a (see Figure 9.10).

This implies that in a year with a measured level x that is lower than or equal to the subscribed level (i.e. $x \leq a$) the cost is $C(x) = T_1 a$, where T_1 is the cost for the subscription per kilowatt per year. During a year when the measured power exceeds the subscribed level (i.e. $x > a$), the cost becomes $C(x) = T_1 a + T_2(x - a)$ instead, where T_2 is the price for excess power per kilowatt per year. The expected cost E_C that weighs the probabilities for different years can be calculated as:

$$
\begin{aligned}
E_C &= \int_{-\infty}^{\infty} C(x)f(x)\mathrm{d}x \\
&= \int_{-\infty}^{a} T_1 a f(x)\mathrm{d}x + \int_{a}^{\infty} [T_1 a + T_2(x - a)]f(x)\mathrm{d}x \\
&= T_1 a \int_{-\infty}^{\infty} f(x)\mathrm{d}x + T_2 \int_{a}^{\infty} xf(x)\mathrm{d}x - T_2 a \int_{a}^{\infty} f(x)\mathrm{d}x \\
&= T_1 a + T_2 \int_{a}^{\infty} xf(x)\mathrm{d}x - T_2 a[1 - F(a)] \\
&= T_1 a + T_2 \{ m[1 - F(a)] + \sigma^2 f(a) \} - T_2 a[1 - F(a)] \\
&= (T_1 - T_2)a + T_2 m + T_2 F(a)(a - m) + T_2 \sigma^2 f(a),
\end{aligned}
\tag{9.20}
$$

where

$f(x)$ is the Gaussian probability function;
$F(x)$ is the Gaussian distribution function.

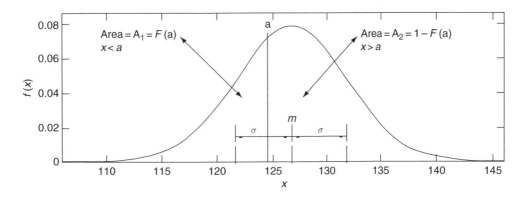

Figure **9.10** Gaussian approximation for the measured power, which defines the power charge

Figure 9.11 shows how the expected cost E_C depends on the selected subscription level.

As shown in Figure 9.11, a higher cost is obtained for too low or too high a subscription level.

The minimum level is obtained when:

$$\Delta a (T_2 - T_1) A_2 = \Delta a T_1 A_1.$$

This implies

$$(T_2 - T_1)[1 - F(a)] = T_1 F(a),$$

which implies

$$F(a) = \frac{T_2 - T_1}{T_2}. \tag{9.21}$$

This solution can be explained in the following way:

1. Assume that the subscription level has been selected to be a. If this level is slightly increased to $a + \Delta a$, the cost for the years with a dimensioning power that is lower than a will increase by $\Delta a T_1 A_1$, where A_1 is the share of years when the dimensioning power is lower than a (see Figure 9.10).
2. During years with a dimensioning power that is higher than a, the cost will instead decrease by $\Delta a (T_2 - T_1) A_2$, where A_2 is the share of years when the dimensioning power is higher than a (see Figure 9.10).
3. The minimal cost is obtained when a small change of a does not affect the cost (i.e. when a cost increase at point 1 equals the cost decrease at point 2), which is the same as Equation (9.21).

Charges as in Table 9.4 make it possible to optimise subscription values. The result is shown in Table 9.5

We now assume that there is wind power on the island of Gotland. This will decrease the amount of power that is supplied from the mainland. The question is now whether there is a new optimal subscription level and if the total expected cost would be reduced.

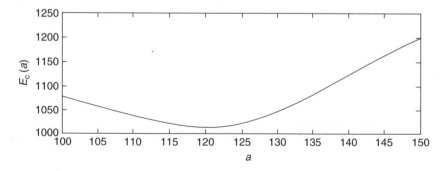

Figure   9.11 Expected cost E_C, thousands of euros, as a function of selected subscription level a

Table   9.5 Optimal subscription levels without wind power

	Tariff	
	Yearly power	High load power
Mean real measured power, m (MW)	126.75	126.75
Standard deviation, σ (MW)	5.1	5.1
Subscription cost per year, T_1 (€ per kW)	3.4	8
Cost of excess power per year, T_2 (€ per kW)	6.9	12
Minimum cost (millions of euros)	0.46	1.03
Subscription level, a (MW)	126.7	124.6

Figure 9.12 shows an example, where the load in Figure 9.10 is decreased by a certain amount of wind power.

As shown in Figure 9.12, wind power will cause the dimensioning power to decrease. In this specific case, the dimensioning power decreased by 4.3 MW, and the installed amount of wind power was 43 MW. At the same time, the dimensioning hour is moved within the week. The dimensioning hour could also be moved to another week with lower wind speeds. On studying more wind situations, we found that the maximum net load during one month was reduced by $m_W = 3.65$ MW, corresponding to 43 MW of wind power. However, the amount was different for different situations, and the standard deviation was 3.1 MW.

In addition to this, one has to consider that the wind speeds differ between different years. With 35 years of data from Visby airport on Gotland, it was found that there is an additional standard deviation of 2.8 MW that has to be considered in order to arrive at a correct estimation of how the maximum net load can vary in different situations.

The net level (power demand – wind power production) for the measured power levels, on which the charges are based, can now be described with a Gaussian distribution according to Figure 9.10. Assuming that there is no statistical dependence between

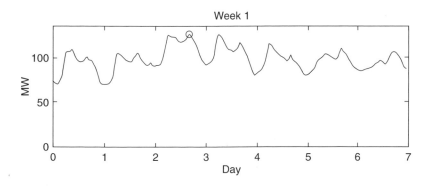

Figure   9.12 Net power consumption (actual load – wind power.)

the load and wind generation in January to February, the parameters for the Gaussian distribution can be calculated as:

$$m_{net} = m_{load} - m_w = 123.1\,MW,$$

$$\sigma_{net} = \sqrt{\sigma_{load}^2 + \sigma_{load-w}^2 + \sigma_{w-year}^2}$$

$$= \sqrt{5.07^2 + 3.1^2 + 2.8^2} = 6.6\,MW.$$

An optimal subscription level can now be obtained by using Equation (9.21), and the expected cost can be estimated by applying Equation (9.20). Table 9.6 presents the results.

A comparison of Tables 9.6 and 9.5 shows that the subscription levels as well as the expected costs decrease depending on the installed amount of wind power. The cost decreases by €0.04 million [(0.46 − 0.44) + (1.03 − 1.01)], which corresponds to (30 k€)/(43,000) = €0.70 per kilowatt per year [(€30 000)/(43 000 kW)] of installed wind power capacity, which lies in the range of 1 % of the total cost of wind power (i.e. it is a small value).

It is important to note that the cost reduction is based on an optimised subscription level in the tariff. It was not part of this chapter to analyse whether the charges that are reduced as a result of use of wind power also correspond to decreased costs in the mainland grid.

9.5 Conclusions

This chapter has shown that the value of wind power in a power system can be divided into the following parts:

- operating cost value;
- capacity credit;
- control value;
- loss reduction value;
- grid investment value.

Table   9.6 Optimal subscription levels for wind power

	Tariff	
	Yearly power	High load power
Mean real measured power, m (MW)	123.1	123.1
Standard deviation, σ (MW)	6.6	6.6
Subscription cost per year, T_1 (€ per kW)	3.4	8
Cost of excess power per year, T_2 (€ per kW)	6.9	12
Minimum cost (millions of euros)	0.44	1.01
Subscription level, a (MW)	123.1	120.3

In Section 9.3, it was shown how these values can be defined as 'true values' (i.e. the cost decrease in an ideal system). Section 9.4 illustrated that payments to wind power owners do not always correspond to these true values, at least not in the current deregulated market in Sweden.

References

[1] IEEE (Institute for Electrical and Electronic Engineers) (1978) 'Reliability Indices for Use in Bulk Power Supply Adequacy Evaluation', *IEEE Transactions on Power Apparatus and systems* **PAS-97** (4) 1097–1103.

[2] Kirby, B., Hirst, E., Parsons, B., Porter K., Cadogan, J. (1997) 'Electric Industry Restructuring, Ancillary Services and the Potential Impact on Wind', American Wind Energy Association, available at: http://www.ornl.gov/ORNL/BTC/Restructuring/pub.htm.

[3] Milligan, M., Porter, K., Parsons, B., Caldwell, J. (2002) 'Wind Energy and Power System Operations: A Survey of Current Research and Regulatory Actions', *The Electricity Journal* (March 2002) 56–67.

[4] Milligan, M. R. (2002) 'Modelling Utility-scale Wind Power Plants, Part 2: Capacity Credit', NREL Report TP-500-29701, available at http://www.nrel.gov/publications/.

[5] Munksgaard, J., Pedersen, M. R., Pedersen, J. R. (1995) 'Economic Value of Wind Power' Report 1, Amternes of Kommunernes Forskningsinstitut (AKF), Coperhagen (in Danish).

[6] Söder, L. (1997) 'Capacity Credit of Wind Power', Elforsk report 97:27, final report from Elforsk project 2061 (in Swedish); copy available from author.

[7] Söder, L. (1992) 'Reserve Margin Planning in a Wind–Hydro–Thermal Power System', 92 WM 168-5 PWRS, presented at IEEE/PES Winter Meeting, New York, 26–30 January 1992; published in *IEEE Transactions on Power Systems* **8**(2) 564–571.

[8] Söder, L. (1988) *Benefit Assessment of Wind Power in Hydro–Thermal Power Systems*, PhD thesis, Royal Institute of Technology, Stockholm.

[9] Söder, L. (1994) 'Integration Study of Small Amounts of Wind Power in the Power System', Electric Power Systems, Royal Institute of Technology, Stockholm.

[10] Söder, L. (1999a) 'The Value of Wind Power for an Owner of a Local Distribution Network', paper presented at 15th International Conference on Electricity Distribution, 1–4 June, 1999, Nice, France; copy available from author.

[11] Söder, L. (1999b) 'Wind Energy Impact on the Energy Reliability of a Hydro–Thermal Power System in a Deregulated Market', For *Proceedings of 13th Power Systems Computation Conference, June 28–July 2, 1999, Trondheim, Norway*; copy available from author.

[12] van Wijk, A. (1990) *Wind Energy and Electricity Production*, PhD thesis, Utrecht University, Utrecht, The Netherlands.

Part B

Power System Integration Experience

10

Wind Power in the Danish Power System

Peter Borre Eriksen and Carl Hilger

10.1 Introduction

The past decade has seen a massive expansion of wind power in Denmark (see Figure 10.1). This trend is the result of targeted political efforts to rely increasingly on renewable energy production by introducing extensive public subsidising schemes.

Denmark is divided into two power systems (Figure 10.1) with no electrical inter-connection:

- The power system in Eastern Denmark (Zealand), with Elkraft as the Transmission System Operator (TSO). This system is part of the Nordic synchronous power system (Nordel).
- The power system in Western Denmark (Jutland–Funen), with Eltra as the TSO. This system is part of the European synchronous power system, the Union for the Coordination of Transmission of Electricity [the Union pour la Coordination du Transport d'Electricité (UCTE)].

Figure 10.2 shows the wind power expansion that has taken place in Denmark since 1985, for both Western Denmark, consisting of Jutland–Funen (with Eltra as TSO) and Eastern Denmark, consisting of Zealand (with Elkraft as TSO). Because there are far better opportunities for placing wind power facilities in the Jutland–Funen area and there is a higher number of suitable sites this part of the country shows the by far largest increase in wind power. There is currently (in mid-2003) a total installed capacity of approximately 3000 MW in Denmark, of which a total of 2400 MW is located in the Eltra area.

Wind Power in Power Systems   Edited by T. Ackermann
© 2005 John Wiley & Sons, Ltd   ISBN: 0-470-85508-8 (HB)

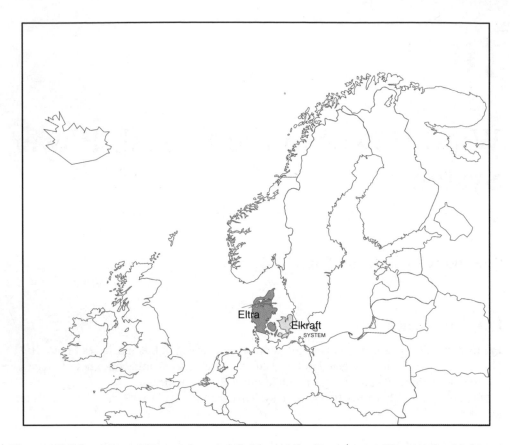

Figure   10.1 Location of Denmark and definition of the Eastern and Western Danish power systems

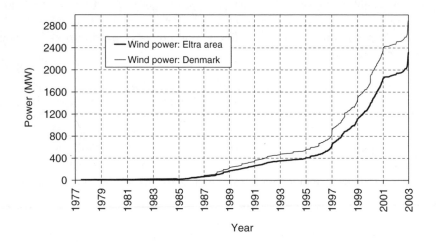

Figure   10.2 Wind power capacity expansion in Denmark

In comparison, in 2002, the maximum hourly consumption reached 2700 MW in the Elkraft area and 3700 MW in the Eltra area. As can be seen, the ratio between installed wind power and consumption is about three times higher in Western Denmark than in Eastern Denmark. In fact, the Eltra area has the largest amount of installed wind power in the world relative to the size of the power system.

As a result, primarily the Western Danish power system has to face the large technical challenges resulting from managing and operating a power system with a significant wind power penetration. This chapter therefore focuses on discussing wind power in the Eltra area. Table 10.1 shows power load and power production capacities in the Eltra area, and Table 10.2 shows the allocation of production in 2002.

The transmission grid in the Eltra area includes 712 km of 400 kV lines and 1739 km of 150 kV lines. The Eltra area is connected to Norway and Sweden by high-voltage direct-current (HVDC) links and to Germany by alternating current (AC) links (400 kV, 220 kV, 150 kV and 60 kV). The connection to the neighbouring countries has a total capacity of about 3000 MW (see Figure 10.3).

The local combined heat and power (CHP) plants (about 700 units) and the wind turbines are located in the distribution grid. The installed distributed generation (DG) capacity thus amounts to about 50 % (see Figure 10.4).

Table 10.1 Power load and production capacities, end of year 2002

	Power (MW)
Local combined heat and power	1596
Wind power	2315
Primary thermal (extraction) units	3107
Peak load	3685
Typical offpeak load, summer	1400
Typical offpeak load, winter	1900
Minimum load	1189

Table 10.2 Distribution of production in the Eltra area, 2002

	Energy (GWh)	Percentage of consumption
Primary (extraction) units	12 928	—
Local combined heat and power	6 723	32.2
Wind power	3 825	18.3
Total production	23 476	—
Foreign exchange	−2 619	—
Consumption	20 858	—

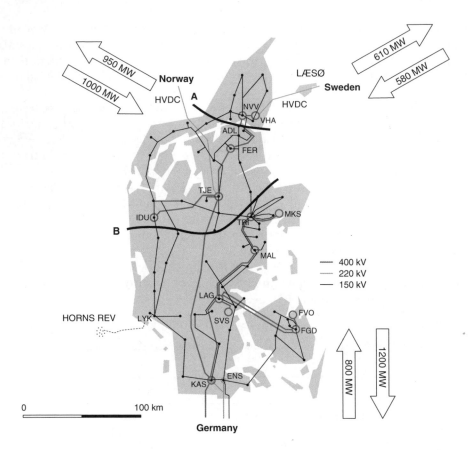

Figure   10.3 Eltra's transmission grid and capacities of foreign transmission links to Norway, Sweden and Germany

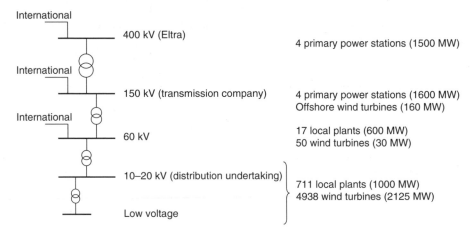

Figure   10.4 Production capacities at each voltage level within the Eltra area

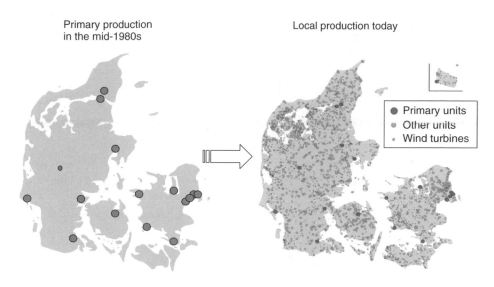

Figure   10.5 Basic changes in the Danish production system: from primary to local production

A variety of DG types are connected to the Eltra system. Some examples are:

- wind turbines, ranging from 11 kW to 2 MW;
- gas-fired CHP units, ranging from 7 kW to 99 MW;
- straw-fired CHP units, ranging from 2 MW to 19 MW;
- coal-fired units, ranging from 18 MW to 44 MW;
- waste-fired units, ranging from 90 kW to 26 MW.

The production system in Denmark has changed radically, as shown in Figure 10.5. The traditional system was based on power stations situated in coastal areas with cooling-water and coal-handling facilities. Now, the system has a great number of local CHP units and wind power plants.

10.2 Operational Issues

Figure 10.2 shows the penetration of wind capacity in the Eltra area. Until the 1990s, wind energy did not pose any major operational problem (i.e. it did not affect the balancing task). Already in the mid-1980s, though, the European Commission invited a number of member countries to participate in a study to investigate the effect of 5 %, 10 % and 15 % wind penetration on power system operation. The study for the Western part of Denmark (EC, 1986) was based on the assumption that the defined geographical area should always be able to balance demand and generation and not depend on neighbouring power systems. The capacity mix for the scenario with wind therefore included gas turbines for situations with no wind.

In the mid-1980s, there were no signs of a Scandinavian-wide electricity market or a European internal electricity market, even though the EU referendum had just been accepted.

Managing an integrated system with substantial wind energy production and, at the same time, taking into account the behaviour of the European wind regime requires a concept that will guarantee the balance of demand and supply in the case of unforeseeable wind speed variations.

The studies in the 1980s assumed that demand and supply would have to be balanced within the Danish system. Today, the TSOs in Europe cooperate in different ways (e.g. on real-time markets) in order to balance production and consumption in real time.

Historically, Nordel and UCTE have evolved differently regarding the balancing of production and demand.

The synchronous Nordic system (Norway, Sweden, Finland and the eastern part of Denmark) applies the following measures:

- In the case of disturbances, half of the primary regulation is to be activated within 5 s and the rest within 30 s. The primary regulation covers the loss of 1100 MW of production and a self-regulation of 100 MW of the load (i.e. the volume must be 1000 MW).
- Frequency regulation reserves of 6000 MW/Hz.
- Manually activated regulating capacity, up or down, within 10 to 15 min.

This balancing philosophy has proved efficient for many years. Recently, however, market activities with hourly schedules have demonstrated poorer frequency qualities at hourly changes, and the TSOs are trying to compensate. An obvious solution would be to switch the schedules to time steps of 15 min instead of one hour, as on the continent.

In the UCTE area (see www.ucte.org), the balancing philosophy is based on the following:

- Primary regulation in the case of disturbances is to be activated within 30 s. The primary regulation covers the loss of 3000 MW of production and a certain degree of self-regulation of the load. Based on the UCTE rules, the primary control capacity within the Eltra area is 35 MW.
- Frequency regulation reserves of 8000 to 15 000 MW/Hz.
- Network controllers covering a certain geographical area: the network controller integrates the total import–export to or from a particular area and controls the schedules agreed between the different TSOs.
- Manually activated regulating capacity up or down in severe situations.

Eltra uses the following facilities:

- Day-ahead market at Nord Pool for balancing wind power production based on the day-ahead forecast.
- Regulating power market (the real-time market) for balancing within the hour prior to operation. In the Eltra area, the secondary control capacity is currently 550 MW

up, and 300 MW down. The manual secondary control consists of 300 MW long-term contracts plus access to the regulating power markets in Sweden and Norway. In general, the cheapest alternative for manual secondary control is used. However, in the case of network congestion, the 300 MW long-term contracts must be used. From 2004 onwards, primary and secondary control will be purchased via a monthly tender procedure instead of long-term contracts.

- Fast-acting production technologies for fine adjusting the balance.
- Network controller that activates the HVDC interconnection to Norway and will replace a more traditional network controller that controls one or several power stations.

Balancing wind production is like driving a big lorry in different situations:

- In the case of stable wind conditions, it is like driving on the motorway at moderate speed – no problem.
- When fronts pass, it is like driving in the countryside, on small winding roads – or even under foggy conditions where you do not know if the road turns to the right or to the left, but you still have to keep up the speed, you cannot stop and from time to time you touch the ditch. This is similar to situations where the manual handling of the regulating power goes wrong.

Within the framework of the European internal electricity market, renewable energy and CHP production can be given a high priority. In the Danish system, end-consumers currently have simply to accept their pro rata shares of priority production. This rule is expected to end in 2004 and legislation is in preparation. Until then, Eltra as the TSO is in charge of balancing the priority production.

In the future, a number of market players are expected to emerge in the context of balancing priority production. Their approach will nevertheless be similar to the way Eltra works today. Therefore, it is interesting to give an example of how Eltra handled the situation up to 2004.

Before explaining the example, we have to describe the Nordic market model.

10.2.1 The Nordic market model for electricity trading

When the Nordic electricity market was set up, the Nordic parliaments of all the countries involved (Sweden, Norway, Finland and Denmark) introduced similar legislation. In each country, a regulator was given an important role in implementing the market opening. All the Nordic countries chose an approach with a TSO. The ownership of the TSOs varies, though, ranging from state ownership via mixed private and state shareholders to cooperatives (i.e. consumer-owned organisations).

Several fields of knowledge and experience have contributed to the development of the Nordic electricity market. A market mechanism has been developed. Power and energy balances still have to be maintained. Striving for competition has to be balanced with maintaining the stability of the system. There must be the appropriate infrastructure to develop the market.

The opening of all networks to all market players and the introduction of a point tariff system were important parts of the market opening. There are several marketplaces. Bilateral trade and the power exchange coexist to their mutual benefit. Brokers operate separate marketplaces, and there is a large number of players in the retail market.

A number of service providers have emerged, such as analysis service providers, information service providers, portfolio managers and balance responsible parties.

For further information, see also Tonjer *et al.* (2000), Granli *et al.* (2002) and Rønningsbakk *et al.* (2001).

10.2.1.1 The Nordic electricity market

The common Nordic electricity market comprises the four countries Denmark, Finland, Norway and Sweden. Iceland is also part of the Nordic community (e.g. it is in the Nordel organisation) but it is not interconnected with the other countries. Table 10.3 shows key figures in the Nordic electricity system in 2001. The peak load of the interconnected system was slightly less than 70 000 MW and, for the same area, the total consumption was 393 TWh.

The population of the common trade area totals 24.0 million people, with 5.4 million in Denmark, 5.2 million in Finland, 4.5 million in Norway and 8.9 million in Sweden.

In 2001 the day-ahead power exchange (Elspot at Nord Pool) reached 111.2 TWh. The same year, the financial trade (Eltermin and Eloption) at Nord Pool totalled 909.8 TWh. Finally, 1748 TWh of bilateral contracts were cleared. The Elspot trade amounted to 28.3 % [(111.2/393) × 100] of the total consumption in the market area.

Table 10.3 Key figures in the Nordic electricity system, 2001

	Denmark	Finland	Iceland	Norway	Sweden	Nordel
Installed capacity (MW)	12 480	16 827	1 427	27 893	31 721	90 348
Consumption (GWh):						
generation	36 009	71 645	8 028	121 872	157 803	395 357
imports	8 603	12 790	—	10 753	11 167	43 313
exports	9 180	2 831	—	7 161	18 458	37 630
total[a]	35 432	81 604	8 028	125 464	150 512	401 040
Breakdown of electricity generation (%):						
hydro power	0	19	82	99	50	55
nuclear power	—	31	—	—	44	23
other thermal power	88	50	0	1	6	20
other renewable power	12	0	18	0	0	2

—No nuclear power production.
[a] generation plus imports minus exports.
Note: 0 indicates less than 0.5 %.

The border between Eltra and Germany is managed in a two-step procedure (for more information on the web, see http://www.eltra.dk or http://www.eon-netz.com):

1. capacity auctioning (yearly, monthly and daily);
2. capacity used to transfer energy across the border based on hourly schedules.

10.2.2 Different markets

The Nordic electricity market consists of several markets: (1) the physical day-ahead (Elspot) trade; (2) the hour-ahead (Elbas) trade and (3) the real-time market. In the following, we will describe the characteristic features of these markets.

10.2.2.1 Elspot market

The basic philosophy of the Nordic day-ahead market (Elspot market) is as follows:

- The power exchange Nord Pool is the common market place for physical Elspot trading in Denmark, Finland, Norway and Sweden.
- The transmission grid is the physical marketplace – the marketplace has a number of players with equal access to the transmission system based on point tariffs.
- The Nordic marketplace is one integrated market.
- The TSO decides the available transfer capacity (ATC) between the potential price areas.
- The ATC is completely supplied to the market (Nord Pool). There are some interconnections between the Nordic marketplace and surrounding systems and markets, which are handled only by the owners of these interconnections. According to a new directive, the 'use-it-or-lose-it' principle has to be implemented for these interconnections.
- Congestion between Elspot areas is handled by market splitting.
- The Nordic system is operated by five TSOs whose aim it is to act, as far as possible, as one TSO.
- In the afternoon before the day of operation, hourly production schedules are set up with the basic intention of balancing the system as a result of the Elspot trading.
- Elbas trade and balancing activities start in the evening before the day of operation.

Table 10.4 shows the activities to be completed during the day before the 24-hour period of operation.

10.2.2.2 Elbas market

The Elbas market is a physical market for power trading in hourly contracts, for delivery the same day and the following day. It allows trading in Finland and Sweden for those hours when Elspot trade has been accepted, covering individual hours up to one hour before delivery. After clearing the Elspot market, the ATC between Finland

Table   10.4 Daily routine

Time	Routine
08:00–09:30	Capacity for market use is calculated based on the 'use-it-or-lose-it' principle across the Danish–German border
10:00–10:30	TSOs inform Nord Pool spot market about the ATC between potential price areas within the Nord Pool market area
12:00	Deadline for bids and offers to Nord Pool Spot
13:00–14:00	Nord Pool Spot clears the market
14:00–19:00	Generators make final production plans
15:00–19:00	Production plans are submitted to the TSOs
16:00–19:00	Final production schedules are determined by the TSOs
16:00–19:30	Bids are made to the regulating power market (there may be changed up to half an hour before the hour of operation)
20:00–24:00	Load forecasts are performed by TSOs

Note: ATC = available transfer capacity; TSO = transmission system operator.

and Sweden for the next day is announced and the market players can start submitting their bids.

There is an ongoing discussion on whether to extend the Elbas market or not.

10.2.2.3 Real-time market

The Nordel Operations Committee has developed a new concept for cooperation between the Nordic countries regarding the balance regulation. The goal has been to create a common regulating power market for all countries. As part of the control cooperation, the part of the system with the lowest control costs will be used for regulation purposes. The rules and prices for the players in the different parts of the system will be as harmonised as possible.

First, the new concept will be implemented in the synchronised system (on the Nordel System, see http://www.nordel.org). Second, Jutland (Western Denmark), which belongs to another control area, will also be part of the cooperation.

10.2.2.4 Main principles

Each TSO receives regulating power bids within its system area and submits the bids to NOIS (Nordic Operational Information System), which is a web-based information system for the exchange of operational information between TSOs. In NOIS, a merit order list of all regulating power bids is composed to form a 'staircase' that is visible to all TSOs.

The balance regulation of the synchronised system is frequency-controlled. Control measures are generally activated according to the order of operational costs, which can be derived from the common list of the regulating power bids. At the end of each hour, the common regulating power price is determined in accordance with the marginal price of operation. This price is used as the reference price. When settling the balance or imbalance, it is included in the calculation of the settlement prices for all the subareas. The subareas continue to use different models for balance settlement, though (one-price or two-price models, marginal prices or middle prices).

In the case of network congestion on the interconnections and within Norway, the regulating power market is divided into different price areas. Bids for regulating power that are 'locked in' are excluded from the 'staircase'. This way, there will be different regulating power prices in different subareas.

The regulating TSO that performs the control action receives compensation for this work. The price of balancing power between the subareas is settled at the common regulating power price or the middle price, if these are different.

In future, Norwegian Statnett and Swedish Svenska Kraftnät will share the main responsibility for frequency control in the Nordel area, in a similar way as today, and will initiate the control procedures in the synchronised system. Western Denmark will be basically managed as today, with a planned supportive power exchange. In that way, this region will participate in the regulating power market. Eltra provides frequency control within the UCTE system and consequently manages the balance control within the Eltra area. The model has been in force since 1 September 2002.

10.2.3 Interaction between technical rules and the market

The procedure for determining the trading capacities is the key to the interaction between market and system operation. A typical course of events may be as follows:

- forecast expected utilisation of the interconnections between Nordel and neighbouring areas;
- forecast load, wind production and other production plans;
- incorporate maintenance plans for production and transmission assets;
- review limiting thermal values based on the $n-1$ criterion ($n-1$ = safe operation even if one component is missing);
- review limiting dynamic values based on the forced outage of the largest unit or dimensioning faults;
- review voltage stability;
- make calculations for certain points of time in the forthcoming 24-hour period;
- determine permissible exchange capacities based on the 24-hour calculations.

Thus, the TSO can ensure security of supply from a technical point of view. In practice, this means that the TSO's operational planning must ensure that:

- it is possible to transport reserves through the system, in the case of forced outages or other events that require a fast activation of production in order to avoid system collapse – the reserves in question are instantaneous reserves that are activated automatically within seconds;
- the balance can be restored in the individual areas after major forced outages, in order to avoid thermal overload or instability – the individual areas must have reserve

capacity at their disposal, and activation of this capacity may take place automatically or manually, according to instructions or otherwise;

• temporary overloading of single components can be contained within their short time ratings.

The transport capacity made available to the market players has to be maximised while also taking into account system reliability aspects.

In Western Denmark, owing to internal congestion, the market capacity depends on the amount of forecasted wind production. A new 400 kV line from mid-Jutland to the north will substantially increase market capacity.

10.2.4 Example of how Eltra handles the balance task

The TSO's major task is to secure a balance between production and demand. This can be done:

• in the range of seconds via automation;
• in the range of minutes via automation and manually (telegraph);
• on an hourly basis via trade in balancing power;
• on a daily basis via Nord Pool Elspot.

Normally, the balancing task is related to:

• variations in demand;
• tripping of power plants;
• random variations in power output from production units.

In view of the increasing share of wind production, the fluctuating output from wind production units creates a major challenge. Figure 10.6 shows a typical aggregated output from all wind turbines in the area on a 15-minute basis.

The power output from wind production fluctuates as a result of the inherent power characteristics and variations of wind speed. Figure 10.7 shows a general power curve for a single turbine. If the wind speed fluctuates between 5 m/s and 15 m/s, this fluctuation will result in a variation of the output from zero to maximum installed capacity.

Wind speed forecasts with reference to physical locations of installed wind capacity are essential. Typically, meteorological institutes focus on forecasts of extreme values and not on variations in the range between 5 m/s and 15 m/s. Considerable efforts have to be made to improve the forecasts (see also Chapter 17).

In order to be able to use the present day-ahead market structure, given the operational requirements of the Elspot market, wind forecasts must define an hourly mean wind speed up to 40 h ahead with reference to a geographical location. Weather-related output from local CHP units must also be evaluated when balancing priority production.

The factual output from a commercial wind turbine is more or less linear in the range between 5 m/s and 13 m/s. Above 13 m/s, production levels out, and at about 20 m/s or

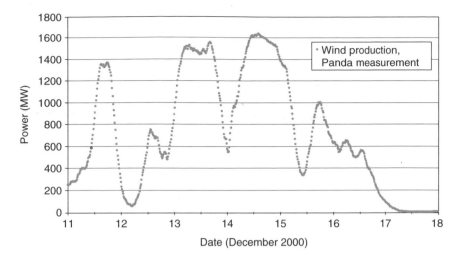

Figure   10.6 Wind production in week 50 2000. *Note*: Panda is a data base for measured 15-minute energy values

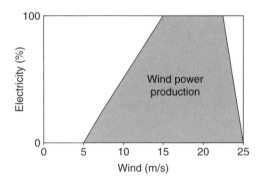

Figure   10.7 Typical wind turbine output

25 m/s the turbines stop and turn out of the wind. Production will be zero below 5 m/s and above 25 m/s.

The wind speed forecast is essential, since in the current Eltra system, a variation of 1 m/s in wind speed can result in a variation of up to 300 MW in wind production.

10.2.5 Balancing via Nord Pool: first step

Eltra is in charge of balancing the priority production (i.e. private wind production and local CHP). Eltra informs the demand side three months in advance about the hourly percentage of priority production that the consumers have to purchase. The three-month forecast has been chosen to enable market participants to act in the power market. This principle is unique to Denmark. The principle question is: 'who is in charge of balancing wind production'. Until now, the TSO has had to take on this responsibility.

Using data on actual weather conditions, Eltra calculates an expected hourly value for wind production and local CHP. The difference between the values of the three-month forecast deducted from the load forecast and the latest forecast is then traded at Nord Pool with a flat bid. Depending on whether the calculated difference is positive or negative, Eltra acts as a seller at price zero or as a buyer at maximum market price.

Figure 10.8 shows 24 hourly values for the three-month forecast and the forecast from the morning of the day prior to the 24-hour operating period. Figure 10.9 illustrates the difference between the two values that Eltra bids to Nord Pool.

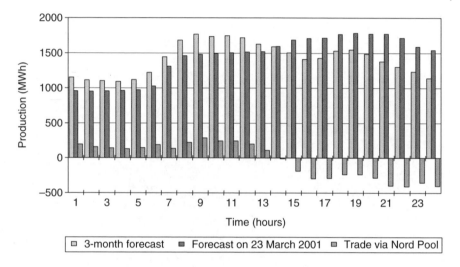

Figure   10.8 Priority production for 23 March 2001

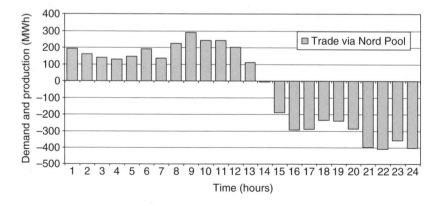

Figure   10.9 Eltra's bid to Nord Pool, 23 March 2001

The operator in the TSO control room will often be aware that corrections of the expected power production will be necessary because of changes in the weather forecast, but once the bids are submitted to Nord Pool at 12:00 they are binding.

10.2.6 The accuracy of the forecasts

Our experience shows that the meteorological forecasts are inaccurate also in normal situations, as can be seen from the following examples of 'good', 'bad' and 'ugly' forecasts from a TSO point of view.

10.2.6.1 The good forecast

Figure 10.10 shows a forecast from the morning of the day prior to the 24-hour operating period, the measured value and the difference. In this case, the forecasted production is consistent with the actual production.

10.2.6.2 The bad forecast

Figure 10.11 is similar to Figure 10.10, but there is a difference of up to 900 MW, which is equivalent to approximately a third of the total load. Owing to this difference, imports from the real-time market were necessary in order to secure the balance. The cost of import determines the prices of the imbalance.

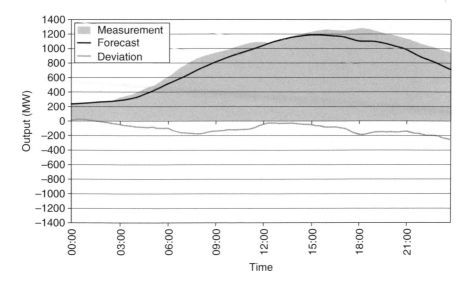

Figure 10.10 The good forecast: priority wind production; average quarter-hour power output on 6 November 2000 and forecast calculated on 5 November at 11:00

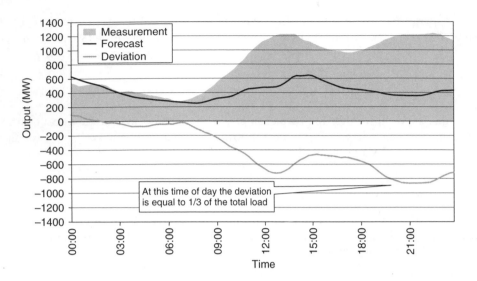

Figure   10.11 The bad forecast: priority wind production; average quarter-hour power output on 25 October 2000 and forecast calculated on 24 October at 11:00

10.2.6.3 The ugly forecast

Figure 10.12 is similar to Figures 10.10 and 10.11, but there is a difference of up to 800 MW. However, the difference changes from a surplus to a deficit, as the wind front is approaching faster than expected.

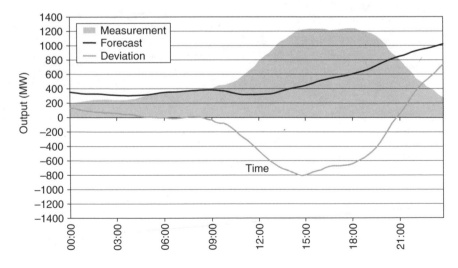

Figure   10.12 The ugly forecast: priority wind production; average quarter-hour power output on 11 December 2000 and forecast calculated on 10 December at 11:00

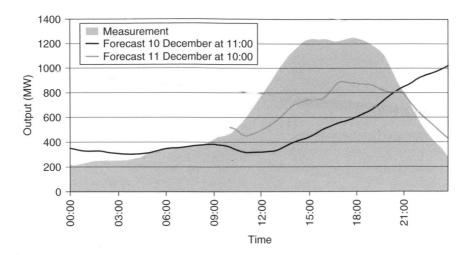

Figure   10.13 Why ugly? Priority wind production; average quarter-hour power output on 11 December 2000 and forecast calculated on 10 December at 11:00 and on 11 December at 10:00

The case is ugly because of the updated forecast at 10:00 hours of the 24-hour operating period. Even the short-term forecast was not able to predict the power surge. Such situations are difficult to handle for TSOs (see also Figure 10.13).

10.2.7 Network controller and instantaneous reserves

Eltra is located inbetween the synchronous Nordic system and the UCTE system. Since Eltra has been operating synchronously within UCTE since the 1960s, all imbalances are automatically placed on interconnections to the German system (see Figure 10.14).

To reduce unscheduled imbalances, Eltra has introduced automatic regulation of the HVDC interconnections to Norway, with certain energy and power limits. This is a type of network controller. This control action lies typically within the range of ±50 MW and is normally active. This function will cease to exist at the end of 2005. In addition to the network controller, power plants in the area supply instantaneous reserves in accordance with UCTE rules.

10.2.8 Balancing prices in the real-time market

Handling imbalances as shown in Figures 10.11 and 10.12 requires the involvement of the real-time market. Eltra purchases regulating power of plus 300 MW and minus 300 MW. As shown, these amounts are not sufficient in all situations. Depending on demand and prices, Eltra will activate regulating power from the Nordic real-time market or from producers south of the Danish–German border. Regulating power affects the pricing of the imbalances. Figure 10.15 shows a price pattern for one month.

We do not include any further description of the real-time market and settlement for balancing power in this chapter. For detailed information on trade mechanisms, the

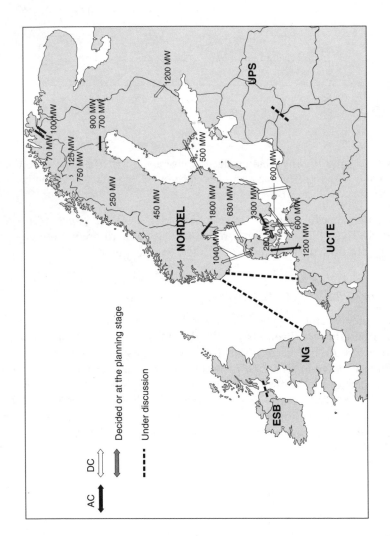

Figure 10.14 Existing, decided and possible cooperative interconnections in Northern Europe

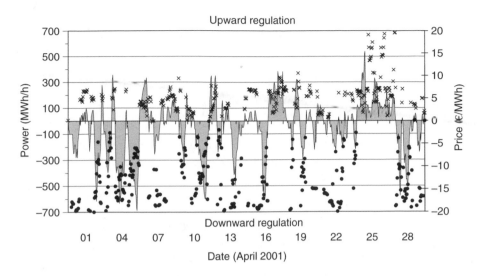

Figure **10.15** Regulating power and prices in the real-time market (April 2001) for the Eltra area. *Note*: the grey areas are the amount of regulating power (MWh/h) for upward and downward regulation; the × and • indicate the prices of the upward and downward regulation, respectively, calculated as additions to and deductions from the area price (right-hand ordinate axis)

regulating power market and the balance philosophy, see the Nordel homepage, at http://www.nordel.org.

10.2.9 Market prices fluctuating with high wind production

Figure 10.16 illustrates how the market price fluctuates with high wind power production during two days in January 2003. It is obvious that market prices can be zero during several hours with high wind energy production and CHP production. The area price for Western Denmark will be between the German price and the area prices in Norway and Sweden.

Figure 10.17 shows a typical day with high wind production and an energy surplus of up to 1500 MWh/h.

10.2.10 Other operational problems

The shift from central production connected to the transmission grid to a pattern of dispersed production connected to the distribution grid influences nearly all electrotechnical design features. The following is only a rough list:

- Previously, load-shedding systems could be activated on 60 kV lines, but now they have to be redesigned and installed in pure load radials (i.e. 10 kV lines).
- Load-shedding signals can no longer rely on local frequency measurements but have to be activated by a frequency signal from the transmission system.

- An effective real-time market is a prerequisite for handling large amounts of renewable production, hence an extended real-time market must be established.
- Local CHP units must be able to contribute to the real-time market using the heat storage capacity for stopping and starting up.
- Local CHP units have to be equipped with additional air coolers to be able to produce even without the normal heat load.
- Local CHP units have to be able to stop and use surplus electricity in electrical heaters.
- Rules for using natural gas must be flexible to allow gas-fired installations to change consumption at short notice (in minutes). The natural storage capacity in the gas transmission system will be used.
- Technical rules for new generators must be rewritten to comply with the new operating conditions in a dispersed production system.
- Black-start conditions for generators and system restoring must be redesigned.
- Reactive power capabilities and balance rules must be redesigned and coordinated with a new general philosophy for reactive balances and voltage control.
- New control concepts must be defined.
- Supervisory control and data acquisition (SCADA) system designs will be changed.
- Load flow tools must have new load and generator models.
- Measurement and estimation of load and generation are key questions.
- Voltage stability must be checked.
- Short-circuit capacity will be reduced.
- Voltage quality will be reduced.
- Working capability of HVDC installations will be reduced for import (gamma angle = the security angle for inverter operation).

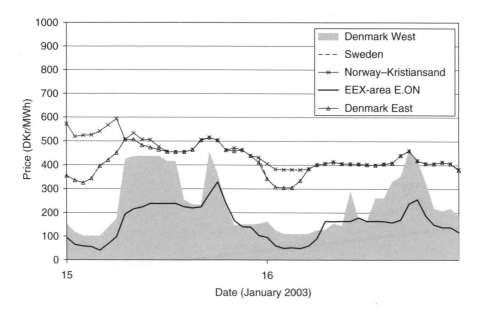

Figure   10.16 Example of market prices, 15–16 January 2003

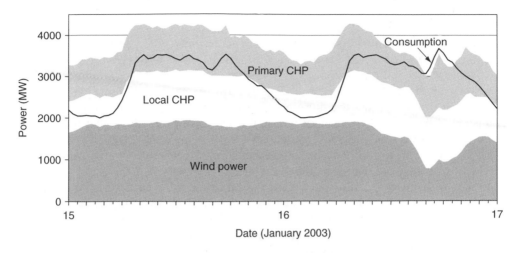

Figure   10.17 High wind power production, local and primary combined heat and power (CHP) and consumption, January 2003

We have not yet seen the full consequences of completely dispersed production when each household installation includes microgeneration based on photovoltaics and fuel cells.

10.3 System Analysis and Modelling Issues

As the transmission system operator, Eltra is responsible for the overall security of supply as well as for the efficient utilisation of the power system. The security of supply involves adequacy of power and energy as well as system security. In addition, Eltra has to see to it that the power market works smoothly.

In order to comply with its tasks as TSO, Eltra has to be able to understand and analyse in detail the physics and technologies of the existing and future power system. Eltra must also be able to analyse the market mechanisms. For that purpose, the application of dedicated mathematical model tools and reliable data are indispensable prerequisites.

10.3.1 Future development of wind power

The expansion plans involving new land-based wind turbines have practically all been stopped because of a general reduction of economic subsidies.

In 2002, the 160 MW Horns Rev offshore wind farm (80 × 2 MW) was commissioned, featuring an AC cable connection to the 150 kV transmission system in the Eltra area. A similar 150 MW offshore wind farm is scheduled for commissioning at Rødsand off Nysted in Eastern Denmark in 2003. Additional future offshore wind farms are under discussion.

10.3.2 Wind regime

Figure 10.18 shows estimated duration curves for the wind power production at Horns Rev in the North Sea and at Læsø in the Kattegat, based on site-specific measurements of offshore wind conditions (from 15 May 1999 to 15 May 2000). The reference offshore wind turbines are 2 MW turbines, where both rotor diameter and hub height are 80 m.

Figure 10.18 also shows the duration curve for wind power production at a windy Danish land-based site (Thy) with a typical turbine layout. It follows that the capacity utilisation factor is much larger offshore (0.45–0.50) than onshore (0.34).

Figure 10.19 illustrates the correlation between synchronous hourly wind power production from wind farms in the Eltra area. As expected, the correlation factor decreases with increased distance between production sites. In addition to this, the correlation is in general higher between the production at two land-based sites than between the generation at a land-based and an offshore site.

In addition, Figure 10.20 shows examples of correlation in time (autocorrelation) for wind power production at Horns Rev and Læsø. The correlation curves are very similar for the two sites; initially, the correlation declines relatively fast and, after about 15 hours, the correlation declines at a slower rate. The location of Horns Rev and Læsø is shown in Figure 10.3.

Finally, Figure 10.21 shows the correlation between the production estimates for Horns Rev and Læsø for varying time displacements. It can be seen that the most significant correlation between the production of the two sites (Horns Rev, located 15 km into the North Sea northwest of Esbjerg, and the Læsø site, located south of Læsø in the Kattegat, east of Jutland) is found by applying a 'delay' of approximately four hours to the production estimates for Læsø, in comparison with Horns Rev. This is consistent with the fact that weather fronts frequently pass Denmark from west to east.

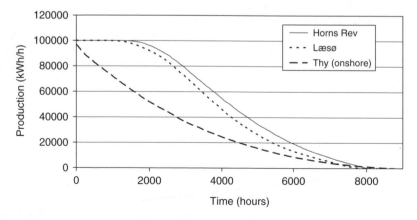

Figure   10.18 Duration curves for estimated offshore and land-based wind power production, based on measurements recorded over a 12-month period (from 15 May 1999 to 15 May 2000): Horns Rev (100 MW, offshore), Læsø (100 MW, offshore), Thy (100 MW, onshore)

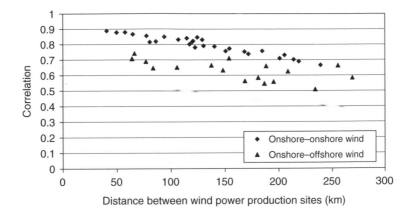

Figure 10.19 Correlation in space between estimated power production at wind farms in the Eltra area, based on 12 months of measurements (15 May 1999 to 15 May 2000)

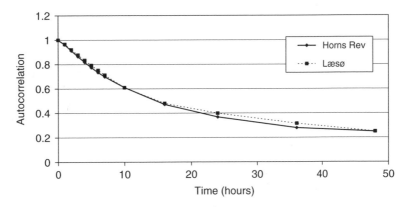

Figure 10.20 Correlation in time for estimated wind power production at two Danish offshore sites, based on 12 months of measurements (from 15 May 1999 to 15 May 2000)

10.3.3 Wind power forecast models

For forecasting wind power, Eltra uses the WPPT (Wind Power Prediction Tool) program package developed by the Department of Informatics and Mathematical Modelling (IMM) at the Technical University of Denmark, in cooperation with Elsam and Eltra (Nielsen, 1999; see also chapter 17)

WPPT predicts the wind power production in Jutland–Funen using online data covering a subset of the total population of wind turbines in the area. The Western Danish Jutland–Funen area is divided into subareas, each covered by a reference wind farm. For each subarea, local online measurements of climate variables as well as meteorological forecasts of wind speed and wind direction are used to predict wind power.

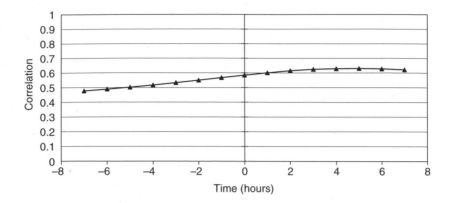

Figure **10.21** Correlation between estimated hourly production at Horns Rev and Læsø, based on 12 months of measurements (from 15 May 1999 to 15 May 2000); positive values on the first axis indicate the number of hours that the Læsø values were 'delayed' compared with the Horns Rev values

The power prediction for the reference wind farm of each subarea is subsequently upscaled to cover all wind turbines in the subarea. Then the predictions for the subareas are compiled to arrive at a total prediction for the entire Eltra area. Table 10.5 outlines the forecast statistics for wind power within the Eltra area for the year 2002.

It follows from the statistics in Table 10.5 that the miscalculation for the year 2002 corresponds to 31 % $\{[(1095\,GWh)/(3478\,GWh)] \times 100\}$ of the total wind energy production. The unpredictable wind power causes large problems with keeping the physical balance and leads to reduced system security. Trading in the Nordic power exchange (Elspot) must be completed at 12 noon of the day prior to the day of operation. The daily miscalculation is thus calculated as the sum of absolute hourly deviations between forecasts of next day's wind power production and the actual production observed later. The daily miscalculations are added up in order to arrive at the yearly miscalculation.

Table **10.5** Wind power statistics within the Eltra area, 2002

Quantity	Value
Installed wind power capacity:	
start of year (MW)	1780
end of year (MW)	2000
Wind energy:	
produced wind energy (GWh)	3478
Miscalculated wind energy (GWh)	1095
Miscalculation (%)	31
Added wind power costs:	
Payment of real-time imbalance power (DKr millions)	68
Added costs of wind power (DKr/kWh)	0.02

Note: in mid-2003, 1 Danish Krone (DKr) = 7.5 euros.

Owing to this miscalculation, in 2002 the added costs of handling wind power amounted to approximately DKr68 million. The added costs correspond to approximately DKr 0.02 per kWh wind power and are related to the purchase of regulating power from the real-time imbalance power market [added costs = (DKr68 million)/ 3478 million kWh)]. The added cost of DKr68 million per year is a substantial amount, and a miscalculation of 31 % leaves great potential for improvement.

The primary reason for the relatively poor wind forecasts is low-quality meteorological data. Eltra is strongly engaged in a number of research and development activities that aim at improving wind power forecasting.

In this respect, the concept of 'ensemble forecasting' looks very promising. The Ensemble Prediction System (EPS) relies on generating several forecasts instead of just one and aims at predicting the correct wind speed instead of providing one generally reliable weather forecast. Initial studies have been conducted in which the meteorological model comes up with 25 wind forecasts instead of a single forecast and then 'selects the best one'. These studies suggest that use of EPS may reduce Eltra's present misforecasts by approximately 20 %. These 25 forecasts have been generated by combining five different vertical diffusion schemes with five different condensation schemes, which results in 25 ensemble members.

In 2003–2004, Eltra will test the 'multi-ensemble concept' on wind power forecasts. For the tests, Eltra will set up a dedicated in-house network (cluster) of PCs to test the method in a real-time setting.

10.3.4 Grid connection

10.3.4.1 Wind farms connected to grids with voltages above 100 kV

Eltra has developed specifications for connecting wind farms, including offshore farms, to the transmission network (Degn, 2000; see also Chapter 7). These specifications, which have become a paradigm in Europe, are the counterpart of the general power station specifications for land-based production facilities.

The most important new requirement is that wind farms – like any other major production plant – are not allowed to lose stability or trip during short circuits in the network that is disconnected by primary network protection. In more popular terms, the turbines must be able to survive a short dead time (about 100 ms) and resume production once the fault has been disconnected and the voltage starts to return.

The specifications require an increased control capability of wind farms according to which a wind farm has to be able to reduce its production from full load to a level between 0 % and 20 % within a few seconds.

The specifications include requirements regarding:

- power and power control, including startup;
- frequencies and voltages (e.g. a wind farm has to be able to operate at deviating frequencies and voltages);
- the wind farm's impact on voltage quality;
- the interaction between power system and wind farm in the case of faults in the power system;
- protection and communication.

10.3.4.2 New wind turbines connected to grids with voltages below 100 kV

The wind turbines that are currently connected to the distribution system were erected in compliance with DEFU KR-111 (DEFU/DTU, 1998) This recommendation applies to the distribution undertakings and includes, in addition to frequency trip limits, very few overall power system requirements.

Based on the experience from applying Degn (2000) and DEFU KR-111 (DEFU/DTU, 1998), the two Danish TSOs have developed a new set of specifications for the connection of wind turbines to grids with voltages below 100 kV. These specifications entered into force in July 2004.

The intention of these specifications is to ensure that the wind turbines have the regulating and dynamic properties that are essential to operate the power system. The control requirements have been set up to allow an increased penetration of wind power and to prepare for a possible distributed control of the power system.

Since these wind turbines usually are connected to the distribution system, the specifications have to balance the local requirements for security of supply (voltage quality, overvoltage, islanding operation, etc.) with the overall system requirements (fault ride-through capability, frequency control, etc.).

10.3.5 Modelling of power systems with large-scale wind power production

Today, distributed generation (DG) in the Eltra system is handled as nondispatchable production. The wind generators produce power according to the wind, and the power production from small-scale CHP units is governed by the local demand for district heating. They are not governed by market demand for power. In addition, when producing the necessary heat, the CHP units today optimise their profits according to statutory power tariffs. The problem is that these tariffs do not reflect the current market value of power.

The present DG in the Eltra area covers about half of the consumption (see Section 10.1), without contributing to regulating power and other ancillary services. Therefore, the present handling of DG causes large operational problems regarding the physical power balance. This leads to reduced system security and increased costs of providing the necessary services from the remaining large central production plants.

The Danish government is preparing a change to the concept of subsidising DG, including the abandoning of its priority status. This means that small-scale CHP units in the future will have to produce according to market signals. The change of concept is expected to start in 2005 for units above 10 MW installed capacity.

10.3.5.1 Wind power and the power market

The variable costs of wind power are only a fraction of the variable costs of thermal production. In this context, wind power is similar to uncontrollable hydropower ('run-of-river' concepts).

Consequently, large quantities of installed wind power in the system will produce a downward pressure on the market price of electricity in the day-ahead market (the spot market) when wind forecasts generate very positive wind power production estimates for the day ahead.

In order to analyse how wind power affects, among other things, market prices and competition within the Nord Pool area, Eltra has developed a new model, MARS (MARket Simulation; Eltra, 2003; Kristoffersen, 2003). MARS is Eltra's new market model for the simulation of prices, production, demand and exchanges in the power market. The model area comprises the Nordic countries (the Nord Pool area) and Northern Germany.

The new feature of this model is that prices, exchanges and so on are calculated on an hourly basis. The model uses the same principles as Nord Pool, including the division of the Nordic countries into price areas with price-dependent bids.

The model is designed for wind power, hydropower and thermal production, including nuclear power. On the demand side, price elasticity is taken into consideration (i.e. that demand varies according to price). Game theory was used to simulate the producers' strategic behaviour (i.e. producers' options for exercising market power).

MARS uses data reported to Nordel by the TSOs for, among others, carrying out system analyses with different model tools. The data are then converted to hourly values in relation to the available information on the distribution of consumption and so on. The data conversion is based on information from the Nord Pool FTP server and on a purchased database of production plants in various countries as well as on various other sources.

As an example of how the MARS model is used, Figure 10.22 shows simulated spot prices in the Eltra area for a week in the winter of 2005. The prices are computed for two scenarios; (1) 'perfect competition' and (2) when the large producers in the Nordic market exercise market power. It can be seen that the prices are low when there is much wind power, and that prices are high when there is little wind power.

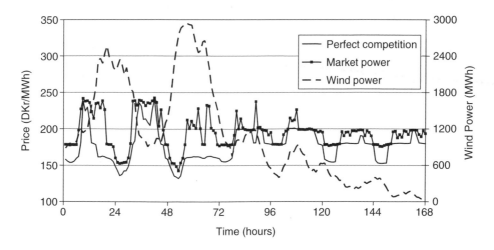

Figure 10.22 Estimated prices in Western Denmark for a week in the winter of 2005 with incidents of high wind power production: perfect competition and market power scenarios

It is also evident that, generally, the potential for exercising market power (i.e. pushing the market price in Western Denmark up through strategic bidding into Nord Pool) is highest during hours with large wind power production in the Eltra area. The reason is that, in such situations, producers can raise the price in Western Denmark to match the price level of its neighbours and hence reap the extra profit.

10.3.6 Wind power and system analysis

10.3.6.1 The Eltra SIVAEL model, including stochastic wind power description

In view of the large share of wind power in the Eltra system, Eltra has updated the SIVAEL simulation model to include the stochastic behaviour of wind power (Pedersen, 2003).

The standard SIVAEL model (Eriksen, 1993, 2001; Pedersen, 1990; Ravn, 1992) solves the problem of how to schedule power production in the best way. It generates the scheduling of power production units, heat production units and CHP units on a weekly basis. The constraints require the power and heat demand and the demand for spinning reserves to be met every hour of the week. The scheduling includes unit commitment (start–stop of units) and load dispatching (unit production rates of power and heat).

SIVAEL handles power and heat generation plants, heat storage systems, wind turbines, foreign exchange, immersion heaters, heat pumps and electricity storage. The scheduling is carried out with the objective of minimising total variable costs, which include variable operational and maintenance costs and startup costs.

By its nature, wind power is largely unpredictable (see Section 10.2.6 and Section 10.3.3). The new and improved SIVAEL model simulates wind power prediction errors as a stochastic process. The parameters in the stochastic model description have been estimated from observations of wind power forecast errors throughout the year within the Eltra system. The work has been carried out in cooperation with IMM at the Technical University of Denmark (DTU).

As a new feature, the updated SIVAEL model can simulate the need for upward and downward regulation of the system, due to changes in wind power in the real-time hour of operation. A change in wind power is by far the most important source of unforeseen regulation in the Eltra system.

With the model it is possible to compare for each hour the need for regulating the system with the available resources, including:

- regulation of primary central units;
- regulation via transmission lines to neighbouring areas;
- regulation of local small-scale CHP units.

The present version does not include the formal optimisation of the use of the regulating resources generated by local CHP units. The optimisation is limited to the two other resources: primary central units and the transmission system. However, the model may be expanded and used for an evaluation of the regulating potential of local CHP units.

Figure 10.23 shows hourly values of simulated wind power prediction errors throughout a year, in the future. The errors are measured as a percentage of the hourly wind power forecasts (with positive figures when production exceeds forecast).

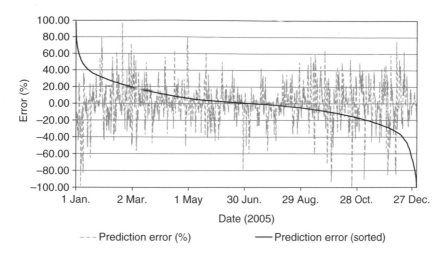

Figure 10.23 Simulation of wind power prediction errors throughout a year, Eltra 2005

10.3.6.2 Active use of small-scale combined heat and power for regulation purposes

As an example of how the SIVAEL model can be applied, Figures 10.24(a) and 10.24(b) present SIVAEL simulations of the Eltra system's demand for regulation in 2005, and how that demand is covered. It is assumed that the demand for regulation is determined by the magnitude of a faulty wind forecast. The 2005 wind forecasts are assumed to be of a similar quality to today's predictions.

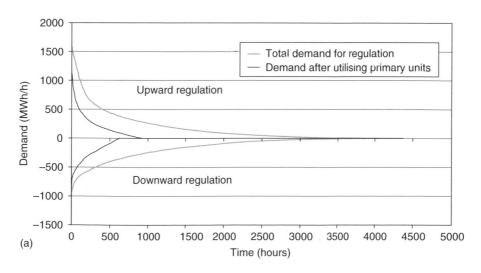

Figure 10.24 Regulating demand and lack of regulating power in the Eltra area, 2005, for the case of high prices in the Nordic market; small-scale combined heat and power (CHP) facilities run according to (a) a three-rate tariff model (as at present) and (b) market signals

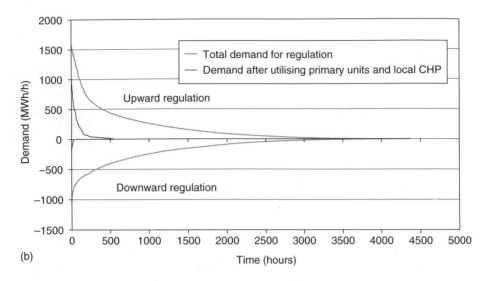

(b)

Figure   10.24 (*continued*)

Simulations were carried out for two scenarios of operating small-scale CHP units:

- Scenario 1: the small-scale CHP units show the same operational patterns as today, with their operation determined by current tariffs and heat requirement (company-specific economic dispatch).
- Scenario 2: the present prioritising has been abandoned, and the operational patterns of the small-scale units follow market signals (socioeconomic load dispatch).

Each of the above scenarios has been analysed relative to a low-price situation, with market prices averaging DKr120 per MWh a year, and a high-price situation, with prices averaging DKr220 per MWh. As an additional feature, the prices have been assumed to vary over the days of the year and the hours of the day.

The results of the high-price situation are compiled in Figure 10.24, which shows that, in the Eltra area, small-scale CHP production has a major potential that can be used for wind power regulation purposes. Activating this resource by opening the market to small-scale CHP production could reduce Eltra's regulation difficulties arising from the unpredictability of wind power. Similar conclusions can be drawn for the low-price situation.

10.3.7 Case study CO_2 reductions according to the Kyoto Protocol

Under the Kyoto Protocol, Denmark has committed itself to reducing carbon dioxide (CO_2) emissions by 21 % compared with 1990. This is equivalent to a total national emission of 55 million tonnes of CO_2 by 2012.

As a result of the Kyoto target, emissions from Danish heat and power production are expected to be less than 23 million tonnes, which corresponds to approximately 14 million tonnes for the Eltra area (Western Denmark).

As part of the efforts to fulfil the Danish Kyoto targets, Eltra regularly studies the technical and emission-related consequences of expanding wind power in Western Denmark. The following sections provide a summary of the assumptions and the findings that Eltra has reached so far. It will serve as a case study of modelling a power system with large-scale wind power production. Eltra's SIVAEL model was used to perform the simulations (see Section 10.3.6).

10.3.7.1 Assumptions

The reference scenario assumes that by 2012 there will be 2500 MW land-based turbines and 760 MW offshore turbines in the Eltra area. Furthermore, the offshore expansion is assumed to take the form of four 150 MW offshore wind farms, in addition to the existing offshore wind farm at Horns Rev.

The Kyoto alternative assumes that by 2012 the Eltra system will include another 1000 MW of offshore wind power at Horns Rev compared with the reference scenario. It is also assumed that immersion heaters will be installed in the district heating systems of the central and natural-gas-fired local CHP areas. Furthermore, it is assumed that some of the local CHP facilities will incorporate air coolers, which would make it possible to generate electricity without generating district heat at the same time.

As a general condition, it is assumed that the local (small-scale) CHP facilities will be operated under market conditions and not, as is the case today (mid-2003), according to fixed tariffs, which do not reflect the current market value of power. This means that the local CHP facilities will not be in service in low-price situations. When the price of electricity is low, peak load boilers and immersion heaters will be used to generate the required heat. Finally, a general CO_2 shadow price of DKr120 per tonne of CO_2 is assumed.

The analyses have been conducted for three different market price scenarios:

- Scenario A: a low-price scenario (average annual electricity price of DKr120 per MWh in the Nordic countries and DKr170 per MWh in Germany).
- Scenario B: a high-price scenario (average annual electricity price of DKr220 per MWh in the Nordic countries and in Germany).
- Scenario C: an extremely-high-price scenario (average annual electricity price of DKr600 per MWh in the Nordic countries and DKr220 per MWh in Germany).

10.3.7.2 Results

Figures 10.25 and 10.26 show selected results of the distribution of power production and emissions for the three price scenarios. They illustrate that an additional 1000 MW offshore wind power would generate 4.2 TWh of power. In the low-price scenario net imports would be reduced by approximately 3.5 TWh, and there would be a similar increase of 2.6 TWh for net exports in the high-price scenario. In addition, another 1000 MW of wind power would lead to minor changes in central and local production.

Figure 10.26 shows that the changed production pattern resulting from an additional 1000 MW wind power would lead to only modest changes in CO_2 emissions in Western

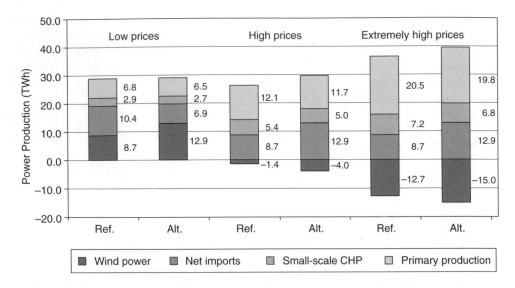

Figure 10.25 Distribution of power production within the Eltra area in the reference case (Ref.) and in the alternative case (Alt.), with an additional 1000 MW of offshore wind power installed by 2012, for three different power price scenarios

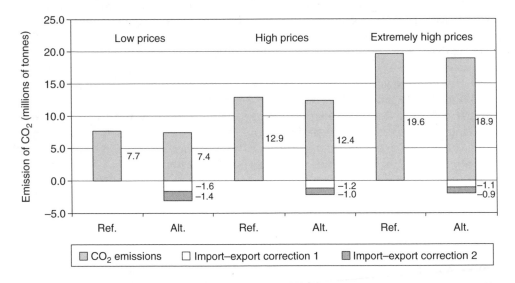

Figure 10.26 Emissions of carbon dioxide (CO_2) within the Eltra area in the reference case (Ref.) and in the alternative case (Alt.), with an additional 1000 MW of offshore wind power installed by 2012, for three different power price scenarios. *Note*: foreign CO_2 emission reductions as a result of an additional 1000 MW of Danish wind power capacity are calculated by two methods of correction (see text)

Denmark (i.e. 0.3 and 0.5 million tonnes CO_2 for the low-price and the high-price scenarios, respectively). This is primarily because of the fact that the Eltra area has already experienced a significant expansion of wind power and that the marginal reduction of the system emissions arising from additional expansions would be modest.

Additional wind power production would, however, reduce nondomestic CO_2 emissions by 1–3 million tonnes, depending on the power price and the CO_2 emissions resulting from the production capacity replaced abroad.

Figure 10.26 gives high and low figures for CO_2 reductions abroad. The low figures are equivalent to replacing natural gas-based power production with an efficiency of 45 % (specific CO_2 emission: 0.456 t/MWh); the high figures are equivalent to a coal-based power production with an efficiency of 40 % (specific CO_2 emission: 0.855 t/MWh). For the low-price scenario, for instance, this would mean a reduction of 1.6 million tonnes and $= 3.0 \ (1.6 + 1.4)$ million tonnes, respectively, in nondomestic CO_2 emissions.

In an international context, it will only be attractive to further expand wind power production in the Eltra area as a means of reducing CO_2 emissions if there is a price on CO_2 and if the CO_2 is traded according to common and harmonised rules.

Another aspect is that further massive wind power expansion in the Eltra area can be realised only if there is additional access to operational reserves and if the transmission network is considerably reinforced. Considerable investments in air coolers, heat storage tanks, immersion heaters and/or heat pumps to decouple the power production from the CHP production should be added to this.

10.4 Conclusions and Lessons Learned

Neighbouring TSOs traditionally offer mutual assistance based on equality and reciprocity. Today, neighbouring assistance implies that the TSOs give each other access to utilise their own 'reserve power market', and the service providers are paid.

There will be further stress on storage and regulation capabilities.

In the future, the real-time market will have a major impact on system operation and prices, and this market will fluctuate considerably. In fact, negative energy prices in this market are already a reality. With this in mind, the following is a wish list for a future electricity market:

- improved wind forecasts;
- enhanced control possibilities for production and load;
- increased interconnection capacity with neighbouring areas;
- enhanced control and resources of Mvar;
- an enhanced regulating power market.

The continuing increase of nondispatchable production and the very dispersed locations on lower voltage levels require the development of a new control philosophy. Studies made in the 1980s regarding the limits for the integration of wind energy were far off the mark in comparison with today's implementation, but two factors have also changed: (1) a pool was the approach chosen for implementing the internal electricity market and

(2) start of the common exchange of regulating power, which was not effective in the 1980s.

We still have to find out what the limits of the system are regarding operation with ever increasing amounts of wind energy. New control systems, flexible load and additional regulating capacity will be some of the new intervention measures used.

References

[1] DEFU/DTU (Danish Utilities Research Association – Technical University of Denmark) (1998) 'Connection of Wind Turbines to Low and Medium Voltage Networks', committee report 111-E, available at http://www.defu.dk.

[2] Degn, P. C. (2000) 'Specifications for Connecting Wind Farms to the Transmission Network', Eltra document 74557; available at http://www.eltra.dk.

[3] EC (European Commission) (1986) 'Wind Penetration in the Elsam Area (Western Denmark)'.

[4] Eltra (2003) 'A Brief Description of Eltra's Market Model MARS', presented at Eltra Seminar on Market Power, Eltra document 156884 v1, available at http://www.eltra.dk.

[5] Eriksen, P. B. (1993) 'A Method for Economically Optimal Reduction of SO_2/NO_x Emissions from Power Stations', in *Proceedings of the IEA/UNIPEDE Conference on Thermal Power Generation and Environment, Hamburg, September 1–3*.

[6] Eriksen, P. B. (2001) 'Economic and Environmental Dispatch of Power/CHP Production Systems', *Electric Power Systems Research* **57**(1) 33–39.

[7] Granli, T., Gjerde, O., Lindström, K., Birck Pedersen, F., Pinzón, T., Wibroe, F. (2002) 'The Transit Solution in the Nordic Electric Power System', feature article, in *Nordel Annual Report 2001*, Nordel, http://www.nordel.org, pp. 37–42.

[8] Kristoffersen, B. B. (2003) 'Impacts of Large-scale Wind Power on the Power Market', in *Proceedings of the Fourth International Workshops on Large-scale Integration of Wind Power and Transmission Networks for Offshore Wind Farms (Session 1), October 2003, Billund, Denmark*, Royal Institute of Technology, Electric Power Systems, Stockholm, Sweden.

[9] Nielsen, T. S. (1999) 'Using Meteorological Forecasts in On-line Predictions of Wind Power', Department of Informatics and Mathematical Modelling, Technical University of Denmark.

[10] Nordel (1992) 'Nordel Network Planning Criteria', Nordel, http://www.nordel.org.

[11] Nordel (2002) 'Nordic System Operation Agreement', available at http://www.nordel.org.

[12] Pedersen, J. (1990) 'SIVAEL – Simulation Program for Combined Heat and Power Production', in *Proceedings of the International Conference on Application of Power Production Simulation, Washington*.

[13] Pedersen, J. (2003) 'Simulation Model Including Stochastic Behaviour of Wind', in *Proceedings of the Fourth International Workshops on Large-scale Integration of Wind Power and Transmission Networks for Offshore Wind Farms (Session 3a), October 2003, Billund, Denmark*, Royal Institute of Technology, Electric Power Systems, Stockholm, Sweden.

[14] Ravn, H. (1992) 'Optimal Scheduling of Combined Power and Heat', IMSOR internal publication, Technical University of Denmark.

[15] Rønningsbakk, K., Gjerde, O., Lindström, K., Birck Pedersen, F., Simón, C., Sletten, T. (2001) 'Congestion Management in the Electric Power System', Nordel Annual Report 2000, feature article in *Nordel Annual Report 2000*, Nordel, http://www.nordel.org, pp. 35–44.

[16] Tonjer, A., Eriksen, E., Gjerde, O., Hilger, C, Lindström, K., Persson, S., Randen, H., Sletten, T. (2000) 'A Free Electricity Market', feature article in *Nordel Annual Report 1999*, Nordel, http://www.nordel.org, pp. 35–40.

11

Wind Power in the German Power System: Current Status and Future Challenges of Maintaining Quality of Supply

Matthias Luther, Uwe Radtke and Wilhelm R. Winter

11.1 Introduction

Political support and, in particular, the fact that the Renewable Energy Sources Act (EEG) in Germany came into force in April 2000 have led to a dramatic increase in the supply of power generated from wind energy. The present situation and the foreseeable offshore developments in the North Sea and Baltic Sea require grid operators to take supplementary measures to meet quality-of-supply requirements in the future. Power supply quality includes quality of voltage and network frequency as well as the interruption frequency and duration, both regarding the power system and customer supply. The integration of renewable energy sources, especially wind energy generation, changes the power system conditions, which does not necessarily have to lead to negative changes in the supply quality.

This chapter summarises the experience with wind power integration in Germany. We will especially focus on the E.ON Netz control area, with an installed wind energy capacity of 5500 MW (as of April 2003). The chapter considers various aspects of future network development and operation with particular reference to system performance on the grid. We will include information on steady-state performance. Additionally, the results of dynamic studies and the resulting requirements regarding wind turbines are presented. It has to be mentioned, that this has been done in close cooperation with

Wind Power in Power Systems Edited by T. Ackermann
© 2005 John Wiley & Sons, Ltd ISBN: 0-470-85508-8 (HB)

manufacturers of wind turbines[1]. Finally, further issues regarding the integration of wind energy into the power system will be discussed.

11.2 Current Performance of Wind Energy in Germany

A look at the wind energy statistics reveals the extent to which this renewable energy source is used in Germany and the dynamic nature of its development. In December 2003, around 14 000 MW were installed in a total of 15 694 wind turbines.

Between April 2002 and December 2003, in a lapse of only 20 months, 3984 additional wind turbines with a total capacity of about 4686 MW were installed and connected to medium-voltage (MV), high-voltage (HV) and extra-high-voltage (EHV) systems. By April 2003, there was a 32 % increase in installed capacity as compared with April 2002 (see Table 11.1). The rate of increase in installed capacity between April 2002 and April 2001 was 44 %.

In the future, there will be no further increase in network load in Germany or in the E.ON Netz area. For the years 2003 and 2004, the rate of increase in newly installed capacity is expected to be reduced to approximately 1500 MW per year. This will be the result of the saturation due to the limited number of areas designated for the use of wind energy.

Figure 11.1 shows a recent forecast of the power supply from wind turbines in Germany up to 2030. The main source of growth here is offshore power generation. At the moment, it is uncertain which projects will ultimately be implemented, and it is also unclear what effect re-powering will have (i.e. the replacement of existing wind turbines by more powerful onshore units). Hence, the planning certainty regarding the future development of the grid is rather low. Several offshore projects are planned in areas with an appropriate water depth in the North Sea and Baltic Sea. Figures 11.2 and 11.3 provide an overview of the projects that are currently under discussion.

The increase in onshore wind power combined with foreseeable offshore developments in the North Sea and Baltic Sea area require further action by the transmission system operator (TSO) in order to guarantee the necessary quality of

Table   11.1 Development of wind energy use in Germany

	Connected wind turbines	Installed capacity (MW)
December 2003	15 694	13 955
April 2003	13 746	12 223
April 2002	11 710	9 269
April 2001	9 525	6 435

---

[1] The authors would like to thank the wind turbine manufacturers organised in the Fördergemeinschaft für Windenergie for fruitful discussion and for their contribution.

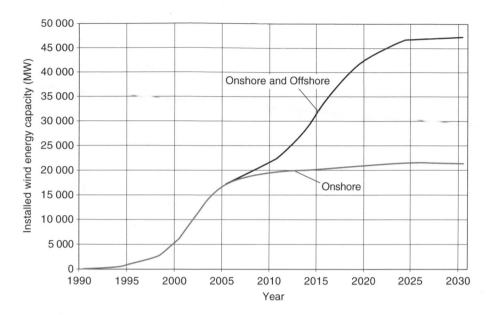

Figure   11.1 Forecast of power supply from wind energy in Germany (www.dewi.com) (Reproduced by permission of DEWI, Germany.)

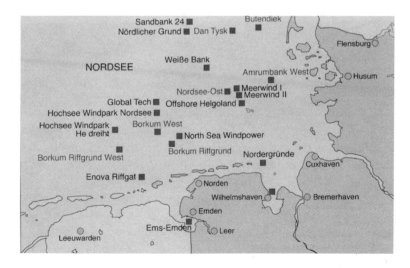

Figure   11.2 Overview of planned offshore wind farms in North and Baltic Sea (Reproduced by permission of DEWI, Germany.)

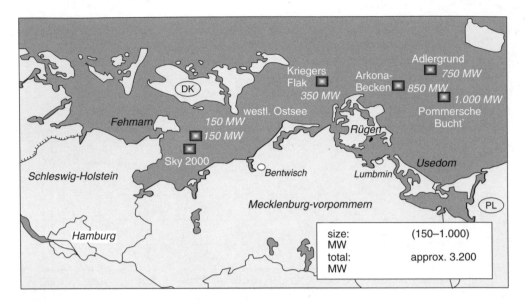

Figure   11.3 Planned offshore wind farms in North and Baltic Sea (Reproduced by permission of DEWI, Germany.)

supply. The German experience is likely to be helpful for further developments in the Union for the Coordination of Transmission of Electricity (UCTE; see also http://www.ucte.org).

11.3 Wind Power Supply in the E.ON Netz Area

The transmission system in Germany is presently controlled by four TSOs. Figure 11.4 shows the areas of the individual TSOs as well as their share of installed wind power generation.

In April 2003, the power supply from wind turbines installed in the E.ON Netz area was about 5500 MW, which represents nearly half of the total installed capacity in Germany. Most of the wind turbines are connected to the MV system in the coastal areas of the North Sea and Baltic Sea (i.e. to the systems that are operated by regional distribution utilities). As the specific capacity of the current wind turbines increases and the turbine operators form groups, there is a growing need for large wind farms to be connected to the HV and EHV system.

Figure 11.5 shows the difference between the wind power fed into the network and the network load during a week of maximum load in the E.ON Netz control area. It is a typical snapshot of the relationship between consumption and supply in this control area. In the Northern part of the power system, the transmission capacity of the HV system reaches its limits during times with low load and high wind power production. Network bottlenecks occur because of the regional generation surplus during low electricity consumption on peak wind power production periods.

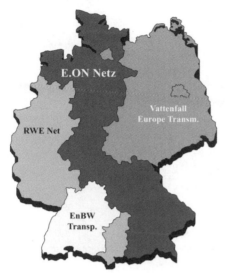

Total of installed wind energy of 12 223 MW :

E.ON Netz: 48 %
Vattenfall Europe Transmission: 37 %
RWE Net: 14 %
EnBW Transportnetze: 1 %

Figure 11.4 Distribution of installed wind capacity in Germany as of April 2003

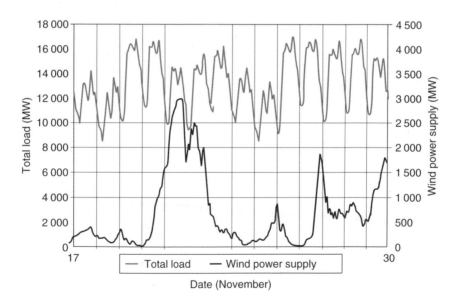

Figure 11.5 Network load and aggregated wind power generation during a week of maximum load in the E.ON Netz control area

11.4 Electricity System Control Requirements

As the meteorological situation can be forecast only to a very limited extent, wind power plants feed in a stochastic form of generation. One major factor of uncertainty here is how precise the forecasts are regarding the exact time of a certain weather condition.

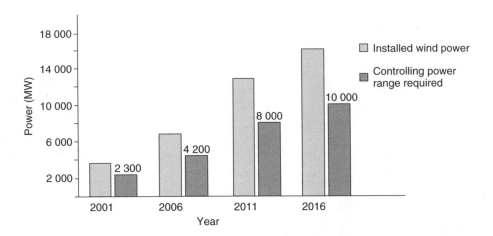

Figure   11.6 Forecast wind power development and controlling power range required (Haubrich, 2003). Reproduced by permission of RWTH, Germany

In the E.ON Netz control area, in high winds the WTs were recorded to feed in a total of about 4800 MW, with fluctuations amounting to 500 MW within a 15-minute period. Occasionally, the controlling power range that is needed during such conditions already exceeds the capacity available in the control area and therefore has to be provided from outside. If this is the case, the controlling power has to be imported from other TSOs via the interconnection lines. For this, the available transfer capacity of the interconnection lines has to be taken in account.

The planned increase in wind power will lead to an increased need for controlling power. Figure 11.6 (Haubrich, 2003) shows the results of a study on how E.ON Netz might be affected by that. The controlling power is required to substitute the positive or negative deviation between actual wind power supply and planned wind power generation; in other words, it has to compensate for the wind forecast failure. The amount of wind power on the grid increases and will continue to require sophisticated congestion management in order to ensure system security, quality of supply and nondiscriminatory access to the grid. Congestion management is carried out by changing control states and, increasingly, by re-dispatching, that is, reducing or relocating generation if limits are reached during a sudden increase in wind power supply, for instance. The TSO has to switch from a daily bottleneck forecast to online cycles in order to match the stochastic phenomenon of wind power generation.

11.5 Network Planning and Connection Requirements

Given the growth rates mentioned above, the priority acceptance of energy from renewable power generation facilities specified in the Renewable Energy Sources (RES) Act will face technical limits. These limits relate to the different grid criteria that apply at the regional level and to the voltage level concerned. For the synchronously connected overall system, the following criteria are essential for secure and reliable operation:

- thermal overloading;
- voltage stability;
- frequency stability.

Thermal overloading occurs when the dispatch of the supply that is fed in exceeds the regional transmission capacity of the system. This can occur, for instance, during times with high wind speeds and low loads on the grids in the coastal regions. The local wind turbines feeding into the MV network then exceed the regional load many times over, which leads to feedbacks and violations of voltage standards in the HV system.

In the medium and long-term, it will be necessary to upgrade the grid of the HV and EHV system in order to remove bottlenecks. This is very well illustrated by the situation in the Schleswig–Holstein region, for which E.ON Netz has worked out an extensive 110 kV upgrading strategy. In view of the current licensing situation, however, it appears doubtful whether grid upgrading will be able to keep pace with the development of wind power. The introduction of generation management could provide temporary relief, however. Generation management is a method of avoiding violations of transmission system standards (e.g. overloading of equipment) by reducing the wind power that is fed in by an adequate portion. This generation management will be limited to the period until additional or new equipment has been installed to allow higher amounts of wind power transfer.

The applications for the necessary grid upgrading schemes have already been submitted. These schemes will be considerably extended, though, by the development of offshore wind power. Figure 11.7 provides an indication of possible developments

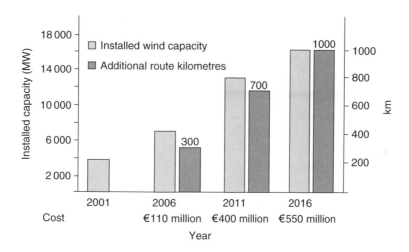

Figure 11.7 Wind power development and grid upgrading (Haubrich, 2003): installed wind capacity and accumulated additional route kilometres, 110 kV and 380 kV (Reproduced by permission of RWTH, Germany.)

within the E.ON Netz control area. Regarding the expected offshore development, additional EHV overhead lines will have to be built in order to transport the power that is generated offshore to the places with high consumption density (e.g. the Ruhr area or the region of Frankfurt-an-Main). Also, in the coastal regions of Germany, the HV networks need to be extended in order to take in the power from the onshore wind farms. Figure 11.7 shows the total additional line kilometres in relation to the increased installed wind generation capacity (Haubrich, 2003).

Modern power systems have evolved into very large systems, stretching out over hundreds and thousands of kilometres. These synchronously interconnected systems must operate stably also for disturbancies. So called interarea oscillations in the low frequency range can affect the complete system. Furthermore, the unbundling of generation, transmission and supply as well as large amounts of wind power fed into the system do not really correspond to the physical nature of synchronously interconnected power systems, which cover a large area where different subnetworks and power plants interact. However, in this new environment, with a possible higher loading of the transmission system, TSOs may be forced to operate the system closer to its stability limits.

Therefore, there are detailed studies regarding the steady-state stability of the UCTE power system. Their goal is to analyse how the reliability of system operation can be maintained given a future environment with further system extensions and an open market (Bachmann *et al.*, 1999, 2002; Breulmann *et al.*, 2000a, 2000b). As a consequence, voltage stability as well as frequency stability and small signal stability are of increasing importance to large interconnected power systems. The frequency of the observed interarea oscillations in the UCTE system lies in the range of 0.22 to 0.26 Hz, and in most cases the damping is sufficient. In some cases, however, there was only poor damping, which has to be considered in relation to reliable system operation. These oscillations in the low frequency range must not excite subsynchronous resonances on the wind turbine shaft, on the one hand. On the other hand, the turbine generator unit control must not excite power oscillations, especially in the low frequency range.

Large-scale wind generation will also have a significant affect on system stability and on frequency and damping of interarea oscillations. Unless appropriate measures are taken, the increasing connection of large-scale wind farms and offshore capacities may lead to additional limiting criteria, because of deficits in reactive power supply or disturbances in dynamic system performance. This may result in critical situations in the overall system, including cascading events and finally the loss of stability. E.ON Netz was the first TSO in Germany to implement additional requirements for the dynamic behaviour of wind turbines in 2003. This was necessary to guarantee system stability in spite of changes in the generation of electrical energy (Lerch *et al.*, 1999).

The results of the investigations show that if the HV and EHV systems are to continue to operate reliably, additional demands will have to be made regarding the connection of wind turbines to the system. These refer especially to providing the required support (Lerch *et al.*, 1999) and involve:

- setting the frequency reduction protection at 47.5 Hz to provide grid support in the event of major disturbances;
- additional requirements of active and reactive power control of wind turbines on the basis of a case-by-case review;

- reactive compensation to avoid system disturbances [e.g. use of static VAR compensators (SVCs) and flexible AC transmission systems (FACTS)];
- operating control through direct integration into implementation and control and protective relaying;
- additional dynamic requirements of wind turbines in the event of faults.

Moreover, the studies show that other neighbouring TSOs have to be involved in further network investigations because fast wind power development affects the secure operation of the entire system. The initial studies in Germany will be followed continuously by investigations covering the whole UCTE area (Deutsche Energieagentur, 2004; E.ON Netz GmbH, 2003).

11.6 Wind Turbines and Dynamic Performance Requirements

The expected expansion of wind power generation in Germany and in Europe calls for additional measures from network operators and also regarding the design of new wind turbines. New requirements are necessary to keep quality and supply standard, both now and in the future.

Energy generation, in general, has been altered by the large number of wind turbines, which have characteristics that are different from those of large power plants. Large thermal power stations require many hours after a breakdown before they can resume operation. As opposed to this, wind turbines are able to resynchronise themselves with the network and provide active power in accordance with the prescribed index values after a short interval (e.g. after a few seconds), following disconnections caused by mechanical failure. However, wind turbines do not yet offer system services on the same scale as conventional power stations. These are actively involved in the network regulation mechanism (e.g. spinning reserve, power frequency regulation, voltage regulation and supply of reactive power). Conventional power stations are able to contribute during several seconds substantially to short-circuit power, because of their rotating centrifugal masses, which both maintains voltage levels and ensures network protection functionality.

In grids where wind energy generation is spread out over a large area and conventional generation from large power stations is simultaneously reduced, severe voltage drops and frequency fluctuations are more likely if the short-circuit capacity is decreased. This leads to both a loss of voltage stability and the selective operation of the network protection systems. Special attention has to be paid to the dynamic behaviour of wind turbines in the event of network failures. These facts make it necessary to define new grid requirements, particularly for wind turbines, in order to include them in the network regulation mechanism. These additional requirements have to take into account the technical standard of decentralised energy supplies, in contrast to conventional power stations. Such additional requirements will help to guarantee a stable and secure grid operation.

11.7 Object of Investigation and Constraints

The installation of wind turbines, especially in northern Germany, has been increasing for years now. In the beginning, single wind turbines were connected to the MV distribution grid. At that time, owing to economical and technical aspects, mostly stall-controlled induction machines without power electronic equipment and control options were installed. Therefore, in case of interactions wind turbines are disconnected if the voltage drops below 80 % of the nominal voltage. After some time, the disconnected turbines are reconnected to the system and return to their operating point.

In recent years, wind farms have mostly been connected to the transmission system (110 kV system and higher voltage levels). Owing to the large amount of installed wind power, the disconnection of wind turbines for an indefinite amount of time in the case of system faults can lead to critical situations, with operation close to stability limits resulting in consequences for the entire synchronously interconnected UCTE system.

E.ON Netz has been the TSO that has been affected the most by the connection of large-scale wind generation to the transmission system. In 2001 it therefore started investigations concerning the maximum outage of installed wind power in the case of system faults (Haubrich, 2003). The results showed that already in 2002 there was a risk that around 3000 MW of wind power would disconnect in the case of heavy system faults located in the northern 380 kV system. Outages of power generation exceeding 3000 MW are defined as dangerous in the UCTE synchronously interconnected system. As a consequence, E.ON Netz defined additional dynamic requirements for wind turbines connected to the transmission system. They have been valid from January 2003. From that time, a wind turbine is not allowed to disconnect if there is a system fault with voltage drops above the line shown in Figure 11.8 (E.ON Netz GmbH, 2001).

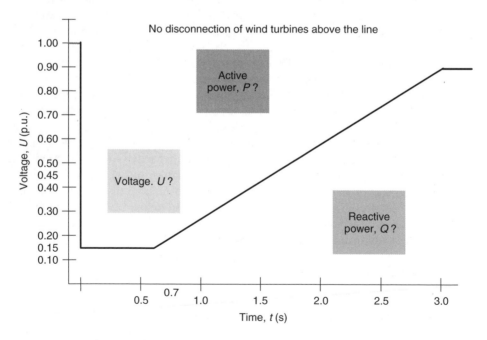

Figure   11.8 Requirements for wind farms valid in the control area of E.ON Netz since January 2003

In addition to the active power balance in the transmission system, the voltage level and the reactive power balance have to be taken into account. Therefore it was necessary to study the dynamic behaviour of wind turbines in the case of system faults in more detail. In particular, the following questions concerning system stability were focused on in this study (Figure 11.9):

- the allowable gradient of active power return of wind farms after fault clearing;
- the limitation of reactive power consumption in the case of system faults;
- the necessary backup voltage operation of wind farms;
- the admissible loading of coupled lines for primary control aspects;
- frequency control in the case of increasing frequency margins.

Initial investigations have shown that additional requirements for wind turbines concerning active power return, backup voltage operation and reactive power balance are absolutely necessary. Otherwise, severe network failures resulting from reactive power consumption of wind turbines after fault clearing could lead to a nonselective disconnection of entire network segments and consumers in the near future.

The goal of the investigation was to determine the constraints to an optimal operation of wind farms in the case of severe system failures concerning active and reactive power balance and the stress to on wind turbines. These additional requirements in combination with the ongoing repowering of older turbine types resulted in future outages of wind power as a result of faults in the EHV network being limited to less than 3000 MW.

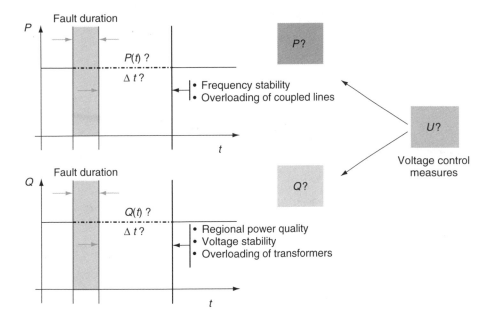

Figure   11.9 Optimal operation of wind turbines concerning active and reactive power balance. *Note*: P = active power; Q = reactive power; U = voltage; t = time

These constraints were added to the Grid Code, which has been valid for the control area of E.ON Netz since August 2003 (E.ON Netz GmbH, 2003).

The requirements were determined without any reference to the type of wind turbine. The requirements have to consist of entirely nondiscriminatory measures that can be applied to all types of wind power installations. This way, the requirements governing the dynamic behaviour in the event of network failures will not cause any distortions in the competition. It may be necessary to add or modify the design of wind turbines and wind farms in order to comply with the system-defined limits.

For the investigation of system interaction, a dynamic model of the complete UCTE system was used. This model is applied in the steady-state analysis and accurately represents all the power system components involved in physical phenomena of system dynamics. Almost the complete 220 and 380 kV transmission grid of UCTE is represented using a specific software package for power system simulation (Lerch *et al.*, 1999). The dynamic model contains:

- 2300 transmission lines;
- 820 transformers 400 kV and 220 kV;
- 500 generators
- 50 wind turbine equivalents in the northern part of the control area of E.ON Netz.

Typical load-flow situations obtained from real-time operation were included in the study. In order to limit the model size, all the units of one generation site are aggregated in one machine, which is valid as far as small-signal stability studies are concerned. The dynamic data, that is the generator characteristics, the excitation controllers, the power system stabilisers, the turbine characteristics and the governors, are mostly described by detailed models in accordance with IEEE standards (IEEE, 1987, 1992). Special attention was paid to modelling wind turbines, wind farms and the 110-kV transmission system in the area bordering the North Sea.

11.8 Simulation Results

11.8.1 Voltage quality

Investigations regarding the effect of wind farms on the entire system have shown that wind farms influence voltage quality, frequency stability and voltage stability, depending on the technology that is available at the time. In particular, wind farms operating with stall-controlled wind turbines may cause lower voltage levels after faults in the transmission system are cleared because of the reactive power consumption of induction machines.

Owing to modern converter technology, limited and full variable-speed wind turbines with asynchronous or synchronous induction generators usually affect the reactive power balance of the system to a lower extent [see Figures 11.10(a) and 11.10(b)]. In the case of severe faults (e.g. deep voltage drops) crowbar firing leads to a disconnection of the converter. During this time, the wind turbines are not controllable, as opposed to synchronous generators using rotating DC-excitation systems. Figure 11.10 shows the grid behaviour of wind turbines in the case of crowbar firing.

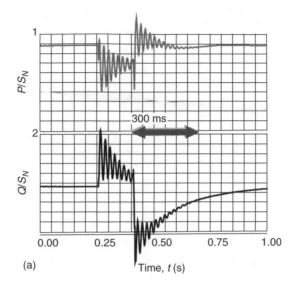

(a)

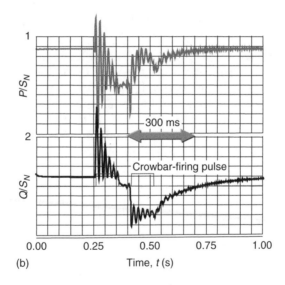

(b)

Figure 11.10 System failure in the case of a three-phase fault close to the wind farm for (a) a wind farm operated with fixed-speed wind turbines (Type A) with an asynchronous induction generator with reactive compensation; (b) a wind farm operated with limited variable-speed wind turbines (Type C) with doubly fed asynchronous induction generators; and (c) a wind farm operated with variable-speed wind turbines (Type D) with synchronous induction generators with full-load frequency converters. In each case: rating = 45 MW; short-circuit power at point of common connection (PCC) = 1000 MVA; nominal voltage at PCC = 110 kV. *Note*: P = active power; Q = reactive power; S_N = nominal three-phase power of induction generator; turbine Types A, C and D are described in Section 4.2.3

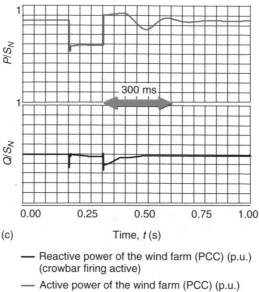

(c)   Time, t (s)

— Reactive power of the wind farm (PCC) (p.u.)
  (crowbar firing active)

— Active power of the wind farm (PCC) (p.u.)

Figure   11.10 (*continued*)

Wind turbines connected to the grid through a full-load frequency converter [Figure 11.10(c)] also have less effect concerning the reactive power balance, because wind generators and the AC network are decoupled.

In contrast to direct AC-coupled synchronous machines, all types of wind turbine can only supply short-circuit currents that are limited to the nominal current. As the investigations have shown, with an increasing amount of wind energy generation connected to the grid, wind turbines have to stay connected to the grid during system faults and have to return quickly to their pre-fault operation point. However, wind turbines that do not comply with the additional dynamic requirements may affect the reactive power balance when remaining connected during the fault. The investigations showed a decreasing voltage level, especially on the 110-kV grid, after fault clearing, because the wind turbines that stayed connected to the grid without complying with the additional dynamic requirements consumed reactive power.

Depending on network status, mainly the Schleswig–Holstein area and the northern part of Lower Saxony would be affected and run a risk of undervoltage-tripping of transmission lines and partial consumer outages.

Modern wind turbines can, depending on their capacity, participate in voltage and frequency regulation in a similar way to conventional power stations. In order to maintain and improve voltage quality, it is practical to switch from constant load power factor ($\cos \varphi$) regulation to backup voltage operation if there are network failures that cause voltage drops of more than 10 % of nominal voltage at individual wind generators. This ensures that terminal voltage does not drop sharply again after fault clearing. Investigation results also show that the quality of the network voltage in the event of failures can be considerably improved by applying this measure (Figure 11.11 and 11.12).

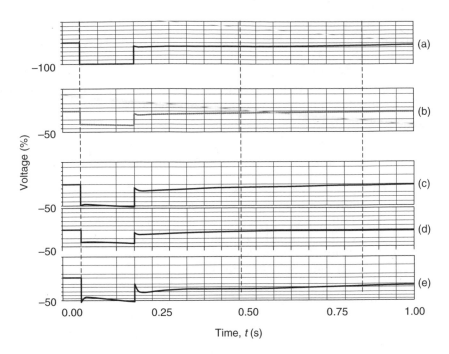

Figure 11.11 Voltage drop (as a percentage of the nominal voltage) after a three-phase fault in the northern part of the E.ON Netz control area (close to the Dollern substation): (a) 380-kV system; (b) 220-kV system; (c) 110-kV system, EMS Elbe; (d) 110-kV system, Weser EMS; (e) 110-kV system, Schleswigs Holstein

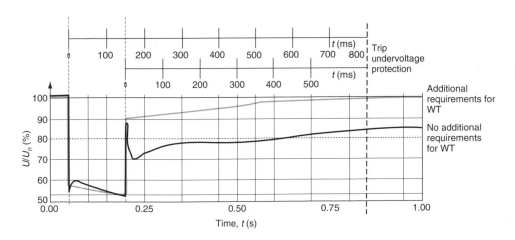

Figure 11.12 Recovery of the grid voltage in the 110 kV system after a three-phase fault in the 380 kV system. *Note*: U = voltage; U_n = nominal phase-to-phase voltage; WT = Wind Turbine

In order to maintain network voltage quality, particularly in 110-kV networks, in the event of network failures it is necessary to prevent as far as possible any repercussions from the wind farms on network voltage. Therefore, wind turbines must return as quickly as possible to their pre-fault operation point. The transient compensation that can occur during failure correction and the return to the voltage must be limited to 400 ms. This additional requirement will prevent long-term reactive power consumption of wind turbines and a voltage decrease to critical levels (Figure 11.12).

11.8.2 Frequency stability

In contrast to conventional power stations, which only experience outages in the case of bus-bar failures, wind turbines set up in large numbers close to one another over wide areas can experience lasting active power outages of up to 2800 MW, even as a result of circuit failures in the EHV network. The outage level is that high because all wind turbines connected before December 2002 disconnect for an indefinite time from the network if generator terminal voltage drops below 80 % of the nominal voltage level. According to simulation results, at present the stationary frequency deviation is expected to be approximately 160 mHz in the case of outages of around 2800 MW in the northern part of the E.ON control area. Figure 11.13 shows the stationary frequency deviation in connection with a three-phase circuit failure near Dollern substation, which is located in Lower Saxony, close to Hamburg.

Additional dynamic requirements in combination with effects from re-powering (i.e. replacing old wind turbines with new ones that comply with network access requirements) have to guarantee within the following years a reduced stationary frequency deviation arising from disturbances in the transmission system. Therefore, it will be important to limit the loss of installed wind power in the event of network failures. In particular, offshore wind power, which will be concentrated to only a few substations

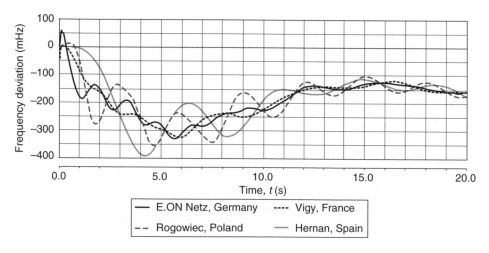

Figure 11.13 Frequency deviations in the UCTE system after an outage of 2800 MW wind power in Northern Germany

close to the sea, must stay connected during severe system faults very close to the point of common coupling (PCC). With modifications of the network connection contract, this can be achieved by upgrading the wind turbines that are already in operation.

Since 1 January 2003 wind turbines that remain connected to the network in the event of network failures have been required to start providing active power immediately after fault clearing and to reach pre-fault active power capacity within 10 s in order to prevent any further increase in the wind power capacity that is temporarily disconnected during such failures. As a result of these measures, the stationary frequency deviation in the event of severe network failures is expected to be less than 200 MHz (about 160 MHz) until 2005, thus remaining below the frequency deviation permitted in the UCTE synchronous integrated network.

Figure 11.13 shows the variations in the increase in active power together with the time at which wind turbines return to their pre-fault operating point, following network failure and failure correction. For the year 2005, approximately 2000 MW of brief active power outages, lasting only a few seconds, are to be expected in the event of severe network failures (in high winds, 90 % wind power), in addition to the 2800 MW of lasting active power outages. The dashed line in Figure 11.14 shows the frequency deviation in the control area of E.ON Netz taking into account a return to full active capacity after one minute.

Even if all newly installed wind turbines increase their active capacity and are able to resume supplying the network with the pre-fault active capacity within 40 s after the fault is cleared, there could still arise critical conditions regarding the allowable frequency deviations and excitation and trip of overcurrent line protection. Therefore, the wind turbines must return to supplying active power as fast as possible. If all newly installed wind turbines increase their active capacity following failure correction at a gradient of 10 % of nominal capacity per second, the network frequency will reach a frequency deviation of 150 MHz within approximately 10 s after the onset of the disturbance.

Disturbances can be divided into two groups: Near-to-generator three-phase short circuits with deep voltage drops and far-from-generator three-phase short circuits.

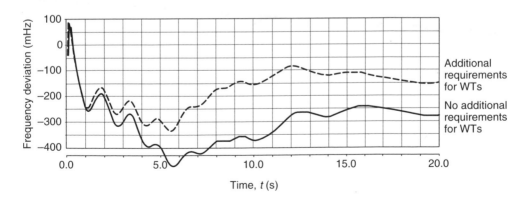

Figure   11.14 Transient and stationary frequency deviation as a function of the gradient of the active power increase of the wind turbines (WTs) following failure correction

Near-to-generator disturbances cause, depending both on the short-circuit ratio at PCC and the type of wind turbine, a varying degree of stress on electrical and mechanical parts of the wind turbines (e.g. gearbox, shaft) that remain connected to the system during the fault. In addition, such deep voltage drops may lead to crowbar firing for doubly fed induction generator (DFIG) wind turbines. In the case of deep voltage drops and higher short-circuit power ratios, induction generators, in general, affect the system by consuming reactive power. In such cases, the active power capacity cannot be raised as quickly as necessary for stable integrated network operation, even if the turbines remain connected to the network. In this kind of situation, there may occur an undesirable compensation between network and wind turbines or wind farm which also causes a high level of mechanical stress on the wind power installation and may considerably shorten its lifetime.

Thus, it seems reasonable in specific cases to exempt wind turbines from the given requirements and allow them briefly to disconnect. Therefore, wind turbines that are affected to varying degrees by network failures are divided into the following two groups, based on voltage drops at the PCC, which increases the active power with different gradients:

- wind turbines affected: in the case of a fault, network voltage collapses to 45 % of residual voltage at the PCC;
- wind turbines stable: in the case of a fault, network voltage remains above 45 % of residual voltage at the PCC.

For this research, for the stable group, a gradient of 20 % of the nominal capacity per second was selected and for the affected group a gradient of 5 % of the nominal capacity per second was selected. Figure 11.15 shows that if in the stable group the time to restore

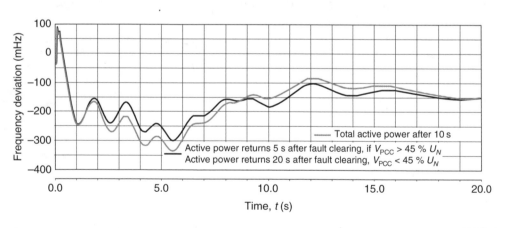

Figure   11.15 Scenario 2005: 2880 MW outage of installed wind power with division of wind turbines into two groups with different active capacity gradients – affected and stable. *Note*: with stable wind turbines the network voltage, V_{pcc}, drops no lower than 45 % of the residual voltage, U_N, at the point of common connection (PCC) in the case of three-phase faults in the transmission system; with affected wind turbines, V drops lower than 45 % of U_N in the case of three-phase faults in the transmission system

the active power levels of all wind turbines in a wind farm is reduced to 5 s and in the affected group to 20 s this constitutes a value equivalent to the requirement presented above regarding active power supply.

If individual wind turbine designs are unable to comply with the requirements, it may be reasonable to switch such turbines selectively into backup voltage operation during a brief period, thereby providing relief to the network voltage. In order to protect regional supply structures and wind turbines during faults with voltage drops below 45 % of the nominal network voltage at the PCC, it may also be reasonable in such cases to permit *a very brief* disconnection of the affected turbines or farm followed by a fast return to the pre-fault operation point.

This, however, requires a rapid resynchronisation of the wind generators to the network after the fault is cleared and a fast return to the normal pre-fault operation in order to ensure that the entire outage does not last longer than 10 s. In this case, the turbines or farm must start supplying active power within 2 s after fault clearing. Active power has to increase at a gradient of at least 10 % of nominal active power. In most cases, the fast return to pre-fault active power levels and frequency stability relating to voltage quality must be given first priority. In cases of slow voltage recovery (e.g. when the voltage remains below 10 % of the nominal voltage after the fault is cleared) voltage stability and local voltage quality will be given first priority for the return of the wind farm to normal operation.

Regarding the forecast for the year 2005, the results of the investigation show that the stable turbine group will have to account for approximately 2 800 MW of installed capacity, which can very rapidly return to its pre-fault operating point after a fault is cleared. There are only about 200 MW in the affected group, which in the case of severe network failures will be allowed a longer time to restore active capacity, because of the behaviour of these turbines. This will not affect stationary frequency deviation, if the stable group increases active power supply with a gradient of 20 % of nominal power.

Overloads on interconnecting feeders and the effect they have also need to be studied. The settings for overcurrent time protection (I very large) have to be included as a reserve protection for the remote protection of the transmission lines, because they can cause a circuit switch-off in the few cases where the current in the protection-end stage (3 s) reaches 1.4 times the nominal current value, unless the conductor current returns to below 0.9 times the nominal current value.

Simulations for the year 2005 show that no trip of the overcurrent time protection on the affected interconnecting lines is to be anticipated through compliance with the described measures (Figure 11.16). Further studies are required to study this phenomenon in Germany (Deutsche Energieagentur, 2004).

If all installed wind farms comply with the requirements presented here, a considerable improvement of the overall behaviour is to be expected in comparison with the current situation. Auxiliary wiring configurations and changes to the regulating design and/or the layout of the wind turbines may be necessary to comply with the requirements.

Wind farms that do not have regulation capabilities will be able to comply with the requirements if they use dynamic compensation facilities (controllable elements, FACTS), or merge with wind turbines or wind farms that can be controlled. An additional increase in wind power through repowering older wind turbines using

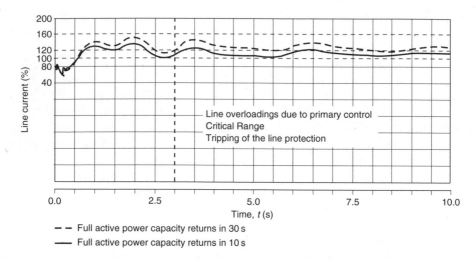

Figure 11.16 Overloading of coupled lines after a three-phase fault in the 380-kV transmission system: line current as a percentage of nominal current as a function of time, *t*

modern wind technology, in combination with the above-mentioned requirements regarding the dynamic behaviour of wind farms, will allow a considerable reduction in the maximum expected wind power outage capacity. This means that, in turn, such system-defined limits as voltage quality, voltage and frequency stability will continue to be fulfilled in the future.

The requirements regarding the dynamic behaviour of wind turbines in the event of network failures are a first step regarding regulations throughout the European Communities as a whole. Over the next few years, large network studies with detailed representations of wind turbines will be carried out to take into account the increasing amount of onshore and offshore wind power and modify the requirements accordingly.

11.9 Additional Dynamic Requirements of Wind Turbines

The dynamic requirements for wind turbines installed in the E.ON control area are based on dynamic network studies. They describe the behaviour of wind turbines in the event of network failures. For the current and future integration of large amounts of wind power into the European transmission system, wind turbines have to comply with the following regulations [14,15,16,17]:

Three-phase short circuits close to power stations shall not as a general rule lead to generation unit instability nor to disconnection from the network above the lower solid line in Figure 11.17. Active power capacity must begin directly after fault recognition and be increased at a gradient of at least 20 % of the nominal capacity per second. The active power increase can take place at a rate of 5 % of the nominal capacity per second within the hatched area in Figure 11.17.

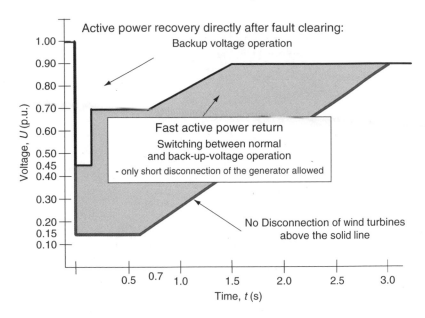

Figure 11.17 Progression of the line-to-line voltage: above the solid line, wind turbines are not allowed to disconnect automatically in the case of grid failures

Short circuits not close to power stations shall not result in a disconnection of the generation unit from the network, even with failure correction in the end stage of network protection of up to 5 s. The stipulations concerning voltage support must be complied with at the same time.

The generation units shall provide voltage support in the event of three-phase short circuits in the network. A switchover to voltage support is to take place in the event of a voltage drop of more than 10 % of the momentary value of the generator terminal voltage. Support of network voltage must take place within 20 ms following failure recognition by supplying reactive current with a factor of 2 % of the nominal current for each percent of the voltage drop. After 3 s, it is allowed to switch back from voltage control to normal operation.

The transient compensation processes in relation to the reactive power consumption following voltage return in quadrant II has to be completed within 400 ms, and the demands regarding voltage support require the exchange of reactive power. Exceptions to these basic principles are possible and have to be agreed upon. The following stipulation applies in such cases:

Within the broken-lined area of Figure 11.18, wind turbines may be disconnected briefly from the transmission system if resynchronisation of the turbines following failure correction does not exceed 2 s and the active power capacity is increased at a gradient of at least 10 % of the installed nominal active capacity per second.

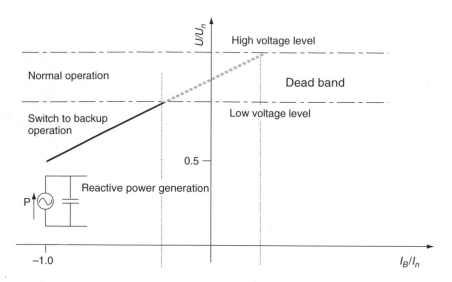

Figure   11.18 Contribution to backup voltage operation in the case of network failures: plot of voltage, U, is a percentage of nominal phase-to-phase voltage, U_n, against reactive current, I_B, as a percentage of rated current, I_n, for voltage drops caused by load flow variation, switching operations and short-circuit faults

11.10 Conclusions

The political support for the use of renewable energies and the promotion of the use of wind power has led to a substantial increase in wind power supply in Germany and especially in the E.ON Netz control area. When transporting and distributing this electricity, the grid reaches its technical limits, and bottlenecks are created on the grid. For a transitional period, these bottlenecks can be handled by using local generation management. In the medium and long term, however, the development of wind power will require a considerable strengthening of the HV and EHV systems in order to ensure an appropriate quality of supply in the future.

The marked growth in wind power supply increases the need for power station reserve capacity, which is required in cases when the actual values of wind power supply differ from predicted wind power supply. The provision and use of such power station reserve capacity generates additional costs to the TSO.

In the future, wind turbines will have to be integrated to a larger extent into the system control mechanism and they will have to comply with requirements that are designed to guarantee secure operation of the interconnected grids. Against this background, E.ON Netz has identified and set up specific requirements for wind turbines and wind farms to provide system services. This has been done in close cooperation with different manufacturers of wind turbines and with other TSOs.

However, further studies are necessary to follow developments in the wind sector. These investigations may lead to a step–by-step upgrade of the requirements regarding the capability of wind turbines and wind farms to support the overall system, especially in the case of offshore wind power integration and its connection to the grid.

References

[1] Bachmann, U., Erlich, I., Grebe, E. (1999) 'Analysis of Interarea Oscillations in the European Electric Power System in Synchronous Parallel Operation with the Central-European Networks', paper BPT99-070-12, presented at the IEEE Power Tech 9 Conference, Budapest, Hungary, 1999.

[2] Bachmann, U., Breulmann, H., Glaunsinger, W., Hoiss, P., Lösing, M., Menze, A., Römelt, S., Zimmermann, U. (2002) 'Verbundstabilität nach dem Synchronanschluß der Netze der Elektrizitätsunternehmen von Bulgarien und Rumänien an den UCTE/CENTREL-Verbund', in *ETG-Tage 1999, ETG-Fachbericht 79*, pp. 251–257.

[3] Bouillon, H. (2002) *Auswirkungen des fluktuierenden Energieangebotes auf den Systembetrieb*, VGB-Tagung, Salzburg, Austria.

[4] Bouillon, H., Dany, G. (2003) 'Ausgleichsleistungs- und energiebedarf in Regelzonen mit hohem Windanteil', *Energiewirtschaftliche Tagesfragen* **53** H9.

[5] Breulmann, H., Borgen, H., Ek, B., Hammerschmidt, T., Kling, W., Knudsen, H., Menze, A., Ross, H., Ring, H., Spaan, F., Winter, W., Witzmann R. (2000a) 'New HVDC Power Links Between UCTE and NORDEL – Analysis of AC/DC Interactions in the Time and Frequency Domains', paper presented at Cigre Conference 2000, Paris.

[6] Breulmann, H., Grebe, E., Lösing, M. Winter, W., Witzmann, R., Dupuis, P., Houry, M., Margotin, P., Zerenyi, J., Dudzik, J., Martin, L., Rodriguez, J. M. (2000b) 'Analysis and Damping of Inter-area Oscillations in the UCTE/CENTREL Power System', paper presented at CIGRE Conference, Paris, 2000.

[7] Deutsche Energieagentur (2004) 'Energiewirtschaftliche Planung für die Netzintegration von Windenergie in Deutschland an Land und Offshore bis zum Jahr 2020, dena-final-report. 2004'. www.deutsche-energie-agentur.de/

[8] E.ON Netz GmbH (2001) 'Ergänzende Netzanschlussregeln für Windenergieanlagen', available at http://www.eon-net2.com.

[9] E.ON Netz GmbH (2003) 'Grid Code High and Extra High Voltage. Status 1. August 2003', available at http://www.eon-netz.com.

[10] Haubrich, H.-J. (2003) 'Technische Grenzen der Einspeisung aus Windenergieanlagen', expert opinion, RWTH, Aachen.

[11] IEEE (Institute of Electrical and Electronic Engineers) (1987) 'Committee Report: Computer Representation of Excitation Systems', *IEEE Transactions PAS-87*, No. 6, 1460–1464.

[12] IEEE (Institute of Electrical and Electronic Engineers) (1992) 'Committee Report: Dynamic Models for Steam and Hydro Turbines in Power System Studies', *IEEE Transactions PAS-92*, No. 6, 1904–1915.

[13] Lerch, E., Kulicke, B. Ruhle, O., Winter, W. (1999) 'NETOMAC – Calculating, Analyzing and Optimizing the Dynamic of Electrical Systems in Time and Frequency Domain', paper presented at 3rd IPST '99, Budapest, Hungary.

[14] Luther, M., Winter, W. (2003a) 'Einbindung großer Windleistungen – Systemverhalten und Maßnahmen zur Erhaltung der Versorgungsqualität im Übertragungsnetz', *ETG-Tagung Zuverlässigkeit in der Stromversorgung*, Mannheim.

[15] Luther, M., Winter, W. (2003b) 'Erweiterte Anforderungen an Windenergieanlagen zur Aufrechterhaltung der Systemstabilität', paper presented at HUSUMWind 2003 – der Kongress, Husum, Germany.

[16] Luther, M., Winter, W. (2003c) 'System Operation and Network Development with Large Scale Wind Generation', paper WG B4:38 presented at the CIGRE Conference on the Simulation of HVDC and FACTS, Nürnberg, Germany.

[17] Winter, W. (2003) 'Grid Code Requirements to Wind Generators', paper WG C6 presented at the CIGRE Conference on Dispersed Generation, Montreal, Canada.

12

Wind Power on Weak Grids in California and the US Midwest

H. M. Romanowitz

12.1 Introduction

The first large-scale deployment of wind energy in wind farms worldwide occurred in California during the period 1982–86. A unique combination of Federal and State incentives, combined with standard offer power purchase agreements (PPAs) created an environment that fostered rapid development with early technology, much of it unproven. The perceived abuses such rapid development fostered brought these incentives to an end by late 1985 and wind energy development stalled. These pioneering efforts demonstrated critical business and technology lessons on a massive scale and laid the foundation for future success.

This chapter looks at the historical practical evolution of wind energy from the perspective of an operator and developer struggling through early problems and uncertainties arising from integrating wind energy into weak rural grids on a large scale. It is important to recognize the early state of grid technology, compared with what is available today. Grid protection assumed load and no generation. Monitoring and communication was poor and slow, leaving grid operators largely unaware of critical operating circumstances in real time. Grid models were primitive and did not accurately reflect the impact of a large proportion of induction machines on the grid. Adequate VAR support was feared, and VAR transport relationships were not well understood. Last, those responsible for grid investment questioned that wind energy would ever be anything other than a passing fad, to be tolerated but certainly not accommodated with investment in adequate facilities.

California has three major wind energy resource areas. The Altamont and San Gorgonio areas are located within the two major California load centres, and adjacent to major 500 kV substations (see Figure 12.1). Tehachapi, the most developed wind

---

Wind Power in Power Systems   Edited by T. Ackermann

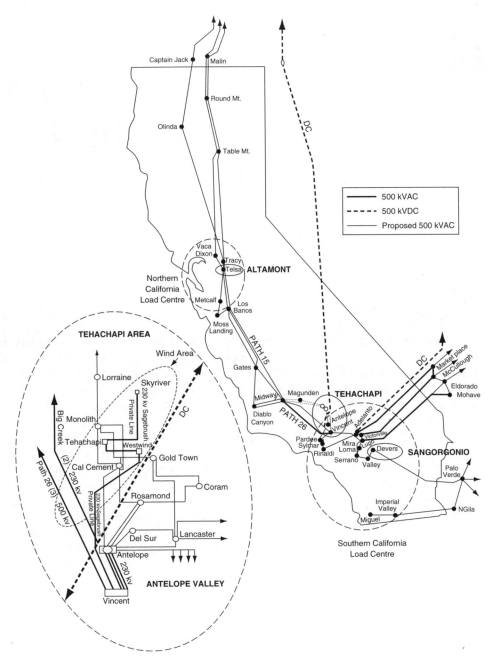

Figure 12.1 California simplified one-line 500 kV grid, major wind resource areas and load centres; Tehachapi grid, including 66 kV subtransmission spread across mountains and desert

resource area, is also the one with by far the greatest potential for additional development. While not remote, the area has not been well integrated into the main grid and is connected by a weak 66 kV system.

Tehachapi presents several interesting and instructive lessons. From necessity, the wind industry has worked closely with the local utility to maximize transfer capability of the existing 66 kV grid with modest upgrades, and may have achieved the greatest transfer capability as a percentage of short-circuit duty of the grid, largely with Type A wind turbines and a small early flexible alternating-current transmission system (FACTS).[1] A small group of Tehachapi developers built a private 230 kV transmission line that has proven to be extremely successful, and presents an expansion opportunity. Tehachapi sits closely adjacent to the major north–south transmission grid of the state (see Figure 12.1) and is an opportunity for substantial effective future integration, if planning and seams interzonal issues can be resolved.

During the later part of the 1990s, wind energy development again began to flourish in the USA, with interesting and important developments in the US Midwest, and with repowers in California. Projects in West Texas and Western Minnesota were developed on weak grids utilizing Type C and D wind turbines with effective results from the enhanced VAR capability of that technology. Projects in Wyoming and North Dakota have utilized high-performance FACTS technology to enhance the weak grids associated with those developments. Repowers in Tehachapi further increased utilization of its weak grid, with Type A and Type B wind turbines, plus added VAR support internal to the projects. The successful achievements from these projects give practical insight into the necessity for and benefit of effective VAR control associated with wind energy projects as they become increasingly large.

12.2 The Early Weak Grid: Background

12.2.1 Tehachapi 66 kV transmission

The Tehachapi Area is served by the Southern California Edison (SCE) company, being on the northern edge of its service territory and immediately adjacent to the Pacific Gas & Electric (PG&E) company, which serves most of Northern California. Service to wind development has been provided by SCE on an existing 66 kV subtransmission and distribution network that covers thousands of square kilometres immediately north of Los Angeles (Figure 12.1).

The Tehachapi area 66 kV system gateway Cal Cement is 40 km from the source Antelope Substation, a 230 kV substation 30 km from the 500 kV Vincent Substation. A network of 66 kV lines winds a web across the desert and mountains of the area. Planned expansion of the grid at 230 kV did not happen for a number of reasons, and currently three 500 kV lines are being planned to serve the area, California's richest wind resource.

SCE was obliged by the state to accommodate renewable energy suppliers, and from 1982 to 1985 signed over 900 MW of standard offer contracts for wind projects in the Tehachapi area, without regard for transmission needs. The primary technical concern

---

[1] For definitions of turbine Types A–D, see Section 4.2.3.

covered rigidly in PPAs was with islanding, causing wind-farm-provided internal VAR support to be severely restricted. Contractual VAR restrictions were aggressively enforced, with the predominant Type A turbines being limited to supplying not more than no-load VARs. In the early years, the interconnection agreement was an appendix to the PPA, with all transmission rights firm.

12.2.2 VARs

VARs are reactive or imaginary power and do no useful work but are essential to properly magnetize and control induction machines, transformers and other inductive devices. VARs, effectively controlled, can cause induction machines to behave much like self-excited synchronous machines. Effective VAR supply can control grid voltage, stabilize the grid and improve grid quality. Chapter 19 has a detailed discussion of VARs.

12.2.3 FACTS devices

FACTS is emerging as a potentially very powerful tool that can materially improve the performance of transmission and distribution systems. It can provide an alternative to 'wire in the air' transmission expansion by unlocking thermal capacity in a line that cannot be used with conventional expansion planning because of the impacts of embedded impedance that is not accommodated conventionally.

Static VAR compensators are at the low end of FACTS technology and have been evolving for nearly 30 years, but only recently have they become widely used. The simplest SVC uses transient free thyristor switching of capacitors to provide automatic VAR support. Such devices can be turned on, off and back on every other cycle and can provide responsive VAR support, including voltage control, to improve grid stability very effectively while supplying the VARs needed by the system. A full-range SVC includes a phase-controlled inductor and can thus also consume VAR, which may be needed where lightly loaded transmission lines are highly capacitive and create unwanted high voltages. The inductor makes the SVC stepless, while the capacitor-only type utilizes one or more steps with granularity dependent on the number of steps. Hybrid SVC is another variation where, in addition, conventional breaker-switched capacitors are added but are controlled in harmony with the fast-switched capacitors to extend the effective capacity at lower cost.

High-end FACTS employ an integrated gate bipolar transistor (IGBT) inverter to couple a capacitor or other energy-storage device to the grid and provide two quadrant stepless control of the VARs, so that they can be full range producing or consuming VARs. An advantage of the IGBT type is that VAR current can be independent of grid voltage. Thus the supplied VARs drop only as the first power of grid voltage instead of as the square of voltage, as with directly connected capacitors or inductors. D-VAR™ is the trade name of one such FACTS device that has had notable success, (see Section 12.3.12).[2]

---

[2] D-VAR™ is a trademark of American Super Conductor.

The IGBT inverters of Type C and D wind turbines are inherently capable of performing a FACTS function very similar to a stand-alone device, and, as will be explained in this chapter, have done so with notable success on weak grids. SVC technology can also be used effectively with Type A and B wind turbines to provide much the same capability, but with VAR support dropping off with voltage more rapidly. As SVC devices may be lower in cost it is more feasible to purchase devices of sufficient size to accommodate voltage drops of as much as 50 %.

Conventional FACTS devices are typically stand-alone products. Most SVCs have been in the form of large custom facilities requiring substantial field construction, but a packaged product is emerging. D-VAR^{TM} type technology has typically been prepackaged and factory built – a product that is delivered, connected and operated but that has been expensive. The effective integration of this technology with wind turbines holds promise for cost efficiencies while performing to a similar level or adequate level according to system needs. The distributed mini SVC discussed in Section 12.3.12 and IGBT Type C or D wind turbine controls allow possible cost efficiencies.

12.2.4 Development of wind energy on the Tehachapi 66 kV grid

The SCE Tehachapi grid of the mid-1980s was designed to serve the Tehachapi–Mojave area along with the outlying High Desert and Mountain areas. The population of the area was less than 20 000 people, with the major served load being several large manufacturing plants, a prison and municipal facilities. The total served load in the entire area was less than 80 MW, and 60 % of that was two large manufacturing plants. The area was served from the Antelope substation by three 40–70 km 66 kV transmission lines and 12 kV local distribution lines.

By early 1986, over 300 MW of wind generation was installed, and SCE stopped further interconnection. The existing grid was more than saturated. Many early wind turbines were not reliable, and a 'shake-out' of good and bad designs quickly eliminated over half the installed capacity. Wind turbine design codes were then generally nonexistent. The strong winds of Tehachapi soon taught many important lessons to the infant industry; within months many wind turbines were scrap, and others seldom ran. Early wind turbines were of small size, and the importance of good siting was not understood, with many turbines being placed in valleys as well as on good ridge-top locations. From the combined effect of early wind turbine unreliability, poor siting, and wind being of different intensity at different locations across the area at any given time, Tehachapi wind generation delivered from the area peaked near only 125 MW in 1986.

The combined generation from all wind turbines in an area at any moment in time is called 'coincident generation' and is a key planning and evaluation measure. Wind blows at different intensities at any instant across a region, and each individual wind turbine produces more or less energy at any moment depending on the wind at its location. The peak coincident generation for the region connected to a transmission grid is the criterium that establishes required facilities since such collected energy is what is transmitted at any moment and is the actual grid need. Such patterns of generation are remarkably consistent and repeatable.

Early grid-related problems were generally associated with unreliability caused by the extensive web of 66 kV lines spread across the desert and mountains, subject to physical damage, lightning strikes and increased loading. Protection schemes were simplistic; control was manual, and operating data were scarce and slowly updated. The skill of a few experienced SCE grid operators contributed substantially to making the marginal 66 kV grid work.

Computer-based wind turbine controllers were uncommon before 1985, and control functions were limited in sophistication and flexibility. Controller reliability generally was low. Utility protection and monitoring schemes were similarly limited in flexibility and were generally designed for operation with large central generation from synchronous generators capable of providing their own excitation and VAR support for the grid.

The utilities were generally unfamiliar with distributed generation and particularly with induction generators. Utility engineers were first mostly concerned about fault clearing and then about self-excitation and islanding in relation to safety issues on the grid. A nearby dramatic resonance event, experienced with transmission line series capacitor VAR support, resulted in major damage to the large Mohave generating station and created excessive caution about VARs generally. There was little experience or understanding about the importance of VAR support for induction generators and of the difficulty of transporting VARs over the transmission system. Contractual rules were thus written severely to restrict wind farms from providing meaningful VAR support, and the foundation for years of difficulty was laid.

12.2.5 Reliable generation

From 1986 through 1991, the US wind industry went through a period of substantial technical and economic challenge, including the bankruptcy or collapse of nearly all wind turbine manufacturers and many wind farms. Warranty obligations were abrogated and the responsibility for the resolution of technical problems shifted from the bankrupt entities to the surviving projects. The local industry worked hard to resolve the technical and economic problems and to get the early turbines to achieve reliable production. Those efforts were quite successful, and by 1991 peak coincident generation from the Tehachapi turbines was well above the then 185 MW grid transfer capability. It became clear that wind energy could provide reliable generation in productive wind resource areas. The utility then made hefty curtailment payments rather than build major transmission, and justified this by the very low cost replacement energy then available from surplus supplies in the Pacific Northwest. Such short-sighted decisions resulting from flawed economic reasoning and self-interest are still being corrected.

Repowering of original projects was tested experimentally in 1996 and 1997 with substantial success. A small number of 500 kW and 600 kW Type A wind turbines replaced original 25 kW to 65 kW wind turbines. Careful siting of the new turbines gave far more productive kilowatt-hours per installed kilowatt, and the new turbines demonstrated substantial reliability. The concept of using tall towers in complex terrain to gain low turbulence and improved suitability was proven. Then, in 1999, there was a substantial repowering of nearly 45 % of the installed capacity on the Tehachapi

66 kV grid. Projects with a capacity factor of 12 % to 20 % or less quickly achieved capacity factors of 35 % to 45 %. Peak coincident generation on the 66 kV grid increased dramatically to 310 MW while total installed capacity remained at 345 MW. (Muljadi *et al.*, 2004).

The increased transfer capability was achieved in progressive steps from 1999 to 2002: although two small transmission upgrades substantially underperformed design expectations, increased windfarm-supplied VARs proved to be extremely effective. In this process, a VAR rationing programme was instituted. The available utility VARs were rationed to each windfarm based on its capacity, and each windfarm had to curtail its output to stay within its VAR allocation. Individual wind-farm transfer capability was thus within the control of each entity, and the economic incentive correctly caused the installation of substantial VARs at the most effective location on the system with dramatically improved transfer capability.

12.2.6 Capacity factor improvement: firming intermittent wind generation

In 2001, an evaluation of the Tehachapi grid showed that while the grid was operating near 100 % capacity for some periods of time, the annual capacity factor for all wind generation on the 66 kV grid was only 27 %. It also showed that if 100 MW of energy storage and wind energy were added, the annual grid capacity factor would be increased to 41 %, while not increasing maximum generation transfer on the grid. Current planning rules in California have not allowed this added capacity, which sits idle for much of the year, to be effectively utilized, this added capacity thus represents a substantial economic opportunity. For intermittent resources, such firming would reduce the environmental footprint of transmission through fewer lines, and increasing its value as a dispatchable, more predictable resource.

In 1992 and 1993 I designed and operated a 2.88 MW 17 280 kWh battery storage facility in the San Gorgonio area, integral with wind generation. The project was the second or third largest battery storage project ever built at the time. It operated successfully for two annual peak seasons for the purpose of meeting firm capacity contract obligations prior to the repowering of an early wind farm. The batteries were surplus submarine batteries, controlled by thyristor inverters, organized in 360 kW modules with six hours storage.

Evolving energy-storage technology will utilize IGBT converters similar to those associated with Type D wind turbines to give a substantial opportunity for enhanced VAR support. When combined with sufficient energy storage capability to create a higher-quality dispatchable wind energy product, transmission facilities can be more fully utilized, costs spread across a larger base, and environmental footprint reduced with fewer transmission lines.

Current grid planning rules provide only firm transmission rights, so that a very substantial opportunity to utilize grid capacity more efficiently is going to waste. Substantial new transmission capacity costs in the range of 30 % to 60 % of the cost of modern efficient wind farms. It is economically efficient to plan to curtail generation for modest periods annually in order better to balance wind farm cost with transmission cost with or without energy storage.

12.3 Voltage Regulation: VAR Support on a Wind-dominated Grid

The Tehachapi 66 kV grid, as noted above, is dominated by over 300 MW of wind energy generation with only 80 MW of local load. Type A wind turbines predominate, with some Type B and a few Type C turbines. In Tehachapi it is necessary to control the voltage through the use of VAR support, and the lessons learned in Tehachapi are important. Chapter 19 presents a good overview and background on voltage control and VAR support generally, whereas the discussion here is focused on specific cases associated with weak grids and wind generation dominating the energy flows. In such cases, VAR support and its effective control is critical to voltage support, and its absence will result in either voltage collapse or grid instability. Tehachapi has experienced both conditions and today is reasonably reliable as a result of the implementation of effective VAR support and selective grid upgrades.

A weak grid can be defined as one where the connected induction machine kilowatt rating exceeds 15 % of the gateway short-circuit S_k. In Tehachapi, wind power in excess of 40 % S_k is successfully being operated. With 345 kW of connected Type A and Type B wind turbines producing 310 MW coincident generation and a gateway short-circuit duty of 560 MVA at Cal Cement, it is clear that Tehachapi can be considered an extremely weak grid.

Another major consideration is that weak grids do not transport VARs well. VARs impact the devices and facilities in their local area to a greater extent. They are partially consumed as they are transported along transmission and distribution lines and as they pass through transformers.

Transmission and distribution lines exhibit a generally capacitive characteristic when lightly loaded and become increasingly inductive as load increases. Some long heavily-loaded lines can be remarkably inductive when heavily loaded. Such characteristic behaviour has a major impact on weak grids and will cause instability or voltage collapse if not effectively dealt with. Tehachapi has experienced both.

12.3.1 Voltage control of a self-excited induction machine

An often talked about but little-used generating machine is the self-excited induction generator. The most common use of this technology has been with small stand-alone generators and a captive fixed load. Capacitors are connected to the induction machine to provide the self-excitation, and the amount of capacitance connected determines the voltage level based on the operating point on the induction machine's saturation curve. The greater the capacitance, the farther up the saturation curve the induction machine operates and the higher the voltage. Increasing load consumes increasing VARs within the induction machine, and stable operation with acceptable regulation can be achieved for applications where the load is reasonably steady and well-defined.

12.3.2 Voltage regulated VAR control

This same technique can be applied to wind energy induction generators associated with Type A and Type B wind turbines. Here, however, the load is not steady, generation is widely variable with wind speed and effective control is much more difficult. In the normal situation of marginal VAR support, grid voltage drops with increasing wind generation. It is possible and practical to use transient-free, fast thyristor control of capacitors to achieve good flat voltage regulation of an induction machine in a wind farm environment and voltage control to dampen out grid oscillations, to the extent that those are desired objectives. This result can also be achieved from the inherent wave-shaping capability of an IGBT inverter as associated with Type C and D wind turbines. In either case, the induction machine can be made to behave very similarly to a synchronous generator, with VAR support for the external grid, and voltage control, and thus be very 'friendly' to the grid.

12.3.3 Typical wind farm PQ operating characteristics

Typical curves of operating induction generators on the Tehachapi grid are shown in Figure 12.2. All four curves are for Type A wind turbines, with VAR support and controls having different VAR support capabilities. One of them shows the effects of less-than-good maintenance. In each case, the data are for an array of wind turbines at the point of interconnection, with no substation transformer included, but with the wind turbine transformers and internal wind farm collection grid included. Voltage regulation causes an approximately 6 % drop from light generation to heavy generation. Reduced grid voltage reduces VAR consumption at heavy generation and increased VARs are

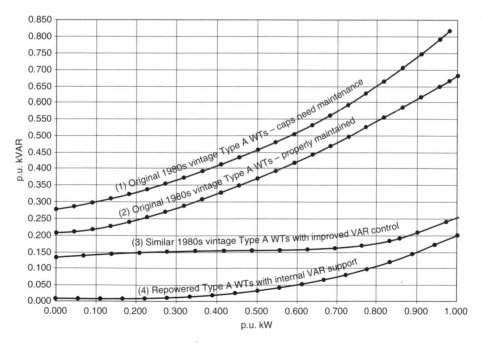

Figure   12.2 *PQ* characteristics of Tehachapi area Type A wind turbine (WT) arrays. *Note:* curve (4) is identical to Curve (4) in Figure 12.3, but note the scale change

required to maintain a flat grid voltage. These curves, as normalized, are quite representative of weak grid conditions experienced and are directly comparable.

Curve (1) of Figure 12.2 is of a wind farm of original 1980s vintage Type A wind turbines with the conventional no-load VAR control of that era but with the VAR control not effectively maintained, a common occurrence with that type of early control system. Curve (2) is of the same wind farm, but with the controls properly maintained in good and fully functional condition. For a 10 MW project, the difference represents a loss of 1.5 MVAR of critical VAR support at the most valuable location for those VARs.

Curve (3) of Figure 12.2 is of a different wind farm with similar 1980s vintage Type A wind turbines, but with improved VAR control. The conventional no-load capacitors have been increased to above-no-load VARs, the wind turbine transformers and wind farm collection grid have been compensated with fixed capacitors and a capacitor bank is switched with generation-level, materially increasing, local VAR support. This level of VAR support at the most valuable location for those VARs for a 10 MW project represents about 6 MVAR of added VARs. This is a substantial and significant increased benefit for improved grid performance and is one of the improvements that have materially raised the transfer capability of Tehachapi.

Curve (4) of Figures 12.2 and 12.3 is of a different wind farm with repowered Type A wind turbines. These turbines have improved VAR support as a part of the turbine

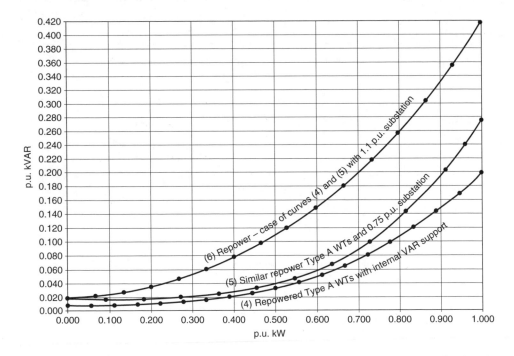

Figure   12.3 PQ characteristics of Tehachapi area Type A wind turbines (WTs) and substation; Curve (4) corresponds to the case where there is no substation and, is identical to Curve (4) in Figure 12.2 (note the scale change)

design with five steps of capacitors switched in and out to maintain approximately zero VARs at the wind turbine main breaker until 75 % of the VAR demand of the turbine is reached. This level of internal VAR support represents about 6.5 MVAR of added VARs at the most valuable location.

With Type A and B wind turbines it is quite feasible to add yet further effective VAR support using the same scheme as in Curve (4) but with further added VARs at the wind turbines, increasing VAR support to provide essentially all of the VARs that the turbines are consuming. Such an arrangement would provide approximately 8.5 MVAR at the most critical location for a 10 MW project. As added VAR support increases, it is important that the control scheme be configured to disconnect the capacitors very rapidly above no-load VARs to avoid driving the induction machine into saturation when an islanding event occurs. The typical speed of response appropriate for such an event is a few tenths of a second (see Section 12.3.9).

Figure 12.3 shows data for two comparable repowered Type A wind turbine projects, but including a project substation, it also shows Curve (4) from Figure 12.2 for comparison. Note that the p.u. kVAR scale is different from that in Figure 12.2. Heavily loaded substation transformers and long internal collection lines require substantial added VARs.

Curve (5) of Figure 12.3 is of a project with Type A wind turbines and is very similar to the situation represented by Curve (4), but with a modestly loaded substation (0.75 p.u.) and some internal supplemental VAR support of the type associated with the Curve (3) project of Figure 12.2, but of lesser relative magnitude.

Curve (6) of Figure 12.3 is of the two projects combined, of both Curve (4) and Curve (5) with different collection, and both fed into the substation associated with Curve (5). The resulting substation loading is at 1.1 p.u. in this case, and that increased loading materially increases VAR consumption. The added VAR support demands created by the heavily loaded substation are material and must be accommodated.

12.3.4 Local voltage change from VAR support

Local voltage changes associated with changes in VAR support can be reasonably predicted. As pointed out in Chapter 19, the impact of VAR support is greater on higher-impedance grids and lesser on low-impedance grids. On weak grids, the relative impedance is usually quite high and the impact of VAR support is usually significant. It is useful to be able to predict the general voltage impact that changes in VAR support will provide. The relationship described by the following formula is useful in making such predictions:

$$\text{percentage voltage change} = \frac{\text{MVAR change} \times 100}{\text{three-phase short-circuit duty, in MVA}} \quad (12.1)$$

From this relationship one can see that VAR changes on a stiff grid have relatively small impact, but on weak grids the impact can be relatively large. The difference between the VAR support of Curves (5) and (6) in Figure 12.3 will cause about 1.2 % voltage change on a normal fully loaded substation transformer, and the impact of Curve (1) of Figure 12.2 will cause about a 6 % voltage change. On a stiff grid, where the wind turbines are a small part of the total connected load, these same VAR impacts will have negligible voltage impact.

12.3.5 Location of supplying VARs within a wind farm

There is a strong advantage in supplying VAR support for induction machines at or
near the induction machine, particularly considering the system topology of typical
wind farms. By providing it at the wind turbine the increased risk of ferro-resonance is
diminished, and harmonic problems and risks are less likely to arise. Also, effective use
can be made of wind turbine power system components, resulting in a lower-cost and
more easily maintained system.

In standard project design, each wind turbine sits behind its own transformer,
having significant impedance, typically 5 % to 6 %, and also behind the wind farm
substation transformer and collection system. Substation transformers typically
have 7 % to 8 % impedance, but since these have a far larger rating than any wind
turbine and may serve many wind turbines their impedance impact at the individual
turbine is relatively small. Some applications of VAR capacitors use dampening
inductance inserted with each bank of capacitors. These typically are in the order of
6 % to 7 % impedance. The available transformer impedance, coupled with careful
system design, can achieve low harmonic currents without the added cost and
complexity of the dampening inductors. Dampening inductors have generally not
been used on most wind farm capacitor installations in the USA and, where they
have been used, they have not dramatically diminished harmonic issues but do
reduce harmonic currents.

In Figure 12.4, the voltage impact of VAR support capacitors is shown for a variety
of locations for the VARs – at the wind turbine, on the wind farm collection system and
externally, on the high voltage (HV) grid, usually supplied by the utility.

Wind turbine case (a) in Figure 12.4 is of a typical Type A or B wind turbine with no-
load VAR support, what has typically been the standard product offering of most
manufacturers. In this configuration, and with a good, low VAR demand generator,
there will be about 1.8 % voltage drop at the wind turbine. This is solely attributable to
inadequate VARs causing a drop in generator voltage with increasing generation.

Wind turbine case (b) in Figure 12.4 has sufficient supplemental VARs at the wind
turbine to carry its share of the VARs needed to provide a 0.95 power factor at the point
of interconnection. This causes a 3.7 % increase in generator voltage at the turbine. One
or two tap adjustments on the wind turbine transformer can accommodate that voltage
rise, and this will also cause the transformer to be less saturated – a desirable situation.
Lower saturation is also helpful with harmonics.

There are other significant issues that must also be addressed with respect to islanding, and one of the consequences of detail oversight is shown later, in Figure 12.7. In the
case illustrated, slow-acting medium-voltage vacuum breakers acted slowly and left
VAR support on an island for far too long. When the energy transfer stopped, this
resulted in a substantial overvoltage condition for a short period (see Section 12.3.9).
Interestingly, Figure 12.9 is of the same event, and at that facility the event was handled
without significant consequence (see Section 12.3.11). On the island, the excess VARs
caused the Type A or B induction machines to overexcite and climb to the maximum of
their saturation curve until the condition cleared.

Good protection and a reasonably rapid speed of response to such conditions is
important. Generally, 0.2 to 0.3 seconds is an adequate and effective time to clear

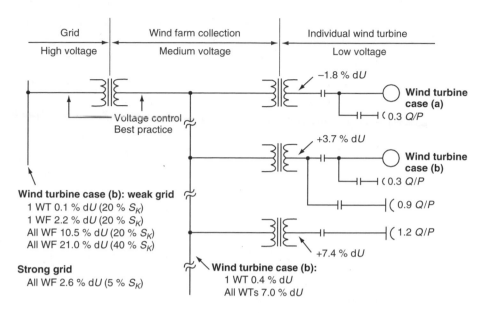

Figure 12.4 Voltage impact of VAR support at alternate locations in the system; the figure shows the voltage impact of VARs on weak and strong grids. Note the potential for small increment control from block switching individual wind turbine VARs. *Note*: WF = wind farm; WT = wind turbine; U = voltage; Q = reactive power; P = active power; S_k = short-circuit power

VAR support in the event of an island and, in the event illustrated in Figure 12.7 would have prevented such an overvoltage condition.

12.3.6 Self-correcting fault condition: VAR starvation

Similar VARs applied solo to the low voltage side of a transformer bank as shown in Case (b) of Figure 12.4 causes double the voltage rise on the low voltage side of the transformer, and must be considered in the design, but can be beneficial.

Capacitors applied on the medium-voltage wind farm collection system, as shown in Figure 12.4, will have a reduced voltage impact because of the lower impedance and higher S_k duty. At this location, the VARs must pass through the wind turbine transformer, and its inductance, before reaching the generator, a critical need location. This location has commonly been used for placing VAR support capacitors since larger increments of VARs can be supplied in small packages, and the convenience is high. However, this location is at higher risk of damaging impacts from harmonic conditions, and these capacitor designs normally do not have dampening inductors feasibly available. Further, some vacuum breakers commonly used for such installations are slow acting and can lead to quite undesirable overvoltage conditions in an islanding event (Section 12.3.9, Figure 12.7).

Utilities commonly supply the VAR support that they provide from the high-voltage grid, usually from capacitors located at utility substations. Such a location is convenient, but is far from the consuming generators and has significant limitations.

In Tehachapi, just over 100 MVAR is supplied from within substations, and another 50 MVAR from the distant Antelope Substation is ineffective. SCE was unable to locate adequate VARs at its facilities, and it was necessary to locate the great majority of the VARs needed by the wind farm within the wind farm.

Block control of supplemental capacitors at a wind turbine can be used as an efficient technique to manage harmonics effectively and to provide a simple and low-cost distributed virtual control scheme. One can achieve effective and fine control of collection voltage and grid voltage by selectively switching one block at a time. A single Case (b) wind turbine will change system voltage at the collection level by only 0.4 %, or at the grid level by only 0.1 %.

12.3.7 Efficient-to-use idle wind turbine component capacity for low-voltage VARs

The effective loading of a wind turbine transformer is not generally impacted significantly by the addition of supplementary VAR capacitors in addition to those supplied with the turbine controller on a Type A or B wind turbine. This is particularly true starting from standard no-load VARs that are about 50 % of the compensation required for a wind farm.

Power system components are generally sized to carry the full load generator current without considering the reduced current with any level of VAR support. In Figure 12.5, this is about 1.16 p.u. current, based on a unity power factor. As can be seen, one could

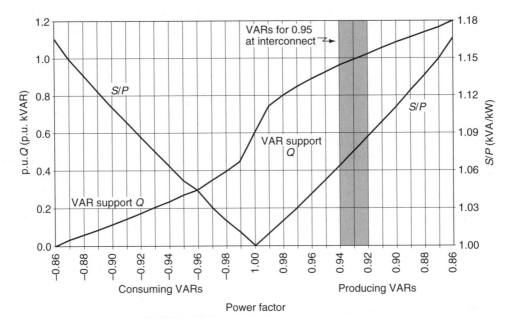

Figure   12.5 VAR impact at wind turbine does not increase major system component size. All of the VARs needed in a wind farm can be supplied efficiently at low voltage. *Note*: Q = reactive power; P = active power; S = apparent power; p.u. = per unit

go to 0.86 power factor producing VARs and still remain within the normal design margin. VARs producing a full generation power factor of about 0.92 or 0.93 will usually give the VARs necessary to achieve a 0.95 power factor at the point of interconnection, the level now required of new installations. In the case illustrated in Figure 12.5, 0.95 can be achieved well within code required design margin, certainly with the extra capacity provided by wind cooling of pad mount transformers.

12.3.8 Harmonics and harmonic resonance: location on grid

When one uses capacitors for VAR supply, it is critical to consider harmonics and harmonic resonance and to design carefully to avoid negative impacts from the introduction of capacitive VARs onto a system. Resonance is the point where the inductive impedance of the connected facilities equals the capacitive impedance of the capacitors, at a given frequency. Thus, at that point, the two impedances cancel each other out, and only the usually small resistance of the network limits current flow. The typical scenario is that resonance is most significant at the third, fifth, or seventh harmonic of system frequency. Although the voltages are low at these harmonics, the very small resistance can produce destructive currents and other undesirable effects.

Ferro-resonance is common with electromagnetic machines, and harmonics exist in all electric systems to a greater or lesser degree. Transformers tend to operate at sufficiently high saturation levels as to generate sufficient harmonics, particularly of the seventh order and below, to create the need for careful attention in the best of systems. Phase-controlled thyristor power controllers, if they exist on a network, often create rich harmonics, and create a need for greater attention. It is good practice to pay substantial attention to harmonics when applying VAR support capacitors to any specific grid, and actual measurement across a full set of conditions is essential. Continuous monitoring is a good idea, as installations and operating conditions change, and the cost of effective monitoring is not great if the monitoring is well organized. Failure to accommodate harmonic issues can lead to premature capacitor failures and, in the worst cases, to other significant consequences.

Figures 12.6(a)–12.6(c) (note the THD scale changes) give a view of the range of harmonic conditions at the wind turbine main terminals, as operating conditions change. Depending on operating point, resonance can be problematic over a fairly broad range. The impedance of an induction machine changes with generation level, along with its VAR requirement. Harmonic resonance does not become significant in most clean systems where VAR support is less than unity. Supplying large blocks of positive VARs from behind the impedance of a transformer and/or dampening inductors is also outside the range of usual clean system resonance and produces low levels of harmonics, particularly at medium and high generation, when VARs are most needed.

The location of the capacitors in the network makes a significant difference. VARs supplied by the utility from capacitors, generators or other sources are generally not as effective nor as efficient in terms of VARs needed to be installed compared with VARs needed at the induction machines. The local collection grid, distribution grid and transmission grid all also consume VARs, particularly at higher energy flows, and this must be accommodated. Needs local to the wind farm can usually be effectively supplied from the wind farm, quite like what is done from a conventional generator, provided the VARs are available. Although this discussion is of the historical opposite condition, it is likely that VAR supply from each wind farm will be required for new interconnection and will improve the system particularly where the control strategy is effective, such as with fast-responding voltage control.

The location of the capacitors and method of control also have a significant impact on harmonic resonance and the range of exposure to resonance. The wind turbine

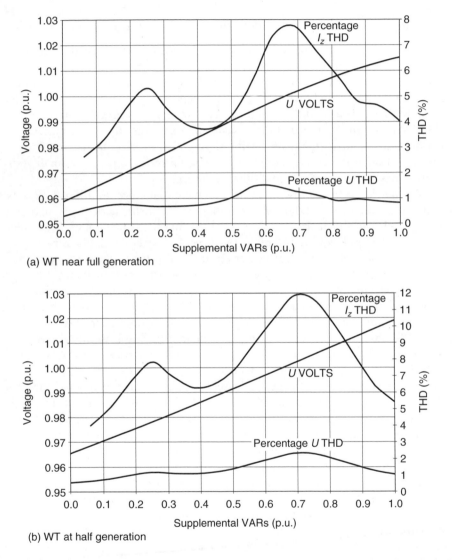

Figure **12.6** Total harmonic distortion (THD) and voltage, U, as a function of per unit (p.u.) supplemental VARs (note the THD scale): VAR capacitors at secondary of wind turbine (WT) transformer and (a) WT near full generation; (b) WT at half generation and (c) WT at 10 % generation. *Note*: I_z – capacitor current

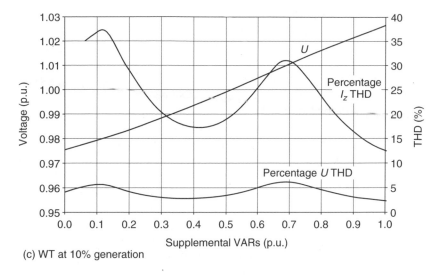

(c) WT at 10% generation

Figure   12.6 (*continued*)

transformer introduces typically 5 % to 6 % impedance and is sized close to the rating of the turbine. A substation transformer typically has 7 % to 8 % impedance, but serves a large number of turbines. It therefore generally has a small impact on the characteristic impedance when seen at any individual wind turbine. Thus, by locating VAR capacitors on the wind turbine side of its transformer, one finds a location less exposed to wide variations in impedance. Also, by using a Delta–Wye or Wye–Delta transformer con-figuration, one eliminates the third harmonic. These factors combined can materially reduce the risk of harmonic resonance in many installations. Effective transformer impedance is substantially increased when the transformer is applied near its maximum capability such as with forced cooled designs.

Lightly loaded transmission lines exhibit a capacitive characteristic. Heavily loaded lines exhibit a substantial inductive characteristic. As line loading passes the point of surge impedance loading (SIL) VAR consumption can increase materially. Similarly, trans-formers when lightly loaded require modest VARs for excitation but when heavily loaded consume substantial VARs. These factors are generally not well evaluated and considered in project design and often lead to unpleasant surprises and impacts on weak grid systems. Conventional transmission network analytical tools have not accurately modelled and predicted grid behaviour associated with substantial installation of induction machines. Only now are improved models beginning to predict grid performance at a reasonable level. Tehachapi has clearly seen such consequences of inadequate models and poor understand-ing of the issues, as have West Texas and Minnesota. As a result, improved models have been developed and reasonably validated, with more to follow, and the result is that others will soon have reasonable tools that can provide good results. However, care is needed, and attention to critical details is quite important to avoid unexpected and unpleasant results.

The medium-voltage collection grid is another common location for VAR support capacitors. That location can provide a low-cost installation for a large block of such capacitors and controls, but caution is indicated, as the range of harmonic exposure is materially increased, and some medium-voltage control devices are slow to operate and thus introduce added exposures that must be accommodated (see Section 12.3.9).

The early design FACTS device applied in Tehachapi is a 14.4 MVAR 12-step capacitor-only system using transient free, fast thyristor switching. This unit is located at the most critical VAR support node and has achieved excellent results, but there are notable issues; 14.4 MVAR is too small for the system needs, and, although it gives excellent performance at light and medium generation, it saturates before high generation when it is needed most. An attempt to add fixed 5 MVAR steps was not successful because of unresolved resonance problems. During light generation, resonance problems when in automatic control have resulted in excessive maintenance and out-of-service events, despite being behind the substantial impedance of dampening inductors and two steps of transformers. It is likely these problems could be readily managed and more effectively operated by using harmonic feedback information and smarter control.

VAR capacitors are designed to work satisfactorily when exposed to relatively high harmonics, with various codes commonly requiring a design margin of 1.3 or 1.35 above nominal current to accommodate unanticipated harmonic currents. The full circuit design must be sized accordingly, including capacitor contactor, breaker, fuses and wiring. Although capacitors can survive in relatively high harmonics, it is good practice to employ control strategies that cause operation away from harmonic resonance and increase reliability and longevity. It is also important to derate capacitors by applying them at below their rated voltage to achieve an appropriately long life with low maintenance costs. Keeping capacitors cool is also important to their long life.

12.3.9 Islanding, self-correcting conditions and speed of response for VAR controls

Islanding is a condition where a wind turbine, a wind farm or a group of wind farms and their associated grid becomes separated from the full grid and may perform as an independent island, if conditions are right to support continued operation without tripping equipment or facilities. There are many risks associated with islanding, and great efforts are taken to prevent this from occurring. Among the greatest risks may be major damage to equipment if an island is reclosed onto the main grid when out of synchronisation, and the safety of repair crews expecting a de-energized grid and shockingly discovering otherwise.

There are other significant issues that must also be addressed with respect to islanding, and one of the consequences of detail oversight is shown in Figure 12.7. There, slow-acting medium-voltage vacuum breakers acted slowly and left VAR support on an island for far too long. When the energy transfer stopped, this resulted in a substantial overvoltage condition.

Good protection and a reasonably rapid speed of response to such conditions is important. Generally, 0.2 to 0.3 seconds is an adequate and effective time to clear VAR support in the event of an island, and in the event illustrated in Figure 12.7 would have prevented such an overvoltage condition.

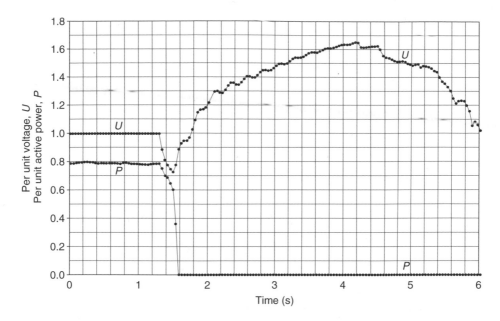

Figure 12.7 Islanding event: overvoltage impacted by slow operating controls for a short period. *Note*: Figure 12.9 is of the same event; at that facility the event was handled without significant consequence. On the island, the excess VARs caused the Type A or B induction machines to overexcite and climb to the maximum of their saturation curve until the condition cleared

12.3.10 Self-correcting fault condition: VAR starvation

The self-correcting concept associated with an unstable line or overloaded grid has been used by planners as a fall-back technique to avoid the need to upgrade facilities to handle maximum generation. Such conditions are commonly associated with Type A or B wind turbines, and the limited VAR support associated with standard wind turbine controllers, and overall system inadequate VAR support compared with need. Wind turbine controllers have historically been set with grid-protection settings that are close to nominal voltage and fast acting. The result has been that, as a voltage collapse event starts to occur, and most such events develop slowly, over a period of some seconds, a significant number of wind turbines will sense low voltage and drop off the grid prior to the collapse fully cascading. As the wind turbines drop off the grid, and if they are consuming VARs, the result is a reduction in VAR consumption and a restoration of grid stability. Figure 12.8 is an example of this condition.

The self-correcting technique does prevent a cascading voltage collapse event due to VAR starvation. It still causes substantial served load grief, though, and is not a good approach for reliable grid operation. Critical served load is also voltage sensitive and also trips at or before the wind turbines trip to clear the condition. This is a condition caused not by the wind turbines, but by inadequate VAR support, and the latter is the condition that should be properly corrected. In the Tehachapi area, such behaviour has caused economic loss to served load, and improved VAR support and/or improved grid facilities are the proper and economically correct solution. The improved VAR support

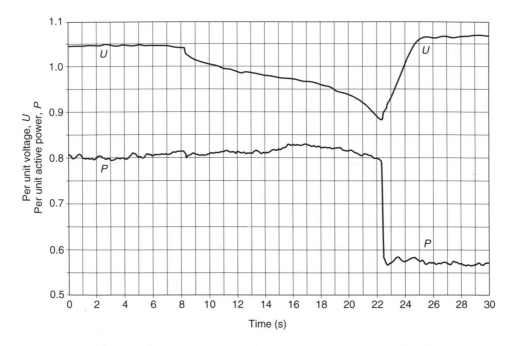

Figure   12.8 Self-correcting feature: VAR starvation grid collapse, self-correcting by WT drop off line

provided by wind farms in recent years has dramatically reduced the occurrence of events such as this, but more could readily be done.

The best new practices for interconnecting wind turbines can readily provide the capability for wind farms to add valuable VAR support to the grid and avoid this sort of problem and to otherwise become at least as good as a synchronous generator in providing grid reliability.

Speed of response for VAR control in such a VAR starvation event can be in the order of a few seconds to catch and correct the event. Thus, controls designed for fast enough performance to handle a critical islanding event would be more than fast enough to effectively handle VAR starvation events.

The behaviour of grid voltage in a typical grid instability situation is of similar speed, and corrective controls responding in 0.2 to 0.3 seconds can provide excellent dampening of the oscillation when voltage-responsive.

12.3.11 Higher-speed grid events: wind turbines that stay connected through grid events

Some grid events, such as the one shown in Figure 12.9 are fast developing and require a different speed of response, or a different approach, to respond effectively to and impact the event. In this event, there is more than a 10 % drop in voltage in about 0.05 s and a nearly 30 % drop in 0.50 s.

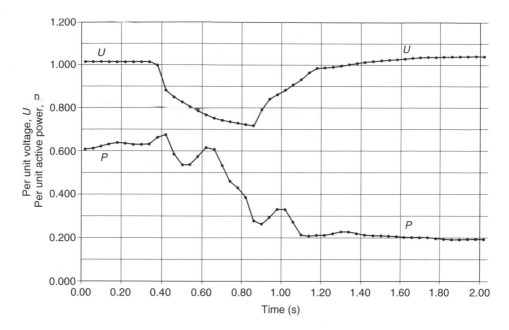

Figure 12.9 The transient event illustrated in Figure 12.7: in this case, the wind farm did not island but instead recovered well; voltage dip can be reduced by fast-acting VAR support

In this wind farm, there are two types of Type A wind turbines. One group consists of reasonably new repower wind turbines, and they tripped offline in approximately 0.2 s after the event exceeded their grid limits, as they should have done, according to their specifications. The other group consists of older Type A wind turbines but with a unique control system. That group nicely stayed online through the event and continued to deliver energy to the grid without interruption, as would a reliable synchronous generator. Interestingly, wind turbines equipped with ride-through capability according to usual standards would also have ridden through this event.

Thus, there are two potential ways in which a wind turbine and wind farm can respond to fast events such as this one. One is by use of very fast acting VAR support. A substantial magnitude of VARs would be required to impact this event materially to keep it below a 10 % drop in voltage. This is possible, but may neither be practical nor desirable as a standard approach. From Figure 12.4 one can see that 1.2 p.u. VARs would not be capable of keeping the voltage to even a 15 % transient drop inside the wind farm. Interestingly, it could do so if all wind farms on the weak grid were able to respond effectively. However, the more appropriate technique is to ride through the event and, in so doing, maintain continuity and reliability of generation. Clearly, new wind turbines effectively equipped can ride through such events, but older wind turbines can also be reasonably equipped to ride through common short-duration events such as the one illustrated in Figure 12.9, provided there is commensurate value in their doing so.

FACTS devices, as well as Type C and D wind turbines equipped with responsive VAR control, and Type A and B wind turbines equipped with fast thyristor VAR control, are all capable of providing an effective contribution to minimize events such as the one shown in Figure 12.9. However, it is clear that substantial VARs are required to assist materially in such an event, and the alternative of ride-through capability appears to be of better value for dealing with such events. When one considers the balance of grid needs, and the way reliability standards work, it seems that somewhat slower, but automatic, VARs combined with ride-through capability would provide essentially the full needs of most grid situations and would be consistent with a cost-effective capability that could be provided by wind turbines and wind farms (see the comments on automatic VAR control in Section 12.3.12).

12.3.12 Use of advanced VAR support technologies on weak grids

Conventional VAR support has historically been provided from the controlled over-excitation of synchronous machines, improving such performance and behaviour through voltage regulators and power system stabilizers. Longer transmission lines and grid segments remote from such generation have used switched capacitors for added VAR support to the extent needed to maintain voltage within limits and to maintain a stable grid. The evolution of various FACTS devices and similar technology has brought the same performance advantages of advanced conventional VAR support to the distributed environments where capacitors have been used historically (Muljadi et al., 2004).

A very early FACTS device used in association with wind turbines and wind farms was the 1991 installation of an ABB SVC in Tehachapi on the SCE grid in the Varwind substation. This device was initially installed at 9.6 MVAR and increased later to 14.4 MVAR. It is composed of 12 steps of transient-free thyristor-switched capacitors, dampening inductors and two steps of transformers. An attempt to add one or two steps of 5 kVAR slow-switched capacitors was not successful because of unresolved resonance issues. This device has at least the capability of voltage control, and observed data have shown that it performs very well in improving grid stability and regulation at the critical node where it is located. However, it clearly has had problems resulting from resonance issues and has been rebuilt at least once. From what can be learned, the resonance issues are severe at low energy flows on the grid. Even the substantial impedance provided in its design was not sufficient to limit harmonic currents adequately during certain low-energy flow conditions, and internal damage resulted from such operation. From these lessons it seems that there is a significant advantage to the fully distributed VAR support model of Figure 12.4 where a single step of operation can be devised for each node that is nicely damped by efficient use of available system components and where those components inherently provide a relatively high per unit impedance at low cost.

Comparable technology to FACTS was developed in the Type D wind turbines of Kenetech, with an installation of 112 turbines of 35 MW in Culbertson County Texas in 1995. There, the VAR support capability of the integrated gate bipolar transformers (IGBT) controllers of these wind turbines materially contributed to the grid stability of a weak grid. The grid in that area was further saturated in 1999 with the addition of 30 MW of Type C turbines with similar capabilities. The combined enhanced VAR capability of these facilities resulted in their successful operation.

At about the same time, several large wind farms were developed in the Lake Benton area of southwestern Minnesota involving similar Type C and Type D turbines and with a similar set of weak-grid problems. Enhanced VAR support provided by these turbines was extensively evaluated by the National Renewable Energy Laboratory (NREL) and the effectiveness of significant performance features is well reported. These evaluations confirm the significant performance advantage of voltage-regulated VAR controls in resolving grid-support and stability problems (Muljadi, Wan and Butterfield, 2002).

A transient-free thyristor-controlled switched-capacitor 1.1 MVAR low-cost distributed VAR support installation was tested in Tehachapi, integrated with the facilities of a megawatt size wind turbine. This unit demonstrated technology for a large virtual SVC-type FACTS device that is modular and of substantial potential capacity based on the size of the wind farm. It is capable of being integrated with a wind farm, centrally controlled through fast-acting distributed control. This arrangement appears to have important harmonic and cost advantages.

FACTS devices constructed in a modular factory built containerised design have been installed and operated successfully to resolve weak grid problems at Foote Creek, Wyoming and Minot, North Dakota. These D-VAR[tm] units utilize an IGBT converter and capacitors providing fast response and generally stepless control of both VAR production and VAR consumption. A further advantage of these particular FACTS devices is that they provide full range control of VARs over a broad voltage range, and thus have superior VAR support capability during a transient event. Capacitors connected directly to the grid do not have this two quadrant capability, and their VAR production decreases as the square of the voltage. The IGBT arrangement, on the other hand, provides for the VARs to decrease only at the first power of voltage. The IGBT converters associated with Type C and D WTs have similar first power of voltage VAR support, and the ability to both deliver and consume VARs. WFs with Type A and B WTs and switched capacitor VAR support also have the inherent capability for wide range production and consumption of VARs since with all of the capacitors turned off, the WTs and WF collection system consume substantial VARs and can be controlled for a wide two quadrant range of control.

12.3.13 Load flow studies on a weak grid and with induction machines

One might wonder why the weak grid areas of California, Minnesota and Texas would have such a difficult time anticipating and resolving the substantial grid capacity and stability issues that have caused such great problems for the utilities, grid customers and the wind industry. A significant factor has been that the models used for load flow studies and transient stability studies simply did not correctly represent induction machines and their impact on the grid. A related factor is that VAR transport issues were not understood by the utilities, and they attempted to maintain self control over VARs and to supply known needed VARs from afar, with poor results. Until the mid-1990s, wind farms were installed by only two utilities in the USA, and Tehachapi was the only large installation with a weak grid. Thus, the problem was not widespread nationally and, until relatively recently, there simply was a lack of attention to the issue first seen in Tehachapi.

In the late 1990s a local College math department did studies of the problem with induction machines on the grid and provided helpful insight. NREL played an extensive role in evaluating the Type C and D wind turbine installations and, as a part of that effort, developed substantial documentation and understanding of the issues. The providers of PSS/E$^{TM(1)}$ and PSLF (general electric positive sequence load flow) load flow programs have recently expended substantial efforts, and their programs have been dramatically improved to handle induction machines and specific wind turbine designs with built-in models. Thus, it appears that a great deal of the technical difficulty of predicting wind turbine, wind farm and grid behaviour may finally be becoming resolved with good tools.

12.4 Private Tehachapi Transmission Line

In 1989 two developers and a consortium of project entities, with 185 MW of stranded unbuilt PPA capacity from the 1985 contracts, elected to build their own private transmission line to carry the new generation out of Tehachapi to the major 500 kV Vincent substation in order to be able to deliver energy from those projects. Another 135 MW of operating PPA capacity on the 66 kV system that was being heavily curtailed was moved to the new private line, for a total installed capacity of 320 MW. This step relieved major curtailment payments and increased economic certainty. This private line is 100 km in total length and was constructed at 230 kV. It has proven to be extremely successful.

The Sagebrush Line at 230 kV, and of approximately similar connected capacity as the SCE 66 kV network, serves as a dramatic comparison, showing what the utility grid could be. Although the Sagebrush Line is considered by the utility to be lower-cost construction, it has performed extremely well and reliably – far more so than the utility grid. The Sagebrush Line has important reliability features that are missing from the 66 kV grid, such as an overhead static ground line that provides substantial lightning protection and good physical protection against vehicle and fire damage, which have so impacted the 66 kV grid. The monitoring and control capabilities of the Sagebrush Line are excellent and provide operators with key operating data that would be most useful on the 66 kV network.

There are limitations to the use of the line. The organizational structure used to allow the private line legally to be built and operated has effectively precluded the addition of additional generators onto the line and prevents use of the line for served customer load. Metering of energy deliveries is based on deliveries at the distant end of the line and thus the generators must absorb approximately 5 % energy losses associated with the line. The line is now 15 years old, and some of the organizational limitations are beginning to be resolved, and the prospect exists for the line to become more widely available for more general use. 60 MW new generation is under construction, soon to be added onto the line, increasing its connected generation to approximately 380 MW, with a prospect for further increased use of the line.

The techniques used to facilitate rapid development and low-cost construction of the line are noteworthy. California utilities have eminent domain rights and can readily acquire rights of way for transmission lines. However, such is not the case

---

[1] Power System Simulation for Engineers; a registered trademark of PTI.

for private developers, as neither special land acquisition rights nor specialized environmental reviews for transmission lines are available. The developers organized a private real estate company and bought up large blocks of land along the desired right of way. The land needed for the transmission rights of way was separated from the residual land. The residual land was then effectively marketed, in some cases by first improving the land for future development and by other value-increasing techniques. In other cases, rights were obtained from governmental agencies such as at the margins of county road rights of way. In one section, the line takes a short detour to avoid property where the seller had very large dollar signs in his eyes and would not 'budge' until too late.

The competing utility has criticized the construction of the 230 kV line along a road right of way, but such is regularly done by the utilities up to at least 115 kV, and there simply does not appear to be a valid argument against the practices used. The concept and execution were outstanding and successful and it was a remarkable achievement to solve a difficult problem that was stalling wind energy development.

The alternative for the developers was to have the utility build the line. Such an approach would likely have resulted in line costs in the order of double of the actual costs and a time of construction several years longer. Such a procedure would have resulted in the 185 MW of projects not being built and the contracts lost. Those contracts were at an unusually high price, which allowed for the construction of the private transmission line, the absorption of transmission losses and the taking of substantial risk to make this happen.

The Sagebrush Line was originally rated for approximately 400 MW, based on the configuration and size of conductor used. A maximum capacity 230 kV line would normally be capable of approximately 800 MW. A double circuit on the same poles would be capable of 1400 MW, limited by grid reliability rules for $N-2$ conditions. It is unfortunate that it was not feasible for that line to have been built at maximum capacity, or upgraded, and be utilized to connect at least some of the current 66 kV wind farms and served customer load more reliably onto a much more reliable grid. If this had been done at the time the line was built, many millions of dollars in curtailment costs would not have been paid by SCE on the 66 kV grid and substantial economic losses due to grid unreliability could have been avoided.

12.5 Conclusions

Experience with weak grid conditions in California and the US Midwest has produced a wealth of knowledge and practical solutions that will be useful as wind industry growth accelerates. VARs are to be respected and not feared. Effective use of VARs can make an induction generator behave much like a conventional synchronous generator and can dampen natural oscillations or instabilities in the grid. While self-correcting schemes have been used to protect old grid systems from cascading overload conditions, proper grid facilities and ride-through capability by wind turbines are essential for a reliable electric system that takes advantage of the low-cost clean energy from wind farms.

Rational, orderly, cost-effective investment in transmission facilities, ahead of the development curve, would allow far more efficient development of wind energy on a large scale and greater reliability to served customers. Increasingly large-scale wind energy development should be better integrated and coordinated with the existing extra-high-voltage grid to take advantage of the efficiencies that would provide.

'Brute-force' techniques such as the more conventional 'wire in the air' have been the method generally used to provide adequate transmission capability, but they can be quite expensive. Alternative techniques of providing transmission capacity, such as aggressive VAR support, planned curtailment of generation at periods of peak coincident generation and use of energy storage as it becomes cost effective offer alternatives to 'wires in the air' for any transmission corridor. Effective coordination of dispatchable generation technologies with clean intermittent resources such as wind can provide an energy supply that reliably matches load. Coordination of generation facilities in a network through use of dispatch priorities or market incentives can increase the production of clean energy, reduce emissions, and reduce the depletion of non renewable fuels while maintaining high power system reliability.

References

[1] Muljadi, E., Wan, Y., Butterfield, C. P. (2002) 'Dynamic Simulation of a Wind Farm with Variable Speed Wind Turbines', paper presented at the AIAA–ASME Wind Energy Symposium, Reno, NV, January 2002.
[2] Muljadi, E., Butterfield, C. P., Yinger, R., Romanowitz H. (2004) 'Energy Storage and Reactive Power Compensator in a Large Wind Farm', paper presented at the AIAA–ASME Wind Energy Symposium, Reno, NV, January 2004.

13

Wind Power on the Swedish Island of Gotland

Christer Liljegren and Thomas Ackermann

13.1 Introduction

This chapter gives an overview of the issues that the local network operator, Gotlands Energi AB (GEAB), has faced on the island of Gotland in connection with increasing wind power penetration. Some of the solutions that have been implemented will be described in a general way. The main focus is on the voltage source converter (VSC) based high-voltage direct-current (HVDC) solution that was installed on the island of Gotland in order to deal with network integration issues related to a very high penetration of Type A wind turbines.[1]

13.1.1 History

The Swedish island of Gotland is situated in the Baltic Sea, about 90 km from the Swedish mainland (see Figure 13.1).

Today's network operator, GEAB, was initially, at the beginning of the last century (1904–20), responsible only for operating electrical lights in the local streets. For this purpose, GEAB developed a small electrical system to supply and operate the local

---

[1] For definitions of wind turbine types A to D, see section 4.2.3.

Wind Power in Power Systems   Edited by T. Ackermann
© 2005 John Wiley & Sons, Ltd   ISBN: 0-470-85508-8 (HB)

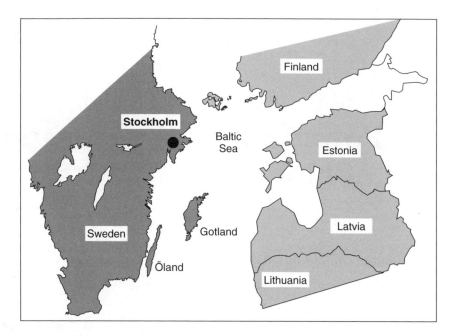

Figure   13.1 Gotland and the Baltic Sea

lights. After the development of metering equipment, GEAB started to sell power to other customers and became the local utility. Today, GEAB is the network operator of the island, which includes the responsibility to provide sufficient power quality to local network customers.

The electrical system was built starting from the village of Slite in the Northern part of the island. The limestone industry that was located there was the first and largest consumer of electricity. Even today the cement industry continues to be the biggest customer and consumes nearly 30 % of the electrical energy on Gotland. In the early 1920s a local coal and oil fired power plant was built on Gotland. In the countryside, small diesel generators were installed for local consumption. In the 1930s overhead lines were built to connect the small power units to the small power system in Slite. This was the beginning of the power system on Gotland.

The next major step was in 1954 with the installation of the first line commutated converter (LCC) based HVDC link in the world, between the mainland of Sweden and Gotland. The rated power of this link was 15 MW. Later on, it was upgraded to 30 MW.

The grid was designed to accommodate a consumption that consisted of small loads in the countryside as well as some villages with small industrial facilities. In 1970 consumers began to use electricity for heating, and the consumption in remote places increased significantly. In 1983, the old LCC based HVDC link was taken out of operation and was replaced by a new LCC based HVDC link with a rated capacity of 150 MW. A few years later, a redundant HVDC link was built.

13.1.2 Description of the local power system

Today, the power network on Gotland consists of approximately 300 km of 70 kV lines, 100 km of 30 kV lines and 2000 km of 10 kV lines, and it has about 36 000 customers. For a simplified single-line diagram of the local 70 kV network, see Figure 13.2.

The maximum load on Gotland is about 160 MW and the minimum is approximately 40 MW. The consumption of electric energy on Gotland is approximately 900 GWh per year. This energy is supplied mainly from the mainland through the LCC based HVDC submarine cables. The HVDC link is also used to regulate the frequency on the island. In order to ensure voltage stability there are also locally installed synchronous generators that play an important role in the operation of the power system. Locally installed gas turbines can be used as backup but produce only marginal amounts of energy per year.

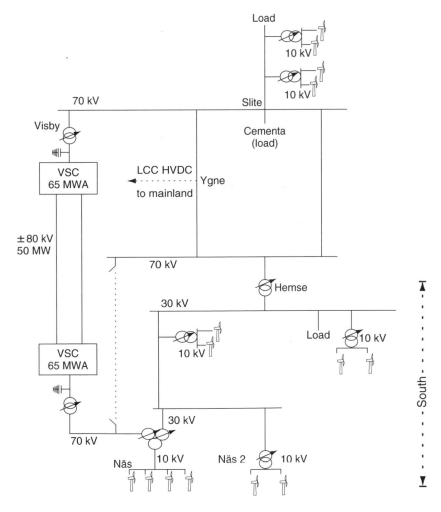

Figure   13.2 Simplified single-line diagram of the power system on Gotland. *Note*: LCC = line commutated converter; HVDC = high-voltage direct-current; VSC = voltage source converter

Good wind conditions, especially at the southern tip of the island, have led to an increased number of wind turbines. In 1984 the total wind power capacity on Gotland was 3 MW, and it increased to 15 MW in 1994. Early in 2003 the number of wind turbines reached 160, with a total installed capacity of 90 MW, generating approximately 200 GWh per year. Wind power production accounts for approximately 22 % of the total consumption on the island. From the perspective of power system operation, it is important to point out that at times with low loads (nights) and high wind speeds, wind power can reach more than twice the minimum load.

For system operation, it is also important that most wind farms are concentrated in the South of Gotland and consist mainly of fixed-speed turbines (Type A). The southern part, in particular the area around Näs, is an important test field for new wind turbines and has played an important part in the history of the development and testing of various wind turbines, including some of the first megawatt turbines in the world. There are plans for further expansion of wind power, particularly in the northern part of the island and offshore.

13.1.3 Power exchange with the mainland

In situations with high wind speeds and low local load, wind power production on the island exceeds local demand. The LCC HVDC link to the mainland, however, was originally designed only to operate in one direction: from the mainland to Gotland.

In 2002 the HVDC link to the mainland was therefore modified in order to allow a change in the direction of the power flow without interfering with frequency control on the network on Gotland. The system is now capable of automatically changing the power flow direction with continuous frequency control.

The approximately 90 MW of wind power on the island lead to an export of approximately 100 MWh of wind power production per year. Hence, for about 40 hours per year the power flow direction that is usually from the mainland to the island must be reversed. With the expected increase of local wind power on Gotland to 150 MW in the near future, the island will export power for about 500 hours per year, which is equal to an export of about 2 GWh.

13.1.4 Wind power in the South of Gotland

The network infrastructure was originally planned according to local demand and did not take into account significant local generation. The North has a strong network that was designed to accommodate large network customers (Visby and Slite). The power system in the South, in contrast, can be considered weak. The local peak demand in southern Gotland is approximately 17 MW, whereas the installed wind power capacity reaches about 60 MW. The situation is even more extreme in the area of Näs, which lies most to the South. There, the peak load is 0.5 MW, with approximately 50 MW of wind power capacity, comprising mainly Type A turbines, installed in the area.

From a technical perspective, this imbalance between load and production makes system operation very difficult. In comparison with a normal distribution system, the short-circuit power, for example, is very low in relation to all the connected equipment.

The solution to the island's power transmission problems was the installation of one of the first VSC based HVDC links in the world.

13.2 The Voltage Source Converter Based High-voltage Direct-current Solution

In the following, the VSC based HVDC solution is presented in more detail.

13.2.1 Choice of technology

Theoretically, there were two alternatives for transmitting the surplus wind power between the southern part of Gotland around the area of Näs to the load centre Visby:

- alternating current (AC; cable or overhead line), or
- direct current (DC; cable, VSC based HVDC technique).

In addition to the transmission lines, the AC alternative requires equipment for the compensation of reactive power, synchronous machines or conventional compensators [static VAR compensators (SVCs)] in order to ensure the required quality of supply for the local network customers. The alternative with synchronous machines was too expensive (in terms of investment, operation and maintenance). An AC three-phase cable was not considered to be economically competitive and an AC overhead line would be difficult to build for environmental reasons.

Being the best available technical alternative, a VSC based HVDC solution was chosen for the project. Common station equipment could be used to meet the power transmission as well as electrical quality requirements regarding the connecting network (for a discussion of VSC based HVDC technology, see also Chapter 22). The VSC HVDC solution was installed in parallel with the existing AC network, which helped to improving the dynamic stability of the entire AC network. Simulation results have shown that a VSC HVDC also helps to maintain the power quality in the northern part of the island.

13.2.2 Description

The VSC based HVDC system installed by Asea Brown Boveri (ABB) in 1999 consists of two converter stations connected by 70 km of double ± 80 kV DC cable. The converters are connected via reactors to the 75 kV AC power system. The transformers are equipped with tap changers to be able to reduce the voltage on the converter side in order to reduce no-load and low-load losses. This special feature was introduced to this installation, since wind power seldom operates at peak production and sometimes does not produce at all. This makes it very important to keep no-load and low-load losses as low as possible, as low-load losses have a much larger impact on the overall project economy than peak-load losses.

The VSCs use pulse-width modulation (PWM) and insulated gate bipolar transistors (IGBTs). With PWM, the converter can operate at almost any phase angle or amplitude.

This can be achieved by changing the PWM pattern, which can be done almost instantaneously. As this allows independent control of both active and reactive power, a PWM VSC comes close to being an almost ideal transmission network component as it is able to control active and reactive power almost instantaneously. This way, the operation characteristics become rather more software-dependent than hardware-related.

13.2.3 Controllability

The control of PWM makes it possible to create any phase angle or amplitude within the ratings. Consequently, control signals to a converter can change the output voltage and current from the converter to the AC network almost instantaneously. Operation can then take place in all four quadrants of the power–reactive-power plane (i.e. active power transmission in any direction can be combined with generation or consumption of reactive power).

The converters can control the transmitted active power in order for it to correspond to the generated power from the wind farms and provide the capability to follow the power output fluctuations from wind power generation. They can also, within certain limits, even out short dips in power generation. This makes it possible to support the frequency control in the power system on Gotland.

The Gotland solution comprises one DC and one AC transmission line in parallel. If the AC line trips, the wind farms will become an isolated AC production area. If the wind turbines produce at maximum, the wind turbine generator speed will increase and eventually the turbine will trip based on high frequency or voltage criteria (individual wind turbine protection systems). The VSC based HVDC link provides the possibility to control this situation by tripping wind turbines and shifting the mode in the Näs converter from active power regulation to frequency regulation.

13.2.4 Reactive power support and control

Reactive power generation and consumption of a VSC can be used for compensating the needs of the connected network, within the rating of the converter. As the rating of the converters is based on maximum current and voltage, the reactive power capabilities of a converter can be traded off for active power capability. The $P–Q$ diagram in Figure 13.3 illustrates the combined active–reactive power capabilities (positive Q is fed into the AC network).

13.2.5 Voltage control

The reactive power capabilities of the VSC are used to control the AC voltages of the network connected to the converter stations. The voltage control of a station constitutes an outer feedback loop and influences the reactive current in order for the set voltage on the network bus to be retained.

At low or no wind power production, the converters will normally switch to a standby position to reduce no-load losses. In the case of a fault with a specific AC voltage

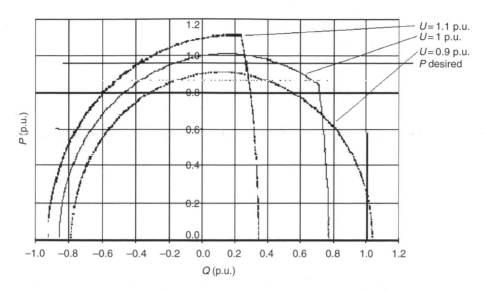

Figure 13.3 Combined active and reactive power capabilities. *Note*: P = active power; Q = reactive power; U = voltage

decrease, the converters could be rapidly (within milliseconds) switched back to operation and assist with voltage support throughout the duration of the fault, thus avoiding severe disturbances for local industries that are sensitive to voltage dips.

The time range for a response to a change in voltage is approximately 50 ms. That means that in the case of a step-order change in the bus voltage, it takes 50 ms to reach the new setting. With this response speed, the AC voltage control will be able to control transients and flicker of up to around 3 Hz as well as other disturbances and keep the AC bus voltage constant. It is thus capable of reducing a considerable part of the wind power generated flicker, caused mainly by Type A wind turbines, at the AC bus.

13.2.6 Protection philosophy

The protection system of the VSC based HVDC is installed to disconnect the equipment from operation in the event of short circuits and other operational abnormalities, especially if they can cause damage to equipment or otherwise interfere with the effective operation of the rest of the system. The protective system is based on the blocking of the IGBT valve switching and/or by AC circuit breakers, which will result in tripping to de-energise the system and thereby eliminate dangerous currents and voltages.

Converter and pole protection systems are especially designed for the VSC based HVDC system and are based on the characteristics of the VSC and its active element, the IGBT transistor. The specific protection systems are designed to handle overcurrents, short circuits across valves, overvoltages, ground faults on the transmission line and the protection of the IGBT valves. Most of the protection systems block the converter. Blocking means that a turnoff pulse is sent to the valves. They will not conduct until they receive a de-block pulse.

During short-circuit faults prevailing for periods of less than a second, the control system will aim at avoiding wind power production tripping. Owing to continuously increased wind power penetration, the grid protection scheme for short circuits must, however, be continuously re-evaluated. It would be advantageous to leave the grid disconnection scheme as well as the flow of fault currents unchanged, to the largest extent possible.

13.2.7 Losses

On Gotland, wind turbines operate most of the time below rated capacity. The average full load hours are around 2000 hours per year for land-based wind turbines, and the total operating time is around 6000 hours out of the 8760 hours in a year.

As the losses in the VSC based HVDC system are higher than those on the parallel AC line, the power flow between the two lines was optimised in order to minimise overall system losses. In this application, it is very important to reduce converter losses also in pure SVC mode and for low power transmission. The DC voltage is the most important parameter for the HVDC light valve losses. The isolation transformer tap changer makes it possible to vary the DC voltage between 95 and 155 kV. This feature is used in combination with a load-dependent DC voltage function. This way, the lowest possible DC voltage is used when active power is low.

13.2.8 Practical experience with the installation

Applying for permits from authorities (including the associated environmental permits) is normally a time-consuming process including questions and demands from various affected parties. With underground DC cables, this process took very little time, primarily because of the limited number of questions raised (related mainly to magnetic fields). Using two underground cables that lie close together and have opposite polarities provides the environmental advantage of comparatively low values for the magnetic field.

A total of 185 property-owners accepted the project within a few weeks. Another contributing factor to the rapid approval was that 50 km of the total 70 km was going to be built on the existing 70 kV line right of way.

The cable was laid with use of a method that is similar to the one applied for midrange voltage cables (i.e. plowing the cable at a depth of at least 650 mm). The joints were planned to be located at convenient places (e.g. at road crossings). At the factory, the cable lengths were adapted to the locations of the joints. About 20 km of the 70 km lead over rocky terrain. In these areas, a rock-milling machine with a width of 120 mm was used. This way, the impact on the environment was kept to a minimum.

The difference from conventional cable laying was the logistics involved in handling 10 000 kg cable drums, each holding approximately 3.5 km of 340 mm^2 cable. Standard trucks were used to transport the 48 cable drums to the sites. The cable lengths were rolled out on the ground before being plowed.

The best machine for the plowing proved to be a timber hauler with an added cable drum stand. When the rock-milling machine had to be used, 300 m of cable were laid per day. Otherwise, a total of several kilometres was laid per day. Roads and watercourses had to be crossed, and telephone cables and power cables had to be moved. However, cable laying as such was not a limiting factor. It was possible to lay cables at temperatures as low as $-10\,°C$.

13.2.9 Tjæreborg Project

In Tjæreborg on the West coast of Denmark, a project was carried out that was similar to the VSC based HVDC project on Gotland. The Tjæreborg demonstration project will be briefly presented below.

In Tjæreborg, an onshore wind farm consisting of four wind turbines is connected to the AC power system via a VSC based HVDC link. This link operates in parallel with a 10 kV AC cable. The HVDC light transmission, with a rating of 7.2 MW and 8 MVA, consists of two VSCs and two ±9 kV DC cables. As the DC cables were laid in parallel to the AC cable, it is possible to operate the wind farm in three different operation modes:

- AC mode via the AC cable only;
- DC mode via the DC cable only;
- parallel mode via the AC and DC cables in parallel.

The option to operate the wind farm in these three different modes has made the demonstration project a study object. It is also very interesting to carry out a comparison of the various operation modes. A detailed discussion of the project, the control models and initial experience is included in Skytt et al. (2001).

13.3 Grid Issues

A successful expansion of wind power generation on Gotland requires an adjustment of the electric system in order to be able to regulate and keep an acceptable voltage quality. In connection with the increasing expansion of the wind power production, GEAB has cooperated with the company Vattenfall to ensure sufficient quality and reliability of power supply. This has resulted in many new ways of looking at network problems such as short-circuit currents, flicker and power flows, among others. In order to solve these problems, new methods have been developed. The basic aspects that were studied are:

- Flicker;
- transient phenomena where faults have been dominating;
- stability in the system with voltage control equipment;
- power flows, reactive power demand and voltage levels in the system;
- calculation of losses in the system with wind power generation;
- instantaneous frequency control with production sources such as gas turbines, but also diesel power plants;
- harmonics.

In the following, the findings regarding flicker, transient phenomena and stability issues with voltage control equipment are briefly presented.

13.3.1 Flicker

The investigations by GEAB showed that Type A wind turbines have different ways of emitting flicker. 'Slow' flicker is due to wind gusts. Repeated startups, connections and disconnections of capacitors are a source of 'fast' flicker. But most of the flicker from wind turbines originates from the so-called '3P' effect, which causes power and voltage fluctuations at the blade–tower passing frequency, which typically gives rise to power fluctuations of around 1–2 Hz. The typical frequencies of voltage fluctuations generated by Type A wind turbines are around 1 and 8 Hz, which are, by coincidence, those to which the human eye is most sensitive.

There are a number of standards and recommendations that set limits for allowed flicker levels (see also Chapters 5 and 7). However, these are often obtained from statistics and assumptions and do not guarantee disturbance-free voltage. GEAB has not used the International Electrotechnical Commission (IEC) norm for flicker when planning the installation of wind farms. Instead, the amplitude of the power fluctuation for different frequencies and limits for the different voltage fluctuations were used. GEAB prefers this method as it provides more information than the statistic value that IEC uses.

Extensive simulation work was carried out to evaluate the impact of the VSC HVDC on flicker levels and to develop a flicker controller that reduces the flicker especially in the range of 1–3 Hz.

GEAB has also noticed that particularly Type A turbines can go into 'synchronous' operation, during specific conditions that depend on the grid. When the above-mentioned flicker controller is applied that phenomenon disappears. Simulation studies have shown that the network angle and the design of the connecting grids are relevant for the phenomenon where the 3P flicker contributions are synchronised. That means that if the network is upgraded the condition changes and this situation can be avoided.

13.3.2 Transient phenomena

Network studies have also shown that the behaviour of asynchronous generators (Type A turbines) is very important during faults in the grid. The subtransient current increases with a larger number of asynchronous generators. If the fault does not last any longer than approximately 200 ms the asynchronous generators consume reactive power from the grid. Thus, the voltage dip at the coupling point is larger and the fault current during the fault becomes lower than without wind power. Overcurrent relays can help to solve this problem. The subtransient fault current and the fault current have to be calculated for a longer period in order to arrive at the right settings for the relay protection.

Simulation studies and measurements have shown that normal synchronous generators control the voltage too slowly and cannot avoid these phenomena. They become a problem once the installed rated power of the asynchronous generators totals around a tenth of the short-circuit power at the coupling point. That was the main reason why GEAB implemented the VSC based HVDC solution. On Gotland, it is more complicated to handle transient phenomena because of the response of the LCC based HVDC link between the mainland and the island. This response differs significantly from that of the VSC HVDC and also of a normal synchronous generator. All this affects the voltage dip during faults. It is important that the voltage dip does not affect the customer voltage, or at least does not exceed a level that is acceptable.

13.3.3 Stability issues with voltage control equipment

The presence of wind turbines alters the normal short-circuit conditions in the network and disturbs the shape of the fault current, which makes it difficult to apply the desired selectivity in the operation of the overcurrent protection. During a fault, the VSC helps to stabilise the voltage.

13.3.3.1 Contribution from VSC-based HVDC

The simulation of a three-phase short circuit applied close to the Näs station with and without the VSC HVDC shows a higher short-circuit current in the first case, without any wind power production (see Figure 13.4).

The converter station supports voltage regulation and thus contributes to stabilising the voltage. During a voltage drop in the AC system, it causes the reactive current to increase up to the capacity limit. The contribution to the short-circuit current decreases

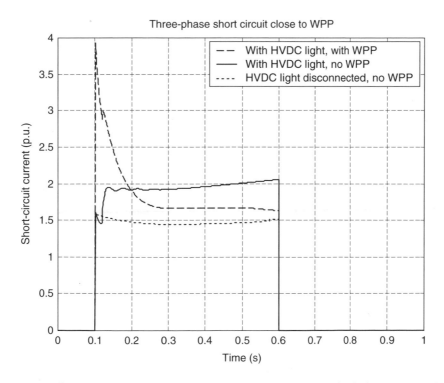

Figure   13.4 Effect of high-voltage direct-current (HVDC) light and wind power production (WPP) on the fault current during short circuit at Näs

with the distance to the fault and the prefault active power transmitted through the link. During a fault with no wind power production (WPP) the short-circuit current is approximately constant if the VSC HVDC is connected.

13.3.3.2 Contribution from wind power production

With WPP, the subtransient short-circuit current can be considerably higher than without WPP, depending on the location of the fault and the level of the connected WPP. Without voltage control, the current decreases over time to the same or lower value than without WPP, owing to the behaviour of the asynchronous generators and capacitors. The voltage control of VSC HVDC helps stabilise the voltage during the fault and therefore the fault current, which is helpful to the operation of the protection system.

13.3.3.3 Protection setting philosophy for the AC system

Owing to the high level of WPP, it is difficult to fulfil the selectivity demands at all production levels, with the normal overcurrent protection being used for the distribution system. However, the operation of the protection system has to be ensured, by setting the protection relays to the minimum values of the short-circuit currents. Computer simulations of short circuits have been used to analyse the proper protection settings.

13.3.4 Validation

GEAB has the goal to maintain a sufficient power quality for its customers even after a large expansion of wind power. Therefore, GEAB has studied four possible fault cases, using computer models with a representation of the entire Gotland power system. The simulation results that included wind power were compared with simulations without wind power.

The most important fault case was named 'Garda'. For this fault case it was possible to compare simulation results with measurements from a real three-phase fault, which was achieved by closing a 10 kV breaker to simulate a solid three-phase short circuit. The sequence consisted only of closing the breaker and letting the overcurrent protection trip the breaker again. The fault lasted only 50 ms, but it showed the response from different equipment. The power response from some wind turbines, the synchronous generators, the LCC HVDC link to the mainland and, of course, from the VSC HVDC was measured.

Voltage dips were measured with a sampling speed of at least 1000 Hz. Voltage dips are defined and evaluated as 20 ms RMS (root mean square) values. Figure 13.5 presents the voltage dips for some key network nodes during a short circuit with a DC voltage at the VSC HVDC of 155 kV. Another short circuit was performed with a VSC HVDC voltage of 96 kV and gave similar results.

Owing to the short fault duration, only the synchronous generators show a physical response and not much response from the controllers can be detected. The same happens in the case with wind power. The simulations and measurements show a very

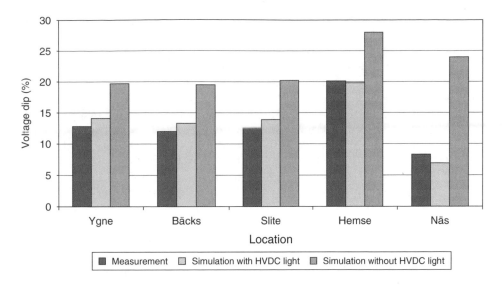

Figure 13.5 Comparison of simulations and measurements of the voltage dip for a short circuit in Garda

similar trend. The VSC HVDC voltage control, however, was slower than in the simulations. The reason was that the gain in the voltage controller was set on a lower value on the actual site, in order to obtain a more stable and safe operation.

Although the short-circuit test worked well, the voltage dip was in most places deeper than the simulations showed. Work on the optimisation of the VSC HVDC voltage control continues and has the goal to achieve similar or better results than the computer simulations.

13.3.5 Power flow

It is common knowledge that wind power production depends strongly on wind speed. On Gotland, local wind power production variations of around 40 % of rated capacity in approximately 10 minutes have been recorded. These comparatively high power output variations are due to the geographical concentration of a large amount of wind power.

The power output variations produce voltage variations in the system, which can affect local power quality. It can be difficult to control the voltage by tap changers and it also affects the maintenance cost of the transformers. A possibility is capacitor on–off switching, but this may affect the voltage quality considerably. The most convenient method of solving voltage control problems with wind power is to use dynamic voltage control. This can be achieved by power electronics, such as SVC, together with intelligent controlling algorithms. On Gotland, the VSC of the HVDC solution is used for the same purpose.

An algorithm was designed for the real-time calculation and management of the reactive power set point as a function of system loads and production, which are provided at regular intervals by the SCADA (supervisory control and data acquisition) system. This can be used to optimise the power flow against low losses and voltage levels.

The algorithm calculates a reactive power set point for the VSC approximately every 5 minutes. The time constants for this slow voltage control must be coordinated with the tap-changer controllers and voltage controllers of the synchronous machines in order to avoid interaction or negative influence on the power system performance. The interaction between dynamic voltage controllers used in tap changers, synchronous machines, power electronics as SVC and other types of voltage control devices in systems with a large wind power penetration is an aspect that must be carefully analysed when setting control parameters since it determines the voltage stability of the system.

The algorithm mentioned before is also used for the real-time calculation of the active power set point in the VSC as a function of the system loads and production given by the SCADA system. The main criterion is total loss reduction, although operation limits, such as maximum line current and so on are taken into account.

13.3.6 Technical responsibility

The ownership of wind power on Gotland varies from case to case. Some wind turbines and wind farms are owned by cooperatives, others by individuals. GEAB as the grid operator has the responsibility to ensure sufficient power quality to all network customers, which includes also local generation.

Even though wind power has a considerable impact on power quality, GEAB has to handle this responsibility without being the owner of the plants. Therefore, GEAB has defined technical requirements for the connection points. These requirements are defined in terms of current or power. There are also some requirements regarding protection and dynamic behaviour. The owners of the wind turbines have the obligation to document the technical capabilities of each wind turbine and also have to follow the requirements set in a contract for the grid connection. The locally developed guidelines have been included in the Swedish guidelines – Anslutning av mindre produktionsanläggningar till elnätet (AMP; see Svensk Energi, 2001) – which apply to all small productions plants to be installed in Sweden.[1]

13.3.7 Future work

Currently, there are applications for more than 300 MW of new wind farms to be connected to the power system on Gotland. GEAB considers that the maximum wind power capacity that can be connected is around 250 MW. However, additional technical issues must be solved before this amount of wind power can be integrated into the Gotland power system. Hence, GEAB is studying further technical solutions for the network integration of wind power generation. Besides continuously studying the impact of increasing wind power penetration on relay protection and voltage control, GEAB wants to find further new solutions for the network integration of wind power.

---

[1] For a more detailed discussion of AMP and other international interconnection rules, see Chapter 7.

13.4 Conclusions

In this chapter we have described the main difficulties that GEAB has faced during the development of wind power on Gotland. The practical experience is likely to coincide with that of other grid systems, even if they do not use VSC based HVDC as a solution. In summary, the main issues are:

- voltage stability related to transient situations, power flows and reactive power demand and the coordination of voltage control in the system: these issues are solved mostly by the installation of the VSC based HVDC system;
- fault situations and relay protection functions as well as selective plans: here, the VSC based HVDC has a large effect and further work has to be done in order to arrive at an optimal solution. Even the technical choice of the protection function and quality has to be taken into consideration.

GEAB has worked with defining requirements that are specific to Gotland but has also cooperated in clarifying standard documents regarding, mostly technical, requirements. During this process, it has become evident that it is much more difficult than expected to deal with these issues. The manufacturing of wind turbines is not standardised, and legal aspects may limit the possibilities to define very specific requirements. It is considered very important for the grid operator, though, to have the option to define very specific requirements for wind farms in order to be able to fulfil its responsibility of operating the power system and ensuring an appropriate quality of power supply.

Further Reading

[1] Castro, A. D., Ellström, R., Jiang Häffner, Y., Liljegren (2000) 'Co-ordination of Parallel AC–DC Systems for Optimum Performance', paper presented at Distributech 2000, September 2000, Madrid, Spain.
[2] Holm, A. (1998) 'DC Distribution, the Distribution for the 21st Century?', paper presented at Distributech 98, November 1998, London, UK.
[3] Liljegren, C., Åberg, M., Eriksson, K., Tollerz, O., Axelsson, U., Holm, A. (2001) 'The Gotland HVDC Light Project – Experiences from Trial and Commercial Operation', paper presented at Cired Conference, Amsterdam, 2001.
[4] Rosvall, T. (2001) *Eletricitet på Gotland*, Ödins Förlag AB, November 2001.
[5] Thiringer, T., Petru, P., Liljegren, C. (2001) 'Power Quality Impact of Sea Located Hybrid Wind Park', *IEEE Transactions on Energy Conversion* 16(2) 123–127.

References

[1] Skytt, A.-K., Holmberg, P., Juhlin, L.-E. (2001) 'HVDC Light for Connection of Wind Farms', in *Proceedings of the Second International Workshop on Transmission Networks for Offshore Wind Farms*, Royal Institute of Technology, Stockholm, Sweden, March 2001.
[2] Svensk Energi (2001) *AMP – Anslutning av mindre produktionsanläggningar till elnätet*, Svensk Energi, Stockholm, Sweden.

14

Isolated Systems with Wind Power

Per Lundsager and E. Ian Baring-Gould

14.1 Introduction

Isolated power supply systems using large amounts of wind and other renewable technologies are emerging as technically reliable options for power supply. Such systems are generally perceived to have their major potential markets as local power supply in first, second and third world countries, but the technology also has considerable potential as a distributed generation component in large utility grids in the developed world.

During the past two decades there have been considerable efforts on the national and international level to implement wind energy in connection with local and regional electric power supply by its integration into small and medium-sized isolated distribution systems. These systems are often, but not necessarily, powered by diesel power plants. Fairly recent reviews made for the Danish International Development Agency, DANIDA (Lundsager and Madsen, 1995) and the Danish Energy Agency (Lundsager *et al.*, 2001) identified about 100 reported and documented wind–diesel installations worldwide.

Much work has been reported in public literature, but since studies of isolated systems with wind power have mostly been case-oriented it has been difficult to extend results from one project to another. This is not least because of the strong individuality that has so far characterised such systems and their implementation, and therefore a consistent and well-developed positive track record has yet to be developed.

This chapter aims to give an account of the basic issues and principles associated with the use of wind energy in isolated power systems and a review of international experience with such systems. The chapter describes a number of recent applications of wind

Wind Power in Power Systems   Edited by T. Ackermann
© 2005 John Wiley & Sons, Ltd   ISBN: 0-470-85508-8 (HB)

power in isolated systems and gives an account of the methodologies and tools now available for the analysis and design of such systems. This chapter also outlines the principles in the successful application of isolated systems with wind power.

14.2 Use of Wind Energy in Isolated Power Systems

In the world of rural electrification, there are two general methods of supplying energy to rural areas: grid extension and the use of diesel generators. In remote areas both options can be exceedingly expensive, grid electrification costing upwards of US$3000 per connection or a continued reliance on expensive and volatile diesel fuel. The inclusion of renewable technologies can lower the lifecycle cost of providing power to rural areas. However, since renewable technologies, other than biomass technologies, are dependent on a resource that is not dispatchable, the combination of a low-cost renewable technology with a more expensive dispatchable technology may provide the most applicable alternative.

Power systems using multiple generation sources can be more accurately described by the term 'hybrid power systems' and can incorporate different components such as production, storage, power conditioning and system control to supply power to a remote community.

The classic hybrid systems include both a direct-current (DC) bus for the battery bank and an alternating-current (AC) bus for the engine generator and distribution; however, recent advances in power electronics and system control are making small single AC bus systems more cost effective. The renewable technology may be attached to either the AC or the DC bus depending on the system size and configuration. Power systems supplying more than single homes or other single point systems usually supply AC power, although some loads may be tapped off the DC bus bar. These power systems can range in size from a few kilowatt-hours (kWh) per day to many megawatt-hours (MWh) per day.

Systems providing smaller loads, less than a few hundred kilowatt-hours per day generally focus around the DC bus whereas larger systems tend to use the AC bus as the main connection point. Recent advances in power electronics and control have allowed the advancement of smaller AC-connected devices where each DC-producing renewable device includes its own dedicated power converter with imbedded control that allows it to coordinate power production. Every configuration, however, can have a great deal of internal variability. Smaller systems will likely use large battery banks, providing up to a few days of storage, and will use smaller renewable generation devices connected to the DC bus. Larger systems focus on the AC bus bar, with all renewable technology designed to be connected to the AC distribution network. In larger systems, the battery bank, if used at all, is generally small and used mainly to cover fluctuations in power production. Larger systems usually contain more and larger equipment that allows for an economy of scale and thus lower power costs.

14.2.1 System concepts and configurations

In systems that focus around the DC bus, the battery bank acts as a large power dampener, smoothing out any short-term or long-term fluctuations in the power flow.

In many ways, it regulates itself based on a few battery-specific parameters. This is not the case with equipment connected to the AC grid, which is much less forgiving. For AC systems the key issues are the balancing of power production, load and voltage regulation on a subcycle time scale. This is done largely with the use of synchronous condensers, dispatchable load banks, storage, power electronics and advanced control systems that carefully monitor the operating conditions of each component to ensure that the result is power with a consistent frequency and voltage. In AC systems, the inclusion of a battery bank can also provide the same power-smoothing functions, although more care must be taken in the control of system power.

In the following subsections three figures are provided to describe the three general types of wind–hybrid power systems. The first, Figure 14.1, illustrates a small conventional DC-based power system providing AC power using a power converter; second, Figure 14.2 shows a small power system focused around the AC bus; last, Figure 14.3 illustrates a larger AC-coupled power system.

Small wind turbines, up to 20 kW, have also been connected to power devices directly; the most common of these is for water pumps in agriculture or for portable water, although other applications such as ice making, battery charging and air compression have been considered. Both mechanical and electric wind turbines are used widely.

14.2.1.1 DC-based hybrid systems for small remote communities

Figure 14.1 illustrates a small conventional DC-based power system providing AC power using a power converter. The use of smaller renewable-based hybrid systems

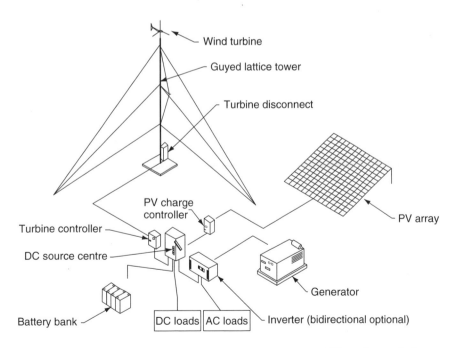

Figure   14.1 DC-based renewable power system. *Note*: PV = photovoltaic

has grown as small wind technology increases in usability and photovoltaic (PV) technology decreases in cost.

Most of these systems use a topography where the DC battery bus is used as a central connection point. Generally, small winds turbines generate variable frequency and voltage AC current that is rectified and applied to the DC bus at the voltage of the battery bank. Energy either is stored in the battery or is converted to AC through an inverter to supply the load. The use of the battery bank smoothes out wind turbine power fluctuations and allows energy generated when there is wind to supply a load at a later point in time. In cases where guaranteed power is required, a dispatchable generator – typically diesel, propane or natural gas – can also be installed to provide the load and charge batteries in the prolonged absence of renewable-based generation.

Control in small DC-based power systems is usually conducted through the voltage of the battery bank. All generators have a voltage limiter, which reduces or shunts any energy generated by the wind turbine if the battery is too full to accept additional power. Recent research conducted at the National Renewable Energy Laboratory (NREL) (Baring-Gould et al., 2001, 1997; Corbus et al., 2002; Newcomb et al., 2003) has shown high potential losses associated with the interaction of DC wind turbines and the battery bank because of premature limiting of energy by the wind turbine controller. For this reason, care must be taken when designing such a power system in order to ensure that proper matching between the different components is achieved. All inverters and load devices also have low-voltage disconnects that stop the discharge of the battery bank if the voltage drops below some present value. The power supplied to the AC bus is controlled through the inverter, which may provide additional system control depending on the unit (for more information on systems using DC architecture, see Allerdice and Rogers, 2000; Baring-Gould et al., 2001, 2003; Jiminez and Lawand, 2000).

14.2.1.2 AC-based hybrid systems for small remote communities

Recent improvements in power electronics, control and power converters have led to the rise of a new system topology, and Figure 14.2 shows a small power system focused around the AC bus. These systems use small, DC or AC generation components, PVs and wind, connected through a dedicated smart inverter to the AC distribution grid. A battery is used to smooth out power fluctuations but also includes its own dedicated power converter.

The prime advantage of this topology is its modularity, allowing the connection or replacement of modules when additional energy is needed. Second, it steps away from the need to co-locate all components where they can be connected to a DC bus, allowing each component to be installed at any location along the micro-grid. These systems generally use system frequency to communicate the power requirements between the different generation and storage modules. The two disadvantages of systems using this topology are cost and the use of sophisticated technology that will be impossible to service in remote areas.

An additional issue is that all energy being stored must pass first from the point of generation to the AC bus and then through the rectifier of the battery dedicated power converter. It must then be inverted again prior to use, resulting in three power conversion cycles, compared with only one for systems using a DC-based topography.

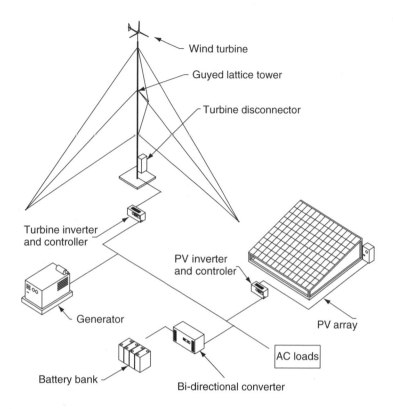

Figure **14.2** AC-based renewable power system. *Note*: PV = photovoltaic

Thus in systems where large amounts of energy storage is expected, such as PV systems designed to provide evening lighting loads, this technology may result in higher system losses (more information on these types of power systems may be located in Cramer *et al.*, 2003; Engler *et al.*, 2003).

14.2.1.3 Wind–diesel systems

Larger power systems are focused around the AC bus and incorporate both AC-connected wind turbines and diesel engines. Figure 14.3 shows a larger AC-coupled wind–diesel power system.

A technically effective wind–diesel system supplies firm power, using wind power to reduce fuel consumption while maintaining acceptable power quality. In order to be economically viable, the investment in the extra equipment that is needed to incorporate wind power, including the wind turbines themselves, must be recouped by the value of the fuel savings and other benefits. As the ratio of the installed wind capacity to the system load increases, the required equipment needed to maintain a stable AC grid also increases, forcing an optimum amount of wind power in a given system. This optimum is defined by limits given by the level of technology used in the system, the complexity of

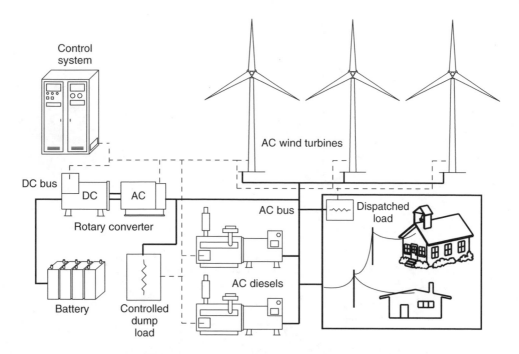

Figure 14.3 A typical large wind–diesel power system

the layout chosen and the power quality required by the user. For this reason the optimal design must be based on careful analysis, not simply the maximum amount of wind energy possible. Although not the focus of this book, it should also be noted that other diesel retrofit options do exist and should be investigated. This includes the use of other renewable technologies, such as biomass, or simply better control of the diesel generators (Baring-Gould *et al.*, 1997).

Because of the large number of isolated diesel mini-grids both in the developed world and in the developing world, the market for retrofitting these systems with lower cost power from renewable sources is substantial. This market represents a substantial international opportunity for the use of wind energy in isolated power supply. To meet this market, an international community has formed, committed to sharing the technical and operating experience to expand recent commercial successes.

A typical isolated diesel power supply system can be characterised by the following:

- It has only one or a few diesel generating sets (gensets). By using a number of diesel gensets of cascading size an optimal loading of diesel gensets may be obtained, thus increasing the efficiency of the diesel plant.
- The existing power system has quite simple system controllers, often only the govern-ors and voltage regulators of the diesel generators, possibly supplemented by load-sharing or self-synchronising devices.
- The local infrastructure may be limited and there may be no readily available resources for operation, maintenance and replacement (OM&R).

- Fuel is generally expensive and is sometimes scarce and prone to delivery and storage problems.
- The diesel engines provide adequate frequency control by the adjustment of power production to meet the load and voltage control by modifying the field on the generator.

14.2.2 Basic considerations and constraints for wind–diesel power stations

In technical terms, an isolated power system for a large community that incorporates wind power will be defined as a wind–diesel system when both the system layout and operation are significantly influenced by the presence of wind power in terms of:

- frequency control, stability of system voltage and limited harmonic distortion;
- the operating conditions of the diesel generators, especially with regard to minimum load;
- provisions for the use of any surplus wind power;
- OM&R of any system components.

Wind–diesel power systems can vary from simple designs in which wind turbines are connected directly to the diesel grid with a minimum of additional features, to more complex systems (Hunter and Elliott, 1994; Lundsager and Bindner, 1994). The important complication of adding wind power to diesel plants is that the production of energy from the wind turbines is controlled by the wind, meaning most turbines cannot control either line frequency or voltage and must rely on other equipment to do so. With only small amounts of wind energy the diesel engines can provide this control function, but with larger amounts of wind energy other equipment is necessary.

With this in mind, two overlapping issues strongly influence system design and its required components: one is the amount of energy that is expected from the renewable sources (system penetration), the other is the ability of the power system to maintain a balance of power between generation and consumption (the primary use for energy storage).

14.2.2.1 Wind penetration

When incorporating renewable-based technologies into isolated power supply systems, the amount of energy that will be obtained from the renewable sources will strongly influence the technical layout, performance and economics of the system. For this reason, it is necessary to explain two new parameters – the instantaneous and average power penetration of wind – as they help define system performance.

Instantaneous penetration, often referred to as power penetration, is defined as the ratio of instantaneous wind power output, P_{wind} (in kW), to instantaneous primary electrical load, P_{load} (in kW):

$$\text{Instantaneous (power) penetration} = \frac{P_{wind}}{P_{load}}. \qquad (14.1)$$

Thus, it is the ratio of how much wind power is being produced at any specific instant. Instantaneous penetration is primarily a technical measure, as it greatly determines the layout, components and control principles to be applied in the system.

Average penetration, often referred to as energy penetration, is defined as the ratio of wind energy output, E_{wind} (in kWh), to average primary electrical load, E_{load} (in kWh), measured over days, months or even years:

$$\text{Average (energy) penetration} = \frac{E_{wind}}{E_{load}}. \tag{14.2}$$

Average penetration is primarily an economic measure as it determines the levellized cost of energy from the system by indicating how much of the total generation will come from the renewable energy device.

14.2.2.2 Power balance

As an illustration of the balancing of generation and consumption of power, Figure 14.4 shows the power balance of a basic wind diesel system, where the instantaneous consumer load is shown as 100 % load. In order to highlight the issues, the system illustrated in Figure 14.4 has an unusually high proportion of wind energy.

At low wind speeds, the diesel generator produces all the power required, but as the wind speed increases the increased power production from the wind turbine is reflected by a decreased production from the diesel generator. Eventually, the diesel generator reaches its minimum load defined by the manufacturer and the surplus power must be taken care of either by limiting the power production from the wind turbine or by diverting the surplus power in a controllable load. If the system configuration and

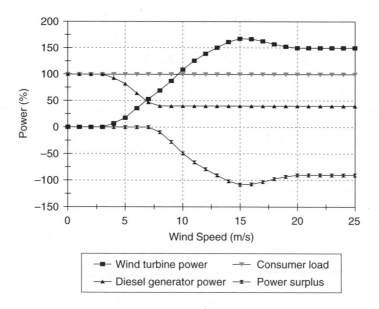

Figure   14.4 Simple power balance of a wind–diesel power system

control allow it, the diesel could also be shut off, allowing the load to be supplied completely by wind energy.

14.2.2.3 Basic system control and operation

The simplest large wind diesel system configuration is shown in Figure 14.5, in which a standard grid-connected wind turbine with an induction generator is connected to the AC bus bar of the system. This represents a simple low-penetration power system.

When the wind turbine power output is far less than the consumer load minus the diesel minimum load, the diesel governor controls the grid frequency and the voltage regulator of the synchronous generator of the diesel set controls voltage.

If the designed wind turbine output is about equal to the consumer load minus the diesel minimum load, controllable resistors, typically referred to as a dump load, may be included to absorb possible surplus power from the wind turbine, in which case the dump load controller helps to regulate the frequency. This would be termed a medium-penetration power system.

If the wind turbine can produce more power than is needed by the load a more complicated system is required, which may also incorporate energy storage to smooth out fluctuations in the wind energy, as was shown in Figure 14.3. In this case, some surplus energy is saved for later release when the need arises. If the wind production is more than the load, it then becomes possible to shut off all diesel generators altogether. Although this maximises fuel savings it requires very tight control of all power system components to ensure system stability. Additionally, the fluctuating nature of the wind power may pose extra problems for the control and regulation of the system. This last configuration would be characteristic of a high-penetration power system.

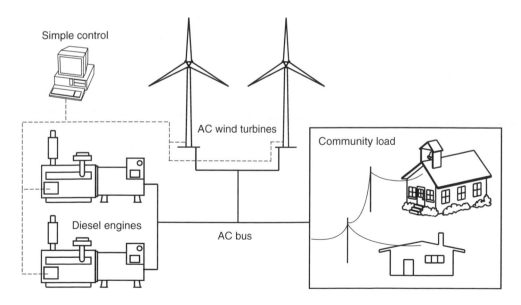

Figure   14.5 Simple, low-penetration wind–diesel power system

As can be intuitively understood, the higher the wind penetration, the more energy that cannot be used by the electric loads is generated. Instead of just dissipating this surplus power in a dump load the power can be used to satisfy additional community loads such as:

- production of fresh water (e.g. by desalination, purification and so on);
- ice making;
- water heating;
- house or district heating;

and much more. Usually, these loads are broken into two categorises:

- Optional loads: loads that will be met only if and when surplus power is available as other energy sources can be used when excess energy is not available (e.g. space heating);
- Deferrable loads: loads that must be met over a fixed period of time, for example on a daily basis, and if not supplied by surplus power will be served by primary bus bar power.

Additionally, some loads can be controlled by the power system instantaneously to reduce the required power demand, thus saving the system from having to start an additional generator to cover what is only a momentary defect of power. This is often referred to as demand-side management.

14.2.2.4 Basic system economics

As with any power system, economics play a key role in the selection of power system design and the technology to be applied. As has been discussed previously (Section 14.2), the reason to consider wind energy as an addition to diesel-based generation is because of its potential lower cost. However, as the level of system penetration increases, we have seen that the complexity of the power system also increases and subsequently so does the cost (Section 14.2.1). Therefore, the optimal level of wind penetration depends on the relative cost difference between increasing the penetration of wind and the difference between generation using wind and diesel technology. We have also seen that if excess energy from the wind can be used in place of other expensive fuels to provide additional services, such as heating, the economics of using alternative sources also improve. Determining this optimal system design is not a trivial problem, but luckily tools that will be introduced later in this chapter (Section 14.6) have been developed to assist with this process.

14.2.2.5 The role of storage in wind–diesel power stations

Storage may be used in a high-penetration power system to smooth out fluctuations in the load and wind power in cases where the diesel generators are allowed to shut off. Additionally, short-term energy storage can also be used to cover short-term lulls in wind energy or allow diesel engines to be started in a controlled manner (Shirazi and Drouilhet, 2003). Generally, energy storage will use either battery storage or flywheel storage. Both of these devices would require the addition of some power conditioning or the use of a bidirectional power converter. Many other forms of storage have been proposed or experimented with. However, the only other successful implementation has been using pumped hydro storage, where excess power is used to pump water into a (high) reservoir to be used later to drive the pump in reverse operation to generate electricity.

The use of hydrogen as a storage medium has received notoriety recently. In this case, an electrolyser uses any excess energy to create hydrogen, which would then be stored for later use. An internal combustion engine or fuel cell would be used to convert the hydrogen into electricity when it was needed. Since the amount of potential storage is dependent only on the size of the storage mechanism, the possibility of long-term or season-based storage makes this option look very attractive. However, the exceedingly high cost of both electrolysers and fuel cells, as well as the very low turn-around efficiency of the system, which approaches 30 % compared with 80 % with lead–acid batteries, still makes this option unattractive from an economic standpoint. Since neither fuel cells nor hydrogen internal combustion engines can be started instantaneously, some form of short-term energy storage will still be required (Cotrell and Pratt, 2003).

The economics of systems with energy storage depend strongly on the reliability and cost both of power electronics, expected to improve significantly with time, and of the storage medium itself. Prediction of the expected life of the battery under site-specific conditions is an important economic and technical issue. The economics of systems without storage are strongly influenced by operational restraints of the diesel system in terms of diesel scheduling, the existence of a mandatory spinning reserve safety threshold and the operational load limits of each generator.

As shown in Figure 14.6, representative for all but very small diesel generators, when diesel generator output decreases, presumably as a result of an increasing instantaneous

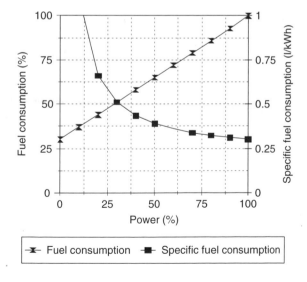

Figure 14.6 Fuel consumption (as a percentage of the maximum) and specific fuel consumption (in litres per kilowatt-hour) as a function of power (as a percentage of the maximum)

penetration, the absolute fuel consumption (in litres per hour) decreases proportionally with diesel power, but it does not reach zero at zero diesel load. Therefore, the specific fuel consumption does not decrease proportionally with decreased load on the diesel. For this reason, it is always better to shut down diesel engines when they are not needed in place of having them operate at very low loading.

14.3 Categorisation of Systems

There is a clear difference between the wind turbines installed in a small isolated system in a rural community and a wind turbine situated in an offshore wind farm in Denmark. Because of the differences in their design characteristics and performance it is useful to introduce a categorisation of power systems according to the installed power capacity. A suggested categorisation is shown in Table 14.1 (Lundsager *et al.*, 2001).

Typically, a micro system uses a small wind turbine with a capacity of less than 1 kW; a village system typically has a capacity between 1 kW and 100 kW, with one or more wind turbine(s) of 1–50 kW; an island power system is typically from 100 kW to 10 MW installed power and with wind turbines in the range from 100 kW to 1 MW; a large interconnected power system typically is larger than 10 MW, with several wind turbines larger than 500 kW installed as wind power plants or wind farms.

The theoretical maximum wind penetration levels of the power systems presented in Table 14.1 are plotted in Figure 14.7 as a function of the total installed capacity. To provide an indicator for very large power systems, the existing conditions of Denmark in 1998 and one projection for Denmark for the year 2030 are used. The dashed line shows the current condition in which the level of wind energy penetration of actual power systems with successful track records decreases as the power system size increases. The dotted line indicates possible future development towards higher penetration levels, which may be achieved in the next 20–30 years. The benchmark used for the future village power system is based on the Frøya Island power system – a Norwegian research system designed to investigate maximum wind penetration (Table 14.2; see Section 14.4.1).

The theoretical feasibility of very high wind energy penetration is seen to change dramatically in the 100 kW to 10 MW system size range. In this range, conventional electricity generation is still based on diesel generation, which has a high cost of energy. The main reasons for the existing low penetration levels in the largest systems is primarily economic, even if at the present stage of development wind energy production cost is on a par with most conventional sources. At any given configuration, there is

Table 14.1 Categorisation of isolated power systems

Installed power (kW)	Category	Description
<1	Micro systems	Single point DC-based system
1–100	Village power systems	Small power system
100–10000	Island power systems	Isolated grid systems
>10000	Large interconnected systems	Large remote power system

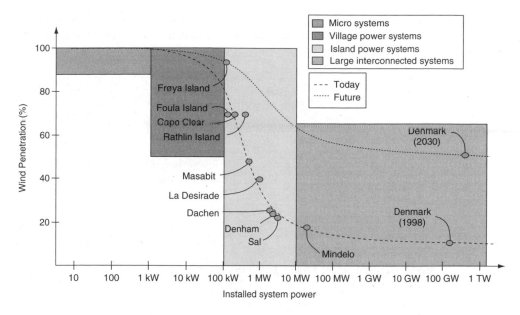

Figure 14.7 Present and expected future development of the wind energy penetration vs. the installed system capacity, Lundsager *et al.* (2001), reproduced with permission

a maximum wind energy penetration, above which the economic return from adding wind power begins to decrease. In addition, managers of larger systems usually take a cautious approach because of generally high fluctuations in wind power.

As indicated by the dotted line in Figure 14.7, the level of wind energy penetration can be expected to increase significantly in the future. Thus the challenge of national (and transnational) systems will be to increase penetration to levels already existing in smaller isolated systems, which themselves seem to be well placed to increase their wind energy penetration to levels typical for just slightly smaller systems.

Obviously, great care has to be taken in the process of introducing wind power into isolated power systems, as many failures have occurred in the past because of over-ambitious system designs with a high degree of complexity and limited background experience in project development. The recommended approach seems to be a gradual increase, starting at the dashed line of Figure 14.7 and moving upwards towards the dotted line in a step-by-step approach applying simple, robust, reliable and well-tested concepts.

14.4 Systems and Experience

The present state-of-the-art of wind turbines is given in Chapter 4 of this book as well as in a number of studies (see for example Hansen *et al.*, 2001). Even though there is rapid development in wind turbine technology there seems to be a pretty clear picture. The present state-of-the-art of power systems with wind energy is more difficult to assess because of the great variety of concepts and applications (for an impression of the

Table **14.2** Selected list of hybrid power systems at research facilities

Laboratory	Country	Year	Rated power (kW)	Special features
NREL	USA	1996	160	Multiple units, AC & DC buses, battery storage
CRES	Greece	1995	45	—
DEWI	Germany	1992	30	Two wind turbines
RAL	UK	1991	85	Flywheel storage
EFI	Norway	1989	50	Wind turbine simulator and wind turbine
RERL–UMASS	USA	1989	15	Rotary converter
IREQ	Canada	1986	35	Rotary condenser
AWTS	Canada	1985	100	Multiple units, rotary condenser
Risø	Denmark	1984	30	Rotary condenser

Source: Pereira, 2000.
Note: NREL = National Renewable Energy Laboratory; CRES = Center for Renewable Energy Resources; DEWI = Deutsches Windenergie-Institut; RAL = Rutherford Appleton Laboratory; EFI = Electric Power Research Institute; RERL–UMASS = Renewable Energy Research Lab at UMass Amherst; University of Massachusetts; IREQ = Institut de Recherche d'Hydro Québec; AWTS = Atlantic Wind Test Site

current state of the technology, see Baring-Gould *et al.*, 2003; Lundsager and Madsen, 1995; Lundsager *et al.*, 2004; Pereira, 2000).

14.4.1 Overview of systems

The source of the system concepts applied can be found in the quite large number of more or less experimental systems installed at research facilities throughout the world. Table 14.2 presents a summary of these systems from Pereira (2000), where a complete overview can be found.

Table 14.3 gives an overview of larger AC-based hybrid power systems installed worldwide over the last decade. All the systems in the table produce power to their community, but most of them are installed in the framework of demonstration or validation projects with some degree of public co-financing (more details of the systems can be found in Baring-Gould *et al.* (2003); Pereira, 2000).

14.4.2 Hybrid power system experience

The most recent account of systems and experience was given at the September 2002 wind diesel conference (Baring-Gould *et al.*, 2003). During the conference, 12 presentations were given providing experience on the operation of more than 17 wind–diesel power systems from around the world. The associated website (DOE/AWEA/Can-WEA, 2002) contains links to all the presentations. We have selected five of these projects that show the varied options for using diesel retrofit technology. These systems

Table   14.3 Selected list of hybrid power systems installed throughout the world in the past decade

Site	Country or region	Reporting period	Diesel power (kW)	Wind power (kW)	Other features	Wind penetration (%)
Wales	Alaska	1995–2003	411	130	Heating, storage	70
St Paul	Alaska	1999	300	225	Heating	—
Alto Baguales	Chile	2001	13 000	1 980	Hydro power	16
Denham	Western Australia	2000	1 970	690	Low load diesel	50
Sal	Cape Verde	1994–2001	2 820	600	Desalination	14
Mindelo	Cape Verde	1994–2001	11 200	900	Desalination	14
Dachen Island	China	1989–2001	10 440	185	Low-load diesel	15
Fuerteventura	Canary Island	1992–2001	150	225	Desalination, ice	—
Foula Island	Shetland Islands	1990–2001	28	60	Heating, hydro power	70
La Desirade	Guadeloupe	1993–2001	880	144	—	40[a]
Marsabit	Kenya	1988–2001	300	150	—	46
Cape Clear	Ireland	1987–1990	72	60	Storage	70[a]
Rathlin Island	Northern Ireland	1992–2001	260	99	Storage	70
Kythnos Island	Greece	1995–2001	2 774	315	Storage, photovoltaic power	—
Frøya Island	Norway	1992–1996	50	55	Storage	94
Denham	Australia	1998–present	1 736	230	—	23
Lemnos Island	Greece	1995–present	10 400	1 140	—	—

[a] Peak.
Sources: Baring-Gould *et al.*, 2003; Pereira, 2000.

also depict installations in very remote locations without easy access to highly developed infrastructure and advanced technical support.

14.4.2.1 Wales, Alaska: wind–diesel high-penetration hybrid power system

The average electric load for this community is about 70 kW, although there are also substantial heating loads for buildings and hot water. The Wales, Alaska, wind–diesel hybrid power system, which underwent final commissioning in March 2002, combines three diesel gensets with a combined power of 411 kW, two 65 kW wind turbines and a 130 Ah battery bank made up of SAFT (SAFT Batteries SA) nickel–cadmium batteries, an NREL-built rotary power converter and various control components. The primary purpose of the system is to meet the village's electric demand with high-quality power, while minimising diesel fuel consumption and diesel engine runtime. The system also directs excess wind power to several thermal loads in the village, thereby saving heating fuel.

Limited data are available, but it is expected that the two wind turbines will provide an average penetration of approximately 70 %, saving 45 % of expected fuel consumption and reducing the operational time of the diesel engines by 25 %. The power system has experienced some service-related technical problems since its installation, which has reduced its performance.

14.4.2.2 St Paul, Alaska: high-penetration wind–diesel power system

In 1999, the first fully privately funded high-penetration, no-storage hybrid power system that maximises wind penetration was installed by TDX Power and Northern Power Systems. The primary components of the St Paul, Alaska, plant include a 225 kW Vestas V27 wind turbine, two 150 kW Volvo diesel engine generators, a synchronous condenser, a 27 000 l insulated hot water tank, approximately 305 m (1000 feet) of hot water piping and a microprocessor-based control system capable of providing fully automatic plant operation.

The primary electrical load for the facility averages about 85 kW, but the system also supplies the primary space heating for the facility, which it does with excess power from the wind generators and thermal energy from the diesel plant. When the wind generation exceeds demand by a specific margin, the engines automatically shut off and the wind turbine meets the load demand with excess power diverted to the hot water tank, which in turn is used to heat the complex. When wind power is insufficient to meet the load, the engines are engaged to provide continuous electric supply as well as provide energy to the hot water system when needed. The total 500 kW wind–diesel co-generation system cost approximately US$1.2 million. Its operation has saved US$200 000 per year in utility electric charges and US$50 000 per year in diesel heating fuel. Since its installation, the load has continued to grow, and additional thermal heating loops have been added to the facilities.

14.4.2.3 Alto Baguales, Chile: wind–hydro–diesel power system in Coyhaique

The system supplies energy to the regional capital of Coyhaique, Chile, providing a maximum power of 13.75 MW. In the autumn of 2001, three 660 kW Vestas wind turbines were installed to supplement the diesel and hydro production. The Alto Baguales wind energy project is expected to provide more than 16 % of the local electric needs and displace about 600 000 liters (158 500 gallons) of diesel fuel per year.

The turbines are operated remotely from the diesel plant, with no additional control capabilities, and operate at around a 50 % capacity factor because of strong winds at the turbine site. To date, the highest recorded penetration, based on 15 minute instantaneous readings taken at the power station, is 22 % of total demand. In the summer of 2003, additional hydro capacity will be installed, allowing the utility to provide the whole load with wind and hydropower, completely eliminating diesel production.

The primary challenge of the project implementation was in obtaining the proper installation equipment, including a crane that had to be brought in over the mountains from Buenos Aires, Argentina.

14.4.2.4 Cape Verde: the three major national power systems

The Archipelago of the Republic of Cape Verde consists of 10 major islands off the Western cost of Africa. Three medium-penetration wind–diesel systems have successfully provided power to the three main communities of Cape Verde – Sal, Mindelo and Praia – since the mid-1990s. These power systems are very simple in nature, containing only a dump load and a wind turbine shutdown function to keep minimum diesel load conditions. The three 300 kW NKT wind turbines at each site are connected to the existing diesel distribution grid in a standard grid-connected fashion. The average loads for the communities vary from 1.15 MW for the smallest, Sal, to 4.5 MW for the largest, Praia (the nation's capital).

At present, the power systems operate at monthly wind energy penetrations of up to 25 %, depending on the system and time of year. Yearly penetrations of up to 14 % for Sal and Mindelo have been experienced. The maximum monthly wind power penetration of 35 % was reached in Sal, with no adverse system impact. The overall experience of the three wind plants has been perceived by the utility ELEKTRA as quite positive and this has resulted in the initiation of a second phase, financed through the World Bank and managed by Cape Verde, in which wind energy penetration of the three power systems will be almost doubled.

Risø National Laboratories of Denmark has conducted a detailed assessment of the phase 2 systems using WINSYS modelling software (Hansen and Tande, 1994). The expansion of these systems is expected to increase the wind energy penetration in the first year of operation to levels of up to 30 % (Mindelo). The average annual diesel fuel consumption is expected to be reduced by an additional 25 %.

14.4.2.5 Australia: wind–diesel power stations in Denham and Mawson

The Denham wind–diesel power station is operated by Western Power and is located on the Western coast of Australia, north of the regional capital of Perth. The power system has a maximum demand of 1200 kW, which is supplied by 690 kW of wind from three 230 kW Enercon E-30 wind turbines, four diesel engines with a total rating of 1720 kW, and a recently installed low-load diesel. The low-load diesel is a standard diesel generator that is modified to allow prolonged operation at very low or negative loads. The device installed in Denham has a load range of +250 kW and −100 kW. The power system was installed by PowerCorp of Australia and is controlled from a highly automated control centre located in the powerhouse using control logic, also developed by PowerCorp. The power system has the capability to operate in a fully automated mode with minimal technical oversight. The system control allows all of the standard diesels to be shut off, resulting in an average penetration of 50 %.

The power system has been operating for more than three years, supplying utility-quality power and saving an estimated 270 000 l (71 300 gallons) of fuel per year.

In a separate but similar project, PowerCorp recently installed two Enercon E30 300 kW wind turbines with a 200 kW dynamic grid interface into the Mawson diesel power system in Antarctica. This system, which does not contain storage, has achieved an average penetration of approximately 60 % since its commissioning in January 2003.

14.5 Wind Power Impact on Power Quality

In systems with turbines connected to the AC bus, the critical consideration is the variation of the power output from wind turbines and its impact on the operation of the power system and on power quality. This impact increases as the level of penetration increases.

The influence on power quality is mainly noted in the level and fluctuation of system voltage and frequency, the shape of the voltage signal (power harmonics) and its ability to manage a reactive power balance. It is also important that the voltage of the distribution network remains constant, especially when interconnected to the wind turbines.

14.5.1 Distribution network voltage levels

Isolated power systems considered in this context are usually characterised by having only one power station. The transmission and distribution network is thus usually quite simple and weak, meaning it is susceptible to low-voltage issues. When the wind turbines are connected to the grid, they will often be at locations where consumers are also connected, not close to the power station.

The voltage at the point of common connection will depend on the output from the wind turbines and on the consumer load. Situations where the voltage becomes high as a result of an overabundance of wind power production and a low consumer load may occur. Measurements completed at the system in Hurghada in Egypt (in the context of the study described in Lundsager *et al.*, 2001) illustrate the dependence of the voltage on wind power production and consumer load. Several other chapters in this book provide examples on the influence of voltage levels of both wind turbine (farm) output and consumer loads, see Chapters 12 and 19 for instance.

14.5.2 System stability and power quality

In all power systems that produce AC power, four critical parameters must be maintained:

- System frequency: the balance of energy being produced and energy being consumed. If more energy is being produced, the frequency will increase.
- Active (real) and reactive power: depending on the types of loads and devices being used, a balance of active and reactive power must be maintained. In addition, some loads will require large amounts of reactive power, such as any inductive motor, and the equipment operating the system will need to provide this reactive power.
- System voltage: stable voltage from the source is required to ensure the proper operation of many common loads. Control is generally maintained by modulating the field of a rotating generator or through the use of power electronics.
- Harmonic distortion of voltage waveform: different generation devices will provide voltage of different harmonic quality. The loads being powered by the device will also impact the harmonic nature of the power, particularly the use of power electronics such as in the ballast of most fluorescent lighting or power converters. Rotating equipment provides the voltage waveform with the least amount of harmonic distortion.

Different devices can be used to control different aspects of system power quality. The need for different devices will depend on the instantaneous penetration of renewable (wind) energy that is expected. The higher the penetration, the more one must worry about how this will impact power quality and the more devices that must be added to the system to ensure high-quality electrical power.

14.5.3 Power and voltage fluctuations

The key driver of many of these power quality issues is the power fluctuations from the wind turbines. The power fluctuations create fluctuations in system voltage and impose fluctuations in the diesel output. Voltage fluctuations can create disturbances such as light flicker and low intensity. The power fluctuations depend not only on the amount of installed wind power capacity but also on the number of wind turbines. The decrease in power fluctuations with increasing number of wind turbines is a result of the stochastic nature of the turbulent wind: the power fluctuations from the individual wind turbines will, to some extent, be independent of each other and they will therefore even out some of the higher frequency fluctuations. Keeping power fluctuations small is a desirable feature in an isolated power system with a high penetration level.

When diesel generators are operating in parallel to wind turbines then, clearly, as the standard deviation of the wind power increases so will the standard deviation of diesel power since the diesel is required to fill any difference between the required load and the wind generation (see Figure 14.8). This is an especially important point when considering the retrofit of existing diesel plants with wind energy as care will have to be taken to ensure that the current diesel controls can cope with the expected wind power fluctuations.

14.5.4 Power system operation

The operation and performance of an isolated system with wind power depends on many factors, too numerous to mention. Generally speaking, however, the complexity

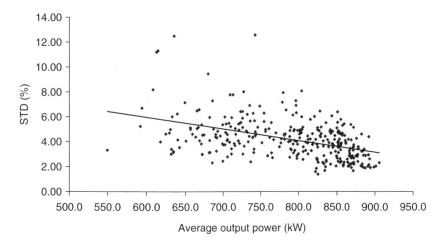

Figure 14.8 Standard deviation (STD) of wind farm power in Mindelo, Cape Verde, as a percentage of rated power
Source: Hansen *et al*., 1999, reproduced with permission.

of operation, as well as the expected reduction of diesel fuel consumption is dependent on the level of wind penetration. The higher the penetration, the higher the fuel savings but also the higher the complexity. Low-penetration systems usually do not require any automated supervisory control whereas high-penetration system must have very sophisticated control systems to ensure proper operation.

The operation of the power system must be carefully considered as part of the power system design. Additionally, the higher the degree of system complexity the more important it is that the system has the ability to operate under simpler structures. This concept of 'failure safety' promotes the use of different levels of operation that can fall back to simpler operation upon the failure of specific components. In the extreme case, the diesels should always be able to supply power to the grid even if everything, including the supervisory control, fails.

One concept that is critical to the operation of remote power systems is that in such systems the levels of penetration that can be reached are much higher than in standard grid-connected wind turbines. For example, systems have been installed where the wind turbine can provide three to four times the existing load, something completely unheard of in the grid-connected market. For this reason, an understanding of the diurnal and seasonal fluctuations in consumer load must be considered in the design of all higher penetration wind–diesel power systems.

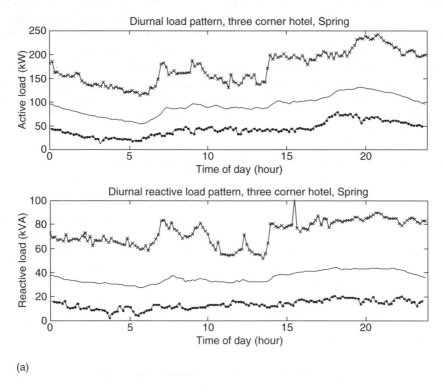

(a)

Figure   14.9 Diurnal patterns for (a) a hotel and (b) residential load, Hurghada, Egypt
Source: Hansen *et al.*, 1999, reproduced with permission.

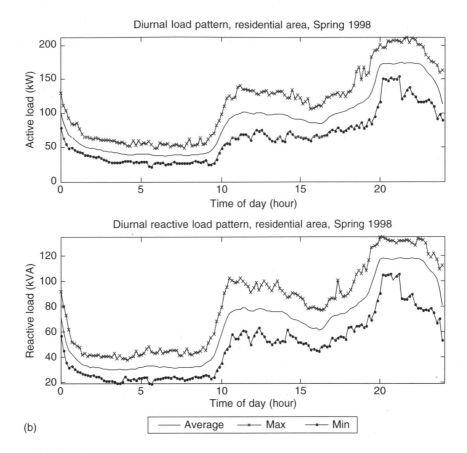

Figure   14.9 (*continued*)

Owing to their impact on the operation strategy of the diesels, several other aspects of the load are also important. Briefly summarising, the minimum load, together with the allowed technical minimum load of the diesel, determines the amount of wind power that can be absorbed by the system without dissipation of surplus power to a dump load or the battery bank. The maximum load determines the necessary installed diesel capacity. The variations in loads and wind power – active and reactive – together with the rate of change of the load and the rate of change of the wind power, determine the necessary spinning capacity in order to ensure adequate power quality and reliability in terms of loss of load probability.

The operation and performance of an isolated system with wind power generally depends on the characteristics of the load. The load is the sum of many individual loads which in some cases may be categorised. Figures 14.9(a) and 14.9(b) show the diurnal load patterns of a hotel and residential load, respectively, measured as part of a study in Egypt (described by Hansen *et al.*, 1999; Lundsager *et al.*, 2001).

Clearly, there is a distinct diurnal pattern for the residential load whereas the hotel load is quite constant over the day. It is also seen that the pattern for the reactive power is very similar to that of the active power. Day-to-day variations of loads shown by the minimum to

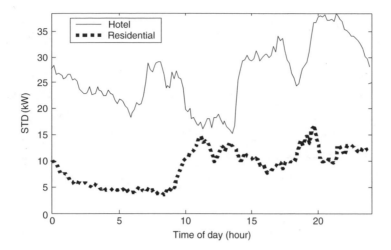

Figure 14.10 Standard deviation (STD) of the load categories in Figures 14.9(a) and 14.9(b)
Source: Hansen *et al.*, 1999, reproduced with permission.

maximum band are small in the residential load case, but in the hotel load case the variation is larger. This is also seen in the standard deviation of the loads shown in Figure 14.10.

Last, in all higher power systems, significant consideration must be given to uses of surplus energy such as space or water heating and water purification. In many cases, the savings of displacing the secondary fuel used in these other processes can play a key role in making a project financially successful.

Owing to their impact on the operation strategy of the diesels, several aspects of the load are important. The minimum consumer load and the allowed technical minimum load of the diesels determine the amount of wind power that can be absorbed by the system without dissipation of surplus power. The maximum consumer load determines the necessary installed diesel capacity. The variations in consumer loads and wind power determine the necessary spinning capacity in order to ensure adequate power quality and reliability in terms of loss of load probability.

14.6 System Modelling Requirements

Numerical modelling is an important part of the design, assessment, implementation and evaluation of isolated power systems with wind power. Usually, the performance of such systems is predicted in terms of the technical performance (power and energy production) and the economical performance, typically the cost of energy (COE) and the internal rate of return (IRR) of the project.

Technically, the performance measures may also include a more detailed design-based analysis of:

- power quality measures;
- load flow criteria;
- grid stability issues;
- scheduling and dispatch of generating units, in particular diesel generators.

Thus a large variety of features are needed to cover the numerical modelling needs in all contexts. No single numerical model is able to provide all features, as they often are in conflict with each other in terms of modelling requirements. This survey of modelling requirements is based on a review of relevant studies that was carried out as part of an earlier study (reported by Lundsager *et al.*, 2001).

14.6.1 Requirements and applications

The requirements of the numerical models depend on the actual application of the modelling. Many factors influence the layout of a numerical model, and the main factors include:

- the objectives of the simulations;
- the time scale of the modelling;
- the modelling principle (deterministic or probabilistic);
- representation of the technical and economic scenarios;
- system configurations, including dispersed compared with single system configurations.

The objectives of the simulations depend on whether the simulations are done in a feasibility study as part of the decision process or as part of the design process. Objectives may concern:

- screening or optimisation of possible system or grid configurations;
- the overall annual performance of selected configurations;
- supervisory control, including scheduling and dispatch of generating units;
- grid modelling, stability or load flow.

Several time scales of system modelling are used, depending on the context and purpose of the simulation. They may focus on:

- transient analysis of electrical transients due to switching, or certain power quality measures: typically a few seconds with a time step of less than 0.001 s;
- dynamic analysis of machine dynamics, power quality or grid stability over specific system state changes: typically around 1 minute with time steps around 0.01 s;
- dispatch analysis of supervisory control, including diesel dispatch: typically a few minutes to an hour at time steps of 0.1 s to 1 s;
- logistic analysis of power flows and seasonal and annual energy production: days, weeks, months, seasons or one year at time steps 10 min to 1 h;
- screening models for logistic analysis that give an overall assessment of the performance of a system, without going into very much detail of the specifics in the operation of the system.

Probabilistic modelling, based on probability distributions such as load duration curves or wind speed probability curves, can work directly with the outputs of typical utility statistics and results from wind resource assessment models such as WAsP

(Mortensen *et al.*, 1993) but they cannot readily represent 'memory' effects such as energy storage, diesel dispatch strategies and deferrable loads. Time series simulation can do that, but to obtain the same statistical significance as probabilistic models Monte Carlo techniques ideally should be applied.

Most models represent technical performance by one year's performance measures (fuel savings and so on) and use an economic lifecycle analysis to establish the economic indicators. This way the technical developments and/or extensions to the system during its lifetime are neglected. This is in fact inadequate to give a realistic representation of the (often) rapid development of isolated systems during the entire technical and economic life of the project, which may be up to 20 years.

A realistic representation of both system configuration and control strategy in the modelling is essential for a reliable and accurate prediction of system performance. Several techniques have been applied to facilitate the flexible modelling of system configuration and connectivity, but the flexible modelling of supervisory control strategy appears to be even more challenging (Pereira *et al.*, 1999).

For dispersed systems, the electric grid becomes part of the system. Deterministic load flow analysis is well established, but probabilistic load flow modelling is more in line with the stochastic nature of wind and loads in isolated systems and the associated probabilistic power quality measures (Bindner and Lundsager, 1994).

Actual grid stability analysis requires real dynamic electromechanical models, and such analysis is frequently outside the scope of isolated system analysis.

14.6.2 Some numerical models for isolated systems

In this section a number of numerical modelling techniques and models are reviewed and, although we do not claim the review is complete, we focus on models for wind and wind–diesel applications that are used for the assessment of the technical–economic performance of the system (for a further review, see Baring-Gould *et al.*, 2003; Clausen *et al.*, 2001; Lundsager *et al.*, 2001). Selected models are briefly described in Table 14.5. Some of the models are not publicly available, but the references provide details of valuable experience with modelling concepts and techniques.

Numerical models are available that can handle practically any requirement for theoretical modelling of isolated power system projects. Thus the availability of models is not a limiting factor for development, but the extent of the efforts involved in applying a complete set of modelling tools for a given isolated community situation may be considerable if not prohibitive.

14.7 Application Issues

One of the major drivers for retrofitting diesel power systems with renewable technology is the relatively high cost of producing power with diesel engines. However, a number of issues have strong influence on the options to be chosen and the success of their implementation.

Table 14.5 Overview of computer models for technical–economical assessment

Model	Description	Source	Availability
HOMER	HOMER is a fast and comprehensive off-grid and distributed micro-power system screening and configuration optimization model. It is widely used and provides a user friendly interface. This model is the state-of-the-art for initial system conceptual analysis.	Lilienthal, Flowers and Rossmann, 1995	Publicly available from NREL, www.nrel.gov/homer
INSEL	INSEL offers almost unlimited flexibility in specifying system configurations by allowing the user to specify the connectivity on a component level. It is intended as an out-of-house-model.	Renewable Energy Group, 1993	Not publicly available
HYBRID2	HYBRID2 is the state-of-the-art time series model for prediction of technical–economical performance of hybrid wind–photovoltaic systems. It offers very high flexibility in specifying the connectivity of systems and is quite widely used.	Baring-Gould, 1996	Publicly available from the University of Massachusetts, www.ecs.umass.edu/mie/labs/rer
SIMENERG	SIMENERG is the only model so far with a very high degree of flexibility in the control and dispatch strategy, using a 'market square' approach, where the economically optimal subset of power sources that satisfy the power demand is dispatched in each time step.	Briozzo *et al.*, 1996	
WINSYS	WINSYS is a spreadsheet (QPW) based model implementing probabilistic representations of resources and demands. WINSYS incorporates the anticipated technical expansions during its lifetime in the technical performance measures, combined with a traditional economic lifecycle cost assessment. Thus WINSYS presents a more realistic lifecycle cost analysis than most other models.	Hansen and Tande, 1994	Not commercially available
WDLTOOLS	WDLTOOLS was developed by Engineering Design Tools for Wind Diesel Systems and is a package containing seven European logistic models: SOMES (The Netherlands), VINDEC (Norway), WDILOG (Denmark), RALMOD (UK) and TKKMOD (Finland). Although somewhat outdated it also includes the modular electromechanical model JODYMOD.	Infield *et al.*, 1994	Not publicly available
RPM-Sim	RPM-Sim is a model that uses the VisSim™ visual environment and allows for the dynamic analysis of wind, PV and diesel systems. Researchers at NREL's National Wind Technology Center developed the model.	Bialasiewicz *et al.*, 2001	Not publicly available

(continued overleaf)

Table   14.5 (*continued*)

Model	Description	Source	Availability
MatLAB	This software uses Simulink^{TM} and the MatLAB^{TM} Power System Blockset to analyse the dynamic response of wind–diesel applications. It was developed by the Hydro-Quebec Research Institute	—	Available through MathWorks, http://www.mathworks.com/
PROLOAD	PROLOAD is a probabilistic load flow analysis code, using Monte Carlo techniques, developed in cooperation with an electrical utility for the dimensioning of distribution systems with wind turbines.	Tande *et al.*, 1999	Not publicly available
RETScreen	RETScreen is a spreadsheet (Microsoft Excel) based analysis and evaluation tool for the assessment of the cost-effectiveness of potential projects with renewable energy technologies. The software package consists of a series of worksheets with a standardised layout as well as an online manual and a weather and cost database. The tool is developed by CEDRL.	RETScreen, 2000	Publicly available from the CEDRL website at www.retscreen.net

Note: CEDRL = CANMET Energy Diversification Research Laboratory; NREL = National Renewable Energy Laboratory (USA).

14.7.1 Cost of energy and economics

The cost of electricity in isolated systems, where existing power supply is typically from diesel power plants, varies over a very wide range, from a low of US$0.20 per kWh to a high of over US$1.00 per kWh.

Clearly, the cost of electricity in existing isolated systems is typically many times higher than in large utility grids, where net production cost of electricity is of the order US$0.04 per kWh. Additionally, the cost for storage of diesel fuel in areas that may receive limited fuel shipments each year may also increase the cost of providing power to remote areas (Clausen *et al.*, 2001).

The cost of electricity from grid-connected wind turbines has decreased to approaching US$0.04 per kWh in large grids and twice to three times this amount for smaller turbines fit for rural applications. The main parameter influencing the COE from wind turbines of a given size is the annual average wind speed.

Electricity production costs from wind–diesel systems are not well documented, and they cover a very wide range. Furthermore, COE from a wind–diesel system may be difficult to ascertain precisely, as the task may be outside the scope of small-scale projects. Therefore, the economic viability of a system will often be assessed by comparing the COE from the wind turbine (including costs of any support technology) with the avoided costs of fuel and diesel service in the existing diesel system. In the case of a retrofit installation of an additional wind turbine in an existing wind–diesel system, this may be a fair assessment; however, in the planning phase of a new isolated system the comparison should be based on total cost of electricity including the capital cost of the alternatives analysed.

14.7.2 Consumer demands in isolated communities

From a consumer's point of view, an isolated power system with wind power should satisfy the same demands as conventional power system solutions. Wind power should, as such, be treated as just another option in accordance with priorities of relevant policies and plans. Since wind energy is resource-driven, in order to be able to assess wind power in comparison with conventional options such as diesel, a supplementary set of information describing the community will most often be necessary.

The feasibility of wind power in small isolated power systems may be particularly sensitive to issues of grid interconnection costs, electricity demand development and diurnal consumer patterns, whether the system supplies electricity 24 hours a day or less. Furthermore, costs for consultants for special studies or power system design add to the investment cost, which may be unacceptable for small communities.

Using Egypt as an example, it was attempted in a study (reported by Lundsager *et al.*, 2001) to acquire a general overview of the potential types of systems in which wind power could be an option. Table 14.6 shows an overview, splitting the communities in three categories.

Hurghada in Egypt has operated as a medium to high wind energy penetration system since 1998, and it has provided demonstration for Egypt of this technology.

14.7.3 Standards, guidelines and project development approaches

A major key to success is the development and application of proper guidelines and standards for these systems. As a market develops, so does the need for developed standards and guidelines for system design, implementation and commissioning. The difference between standards and guidelines are illustrated by the two types of

Table 14.6 Overview of isolated power systems in Egypt

Category	Description	Estimated number in Egypt
1	Micro grids currently not electrified or nongrid communities that get electricity from buying or charging batteries or from individual household systems	>100
2	Village power and island power systems, presently powered by diesel systems of sizes between 50–1000 kW	50
3	Island and large interconnected power systems – isolated or end-of-line systems with 1–150 MW installed capacity of diesel or gas turbine supplied loads	5–10

Source: Lundsager *et al.*, 2001, reproduced with permission.

documents that are now being developed in the framework of the International Electro-technical Commission (IEC), described below.

Standards deal with the specific technical elements of energy (DRE) systems and are the responsibility of individual technical committees in the national and international standardisation framework that include the IEC, the Comité Européen de Normal-isation Electrotechnique (CENELEC) and other bodies. The standards are unique, and all project implementation guidelines should refer to one and the same set of renewable energy system standards. This set-up is adopted in the development of the IEC/publicly available specification (PAS) guidelines IEC PAS 62111 (IEC, 1999) for small DRE systems for rural electrification in third world countries. This document is currently being developed into an IEC guideline under the title of 'Recommendations for Small Renewable Energy and Hybrid Systems for Rural Electrification', IEC 62257 (IEC, 0.000). The development of this standard, however, focuses on smaller power systems, typically under 50 kW total output power, and is thus not applicable to larger wind–diesel applications.

In 'Isolated Systems with Wind Power – An Implementation Guideline', by Clausen *et al.* (2001), a unified and generally applicable approach to project implementation is attempted to support a fair assessment of the technical and economical feasibility of isolated power supply systems with wind energy.

General guidelines and checklists on which information and data are needed to carry out a project feasibility analysis are presented as well as guidelines of how to carry out the project feasibility study and environmental analysis. The main process detailed by Clausen *et al.* (2001) includes:

- fact finding;
- project feasibility analysis;
- environmental impact analysis;
- institutional and legal framework;
- financing;
- implementation.

At the initial stage of development this can be summarised in an approach where the following are combined in the project preparation:

- site evaluation based on resource assessment modelling;
- technical assessments based on logistic modelling techniques;
- economic evaluations based on lifecycle costing principles.

These three basically computational techniques should be coupled to a thorough ana-lysis of nontechnical (institutional) issues.

As is commonly stated, the key to success in the application of wind power in isolated power supply systems is proper project preparation. In this chapter we have described some of the factors related specifically to isolated systems; other factors, such as wind resource assessment, wind turbine characteristics and grid integration issues, are dealt with in other chapters of this book.

14.8 Conclusions and Recommendations

A number of issues have been identified that should be considered when developing a wind power project as part of an isolated power system, be it a small system of a few kilowatts or a large megawatt-size system. A successful project will often have the following characteristics:

- It will use updated versions of relevant international standards, including the that for decentralised power systems with renewable energies now in progress within the IEC.
- Best practice guidelines for project implementation will be applied, including common references and relevant experience from recent projects.
- The wind power project in the isolated system in question will be part of a concerted action in a national and international programme rather than an individual project.
- The wind power technology applied in a small to medium-size system will follow simple and proven approaches, for instance by repeating and/or scaling pilot and demonstration systems with positive track records, which may have been developed from filtering down from large-scale systems any technological achievements adaptable to smaller systems.
- Small systems will be developed to apply rugged technology suitable for remote communities.
- No experimental systems will be installed at rural remote communities unless previously thoroughly tested and documented at test benches dedicated to serve as experimental facilities. Any pilot or proof-of-concept system should be implemented only with the consensual understudying that the system will likely need higher levels of support for the life of the system, not only the life of the expected research project.
- Ownership will be well defined, with a built-in interest identified to ensure long-term interest and funding of operation, maintenance and reinvestments when needed.
- An organisation will be established with the necessary capacity and capability for implementation, operation and maintenance, preferably including the backup from a relevant national or regional knowledge centre.
- A sufficiently detailed feasibility study will have been performed.
- Modelling assumptions, input data and methodology applied for the feasibility study and system design will reflect the true hardware reality for the types of systems in question.

The technical capacity to design, build and operate isolated power systems with a high penetration of wind power exists, but the mature product and the market have not yet met. Nevertheless, there is today an industry offering small wind turbines (10–300 kW) and associated equipment for hybrid system applications with a long-term commitment. This indicates a belief that a market is emerging and that interest from some large wind turbine manufacturers can be expected.

However, the main barrier is the lack of a positive, well-documented track record for isolated systems with wind power. Some of this has been a result of early technical failures in complex systems but, more generally, the technology has yet to supply enough successful examples to be accepted by the mainstream markets. Therefore, in addition to, of course, ensuring that any system implemented in an isolated community works reliably and economically, it is important also to ensure that proper follow-up and monitoring of system performance is carried out. Only then will a proper and well-documented track record be established. Additionally, in order for this potential market to develop, a simple approach should be found that can be applied at low cost and at a risk comparable to the risk involved when designing a diesel-powered system.

The above recommendations are seen as steps that could open and extend access to electricity for the benefit of, for example, small rural communities but, in a broader scope, the process could lead in the direction of an increased use of wind power in the whole range of isolated power systems from a few kilowatts to several megawatts.

References

[1] Allderdice, A., Rogers, J. H. (2000) 'Renewable Energy for Microenterprise', booklet, Report BK-500-26188, National Renewable Energy Laboratory, Golden, Co.

[2] Baring-Gould, E. I. (1996) 'Hybrid2: The Hybrid System Simulation Model', Report TP-440-21272, National Renewable Energy Laboratory, Golden, Co.

[3] Baring-Gould, E. I., Castillo, J. Flowers, L. (2003) 'The Regional Replication of Hybrid Power Systems in Rural Chile', in *Proceedings of the 2003 AWEA Conference, Austin, TX, May, 2003*, American Wind Energy Association.

[4] Baring-Gould, E. I., Barley, C. D. Drouilhet, S., Flowers, L., Jimenez T., Lilienthal, P., Weingart, J. (1997). 'Diesel Plant Retrofitting Options to Enhance Decentralized Electricity Supply in Indonesia', in *Proceedings of the 1997 AWEA Conference, Austin, TX, June 1997*, American Wind Energy Association.

[5] Baring-Gould, E. I., Flowers, L., Lundsager, P., Mott, L., Shirazi, M., Zimmermann, J. (2003) 'World Status of Wind Diesel Applications', in *Proceedings of the 2003 AWEA Conference, Austin, TX*, American Wind Energy Association.

[6] Baring-Gould, E. I., Newcomb, C., Corbus, D., Kalidas, R. (2001) 'Field Performance of Hybrid Power Systems', Report CP-500-30566, National Renewable Energy Laboratory, Golden, Co.

[7] Bialasiewicz, J. T., Muljadi, E., Nix, R. G., Drouilhet, S. (2001) 'Renewable Energy Power System Modular Simulator: RPM-SIM User's Guide (Supersedes October 1999)', Report TP-500-29721, National Renewable Energy Laboratory, Golden, Co.

[8] Bindner, H., Lundsager, P. (1994) 'On Power Quality Measures for Wind–Diesel Systems: A Conceptual Framework and a Case Study', paper presented at European Wind Energy Conference, Thessaloniki, Greece, June 1994.

[9] Briozzo, C., Casaravilla, G., Chaer, R., Oliver, J. P. (1996) 'SIMENERG: The Design of Autonomous Systems', pp. 2070–2073. in *Proceedings of the World Renewable Energy Congress III, Reading, UK, 11–16 September 1994*, pp. 2070–7073.

[10] Clausen, N-E., Bindner, H., Frandsen, S., Hansen, J.C., Hansen, L.H., Lundsager, P. (2001) 'Isolated Systems with Wind Power – An Implementation Guideline', publication R-1257 (EN), June 2001, Risø National Laboratory, Roshilde, Denmark.

[11] Corbus, D., Newcomb, C., Baring-Gould, I., Friedly, S. (2002) 'Battery Voltage Stability Effects on Small Wind Turbine Energy Capture', preprint, Report CP-500-32511, National Renewable Energy Laboratory, Golden, Co.

[12] Cotrell, J., Pratt, W. (2003) 'Modeling the Feasibility of Using Fuel Cells and Hydrogen Internal Combustion Engines in Remote Renewable Energy Systems', Report TP-500-34648, National Renewable Energy Laboratory, Golden, Co.

[13] Cramer, G., Reekers, J., Rothert, M., Wollny, M. (2003) 'The Future of Village Electrification: More than Two Years of Experience with AC-coupled Hybrid Systems', in *Proceedings of the 2nd European PV Hybrid and Mini-Grid Conference, Kassel, Germany*, pp. 145–150, Boxer, Germany.

[14] DOE/AWEA/CanWEA (2002) *Proceedings of the 2002 DOE/AWEA/CanWEA Wind–Diesel Workshop*, http://www.eere.energy.gov/windpoweringamerica/wkshp_2002_wind_diesel.html.

[15] Engler, A., Hardt, C., Bechtel, N., Rothert, M. (2003) 'Next Generation of AC Coupled Hybrid Systems: 3 Phase Paralled Operation of Grid Forming Battery Inverters', in *Proceedings of the 2nd European PV Hybrid and Mini-Grid Conference, Kassel, Germany*, pp. 85–90, Boxer, Germany.

[16] Hansen, J. C., Tande, J. O. G. (1994) 'High Wind Energy Penetration Systems Planning', paper presented at the European Wind Energy Conference, Thessaloniki, Greece, June 1994.

[17] Hansen, J. C., Lundsager, P., Bindner, H., Hansen, L., Frandsen, S. (1999) 'Keys to Success for Wind Power in Isolated Power Systems', paper presented at the European Wind Energy Conference, Nice, France, March 1999.

[18] Hansen, L. H., Helle L., Blaabjerg F., Ritchie E., Munk-Nielsen S., Bindner, H., Sørensen, P., Bak-Jensen, B. (2001) 'Conceptual survey of Generators and Power Electronics for Wind Turbines', publication R-1205 (EN), Risø National Laboratory, Roskilde, Denmark.

[19] Hunter, R., Elliot, G. (Eds) (1994) *Wind–Diesel Systems*, Cambridge University Press, Cambridge, UK.

[20] IEC (International Electrotechnical Committee) (1999) 'Publicly Available Specifications on "Decentralised Renewable Energy Systems" (DRES)', IEC PAS 62111, July 1999, IEC, Geneva, Switzerland.

[21] IEC (International Electrotechnical Committee) (in development) 'Recommendations for Small Renewable Energy and Hybrid Systems for Rural Electrification', IEC 62257, Geneva, Switzerland.

[22] Infield, D., Scotney, A., Lundsager, P., Bindner, H., Pierik, J., Uhlen, K., Toftevaag, T., Falchetta, M., Manninen, L., Dijk, V. V. (1994) 'WDLTools – Engineering Design Tools for Wind Diesel Systems', Energy Research Unit, Rutherford Appleton Laboratory.

[23] Jimenez, A. C., Lawand, T. (2000) 'Renewable Energy for Rural Schools', booklet, Report BK-500-26222, National Renewable Energy Laboratory, Golden, Co.

[24] Lilienthal, P., Flowers, L., Rossmann, C. (1995) 'HOMER: The Hybrid Optimization Model for Electric Renewable', in *Windpower '95 – Procedings*, American Wind Energy Association, pp. 475–480.

[25] Lundsager, P., Madsen, B. T. (1995) 'Wind Diesel and Stand Alone Wind Power Systems', final report, May 1995, BTM Consult and Darup Associates ApS.

[26] Lundsager, P., Bindner, H., Clausen, N-E., Frandsen, S., Hansen, L. H., Hansen, J. C. (2001) 'Isolated Systems with Wind Power – Main Report', publication R-1256(EN), June 2001, Risø National Laboratory, Roskilde, Denmark.

[27] Lundsager, P., Bindner, H. (1994) 'A Simple, Robust, and Reliable Wind–Diesel Concept for Remote Power Supply', paper presented at the World Renewable Energy Congress III, Reading, UK, 11–16 September 1994.

[28] Mortensen, N.G., Petersen, E. L., Frandsen, S. (1993) 'Wind Atlas Analysis and Application Program (WAsP), User's Guide', Risø National Laboratory, Roskilde, Denmark, http://www.wasp.dk/.

[29] Newcomb, C., Baring-Gould, I., Corbus, D., Friedly, S. (2002) 'Analysis of Reduced Energy Capture Mechanisms for Small Wind Systems', in *Proceedings of the 2002 AWEA Conference, June 3–5, 2002, Portland, Oregon, USA*, American Wind Energy Association.

[30] Pereira, A. de Lemos, Bindner, H., Lundsager, P., Jannerup, O. (1999) 'Modelling Supervisory Controller for Hybrid Systems', paper presented at the European Wind Energy Conference, Nice, March 1999.

[31] Pereira, A. de Lemos (2000) *Modular Supervisory Controller for Hybrid Power Systems*, PhD thesis, publication R-1202 (EN), June 2000, Risø National Laboratory, Roskilde, Denmark.

[32] Renewable Energy Group (1993) 'INSEL Reference Manual', version 4.80 ed., Department of Physics, University of Oldenburg, Oldenburg.

[33] RETScreen (2000) 'Pre-feasibility Analysis Software', CANMET Energy Diversification Research Laboratory (CEDRL), www.retscreen.gc.ca.

[34] Shirazi, M., Drouilhet, S. (2003) 'Analysis of the Performance Benefits of Short-term Energy Storage in Wind–Diesel Hybrid Power Systems', Report CP-440-22108, National Renewable Energy Laboratory, Golden, Co.

[35] Tande, J. O. G., Palsson, M. Th., Toftevaag, T., Uhlen, K. (1999) 'Power Quality and Grid Connection of Wind Turbines. Part 1: Stationary Voltages'.

15

Wind Farms in Weak Power Networks in India

Poul Sørensen

15.1 Introduction

Several countries support wind energy in developing countries in order to ensure sustainable development that can meet the rapidly growing demand for power in these countries. India is the developing country that has experienced by far the fastest expansion of wind energy. It is therefore selected as a case study in this chapter. India is also an interesting and typical case, as the networks are weak and in the process of being built. The present chapter is based on a study of wind farms connected to weak grids in India (Sørensen *et al.*, 2001).

India has experienced a fast increase in wind energy, especially in the mid-1990s, when subsidies and tax allowances for private investors in wind farms promoted the installation of wind farms. The State of Tamil Nadu in the southern part of India was leading this development, but the State of Gujarat also installed a substantial capacity of wind energy. Figure 15.1 shows the centres of this development: the Muppandal area in Tamil Nadu and the Lamba area in Gujarat.

The total wind energy capacity installed in Tamil Nadu during the 1990s amounted to approximately 750 MW (Ponnappapillai and Venugopal, 1999), 20 MW of which were funded by the government as demonstration projects. The remaining 730 MW were funded by private investors. According to the load dispatch centre of Tamil Nadu Electricity Board (TNEB), the total conventional power supply capacity in Tamil Nadu amounts to 7804 MW and comprises thermal, nuclear and hydro power stations. That means that in Tamil Nadu wind power capacity corresponds to 8.8 % of total power

Wind Power in Power Systems Edited by T. Ackermann
© 2005 John Wiley & Sons, Ltd ISBN: 0-470-85508-8 (HB)

Figure   15.1 Map of India, 2001, showing Muppandal (State of Tamil Nadu), and Lamba (State of Gujarat)

supply capacity. In 1999–2000, the wind power production in Tamil Nadu was 1157 GWh, supplying 3.1 % of total consumption (37 159 GWh).

In Gujarat, the utilisation of wind energy started with government demonstration projects in 1985. The total installed wind energy capacity is 166 MW: 16 MW have been funded as government demonstration projects, and the remaining 150 MW have been funded by private investors. In 2000, the capacity of conventional power supply in Gujarat amounted to 8093 MW, including private sector and national grid allotment. Consequently, wind power capacity corresponded to 2.0 % of the total power supply capacity in Gujarat.

In comparison, the Western part of Denmark had approximately 1900 MW of installed wind power by early 2001, according to Eltra (http://www.eltra.dk) which is the transmission system operator (TSO) in charge of this region. Other power plants had a capacity of 4936 MW. Wind energy capacity corresponded to 28 % of total installed power capacity. In 2000, the wind power production in the Eltra area was 3398 GWh, corresponding to 16.5 % of the total consumption (20 604 GWh).

Table 15.1 summarises the above-mentioned numbers. Even though the periods the figures refer to do not exactly coincide, a comparison is still reasonable. The figures indicate that the wind energy penetration level in Tamil Nadu is high, though not exceptionally high, whereas in Gujarat, the wind energy penetration is moderate.

Table 15.1 Penetrations in Gujarat, Tamil Nadu and the Eltra area (Denmark) in 2000

	Area		
	Tamil Nadu	Gujarat	Eltra
Wind capacity (MW)	750	166	1 900
Conventional capacity (MW)	7 804	8 093	4 936
Total capacity (MW)	8 554	8 259	6 836
Wind capacity penetration (%)	8.8	2.0	27.8
Wind production (GWh)	1 157	—	3 398
Consumption (GWh)	37 159	—	20 604
Wind production penetration (%)	3.1	—	16.5

— Production and consumption data for Gujarat are not available

Evaluating the penetration levels, it has to be taken into account that the power systems in Tamil Nadu and Gujarat are both parts of larger power systems, similar to the power system in the western part of Denmark, which is part of the strong central European power system. The power system in India is divided into four subsystems. These are interconnected only through DC-links and consequently do not operate synchronously (Hammons *et al.*, 2000). Tamil Nadu is part of the power system in the southern part of India, together with the three neighbouring States of Kerala, Karnataka and Andhra Pradesh. Gujarat, however, is included in the power system of western India together with the two neighbouring States of Maharastra and Madhya Pradesh.

In India, the power systems are publicly owned. Each State has its own electricity board, [e.g. the Tamil Nadu Electricity Board (TNEB) and the Gujarat Electricity Board (GEB)]. Each State also has its own development agency for planning wind energy development and for handling the grid connection [e.g. the Gujarat Energy Development Agency (GEDA)]. In Gujarat, GEDA owns and operates dedicated wind farm substations, whereas in Tamil Nadu, the substations are owned and operated by TNEB and are not completely dedicated to the wind farms.

In India, as in other countries in eastern Asia, industrial and economic development is progressing very quickly and, at the same time, the population is growing very quickly too. This results in fast growing demands on power systems. The electricity boards have not been able to develop the necessary power system capacities for supply, transmission and distribution accordingly. As a consequence, the power systems have in several ways been weakened over the years.

The major part of the wind energy installations is concentrated in a few rural areas, where the existing distribution and transmission grids are particularly weak. As will be shown next, the power quality at the connection points of the wind turbines is mostly poor, which affects the operation of the wind turbines. To some extent, the connection of wind turbines to these grids has even worsened the power quality, because grid reinforcements have not been sufficient to integrate the wind energy into the power systems.

In this chapter I describe the interrelationship between wind turbines and power quality in the Indian power systems. First, I will characterise the grids in the areas where the wind turbines are connected and describe the overall electrical design of the grid connections of the wind turbines. Referring to the characteristics of grids and wind turbines, I will describe the impact of the wind turbines on power quality, on the one hand, and the effect that the power quality has on the wind turbines, on the other hand.

15.2 Network Characteristics

This section summarises the main characteristics of the Indian grids, which are important with regard to the interrelationship between wind turbines and grids. I will focus on the grid in the Muppandal region, where more than half of the wind turbines in Tamil Nadu are connected. In addition, the grid in the coastal region around Lamba in Gujarat is described.

15.2.1 Transmission capacity

One of the most critical issues for the development of wind energy in India has been the transmission capacity of the grid in the areas where the wind farms were built. Similar to many other countries, wind farms are concentrated in rural areas, where the existing transmission grids are very weak. In addition, in India the new wind farms were built during a comparatively short period of time and were restricted to a few areas, and the reinforcement of the transmission system in these areas has lagged behind the fast development of wind energy.

In Tamil Nadu, almost 600 MW of wind energy were installed over the three-year period from 1994 to 1997. Approximately 400 MW of the wind energy capacity are located in the Muppandal area. Figure 15.2 shows the grid connection diagram for that area. The 110 kV Arumuganeri–Kodayar ring mains was constructed to supply the Muppandal area with power. The capacity of the ring mains itself is only 200 MVA. To improve transmission capacity, an additional 110 kV line to Melak Kalloor and two 230 kV lines from the S. R. Pudur substation to the Tutucorin 230/110 kV substation and the Kayathar 230/110/11 kV substation, respectively, have now been installed as a supplement to the ring mains.

Figure 15.2 indicates the short-circuit levels at selected points of the 110 kV and 230 kV grid. The short-circuit power at the wind farm substations of the 110 kV ring mains is about 800 MVA. This is very low considering that approximately 200 MVA of wind power are connected directly to the ring.

In Gujarat, there are 60 MW of wind power capacity in the coastal area around Lamba. The wind farms are connected to three dedicated substations, as shown in Figure 15.3. The power from the wind farms is evacuated through a 66 kV line with a capacity of 30 MVA, and a 132 kV double line with a total capacity of 60 MVA.

In June 1998, the capacity installed at these substations amounted to 75 MW. However, in June 1998, a capacity of approximately 40 MW was damaged by a cyclone, which also damaged the overhead lines that were used for the transmission of the power from the wind farms. Before the cyclone, the only transmission line was a 66 kV line with a capacity of 30 MVA. This configuration led to grid outages as a result of load shedding and grid breakdowns after a 66 kV conductor snapped because of age and overloading. The process of ageing of the line has been accelerated by the saline atmosphere close to the sea.

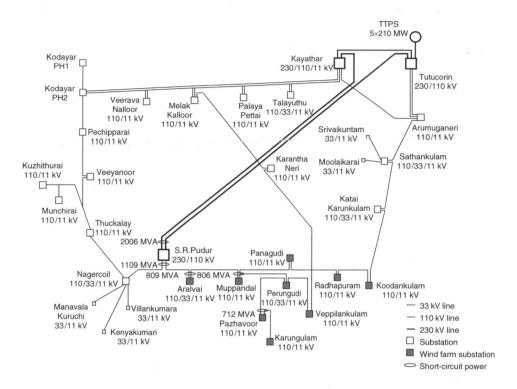

Figure   15.2 Muppandal grid with the original 110 kV Arumuganeri–Kodayar ring mains and 230 kV line in Muppandal, Tamil Nadu, based on drawings by an engineer of the Muppandal office of the Tamil Nadu Electricity Board

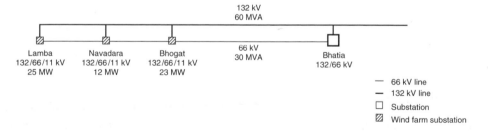

Figure   15.3 Wind farms and substations on the costal lines of Porbandar, Gujarat: dedicated lines for transmission and no tapping for rural distribution

15.2.2 Steady-state voltage and outages

The insufficient capacity of the power systems causes large variations in the steady-state voltage, and outages on the grids. Figure 15.4 shows measurements of the steady-state voltage on the primary side (110 kV) of the wind farm substation of Radhapuram in 1999 (Sørensen *et al.*, 2000). Some of the measurements are from the windy season in May and August, others are from the less windy season in October. The Radhapuram

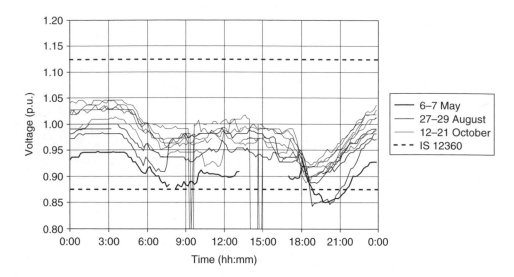

Figure   15.4 Steady-state voltage measured in Radhapuram substation. *Note*: IS 12360 = Indian Standard 12360 (Indian Standard 1988)

substation is one of the eight wind farm substations connected to the 110 kV ring mains in Muppandal (Figure 15.2).

The curves in Figure 15.4 show distinct diurnal variations in the voltages, with typical low voltages during the peak load period in the evening and high voltages at night. During two of the days, the steady-state voltage is about 15 % below the rated voltage for approximately 2 h. This is below the tolerance of ±12.5 % specified in the Indian Standard (IS) 123600 (Indian Standard, 1988). The voltage variations on the 110 kV mains indirectly affect the voltages at the wind turbine terminals. The transformers in the substations in Muppandal have no automatic voltage regulation to compensate for this. Remote voltage regulation is carried out manually from the substation by observation of the voltages in the substation.

Figure 15.4 also shows four outages with a total duration of approximately 1 h, during approximately 200 h of measurements. Since the measurements are taken on the primary side of the substation, these outages only include outages on the 110 kV level or higher. The number of outages is expected to be higher on the low voltage side, where the wind turbine terminals are connected to the grid. According to TNEB, the grid availability of the dedicated wind farm feeders during peak production is in the range of 95–96 %, with an annual average of 98 %.

The main reason for the voltage variations in Figure 15.4 is the load variation, which can be observed by the regular diurnal variation. The wind farms will, however, also affect voltage, both as a result of power production (which increases voltage) and as a result of reactive power consumption (which decreases voltage).

In Gujarat, the interruption time on the line to Lamba was extremely long, before the cyclone and the reinforcement of the 66 kV line with a 132 kV line. Rajsekhar, Hulle and Gupta (1998) concluded that the total operational time of the wind turbines in Lamba was reduced by 4–5 % as GEDA disconnected 11 kV feeders in order to avoid the 66 kV

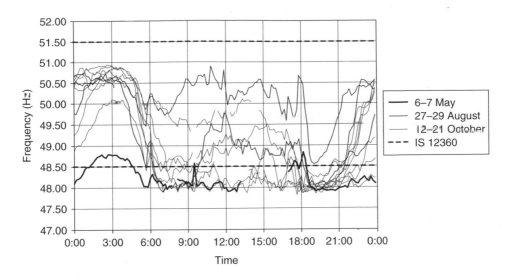

Figure 15.5 Frequency measured at Radhapuram substation. *Note*: IS 12360 = Indian Standard 12360 (Indian Standard, 1988)

line tripping, and by another 474 h per year (5 %) in 66 kV outages due to faults in the GEB system. After the installation of the 132 kV line, GEB has reduced the number of interruptions to two per month, with a duration of 6–8 h per month.

15.2.3 Frequency

In the Indian power systems, the frequency varies significantly compared with European systems. Figure 15.5 shows the frequency measured simultaneously with the voltage.

The curves in Figure 15.5 have diurnal variations in frequency which are similar to the voltage variations shown in Figure 15.4, with frequencies of about 51 Hz at night and down to 48 Hz during the day. According to IS 12360, the frequency should be 50 Hz ± 3 % (i.e. in the interval of 48.5 Hz to 51.5 Hz). It appears from the curves that the frequency is floating until a lower limit of 48 Hz is reached. Then the frequency is regulated, presumably by load shedding, in order to minimise the gap between supply and demand.

The wind farms are not a major source of frequency variation, as the installed wind power capacity is relatively small compared with the total capacity of the interconnected system in south India.

15.2.4 Harmonic and interharmonic distortions

With decreasing prices, the application of power electronics for power supplies and motor drives, for instance, is steadily increasing all over the world. The increased load from power electronics causes harmonic and interharmonic distortions in the power systems.

Standard power supplies for appliances emit only low orders of odd harmonics. Since 3rd-order harmonics are cancelled out by star–delta connected transformers, the main emission to the medium-voltage grid from power supplies are 5th and 7th harmonics. Modern motor drives emit higher-order harmonics and interharmonics, typically with frequencies above 1 kHz.

Measurements at the wind farm substations have shown only moderate harmonic and interharmonic distortions of the voltages at the grid connection points on the primary sides (110 kV and 66 kV), except for distortions at Radhapuram substation, caused by wind turbines with power converters.

15.2.5 Reactive power consumption

The loads on the power systems in India consume a significant amount of reactive power, mainly as a result of the use of agricultural pumps. The reactive power consumption affects the power system by:

- increasing losses in the transmission and distribution grids;
- reducing thermal capacity for the transmission of active power in the distribution and transmission grids;
- reducing the active power capacity of the synchronous generators in the other power plants in the system;
- raising the risk of voltage instability in the power system (Taylor, 1994).

Usually, the reduction of the active power capacity of the synchronous generators is not a critical issue. In India, however, it is important because production capacity is insufficient to supply the demand.

15.2.6 Voltage imbalance

The voltage imbalance can be high on rural feeders that are dominated by unbalanced single-phase loads. Experience with wind turbines connected to rural load feeders in India indicates that the electricity boards perform load shedding on single phases of the rural feeders, which causes severe voltage imbalances. However, most of the wind turbines in India are connected to the substations by dedicated wind farm feeders. Therefore, measurements on load feeders to quantify this issue have not been carried out.

Voltage imbalances have been measured on the primary sides of five wind farm substations in Tamil Nadu and Gujarat (Sørensen, Unnikrishnan and Lakaparampil, 1999), and the results show low voltage imbalances. IS 12360 does not include limits for voltage imbalance. The European standard for voltage quality, EN 50160 (European Standard, 1999) sets a limit of 2 %. This limit was not exceeded by any of the measurements at the substations.

15.3 Wind Turbine Characteristics

Many of the wind turbines in India have a rated capacity of between 200 kW and 250 kW, but there is the whole range, from 55 kW to 500 kW. Most of the wind turbines are produced in India, in cooperation with the western wind turbine industry.

The majority of wind turbines installed in India use induction generators directly connected to the grid (wind turbine Type A)[1]. Both stall control (Type A0) and pitch control (Type A1) are used. The capacitor banks in these wind turbines are typically designed to compensate only the induction generators' no-load consumption of reactive power. Owing to a substantial increment in the fees for reactive power in Tamil Nadu in 2000, some manufacturers now compensate the full reactive power consumption.

In India, another widely used grid connection concept for wind turbines is to connect the generator through a full-scale power converter (Type D). Most of these wind turbines use pitch control (D1). However, the wind turbines with power converters use different types of components. Both induction generators and synchronous generators are used, and the power converters use either thyristor switches or integrated gate bipolar transistor (IGBT) switches. Power converters using thyristor switches also consume reactive power, whereas power converters with IGBT switches can control the power factor fully at the wind turbine terminals. If the power factor is controlled to unity at the wind turbine terminals, only the transformer uses reactive power.

Type C wind turbines [doubly fed induction generators (DFIGs)] are not installed in India yet. When wind energy development in India peaked during the mid-1990s, this type of wind turbine was not very common.

15.4 Wind Turbine Influence on Grids

This section summarises the influence of wind turbines on the power quality of the weak grids to which they are connected.

15.4.1 Steady-state voltage

In many countries, the influence on steady-state voltage is the main design criteria for the grid connection of wind turbines to the medium-voltage distribution system. In particular, if wind turbines are connected to existing rural load feeders, the effect of the wind turbines on the steady-state voltage is important (Tande *et al.*, 1996; DEFU, 1998).

According to IS 12360, the tolerance of the voltage on the 11/33 kV level is +6/−9%. Voltage variations must be kept within these limits if wind turbines are connected to rural feeders. If the wind turbines are connected to dedicated feeders, the voltage limits are set by the official wind farm planning considerations (Government of India, 1998). These specify that the voltage increase at each wind turbine must not exceed 13 % of rated voltage.

15.4.2 Reactive power consumption

The reactive power consumption from wind farms adds to the reactive component originating from agricultural loads. In particular, Type A wind farms consume reactive power, but even Type D wind farms that produce at unity power factor consume

---

[1] For definitions of wind turbine Types A to D see Section 4.2.3.

reactive power for the transformers in wind farms and substations and in the overhead lines. At the same time, wind farms feed active power into the grid. These two factors together reduce the power factor at the conventional power stations.

Looking at the primary side of the wind farm substations, the wind farms affect the reactive power of:

- wind turbines themselves, especially wind turbines with directly grid connected induction generators;
- step-up transformers between wind turbines and wind farm feeders;
- wind farm feeders;
- substation transformers.

Generally, these components consume reactive power, even though some wind turbines may operate at a leading power factor and wind farm feeders produce reactive power at low loading.

The reactive power on the primary side of five wind farm substations was measured. Figure 15.6(a) plots the measured 10-minute-average values of reactive power against active power at three substations in Tamil Nadu, whereas Figure 15.6(b) illustrates similar measurements at two substations in Gujarat.

Figures 15.6(a) and Figure 15.6(b) indicate that the reactive power consumption of the wind farm substations is lower in Tamil Nadu than in Gujarat. The main technical reasons for the differences in reactive power consumption are that:

- in Gujarat, approximately half of the capacitors in the wind turbines are defective, whereas in Tamil Nadu most of the wind turbine capacitors are working;
- the TNEB has installed capacitors on the secondary sides of the wind farm substations, which connect wind turbines that consume reactive power.

The high rate of defective wind turbine capacitors in Gujarat is mainly owing to the fact that wind farm operators in Gujarat have no incentive to replace the defective capacitors. In Tamil Nadu, in contrast, wind farm operators have to pay for reactive power consumption. Another aspect could be that the environment is more aggressive in Gujarat than it is in Tamil Nadu. However, the climate may only be a relevant aspect at Lamba substation, which collects power from wind turbines installed on the coastline. The wind turbines in Dhank are situated inland, in a less aggressive environment.

The three substations in Tamil Nadu represent different technologies of connecting wind turbines to the grid. Aralvai connects only wind turbines with directly grid connected induction generators (Type A), Radhapuram connects only wind turbines with IGBT-based power converters (Type D) and, finally, Karunkulam connects different types, including wind turbines with thyristor-based power converters (also Type D). Figure 15.6(a) indicates that the reactive power consumption of the wind farms can be kept low in the lower power range, independent of the wind turbine technology. There are no measurements available regarding the upper half of the power range. At rated power production, though, reactive power consumption is expected to increase to a higher level in Aralvai and Karunkulam than in Radhapuram because of the different technologies.

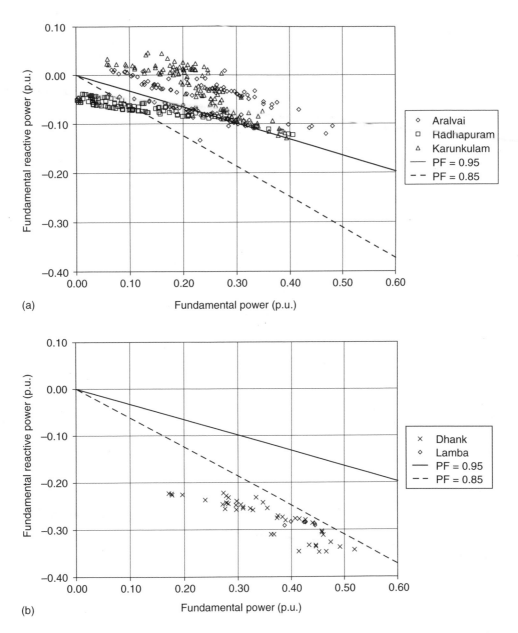

Figure 15.6 Reactive power consumption measured on the primary side of wind farm substations in (a) Tamil Nadu and (b) Gujarat. *Note*: PF = power factor

The wind turbines with induction generators (Type A in Aralvai and Karunkulam) will increase reactive power consumption because of the leakage reactance in the induction generators. Wind turbines with thyristor-based converters (Type D in Karunkulam), will also have an increased reactive power consumption at rated power.

The wind turbines with IGBT-based converters (Type D, in Radhapuram) will not consume reactive power, not even at rated power. However, as mentioned above, in all wind farms, transformers and cables will consume reactive power, and this consumption will increase accordingly if power increases to rated power.

15.4.3 Harmonic and interharmonic emission

According to the International Electrotechnical Commission (IEC) standard on the measurement and assessment of power quality characteristics for grid connected wind turbines, IEC 61400-21 (IEC, 2001), harmonic emissions have only to be measured and assessed for wind turbines with power converters. There are reports regarding harmonic emissions from a few wind turbines with directly grid connected induction generators, but there is no agreed procedure for measuring harmonic emissions from induction machines. Further, there is no reported instance of customer dissatisfaction or damage to equipment as a result of harmonic emissions from such wind turbines. Therefore, IEC 61400-21 does not require measurements of harmonic emissions from such wind turbines.

Measurements at the wind farm feeders of Dhank substation showed 5th and 7th harmonic currents from wind turbines with directly grid connected induction generators (Sørensen 2000). It was concluded that these harmonics are likely to originate from induction generators with irregular windings. Another explanation is that the capacitors together with the induction generator and grid reactance cause a resonance, typically with an eigenfrequency around the 5th and 7th harmonic frequencies. These harmonics have also been observed in other grids outside India.

Analyses of the harmonic power flow indicate that only one type of wind turbine emits harmonics. IS 12360 does not include limits for harmonics, but the limits in the European voltage quality standard EN 50160 are not exceeded. The standards do not include absolute limits for current harmonics, because the grid conditions influence the acceptable level. However, IEC 61000-3-6 (IEC, 1996a) gives some example values, and it was concluded that the harmonic emission from the induction generators does not exceed these limits, with the 5th harmonic current from one feeder reaching the limit.

Measurements at the 110 kV level of Radhapuram substation have indicated a significant emission of harmonics and interharmonics to that high voltage level. Figure 15.7 shows the spectrum of the amplitudes of each of the harmonics up to the 200th harmonic (i.e. approximately 10 kHz). The majority of the wind turbines in Radhapuram use a power converter with a switching frequency of 5 kHz. The distribution of the distortions over frequencies in Figure 15.7 indicates that the source of the distortion is the power converters in the wind farm. Distortions occur at around multiples of the shifting frequency, but apparently the shifting frequency is slightly below the stated 5 kHz.

Both voltage and current distortions measured at Radhapuram substation amount to approximately 6 %. However, measurements with such a high bandwidth and on such a high voltage level are subject to inaccuracies. The figure also shows that the distortions are above the 50th harmonic and, consequently, it does not affect the total harmonic distortion as defined in EN 50160. TNEB in Muppandal has not experienced any problems with the distortions at Radhapuram substation.

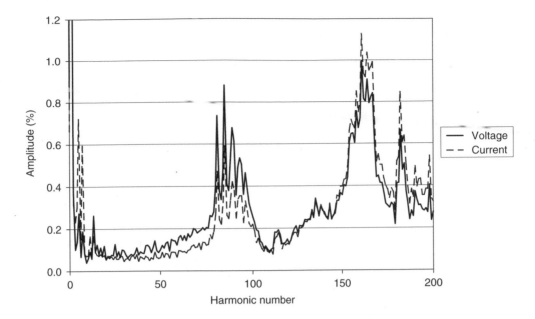

Figure 15.7 Harmonic amplitudes measured at the 110 kV level of Radhapuram substation, where 79 410 kW wind turbines with integrated gate bipolar transistor based power converters are connected

The measurements at neighbouring substations at the 110 kV level show that the harmonic distortions at Radhapuram substation have only a very weak effect on them. This indicates that the distortion is filtered out by the capacitances of the 110 kV lines.

15.5 Grid Influence on Wind Turbines

Connecting wind turbines to weak grids affects the most essential aspects of wind turbine operation, such as performance and safety. According to Rajsekhar, Hulle and Gupta (1998), the inadequate transmission capacity from the wind farms in the Lamba region has strongly affected the wind turbines' performance and maintenance, for instance. This section summarises the effect the connection to weak grids has on wind turbines in India, regarding these and other aspects.

15.5.1 Power performance

The power performance of the wind turbines connected to the weak grids in India is influenced by the following grid parameters:

- outages;
- frequency;
- voltage imbalances;
- steady-state voltage.

Grid outages reduce the performance simply because wind turbines shut down during outages. The frequency variations in the grid mainly affect the performance of stall-regulated wind turbines. The performance is modified because changes in the grid frequency cause changes in the rotor speed of the wind turbine. This changes the angle of attack of the relative wind speed as seen from the rotating blades, which has a decisive influence on the performance of the wind turbine.

Figure 15.8(a) shows calculations of how the conventional 50 Hz power curve of a stall-regulated wind turbine changes in the case of 48 Hz and 51 Hz grid frequency, respectively. The figure shows that up to 10 m/s wind speed the performance is very similar for the three frequencies. If the wind speed exceeds 10 m/s, the wind turbine on the 48 Hz grid starts to stall. If the frequency, and consequently the rotor speed, is higher, the stall and the maximum power point will shift to a higher wind speed, and the maximum power will reach a higher value.

Figure 15.8(a) shows that for frequency variations from 48 Hz to 51 Hz the difference between the maximum values of the power curves is approximately 20 %. This is a normal frequency range in Indian grids, according to the measurement shown in Figure 15.5.

A stall-regulated wind turbine designed for a 50 Hz grid connection can pitch its blades to avoid production above the rated 51 Hz. Alternatively, the wind turbine may trip, as the generator gets too hot if power production is too high. In both cases, the performance of the wind turbine will be affected.

Figure 15.8(b) shows similar performance calculations for a pitch-regulated wind turbine. It shows how the blade angle control ensures a maximum steady-state power of 1 p.u. The 51 Hz power curve for the pitch-regulated wind turbine is almost identical to the 50 Hz power curve, whereas the 48 Hz power curve lies slightly below the 50 Hz power curve. Consequently, the performance of pitch-regulated wind turbines is expected to be slightly inferior as a result of the frequency variations. However, the influence of the frequency on pitch-regulated wind turbines is much smaller than the effect it has on stall-regulated wind turbines.

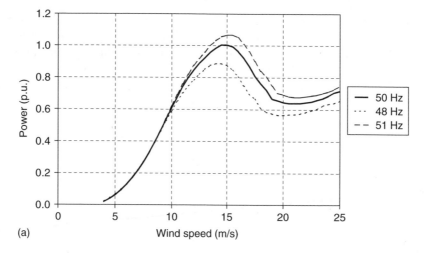

(a)

Figure 15.8 Power curves for (a) a stall-regulated and (b) a pitch-regulated wind turbine, at different frequencies

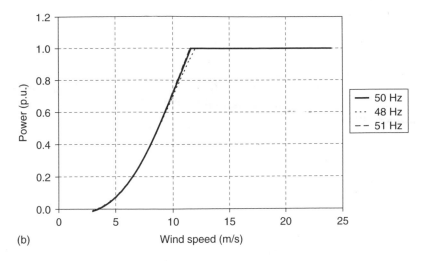

(b)

Figure   15.8 (*continued*)

The performance of wind turbines with power converters is in principle not affected by frequency variations on the grid, because the frequency on the generator side of the converter is controlled to optimise rotor speed, independent of grid frequency.

There are also other grid aspects that influence the performance of wind turbines. A large voltage imbalance will increase losses in induction generators (Lakervi and Holmes, 1995). The losses themselves cause a minor reduction in the produced power, but the losses also cause a heating of the generator, which in some cases forces the wind turbines to shut down.

Low voltage on the wind turbine terminals can cause the voltage relays to trip the wind turbine. Low voltages also cause increased losses, which has only a small influence on the performance. However, the losses also cause heating of the generator, which may, in turn, trip the wind turbine.

15.5.2 Safety

Connection to weak grids also influences the safety of wind turbines:

- The large number of grid outages imply an increased frequency of faults, which reduces the safety.
- Grid abnormalities, such as voltage and frequency variations or voltage imbalances, can trip relays or result in the heating of the generator, for instance, which also increases the risk of faults.
- Low voltages are also critical because they reduce the maximum torque of the induction generator, which is essential to the safety of wind turbines with directly grid connected induction generators. Example calculations have shown that a 10 % voltage drop reduces the maximum torque from approximately 2 times the rated torque to approximately 1.5 times the rated torque (Vikkelsø, Jensen and Sørensen, 1999).

15.5.3 Structural lifetime

The lifetime of the mechanical components is also affected by connection to a weak grid. In particular, the drive train is exposed. If a wind turbine stops, the generator normally remains connected to the grid until the brakes have reduced the power. That way the generator serves as an electrical brake. In the event of a grid outage, the wind turbine will first accelerate, and then the mechanical brakes will have to provide full braking. Wind turbines with pitch control or active stall control will be able to reduce the stress by using the pitch to decelerate and stop the wind turbine.

Also, startups in very high winds will put stress on the mechanical system. In particular, for stall-controlled wind turbines with directly grid connected induction generators (Type A0), the load on the drive train can be high during connection at high wind speeds, and this can affect the lifetime of the gearbox.

15.5.4 Stress on electric components

Switching operations of the capacitor banks in wind turbines with directly grid connected induction generators cause transient overvoltages, which stress the capacitors. Consequently, the number of switching operations can affect the lifetime of the capacitors.

IEC 831-1 (IEC, 1996b) and IEC 931-1 (IEC, 1996c) include the capacitor design limits for transient switching voltages. They are limited to a maximum of 5000 switching operations per year, each with a maximum peak voltage of twice the rated voltage amplitude. Simulations of switching 100 kvar capacitors to a 225 kW wind turbine indicate that the peak value of the voltage does not exceed the specified value of twice the rated amplitude. The number of 5000 switching operations per year is a very high limit for a wind turbine, which only switches in capacitors during startup. Capacitors designed according to standards should therefore not suffer from ageing due to grid abnormalities, if only no-load compensation is used. Also, the temperature variations in India are within the limits specified in IEC 831-1 and IEC 931-1.

Poor voltage quality puts stress on directly grid connected induction generators, too. Voltage imbalances and voltage drops cause increased losses in the generator, which will heat the generator. In addition, experience with induction motors shows that unbalanced currents in the generator cause reverse torque, which tends to retard the generators.

15.5.5 Reactive power compensation

As stated above, the reactive power consumption of wind farms has an important influence on the power system. However, the power system also affects the reactive power consumption of the wind farms. This is particularly the case for wind turbines with directly grid connected induction generators and capacitor compensation (mainly Type A). Mainly the voltage, but also the frequency, of the grid will affect the power factor. Example calculations have shown that a 10 % voltage drop reduces the power factor from 0.96 to 0.94, whereas the influence of the frequency on the power factor is insignificant.

15.6 Conclusions

The findings in India show that wind turbines connected to weak grids have an important influence on the power system, but also that weak grids have an important influence on wind turbines. These findings apply in general to wind farms connected to weak grids, but the Indian case is pronounced regarding the weakness of the grids as well as the wind energy penetration level.

The main concern related to the effect that wind turbines have on the grid is the variation in the steady-state voltage and the reactive power consumption, as well as the influence on voltage fluctuation, and harmonic and interharmonic distortions on the grid.

The weak grid in India is characterised by large voltage and frequency variations, which affects wind turbines regarding their power performance, safety and the lifetime of mechanical and electrical components. Moreover, grid variations modify the reactive power consumption of the wind farms. Hence, it is not only grid companies and customers that should have an interest in strong networks, but also wind farm operators.

References

[1] DEFU (Danish Utilities Research Association) (1998) 'Connection of Wind Turbines to Low and Medium Voltage Networks. Elteknikkommiteen', DEFU KR 111-E, DEFU, Lyngby, Denmark.

[2] European Standard (1999) 'Voltage Characteristics of Electricity Supplied by Public Distribution Systems', EN 50160, European Committee for Electrotechnical Standardisation (CENELEC).

[3] Government of India (1998) 'Wind Power Development in India: Towards Global Leadership', Ministry of Non-conventional Energy Sources, Delhi, India.

[4] Hammons, T. J., Woodford, D., Loughtan, J., Chamia, M., Donahoe, J., Povh, D., Bisewski, B., Long, W. (2000) 'Role of HVDC Transmission in Future Energy Development', *IEEE Power Engineering Review* Volume 20, p. 10–25.

[5] IEC (International Electrotechnical Commission) (1996a) 'Electromagnetic compatibility (EMC). Part 3: Limits. Section 6: Assessment of emission limits for distorting loads in MV and HV power systems – Basic EMC publication', IEC 61000-3-6, IEC, Geneva, Switzerland.

[6] IEC (International Electrotechnical Commission) (1996b) 'Shunt Power Capacitors of the Self-healing Type for A.C. Systems Having a Rated Voltage up to and Including 1000 V, General Performance, Testing and Rating – Safety Requirements – Guide for Installation and Operation', IEC 831-1, IEC, Geneva, Switzerland.

[7] IEC (International Electrotechnical Commission) (1996c) 'Shunt Power Capacitors of the Non-self-healing Type for A.C. Systems Having a Rated Voltage up to and Including 1000 V, General Performance, Testing and Rating – Safety Requirements – Guide for Installation and Operation', IEC 931-1, IEC, Geneva, Switzerland.

[8] IEC (International Electrotechnical Commission) (2001) 'Measurement and Assessment of Power Quality of Grid Connected Wind Turbines', IEC 61400-21, IEC, Geneva, Switzerland.

[9] Indian Standard (1988) 'Voltage Bands for Electrical Installations Including Preferred Voltages and Frequency', IS 12360, Delhi, India.

[10] Lakervi, E., Holmes, E. J. (1995) 'Electricity Distribution Network Design', *IEE Power Engineering Series 21*, 2nd Edition. Peter Peregrinus Ltd, UK 1995.

[11] Ponnappapillai, S., Venugopal, M. R. (1999) 'Policy of TNEB in Developing Wind Farms in Tamil Nadu', presented at the Workshop on Wind Power Generation and Power Quality Issues, Trivandrum, India.

[12] Rajsekhar, B., Hulle, F. V., Gupta, D. (1998) 'Influence of Weak Grids on Wind Turbines and Economics of Wind Power Plants in India', *Wind Engineering* **22(3)** p. 171–181.

[13] Sørensen, P. (2000) 'Harmonic and Interharmonic Emission from Wind Turbines', presented at the Nordic Wind Power Conference (NWPC), NWPC 2000, Trondheim, Norway.

[14] Sørensen, P., Unnikrishnan, A. K., Lakaparampil, Z. V. (1999) 'Measurements of Power Quality of Wind Farms in Tamil Nadu and Gujarat', presented at the Workshop on Wind Power Generation and Power Quality Issues, Trivandrum, India.

[15] Sørensen, P. E., Unnikrishnan, A. K., Mathew, S. A. (2001) 'Wind Farms Connected to Weak Grids in India', *Wind Energy* **4** 137–149.

[16] Sørensen, P., Madsen, P. H., Vikkelsø, A., Jensen, K. K., Fathima, K. A., Unnikrishnan, A. K., Lakaparampil, Z. V. (2000) 'Power Quality and Integration of Wind Farms in Weak Grids in India', publication R-1172(EN), Risø National Laboratory, Roskilde, Denmark.

[17] Tande, J. O., Nørgård, P., Sørensen, P., Søndergård, L., Jørgensen, P., Vikkelsø, A., Kledal, J. D., Christensen, J. S. (1996) 'Power Quality and Grid Connection of Wind Turbines', Summary Report, publication Risø-R-853(Summ.) (EN), DEFU-TR-362 E, Risø National Laboratory, Roskilde, Denmark.

[18] Taylor, C. W. (1994) 'Power System Voltage Stability', California, USA McGraw-Hill, 1994.

[19] Vikkelsø, A., Jensen, K. K., Sørensen, P. (1999) 'Power Factor Correction in Weak Grids using Power Capacitors', presented at the Workshop on Wind Power Generation and Power Quality Issues, Trivandrum, India.

16

Practical Experience with Power Quality and Wind Power

Åke Larsson

16.1 Introduction

Perfect power quality means that the voltage is continuous and sinusoidal, with a constant amplitude and frequency. Power quality can be expressed in terms of physical characteristics and electrical properties. It is most often described in terms of voltage, frequency and interruptions. The quality of the voltage must comply with the requirements stipulated in national and international standards. In Europe, voltage quality is specified in EN 50160 (CENELEC, 1994). In these standards, voltage disturbances are subdivided into voltage variations, flicker, transients and harmonic distortion. Figure 16.1 shows a classification of different power quality phenomena.

Grid-connected wind turbines may affect power quality. The power quality depends on the interaction between the grid and the wind turbine. This chapter is subdivided into five sections. Four sections focus on different aspects of voltage disturbances, such as voltage variations (Section 16.2), flicker (Section 16.3), harmonics (Section 16.4) and transients (Section 16.5). Section 16.6 focuses on frequency variations. A wind turbine normally will not cause any interruptions on a high-voltage grid. Interruptions therefore will not be considered.

16.2 Voltage Variations

Voltage variations can be defined as changes in the RMS (root mean square) value of the voltage during a short period of time, mostly a few minutes. Voltage variations on the grid are caused mainly by variations in load and power production units. If wind

Wind Power in Power Systems   Edited by T. Ackermann
© 2005 John Wiley & Sons, Ltd   ISBN: 0-470-85508-8 (HB)

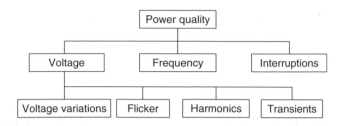

Figure   16.1 Classification of different power quality phenomena

power is introduced, voltage variations also emanate from the power produced by the turbine. All types of wind turbine cause voltage variations. The power production from wind turbines may vary widely and not only as a result of variations in the wind. It may also momentarily drop from full to zero power production in the event of an emergency stop or startup in high winds.

Several methods are used to calculate voltage variations. There are several computer programs for load-flow calculations available on the market. Utility companies use such software for predicting voltage variations caused by load variations. Load-flow calculations can advantageously be used to calculate variations in the voltage caused by wind turbines. Another analytical method is simply to calculate the voltage variation caused by the grid impedance Z, the active power P and the reactive power Q (Ballard and Swansborough, 1984). The analytical method uses a simple impedance model, shown in Figure 16.2. U_1 is the fixed voltage at the end of the power system and U_2 is the voltage at the point of common connection (PCC).

The voltage at the PCC can be expressed as

$$U_2 = \left[a + (a^2 - b)^{1/2} \right]^{1/2}, \tag{16.1}$$

where

$$a = \frac{U_1}{2} - (RP - XQ),$$
$$b = (P^2 + Q^2)Z^2.$$

and R is the resistance and X is the reactance.

Figure 16.3 shows the calculated voltage of the grid at the PCC for different X/R ratios and for a constant short-circuit ratio. The short-circuit ratio is defined as the ratio between the short-circuit power of the grid at the PCC and the rated power of the wind

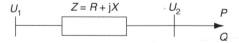

Figure   16.2 Simple impedance model. *Note*: U_1 = fixed voltage at the end of the power system; U_2 = fixed voltage at the point of common connection; Z = grid impedance; R = resistance; X = reactance; P = active power; Q = reactive power

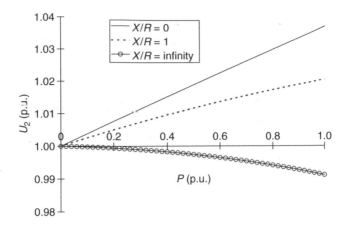

Figure 16.3 Voltage variations at different ratios of reactance, X, to resistance, R. The short-circuit ratio is constant. *Note*: $U_2 =$ fixed voltage at point of common connection; $P =$ active power

turbine. As can be seen in Figure 16.3, a low X/R ratio will increase the voltage at the PCC whereas a high X/R ratio will lower the voltage. Low X/R ratios are found mainly in weak grids in rural areas where long overhead lines with a voltage level below 40 kV are used.

Figure 16.4 presents the measured voltage variation of a wind turbine at Risholmen, Sweden. The calculated voltage variation obtained by Equation (16.1) is also shown. At this specific site, the short-circuit ratio is 26 and the X/R ratio at the PCC is approximately 5.5. As can be seen in Figure 16.4, the voltage variations are within tight limits, at this specific site, because of a high X/R ratio at the PCC.

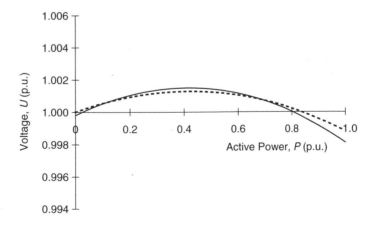

Figure 16.4 Measured (solid line) and calculated (dashed line) voltage variations of the wind turbine site at Risholmen, Sweden

16.3 Flicker

Flicker is the traditional way of quantifying voltage fluctuations. The method is based on measurements of variations in the voltage amplitude (i.e. the duration and magnitude of the variations). Flicker is included in standard IEC 60868 and Amendment 1 (IEC, 1990a; IEC, 1990b). Figure 16.5 shows the magnitude of maximum permissible voltage changes with respect to the number of voltage changes per second, according to IEC 60868.

The fluctuations are weighted by two different filters. One filter corresponds to the response of a 60 W light bulb and the other filter corresponds to the response of the human eye and brain to variations in the luminance of the light bulb (Walker, 1989).

Flicker from grid-connected wind turbines has been the subject of several investigations (Gardner, 1993; Saad-Saoud and Jenkins, 1999; Sørensen *et al.*, 1996). Flicker from wind turbines occurs in two different modes of operation: continuous operation and switching operations.

16.3.1 Continuous operation

Flicker produced during continuous operation is caused by power fluctuations. Power fluctuations mainly emanate from variations in the wind speed, the tower shadow effect and mechanical properties of the wind turbine.

Pitch-controlled turbines will also have power fluctuations caused by the limited bandwidth of the pitch mechanism, in addition to fluctuations caused by the tower shadow. This is valid for all pitch-controlled turbines. The method to overcome this drawback and to prevent such power fluctuations from being transferred to the grid is to use a variable-speed generator in the wind turbine. The power of pitch-controlled turbines is controlled by the angle of the blades. This means that at high wind speeds, normally between 12–14 m/s, and the cut-off wind speed, at 20–25 m/s, the steady-state

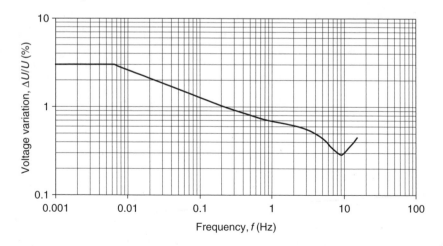

Figure   **16.5** Flicker curve according to IEC 60868 (IEC, 1990a)

value of the power output should be kept close to the rated power of the generator. This is achieved by pitching the blades. Figure 16.6(a) shows the steady-state value of the power (solid line) as a function of wind speed. At wind speeds exceeding 12 m/s the steady-state value of the power is kept equal to rated power. However, pitching the blades implies that the power curve shifts. This is illustrated in Figure 16.6(a) where the dotted line shows the instantaneous power curve, when at a wind speed of 15 m/s the blades are pitched for operation at rated power.

Unfortunately, the wind speed is not constant but varies all the time. Hence, instantaneous power will fluctuate around the rated mean value of the power as a result of gusts and the speed of the pitch mechanism (i.e. limited bandwidth). As can be seen in Figure 16.6(a), variations in wind speed of ±1 m/s give power fluctuations with a magnitude of ±20 %. Figure 16.6(b) shows the power from a stall-regulated turbine under the same conditions as the pitch-controlled turbine in Figure 16.6(a). Variations

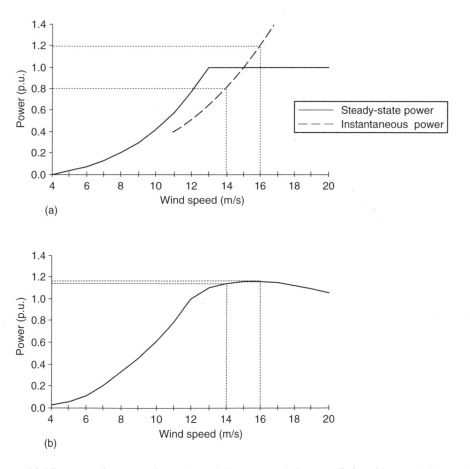

Figure 16.6 Power as function of wind speed from (a) a pitch-controlled turbine and (b) a stall-regulated turbine. Reprinted, with permission, from A. Larsson, 'Flicker Emission of Wind Turbines During Continuous Operation', *IEEE Transactions on Energy Conversion* 17(1) 114–118 © 2002 IEEE

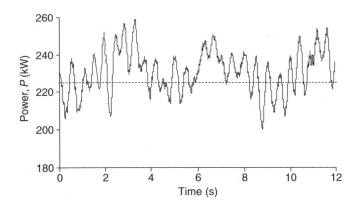

Figure 16.7 Measured power during normal operation of a Type A1 wind turbine (solid line). In the figure, the steady-state power is also plotted (dotted line). Reprinted, with permission, from A. Larsson, Flicker 'Emission of Wind Turbines During Continuous Operation', *IEEE Transactions on Energy Conversion* 17(1) 114–118 © 2002 IEEE

in wind speed cause also power fluctuations in the stall-regulated turbine, but they are small in comparison with those in a pitch-controlled turbine.

Figure 16.7 shows the measured power of a Type A1 wind turbine with a rated power of 225 kW in high winds. The figure shows variations in the power produced by the turbine. As previously mentioned, Type A wind turbines produce power pulsations as a result of wind speed gradients and the tower shadow.[1] The frequency of the power pulsations is equal to the number of blades multiplied by the rotational speed of the turbine (i.e. the 3-p frequency). The figure also indicates the power fluctuations caused by gusts and the speed of the pitch mechanism.

Measurements are required to determine flicker emission produced during the continuous operation of a wind turbine. IEC 61400-21 warns that flicker emission should not be determined from voltage measurements, as this method will be influenced by the background flicker of the grid (IEC, 1998). The method proposed to overcome this problem is based on measurements of current and voltage. The short-term flicker emission from the wind turbine should be calculated based on a reference grid using the measured active and reactive power as the only load on the grid.

16.3.2 Switching operations

Switching operations will also produce flicker. Typical switching operations are the startup and shutdown of a wind turbine. Startup, stop and switching between generators or generator windings will cause a change in power production. The change in power production will lead to voltage changes at the PCC. These voltage changes will, in turn, cause flicker.

---

[1] For definitions of wind turbine Types A to D, see Section 4.2.3.

16.3.2.1 Startup

The startup sequences of variable-speed wind turbines as well as fixed-speed wind turbines are all different. Variable speed wind turbines are normally equipped with pitch control. Generally, owing to the controllable speed of the turbine and the pitch control, the starting sequence of variable-speed wind turbines is smoother than that of fixed-speed wind turbines.

Fixed-speed wind turbines

Figure 16.8(a) shows the measured power during the startup of a Type A1 wind turbine. The startup of the turbine occurs at time $t = 30$ s. As can be seen, the wind turbine consumes reactive power in order to magnetise the generator. The soft-starter operates for two or three seconds in order to limit the current to the rated value. The reactive power is then compensated by shunt capacitor banks. It can be seen that the capacitors are switched in four steps, at time intervals of approximately 1 s. Once all capacitor banks have been switched in at approximately $t = 35$ s, the blades of the turbine are pitched, which increases power production. The power production also affects the reactive power consumption.

Figure 16.8(b) shows the corresponding terminal voltage of the wind turbine. The voltage change caused by the startup of the turbine can be divided into two parts. The first part is caused by the reactive power consumption of the generator. As can be seen, the reactive power consumption causes a voltage drop. Once the capacitors are connected and the reactive power consumption returns to zero, the voltage level is restored. The second part is caused by the power production. As the power production increases, the voltage level begins to rise.

Variable-speed wind turbine

Figure 16.9 shows the active and reactive power during the startup of a Type D1 wind turbine in high winds. At time $t = 30$ s the generator is connected to the grid via the converter. The active power increases smoothly from zero to half the rated power in 30 s. During this period the active power increases; the reactive power is controlled in order to keep the power factor constant. At this particular site, a power factor of 0.98 was chosen. Again, the starting sequence of variable-speed wind turbines is smoother than that of fixed-speed wind turbines.

16.3.2.2 Shutting down

If the wind speed becomes either too low or too high, the wind turbine will stop automatically. In the first case the turbine will be stopped in order to avoid a negative power flow. In the second case it will be stopped to avoid high mechanical loads. At low wind speeds (3–4 m/s) the active power is almost zero. The stop will be rather soft and the impact on the voltage at the PCC will be small at such low wind speeds. The impact may be larger at high wind speeds (greater than 25 m/s) since the turbine produces at rated power on these occasions. If the turbine is stopped and the power decreases from rated power to zero production, the voltage at the PCC will be affected.

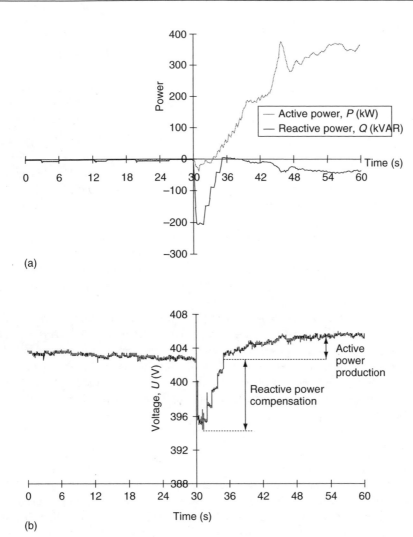

Figure 16.8 (a) Measured power and (b) Measured voltage during startup of a Type A1 wind turbine with a rated power of 600 kW Reprinted, with permission, from A. Larsson, 'Flicker Emission of Wind Turbines caused by Switching Operations', *IEEE Transactions on Energy Conversion* 17(1) 119–123 © 2002 IEEE

Fixed-speed wind turbines

Figure 16.10 shows the stop of a Type A0 wind turbine with a rated power of 600 kW. The wind turbine operates at approximately half the rated power when the turbine is stopped. Once the turbine is stopped, the capacitor bank for reactive power compensation is switched off. After a couple of seconds, brakes reduce the turbine speed. Figure 16.10 illustrates that the turbine's speed is reduced by braking at time $t = 15$ s. In order

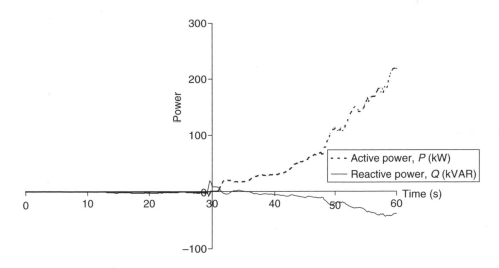

Figure 16.9 Measured power during startup of a Type D1 wind turbine with a rated power of 500 kW Reprinted, with permission, from A. Larsson, 'Flicker Emission of Wind Turbines caused by Switching Operations', *IEEE Transactions on Energy Conversion* 17(1) 119–123 © 2002 IEEE

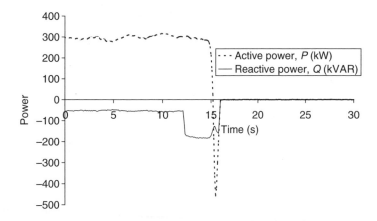

Figure 16.10 Measured power during stop of a Type A0 wind turbine with a rated power of 600 kW Reprinted, with permission, from A. Larsson, 'Flicker Emission of Wind Turbines caused by Switching Operations', *IEEE Transactions on Energy Conversion* 17(1) 119–123 © 2002 IEEE

to make sure that the turbine is actually stopped, the generator is disconnected when the power is reversed.

Variable-speed wind turbines

Figure 16.11 shows the stop of a Type D1 wind turbine in high winds. The turbine operates at rated power. The power from the wind turbine begins to decrease at time $t = 6$ s. Four seconds later, the power has dropped from rated power to zero. Similar to the startup, the stop is very gentle and smooth.

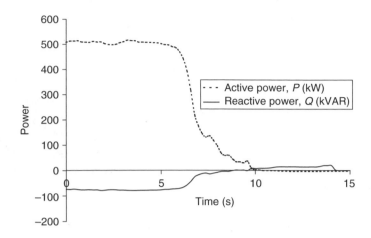

Figure 16.11 Measured power during stop of a Type D1 wind turbine with a rated power of 500 kW Reprinted, with permission, from A. Larsson, 'Flicker Emission of Wind Turbines caused by Switching Operations', *IEEE Transactions on Energy Conversion* 17(1) 119–123 © 2002 IEEE

According to IEC 61400-21, switching operations have to be measured during the cut-in of a wind turbine and for switching operations between generators. Switching operations between generators are only relevant for wind turbines with more than one generator or a generator with multiple windings.

16.4 Harmonics

Voltage harmonics are virtually always present on the utility grid. Nonlinear loads, power electronic loads, and rectifiers and inverters in motor drives are some sources that produce harmonics. The effects of the harmonics include overheating and equipment failure, faulty operation of protective devices, nuisance tripping of a sensitive load and interference with communication circuits (Reid, 1996).

Harmonics and interharmonics are defined in IEC 61000-4-7 and Amendment 1 (IEC 1991, 1997). Harmonics are components with frequencies that are multiples of the supply frequency (i.e. 100 Hz, 150 Hz, 200 Hz, etc.). Interharmonics are defined as components having frequencies located between the harmonics of the supply frequency.

The signal that is to be analysed is sampled, A/D-converted and stored. These samples form a window of time ('window width') on which a discrete Fourier transformation is performed. The window width, according to the standard, has to be 10 line-periods in a 50 Hz system. This window width will provide (the) distance between two consecutive interharmonic components of 5 Hz. Figure 16.12 shows the interharmonic components between 1250 Hz and 1300 Hz of the measured current from a variable-speed wind turbine. The current has been analysed in accordance with IEC 61000-4-7.

Fixed-speed wind turbines are not expected to cause significant harmonics and inter-harmonics. IEC 61400-21 does not include any specifications of harmonics and inter-harmonics for this type of turbine. For variable-speed wind turbines equipped with a

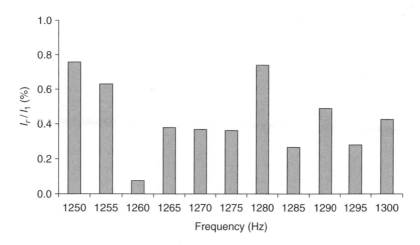

Figure 16.12 Current interharmonic content between 1250 Hz and 1300 Hz. *Note*: I_n = harmonic current of order n

converter, the emission of harmonic currents during continuous operation has to be specified. The emission of the individual harmonics for frequencies of up to 50 times the fundamental grid frequency are to be specified, as well as the total harmonic distortion. According to IEC 61000-4-7, the following equation has to be applied to determine the harmonic currents that originate from more than one source connected to a common point:

$$I_n = \left(\sum_k I_{n,k}^{\alpha} \right)^{1/\alpha},$$
(16.2)

where I_n is the harmonic current of order n; $I_{n,k}$ is the harmonic current of order n from source k; and α is an exponent, chosen from Table 16.1. This recommendation is valid for wind farm applications.

Figure 16.13 shows the harmonic content of two Type D1 wind turbine equipped with forced-commutated inverters. The harmonic content is obtained in two different ways. One way is to use Equation (16.2) and the measured time series from each wind turbine. The other way is to use the measured time series of the total current from both wind turbines.

Table 16.1 Exponent, α, for harmonics

α	Harmonic number, n
1	$n < 5$
1.4	$5 \leq n \leq 10$
2	$n > 10$

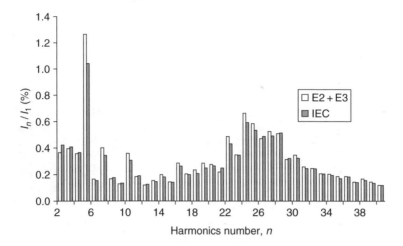

Figure 16.13 Calculated harmonic content from two Type D1 wind turbines equipped with forced-commutated inverters (i) by using the measured time series of current from each wind turbine (E2 + E3) and by using Equation (16.2) in text and (ii) by using the measured time series of the total current from both wind turbines (IEC). *Note*: I_n = harmonic current of order n; IEC = standard IEC 61400-21 (IEC, 1991, 1997)

16.5 Transients

Transients seem to occur mainly during the startup and shutdown of fixed-speed wind turbines (Demoulias and Dokopoulos 1996). The startup sequence of a Type A wind turbine is performed in two steps. First, the generator is switched on. A soft-starter is used to avoid a high in-rush current. Once the soft-starter begins to operate and the generator is connected to the grid, the shunt capacitor banks are switched in. The shunt capacitor banks are switched directly to the grid without any soft switching devices. Once the shunt capacitor banks are connected, a high current peak occurs (see Figure 16.14). This transient may reach a value of twice the rated wind-turbine current and may

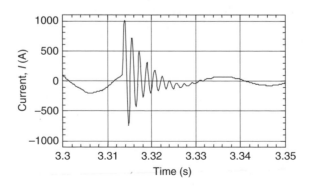

Figure 16.14 Measured oscillating current when connecting shunt capacitors during the startup sequence of a 225 kW wind turbine of Type A1

substantially affect the voltage of the low-voltage grid. The voltage transient can disturb sensitive equipment connected to the same part of the grid.

The amplitude of the current originating from switching in an unloaded capacitor is determined by the impedance of the grid and the capacitance of the capacitor. The frequency, f, of the transient can be determined approximately by:

$$f = \frac{1}{2\pi}\left(\frac{1}{LC}\right),\qquad(16.3)$$

where L is the inductance of the grid and C is the capacitance of the capacitor.

In order to improve the calculations of the connecting current and voltage, a more detailed model must be used. The Electro Magnetic Transient Program (EMTP) makes it possible to use frequency-dependent parameters. Calculations of switching transients on a low-voltage grid equipped with two wind turbines are presented in Larsson and Thiringer (1995).

16.6 Frequency

On the one hand, the incorporation of a relatively small amount of wind power into the utility grid normally does not cause any interfacing or operational problems. The intermittent power production from wind turbines is balanced by other production units (Hunter and Elliot, 1994). On the other hand, the effect of wind power is very important in autonomous power systems (see also Chapter 14). In an autonomous grid, the spinning reserve is supplied by diesel engines, and it is usually small. This small spinning reserve will give rise to frequency fluctuations in case of a sudden wind rise or wind drop. Hence, in a wind–diesel system, voltage and frequency fluctuations will be considerably larger than in an ordinary utility grid.

Over the past decade, different types of wind turbines and wind–diesel systems for autonomous grids have been tested. Type A wind turbines used to be the most common.

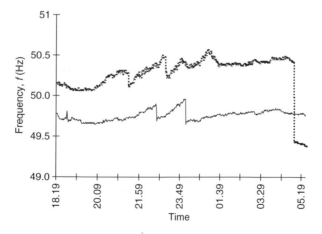

Figure **16.15** Frequency variations during two nights. During one night the wind turbines (WTs) operated (gray line) and during the other night they were shut off owing to low winds (black line)

Today, an increasing number of D1 types are used. Figure 16.15 shows measurements at a wind–diesel system with a relatively small amount of wind power, during two nights.

The installed wind power corresponded to approximately 10 % of the total diesel power installed on the island. The frequency on the grid was measured during two nights, one night with wind turbines and one night without. There were two frequency drops during the night when the wind turbines were not operating (black line). These two drops were most likely caused by the stop of one or several diesel engines. The other curve (gray line), which represents the frequency with operating wind turbines, shows an increase in frequency. The frequency exceeded 50 Hz throughout the night, which indicates that some diesel engines ran at low load. A likely explanation is that the utility company was afraid to stop too many diesel engines in case the wind suddenly dropped. If the portion of wind power is to be increased further (i.e. if the wind–diesel system is to operate solely on wind power under high-wind conditions) the power from the wind turbines must be controllable. Figure 16.16 shows measurements from such a specially designed wind–diesel system, which uses a Type D1 wind turbine.

The figure shows the power from the wind turbine and the frequency measured during one night. The figure illustrates that the turbine was switched off and did not produce any power during the first 1.5 h. During this time, the plant operated in diesel mode. The plant then worked in a mixed mode and the wind turbine operated together with the diesel for approximately 4 h. After this, and for the rest of the night, the wind speed was high enough for the wind turbine to operate alone. The total power consumption was relatively constant, slightly exceeding 15 kW. The criterion for operating the plant in wind mode is that the rotational speed of the wind turbine exceeds a predetermined value, in this case 60 rpm.

The frequency rose from approximately 48 Hz in the diesel mode to 50 Hz in the mixed and wind modes. The diesel seems to have a governor with a frequency of 52 Hz at no load and up to 48 Hz at full load. For the rest of the night, the plant ran in the

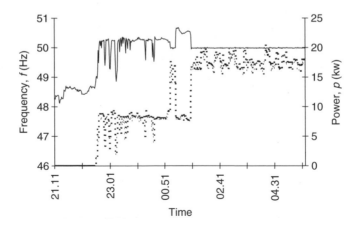

Figure   16.16 Frequency variations (black line) and power output from a Type D1 wind turbine (gray line) during one night

wind mode. As can be seen, the frequency was very stable during that time. In fact, the frequency was much more stable in the wind mode than in the other two modes.

16.7 Conclusions

This chapter investigated the power quality of grid-connected wind turbines. Furthermore, it focused on the characteristics of fixed-speed and variable-speed wind turbines. From an electrical point of view, wind turbines may be divided into two main groups – fixed-speed and variable-speed. Both groups of wind turbines have advantages and disadvantages regarding their interaction with the grid and power quality. Wind turbines have an uneven power production following the natural variations in the wind. Uneven power production is the same for all kinds of wind turbines. Each time a turbine blade passes the tower, it enters into the tower shadow. If the turbine operates at fixed speed, the tower shadow and wind speed gradients will result in fluctuating power. Both uneven power production and power fluctuations cause voltage variations. Load-flow calculations can be used to calculate slow variations in the voltage caused by the uneven power production of wind turbines. The power fluctuations of the wind turbine may cause flicker disturbances. In order to calculate the impact of flicker, measurements and subsequent flicker calculations must be performed.

Apart from possible oscillations between the grid impedance and the shunt capacitor banks for power factor correction, fixed-speed wind turbines do not produce any harmonics. When it comes to variable-speed wind turbines, however, the situation is different. Depending on the type of inverter used, different orders of harmonics are produced.

Transients seem to occur mainly when wind turbines are started and stopped. A large in-rush current and thereby a voltage dip can be avoided if the wind turbine is equipped with a soft-starter. When the shunt capacitor bank is switched in, a large current peak occurs. The current peak may substantially affect the voltage on the low-voltage side of the transformer.

In an autonomous grid supplied by diesel engines, the spinning reserve is limited and gives rise to frequency fluctuations when fast load changes occur. Hence, the frequency of an autonomous grid is normally not as stable as that of a large grid. When wind power is incorporated into an autonomous grid, a sudden rise or drop in wind will affect the power balance, with frequency variations as a result. The use of sophisticated variable-speed wind turbines can eliminate this problem and actually improve the frequency balance.

References

[1] Ballard, L. J., Swansborough, R. H. (Eds) (1984) *Recommended Practices for Wind Turbine Testing: 7. Quality of Power. Single Grid-connected WECS*, 1st edn, Elsgaards bogtrykkeri.

[2] CENELEC (European Committee for Electrotechnical Standardisation) (1994) Voltage Characteristics of Electricity Supplied by Public Distribution Systems, EN 50 160, CENELEC.

[3] Demoulias, C. S., Dokopoulos, P. (1996) 'Electrical Transients of WT in a Small Power Grid', *IEEE Transactions on Energy Conversion* 11(3) 636–642.

[4] Gardner, P. (1993) Flicker from Wind Farms, in *Proceedings of the BWEA/SERC RAL Workshop on Wind Energy Penetration into Weak Electricity Network, Rutherford, UK, June*, pp. 27–37.

[5] Hunter, R., Elliot, G. (1994) *Wind–Diesel Systems*, Cambridge University Press, Cambridge, UK.

[6] IEC (International Electrotechnical Commission) (1990a) 'Flickermeter – Functional and Design Specifications', IEC 60868, IEC.

[7] IEC (International Electrotechnical Commission) (1990b) 'Amendment 1 to Publication 60868, Flickermeter – Functional and Design Specifications', IEC.

[8] IEC (International Electrotechnical Commission) (1991) 'Electromagnetic Compatibility, General Guide on Harmonics and Inter-harmonics Measurements and Instrumentation', IEC 61000-4-7, IEC.

[9] IEC (International Electrotechnical Commission) (1997) 'Amendment 1 to Publication 61000-4-7, Electromagnetic Compatibility, General Guide on Harmonics and Inter-harmonics Measurements and Instrumentation', IEC.

[10] IEC (International Electrotechnical Committee) (1998) 'Power Quality Requirements for Grid Connected WT', IEC 61400-21, Committee Draft, IEC.

[11] Larsson, A. (2002a) 'Flicker Emission of Wind Turbines During Continuous Operation', *IEEE Transactions on Energy Conversion* 17(1) 114–118.

[12] Larsson, A. (2002b) 'Flicker Emission of Wind Turbines Caused by Switching Operations', *IEEE Transactions on Energy Conversion* 17(1) 119–123.

[13] Larsson, Å., Thiringer, T. (1995) 'Measurements on and Modelling of Capacitor-connecting Transients on a Low-voltage Grid Equipped with Two WT', in *Proceedings of the International Conference on Power System Transients* (IPST '95), Lisbon, Portugal, September, pp. 184–188.

[14] Reid, W. E. (1996) 'Power Quality Issues – Standards and Guidelines', *IEEE Transactions on Industry Applications* 32(3) 625–632.

[15] Saad-Saoud, Z., Jenkins, N. (1999) 'Models for Predicting Flicker Induced by Large WT', *IEEE Transactions on Energy Conversion* 14(3) 743–748.

[16] Sørensen, P., Tande, J. O., Søndergaard, L. M., Kledal, J. D. (1996) 'Flicker Emission Levels from WT', *Wind Engineering* 20(1) 39–45.

[17] Walker, M. K. (1989) 'Electric Utility Flicker Limitation', in *Proceedings of the Power Quality Conference, Long Beach, California, October*, pp. 602–613.

17

Wind Power Forecast for the German and Danish Networks

Bernhard Ernst

17.1 Introduction

Wind-generated power now constitutes a noticeable percentage of the total electrical power consumed and in some utility areas it even exceeds the base load on the network. This indicates that wind is becoming a major factor in electricity supply and in balancing consumer demand with power production. A major barrier to the integration of wind power into the grid is its variability. Because of its dependence on the weather, the output cannot be guaranteed at any particular time. This makes planning the overall balance of the grid difficult and biases utilities against using wind power. Accurate forecasting of power inputs from wind farms into the grid could improve the image of wind power by reducing network operation issues caused by fluctuating wind power.

The variability of wind power can also affect the price that is paid for wind-generated electricity. Some countries give a subsidised price, higher than the pool price, to wind generators. The trend, however, is towards deregulated electricity markets. Under these regimes, producers may have to contract to provide firm power and might be penalised if they overproduce or underproduce. This reduces the value of the energy they sell which in turn reduces the incentive to invest in wind energy. By aggregating the power output of wind farms and obtaining accurate predictions of the power that will be fed into the grid it may be possible to raise the achieved price, and this would make wind power a more favourable investment. Higher prices would also mean that more sites will become feasible, as those with lower average wind speeds would become economic.

Wind Power in Power Systems   Edited by T. Ackermann
© 2005 John Wiley & Sons, Ltd   ISBN: 0-470-85508-8 (HB)

The requirements regarding wind power prediction systems vary considerably depending on the specific market and the distribution of wind turbines. The French utility Electricité de France (EDF), for instance, wants a day-ahead as well as a week-ahead forecast. The value to be forecasted is the mean wind power production over thirty minutes, on a national scale (France and other European countries) and on a regional scale (Dispower, 2002). The British New Electricity Trading Arrangement (NETA) includes completely different requirements. The spot market closes only one hour before the time of delivery. However, generators find two-hour predictions most useful because of the work that has to be done prior to the delivery. In Denmark and Germany, predictions are used mainly for the day-ahead market, which closes at noon in Denmark (i.e. up to 36 hours before time of delivery) and at 3 p.m. in Germany (up to 33 hours before the time in question). Hence, the requirements for wind power forecast systems are usually set by market operation constraints, not by technical or physical constraints. The hydro-based system in Scandinavia, for instance, technically does not require a 36-hour forecast as the hydro system can adjust generation in a fraction of an hour. However, market operation requires a 36-hour forecast.

In addition to the different timespans that forecasts cover, different spatial resolutions are required. In Denmark and in Germany, for instance, wind turbines are spread all over the country. Therefore, the wind power is forecasted mainly for large areas (see also Chapters 2, 10 and 11). In the USA and in Spain, the installed wind power is limited to a number of large wind farms. This requires a forecast for single wind farms instead of wide areas. However, most of the prediction systems tend to cover all possible needs.

17.2 Current Development and Use of Wind Power Prediction Tools

Meteorological and research and development (R&D) institutes initiated and developed the short-term prediction of wind power. The Risø National Laboratory and the Technical University of Denmark were first in this area. They have used detailed, area-specific, three-dimensional weather models and have worked with numerical weather prediction (NWP) models, such as HIRLAM (High Resolution Limited Area Model), the UK MESO (UK Meteorological Office Meso-scale model) or the LM (Lokal-Modell of the German Weather Service). These systems work like a weather forecast, predicting wind speeds and directions for all the wind farms in a given area. On this basis, they calculate the output power using physical equations such as in WAsP (Wind Atlas Analysis and Application Program, http://www.wasp.dk) or similar programs. The models make it possible to produce predictions for up to 48 hours ahead. They have been designed primarily to provide information for network companies to be able to operate their power systems. Similar projects have been carried out

- in Norway at the Norwegian Meteorological Institute;
- in the USA at Wind Economics and Technology (WECTEC) and at TrueWind Solutions;

- in Germany at the Institute of Solar Energy Supply Systems (ISET), at Oldenburg University and at Fachhochschule Magdeburg;
- in Japan, at Mie University and the Toyohashi University of Technology;
- in Ireland at the University College Cork.

Some of the systems use statistics, artificial neural networks (ANNs) or fuzzy logic instead of physical equations. Their advantage over standard computing is that they are able to 'learn' from experience. The disadvantage is that they need a large dataset to be trained with before the system works properly.

In the following, several systems will be described in more detail. The main focus is on two systems used by the transmission system operators (TSOs) in Denmark and in Germany. In Denmark, the Wind Power Prediction Tool (WPPT) was developed by the University of Copenhagen and is used by the TSO in Western Denmark, Eltra, whereas the TSO in Eastern Denmark Elkraft uses a similar model that was developed in-house. In Germany, E.ON, RWE Net and VET use the Advanced Wind Power Prediction Tool (AWPT), which was developed by ISET and predicts 95 % of German wind power.

17.3 Current Wind Power Prediction Tools

17.3.1 Prediktor

The Prediktor system has been developed by Risø National Laboratory, Denmark, and it uses, as far as possible, physical models. Large-scale flow is simulated by an NWP model – HIRLAM in this case. The wind is transformed to the surface using the geostrophical drag law and the logarithmic wind profile. When zooming in on the site, more and more detail is required. Such details is provided by the Risø WAsP program. WAsP takes into account local effects (lee from obstacles, the effect of roughness and roughness changes, and speed-up or down of hills and valleys). The Risø PARK program is used to include the shadowing effects of turbines in a wind farm. Finally, two model output statistics (MOS) modules are used to take into account any effects not modelled by the physical models and general errors of the method.

The model runs twice a day generating 36-hour predictions for several wind farms in the supply area. For each wind farm, power predictions can also be scaled up in order to represent a region instead of only the wind farm. In order to evaluate the model, the predictions are compared with those of the *persistence* model, which is a very simple model stating that

$$P(t + l) = P(t), \tag{17.1}$$

where $P(t)$ is the production at time t, and l is the 'look-ahead time'. This model could popularly be called the 'what-you-see-is-what-you-get' model. Despite its apparent simplicity, this model describes the flow in the atmosphere rather well, because of the characteristic time scale of weather systems; the weather in the afternoon often is the same as it was in the morning. For a short prediction horizon of only a few hours, this model is hard to beat.

A comparison of the two models gives the following results (Landberg, 2000; Landberg *et al.*, 2000b):

- The prediction model outperforms the persistence model after six hours.
- The mean absolute error of the prediction model is around 15 % of the installed capacity. A well-predicted wind farm has a scatter of as little as 10 %, and a not so well predicted wind farm has a scatter of up to 20 %.
- The performance of the prediction model deteriorates very gently – during 36 hours the error increases by less than 10 %.

17.3.2 Wind Power Prediction Tool

WTTP implements another philosophy for predicting wind power. It was developed as a cooperation between Eltra and Elsam and the Department of Informatics and Mathematical Modelling (IMM) at the Technical University of Denmark (DTU). WPPT applies statistical methods for predicting the wind power production in larger areas. It uses online data that cover only a subset of the total population of wind turbines in the area. The relevant area is divided into subareas, each covered by a reference wind farm. Then local measurements of climatic variables as well as meteorological forecasts of wind speed and direction are used to predict the wind power for periods of time ranging from half an hour to 36 hours. The wind farm power prediction for each subarea is subsequently upscaled to cover all wind turbines in the subarea; and then the predictions for subareas are summarised to arrive at a prediction for the entire area.

Input data to WPPT are wind speed, wind direction, air temperature and the power output at the reference wind farms. They are sampled as 5 minute mean values. In addition to that, meteorological forecasts are used as an input to the models. The forecasts are updated every 6 hours and cover a period of 48 hours ahead with a 1 hour resolution. In general, WPPT uses statistical methods to determine the optimal weight between online measurements and meteorological forecasted variables.

The first version of WPPT went into operation in October 1994. This version did not use any NWP data. Predictions were based on the statistical analysis of measurements from only seven wind farms. These predictions were not suitable for forecast horizons that exceeded 8 hours:

- because of the poor reliability of measurement equipment;
- because too few wind farms were included;
- because of a lack of meteorological forecasts.

A new project was started in 1996. Its aim was to obtain better predictions by including more wind farms, by improving the reliability of the equipment and by getting good meteorological forecasts. The ELTRA dispatch centres use this system with good results. There, it provides forecasts for more than 2 GW installed wind power capacity.

Figure 17.1 shows the distribution of the error of 36 hour predictions for the year 2001. At that time, installed capacity was 1900 MW. Some 37 % of the errors are within ±100 MW; large errors over 500 MW occur only 7 % of the time. Figure 17.2 shows the

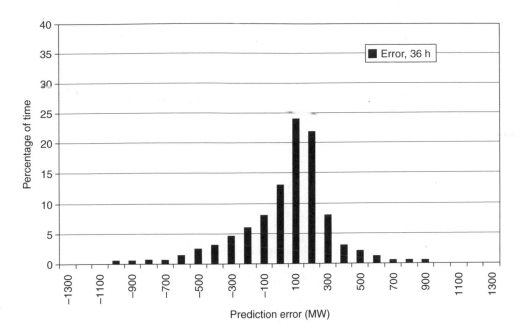

Figure 17.1 Distribution of prediction errors for 1900 MW of installed wind power with a prediction horizon of 36 hours
Source: Holttinen, Nielsen and Giebel, 2002. (Reproduced by permission of KTH, Sweden.)

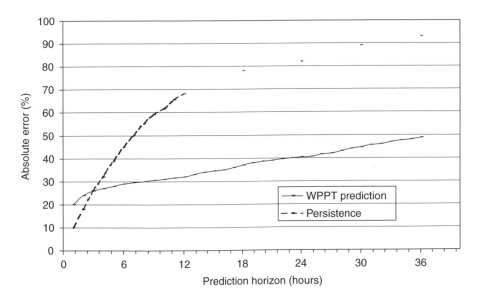

Figure 17.2 The total absolute prediction error during one year for different prediction horizons, as a percentage of the total production in 2001. *Note*: WPPT = Wind Power Prediction Tool
Source: Holttinen, Nielsen and Giebel, 2002. (Reproduced by permission of KTH, Sweden.)

total prediction error depending on the forecast horizon. For forecast horizons of more than 3 hours, WPPT outperforms the persistence model. The graphs were produced with the 1997 version of WPPT. In 2003, a new, improved, version was installed, (Holttinen, Nielsen and Giebel, 2002; Nielsen, Madsen and Tofting, 2000).

17.3.3 Zephyr

In Denmark, a new program for short-term predictions of wind energy is being developed. Risø National Laboratory and IMM jointly develop the system, and all Danish utilities will use the system. The new software is called Zephyr and, being based on Java2, it offers a high flexibility. New prediction models have become possible because new input data are available. Zephyr merges the two Danish models Prediktor and WPPT. This combination will ensure that there are reliable forecasts available for all prediction horizons, from short-range (0–9 hours) to long-range (36–48 hours) predictions, as IMM uses online data and advanced statistical methods, which provide good results for the short-range prediction horizon. In addition to that, the use of meteorological models such as the HIRLAM model of the Danish Meteorological Institute benefits the long-term forecasts. An intelligent weighting of the results of both modelling branches then gives the result for the total supply area of the customer (Landberg, 2000a).

17.3.4 Previento

Previento is the name of a forecast system, that has been developed recently at Oldenburg University. It provides power predictions for a larger area (e.g. the whole of Germany) for the following two days. Its approach is similar to that of the Prediktor system. The German version uses data of the German Weather Service instead of HIRLAM data. Furthermore, it provides an estimation of the possible error depending on the weather situation (Beyer et al., 1999; Focken, Lange and Waldl, 2001).

17.3.5 eWind

The eWind[TM] system was developed by True Wind Solutions, USA, and runs now at Southern California Edison (SCE) Company, with about 1000 MW of installed wind power. The system is composed of four basic components:

- a set of high-resolution three-dimensional physics-based atmospheric numerical models;
- adaptive statistical models;
- plant output models;
- a forecast delivery system.

The physics-based atmospheric models are a set of mathematical equations that represent the basic physical principles of conservation of mass, momentum and energy and the equation of state for moist air. These models are similar to those used by operational weather forecast centres throughout the world. However, the eWind[TM] models are run at a higher resolution (i.e. with smaller grid cells), and the physics and data incorporated into the models are specifically configured for high-resolution simulations for wind forecasting applications. The current eWind™ configuration uses a single numerical model [the Mesoscale Atmospheric Simulation System (MASS) model] to generate the forecasts.

The statistical models are a set of empirical relationships between the output of the physics-based atmospheric models and specific parameters that are to be forecasted for a particular location. In this application, the specific parameters are the wind speed and direction and air density at the location of the wind turbines that provide power to each of the five substations. The role of the statistical model is to adjust the output of the physics-based model in order to account for subgrid scale and other processes that cannot be resolved or otherwise adequately simulated on the grids that are used by the physical model.

The third component of the system is a wind turbine output model. This model is a relationship between the atmospheric variables and the wind turbine output. The wind turbine output can be either a fixed relationship that applies to a particular wind turbine configuration or it can be an empirical statistical relationship derived from recent (e.g. 30-day) atmospheric data and wind turbine output data. The final piece is the forecast delivery system. The user has the option of receiving the forecast information via email, an FTP transmission, a faxed page or on a web page display.

For the SCE day-ahead power generation, the system is configured to deliver two forecasts per day: an afternoon forecast at 5 p.m. PT, and a morning forecast at 5:30 a.m. PT. Each cycle consists of the following three parts (Bailey, Brower and Zack, 2000; Brower Bailey and Zack, 2001):

- execution of the physical model;
- reconstruction of the statistical model equations based on the previous 30 days of physical model forecasts and measured data;
- evaluation of these statistical equations for each forecast hour in the current cycle.

17.3.6 SIPREÓLICO

SIPREÓLICO is a statistics-based prediction tool developed by the University Carlos III, Madrid, Spain, and the transmission system operator Red Eléctrica de España (REE). It is presently a prototype that provides hourly predictions with a prediction horizon of up to 36 hours. They are generated by using meteorological forecasts, as well as online power measurements, as input data to time series analysis algorithms. REE already uses these predictions for its online system operation. First, SIPREÓLICO produces predictions for single wind farms. Once there are predictions for each wind farm, they are aggregated in zones. Finally, the production forecast for the whole of Spain is generated. For a given wind farm, SIPREÓLICO uses four types of input: the characteristics of the wind farm; historical records of incoming wind and output power, to arrive at a real power curve; online measurements of power output; and meteorological predictions provided by HIRLAM. The algorithms that SIPREÓLICO uses to generate the predictions depend on what type of input is available. Input data may be basic, additional or complete. Basic data are those that are available for every wind farm, and consist of the standard power curve and meteorological forecasts. Additional data are the basic data plus the real power curve. Complete data are the basic data plus online measurements of generated energy.

If only basic data are available, a prediction is based on the wind speed predictions and the standard power curve which aggregates the single power curves of the existing turbines. In the case of additional data, there are historical records of incoming wind and output power for a wind farm. A real wind farm power curve can be produced, and the predictions will be more accurate. In this case, wind direction forecasts as well as wind speed can be taken into account. If online measurements are available, a statistical time series analysis is performed. This method is the most accurate and it is the main part of SIPREÓLICO. Currently, more than 80 % of the wind farms connected to the grid supply online information. That means that there are accurate predictions for most of the wind farms. The performance of SIPREÓLICO in the Andalucía area, Southern Spain, is above average. This outstanding performance is attributable to the quality of the HIRLAM data for that area. In contrast, the performance of SIPREÓLICO in the region of Navarra is below average. The wind farms in Navarra are located far away from the coast and the terrain is very complex, too. This is probably the reason for the large differences between wind speeds that were measured by the wind farms' anemometers and the HIRLAM predictions. The location of the wind farms results in poor forecasts for prediction horizons that exceed 12 hours. The performance for the whole of Spain is satisfactory, though. The aggregation of individual predictions has a positive effect on the performance of SIPREÓLICO (Sánchez *et al.*, 2002)

17.3.7 Advanced Wind Power Prediction Tool

17.3.7.1 Online monitoring of wind power generation

AWPT is a part of ISET's Wind Power Management System (WPMS) that includes an online monitoring system, a short-term prediction system (1 to 8 hours ahead) and a day-ahead forecast. First, we will give a brief overview of the WPMS, starting with the online monitoring system. The most precise method for obtaining data for the generation schedule and grid balance is to monitor online the power output of each wind turbine in a certain supply area. In Germany, however, wind turbines are scattered throughout the country, and it is hardly realistic to equip all wind turbines with an online monitoring system (Ensslin *et al.*, 1999).

Online monitoring requires an evaluation model, which makes it possible to extrapolate the total power fed in by the wind turbines of a larger grid area or control area from the observed time series of power output of representative wind farms (see Figure 17.3). In cooperation with the German TSO E.ON Netz GmbH, ISET has successfully developed an online monitoring system that is able to provide data on the wind power generation of about 6 GW from all the wind turbines of the entire utility supply area (EWEA, 2000). This model transforms the measured power output from 25 representative wind farms with a total capacity of 1 GW into an aggregated wind power production for approximately 6 GW of installed wind power capacity. The selection of the sampled wind farms and the development of the transformation algorithms are based on the long-term experience of the '250 MW Wind' programme and on its extensive

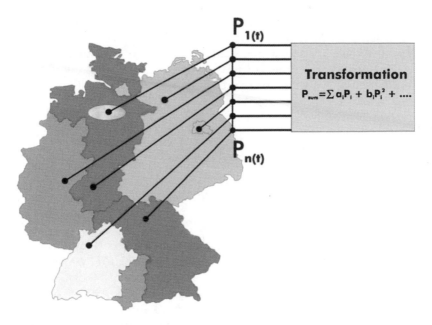

$P_{1(t)}$

Transformation

$P_{sum} = \sum a_i P_i + b_i P_i^2 + \dots.$

$P_{n(t)}$

Figure 17.3 Representative wind farms providing input data for the online model in the Wind Power Management System of the Institut für Solare Energieversorgungstechnik

measurement data and evaluation records (IRENE, 2001). The current wind power production is calculated by extensive equation systems and parameters, which consider various conditions, such as the spatial distribution of the wind turbines and environmental influences. The data collected at the selected wind farms are transmitted online to the control centre.

The online model is a basic part of ISET's WPMS, which consists of three levels. The second and third level (the day-ahead forecast and the short-term prediction) are based on ISET's wind power prediction tool AWPT.

17.3.7.2 Short-term prediction

In cooperation with E.ON Netz, Vattenfall Europe Transmission, Aktiv Technology and the German Weather Service, ISET has developed a new wind power prediction model, the AWPT. This model uses a mixture of three successful approaches:

- accurate numerical weather prediction, provided by the German Weather Service;
- the determination of the wind farm power output, based on ANNs;
- the online model, used to extrapolate the total power that is fed into the utilities' grid from the predicted power.

The meteorological component of the prediction tool is based on operational weather forecasting. For this purpose, the routine updates from the NWP model for the respective area (i.e. the Lokal-Modell of the German Weather Service) are used. The

Lokal-Modell is the latest generation model of the German Weather Service and is specifically designed for handling the typical small-scale circulation patterns in German inland areas. It provides results with a spatial resolution of $7 \times 7\,km^2$ and in an hourly sequence. The updates are calculated twice a day. The following output data of the Lokal-Modell are used for wind power prediction (see also Figure 17.4):

- wind speed at 30 m above ground;
- wind direction;
- air pressure and temperature;
- humidity;
- cloud coverage;
- temps (plot of air temprature in different altitudes), which are used for the determination of the atmospheric stability class.

For selected, representative wind farm sites, the routine forecast updates are evaluated and the corresponding power output of the wind farm is calculated by ANN. The medium points of these sites may be used as the grid points of the Lokal-Modell, for instance.

Several institutes had already examined whether and how ANNs could be used to predict the power output of wind turbines (Tande and Landberg, 1993). Unlike previous projects, ISET not only uses measurement data from the past few hours for its power predictions but also applies ANN modules to understand the relationship between the meteorological data and the wind power output, using past wind and power data. By comparing the computed results with the observed power data, the optimal configuration of the ANN modules is determined. The advantage of this approach is that it uses

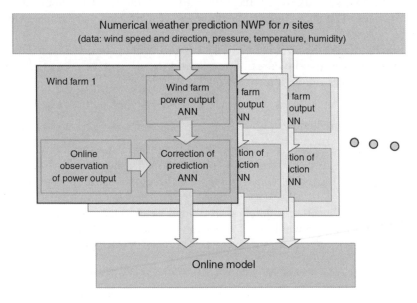

Figure   17.4 The numerical weather prediction (NWP) model. *Note*: ANN = artificial neural network

observed data to determine the coherence between meteorological data and wind power output. It is not really possible to describe sufficiently the real relationship between meteorological data and wind farm power output by physical models. Moreover, incorporation of further parameters does not require expensive modifications of the model. These trained networks compute the predicted wind power output of the representative wind farms. The predicted power output is then used as the input to the transformation algorithm of the online model. The online model requires only a limited number of locations with predicted wind speed in order to predict the total wind power generated in large utility supply areas. This tool represents the second tier of the WPMS and provides the wind power output for the control area or for selected subareas. The resolution is 1 hour and the prediction schedule is 72 hours. The NPW at the German Weather Service runs twice a day, starting at midnight and at noon and finishing at 8 a.m. and 8 p.m., respectively. The day-ahead forecast (typically used for the load management) is based on the results of the morning run of the NWP and provides the wind power generation for the entire next day, without any update. The forecast horizon is therefore between 24 hours and 48 hours. For this forecast period, the average error [root mean square error (RMSE)] between predicted and observed power is about 9.6 % of the installed capacity.

Figure 17.5 shows the frequency distribution of the forecast error for the same time period. Large errors, over 20 %, occur only 3 % of the time; 86 % of the time the forecast errors range from −10 % to +10 %.

The most frequent cause of forecast errors is attributable to the wrong timing of significant, large variations of weather situations. Wind power prediction models that are based only on operational weather forecasts are not able to correct these deviations. Therefore, another module, the third tier, uses the predicted wind farm power output, computed by the second tier in combination with measured wind farm power output of

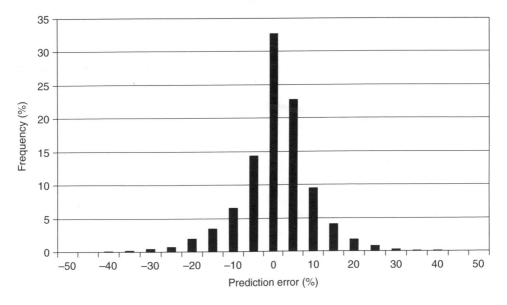

Figure   17.5 Frequency distribution of the forecast error (as a percentage of installed power) of the day-ahead forecast (prediction horizon of between 24 and 48 hours)

Table 17.1 Forecast errors RMSE and correlation between measurement and prediction

Forecast time (h)	ISET AWPT 3		Persistence	
	RMSE (%)	Correlation	RMSE (%)	Correlation
3	5.2	0.95	6.5	0.92
4	5.7	0.94	8.0	0.88
5	6.1	0.93	9.4	0.84
6	6.3	0.93	10.5	0.80

Note: AWPT = Advanced Wind Power Prediction Tool; ISET = Institüt für Solare Energieversorgungstechnik; RMSE = root mean square error.

the recent past, in order to provide topical updates and adjustments of the predictions computed by the second layer. These updates and adjustments of the predicted power output for the following 6 hours are also computed by ANN and can be carried out at any time. Table 17.1 shows the accuracy of the 3–6 hour forecasts in comparison to the persistence model.

The advantages of this model can be summarised as follows:

- the model architecture and the combination of online monitoring and prediction modelling make it universally applicable;
- it provides high precision with minimum computation time;
- it is easily adaptable to other renewable energy sources.

E.ON has been using the model since July 2001 and, in 2003, the model was installed at two other German TSOs, RWE Net and Vattenfall Europe Transmission GmbH. It is also used to calculate the horizontal wind energy exchange between the TSOs in Germany. Although already in operation, the model is continuously being improved. The next important feature is to provide a confidence interval for each prediction (Ensslin *et al.*, 2003).

17.3.8 HONEYMOON project

HONEYMOON (http://www.honeymoon-windpower.net) is being developed by University College Cork, Ireland and is funded by the European Commission. It is scheduled to be completed by 2004. Its approach differs from other wind power forecasting in that a wind power module is directly coupled to an NWP model in order to obtain the wind power from the NWP. Offshore winds and waves can be predicted, too, by coupling an atmospheric model to wave and ocean models. Offshore wind farms not only require forecasts that include the ocean model and take into account the direct impact of the sea on wind power prediction but also require predictions of waves and currents for maintenance purposes. With this coupling, customers may be provided with offshore predictions for the maintenance of offshore turbines and towers.

HONEYMOON uses ensemble predictions to supply an uncertainty estimate of the power predictions.

17.4 Conclusions and Outlook

17.4.1 Conclusions

The described methods can be roughly divided into two classes: those that apply NWP and use, as far as possible, physical equations in order to calculate the power output of wind farms, and those that use statistical methods to obtain the power output from NWP results. In addition to NWP data, the statistical systems incorporate online measurements to optimise short-time predictions of a few hours. However, the physical systems use some model output statistic (MOS) modules to correct errors systematically. In general, the results show that NWP-based forecasts are better at supplying forecasts for longer periods of time, between 6 to 48 hours, with systems providing very short-term predictions (0.5 h to 6 h) needing additional measurements to provide accurate results. The first version of WPPT, for instance, used only measurement input data. It provided satisfactory predictions for short horizons of up to 8 hours, but was not useful for longer prediction horizons. Its second version soon followed and also used NWP data as input data. This version gave much better results for longer prediction times.

Prediktor and Previento use only NWP data. The systems require measurement data to optimise the systems with MOS, but the data do not have to be collected online. This makes the data collection much easier.

Zephyr, SIPREÓLICO and AWPT use NWP and online measurements as input data. These systems combine the advantages of both approaches: the high accuracy of the NWP for a longer prediction horizon (up to 48 hours) and the advantage of online measurements for forecasts in the short-term range. However, the required online measurements cause additional costs. The measured wind farms have to be selected carefully in order to be representative of the whole area, and the equipment has to be very reliable.

For calculating the prediction error, we first need the measured wind power. These data do not have to be available online, but they have to have a high accuracy. A single wind farm is easily observed, but to compute the sum power of a large supply area usually involves some upscaling algorithm. This depends, however, on the structure of the wind farms. In Denmark and Germany, for example, there are many small wind farms as well as single wind turbines. It is therefore difficult to arrive at the sum power through measurements. In Spain and the USA, there are fewer but larger wind farms, which makes it easier to obtain real measurement data without any upscaling. It is evident that the higher the ratio between measured and total power, the higher the upscaling accuracy, on the one hand, but the more costly the data collection, on the other hand. The Zephyr system is planned to obtain measurement data from all wind farms in the area, whereas Prediktor and WPPT use only a couple of dozen large wind farms in the area. AWTP uses 25 wind farms in the E.ON area and about 100 measurements from single turbines in the '250 MW Programme'. Previento includes about 50 measurements of single turbines for confirming the prediction for the whole of Germany.

Various error functions are used to calculate the prediction error. The most common is the RMSE normalised to the installed wind power. The correlation coefficient is very useful for this purpose, too. (For a summary of the prediction systems described in this chapter, see Table 17.2).

Table 17.2 Overview of prediction systems

Name	Developer	Method		Input data		In operation by July 2003?
		Physical	Statistical	NWP	Measurement	
Prediktor	Risø National Laboratory, Denmark	✓	X	✓	X	Yes, for several wind farms (approximately 2 GW) in Denmark
Wind Power Prediction Tool	IMM, and University of Copenhagen, Denmark	X	✓	✓	✓	Yes, in Denmark
Zephyr	Risø National Laboratory and IMM, Denmark	✓	✓	✓	✓	Under development
Previento	Oldenburg University Germany	✓	X	✓	X	Yes, for several wind farms in Germany
e Wind™	True Wind Solutions, USA	✓	X	✓	X	Yes, for several wind farms (about 1 GW)
SIPREÓLICO	University Carlos III, Madrid, and Red Eléctrica de España	X	✓	✓	✓	Yes, in Spain (about 4 GW)
Advanced Wind Power Prediction Tool	ISET, Germany	X	✓ (ANN)	✓	✓	Yes, in Germany (about 12 GW)
HONEYMOON	University College Cork, Ireland	✓	X	✓	X	Under development

Note: ANN = artificial neural network; IMM = Department of Informatics and Mathematical Modelling, Technical University of Denmark; ISET = Institut für Solare Energieversorgungstechnik.

The utilities have other criteria to evaluate the quality of predictions. A too optimistic forecast of wind power, for examples, causes a lack of power in the system and the utility has to buy expensive regulation power. As opposed to this, a too pessimistic prediction is not as expensive for the utility. However, unless the predicted and the true wind power exceeds a certain amount (a couple of hundred megawatts in a TSO area) the error is negligible. The goal for the development of forecast systems is to achieve a minimum RMSE and a maximum correlation. Optimisation of the systems for the economic needs of the utilities is then a subsequent step.

All the systems first calculate the power output of single wind farms. They then apply an upscaling algorithm to obtain the sum power of the entire supply area. Regardless of the applied system, the RMSE for a single wind farm is between 10 % and 20 %. After upscaling to the sum power, the RMSE drops under 10 % because of the smoothing effects produced by adding different signals. The larger the area the better the overall prediction (for further information, see Holttinen, Nielsen and Giebel, 2002).

The systems are difficult to compare because they are applied in different areas. The Danish systems run in a mostly flat terrain, which makes obtaining an accurate NWP much easier, There are fewer slack periods, though, and the average wind power output is higher, which increases the overall error. In Germany, it is much more difficult to arrive at an accurate NWP, especially in the lower mountain areas, but periods of low wind power, especially in the summer, decrease the overall RMSE.

It is impossible to decide which of the described systems is best, unless they are compared under the same circumstances. It all depends on the user's needs and his or her abilities of providing data. Zephyr, AWPT and SIPREÓLICO are probably superior regarding the entire prediction horizon from 1 to 48 hours. However, the costs related to setting up the system are high, especially for collecting online data.

Systems using physical equations, such as Prediktor or Previento, require the exact location and environment of the wind farms they predict. They also need a long computation time to transform the geostrophic wind speed down to the hub heights of the wind turbines, with WAsP or similar programs. Statistical models, such as WPPT and AWPT, however, need time to learn the correlation between wind and power when they are being set up, but they require only a minimum of computing time. The ANEMOS project, funded by the European Commission, compares different approaches of forecasting and is developing a new system that is expected to outperform the existing systems (http://anemos.cma.fr).

All prediction systems have produced unsatisfactory forecasts in the following situations:

- during very-fast-changing weather conditions;
- in very high wind situations (storms), when wind turbines cut out for safety reasons;
- when very local weather changes (thunderstorms) cause problems to the upscaling;
- when bad measurements are produced that are not automatically detected (only for systems using online data);
- when weather service forecasts are delayed or missing;
- when meteorological forecast data are poor (very often, a time shift between forecast and real data causes a large error).

17.4.2 Outlook

The largest potential for achieving better prediction accuracy lies in an improved NWP model. A wind power model cannot be better than its weather input data. Running the NWP models repeatedly with slightly different input data (ensemble forecasting) will give the most probable result. Furthermore, the spread of the single results of the ensemble will provide an estimation of the uncertainty, which is also very helpful for users of the model.

The systems using online data can improve the NWP-based forecast, but only for a small prediction horizon of a few hours ahead. Nevertheless, a lot of research will be dedicated to these forecasts, because they are very interesting from a technical point of view. Many of the new power plants are fast adjustable, such as gas turbines or small combined heat and power plants with heat storage. With a very accurate and reliable forecast for the next few hours, these plants do not have to idle but can be switched off if there is sufficient wind. This will help to save fossil fuel and achieve maximum benefits from wind power regarding energy saving.

Another major field of future development is prediction for offshore wind farms. Although weather prediction for offshore sites seems at first sight to be easy because of the flat orography, the interaction between wind and waves and coastal effects requires further work.

In addition to short-term predictions, power plant operators have an increased interest in seasonal forecasts in order to be able to plan their fuel stock accurately.

References

[1] Bailey, B., Brower, M., Zack, J. (2000) 'Wind Forecast: Development and Applications of a Mesoscale Model', in *Wind Forecasting Techniques: 33 Meeting of Experts, Technical Report from the International Energy Agency, R&D Wind*, Ed. S.-E. Thor, FFA, Sweden, pp. 93–116.

[2] Beyer, H. G., Heinemann, D., Mellinghof, H., Mönnich, K., Waldl, H.-P. (1999) 'Forecast of Regional Power Output of Wind Turbines', presented at the 1999 European Wind Energy Conference and Exhibition, Nice, France, March 1999.

[3] Brower, M., Bailey, B., Zack, J. (2001) 'Applications and Validations of the Mesomap Wind Mapping System in Different Climatic Regimes', presented at Windpower 2001, Washington, DC, June 2001.

[4] Dispower (2002) 'Progress Report for DISPOWER', reporting period 1 January 2002 to 31 December 2002; project founded by the European Commission and the 5th (EC) RTD Framework Programme (1998–2002) within the thematic programme 'Energy, Environment and Sustainable Development', Contract ENK5-CT-2001-00522.

[5] Ensslin, C., Ernst, B., Hoppe-Kilpper, M., Kleinkauf, W., Rohrig, K. (1999) 'Online Monitoring of 1700 MW Wind Capacity in a Utility Supply Area', presented at the European Wind Power Conference and Exhibition, Nice, France, March 1999.

[6] Ensslin, C., Ernst, B., Rohrig, K., Schlögl, F. (2003) 'Online Monitoring and Prediction of Wind Power in German Transmission System Operation Centers', presented at the European Wind Power Conference and Exhibition, Madrid, Spain, 2003.

[7] EWEA (European Wind Energy Association) (2000) 'Press Release 24', January 2000.

[8] Focken, U., Lange, M., Waldl, H.-P. (2001) 'Previento – A Wind Power Prediction System with an Innovative Upscaling Algorithm', presented at the 2001 European Wind Energy Conference and Exhibition, Bella Center, Copenhagen, Denmark, 2–6 July 2001.

[9] Holttinen, H., Nielsen, T.S., Giebel, G. (2002) 'Wind Energy in the Liberalised Market – Forecast Errors in a Day Ahead Market Compared to a More Flexible Market Mechanism', presented at the 2nd Inter-

national Symposium on Distributed Generation: Power System and Market Aspects, Stockholm, 2–4 October 2002.

[10] IRENE (2001) 'Integration of Renewable Energy in the Electrical Network in 2010 – IRENE', EC funded project, ALTENER XVII/4.1030/Z/99-115.

[11] Landberg, L. (2000) 'Prediktor – an On-line Prediction System', presented at Wind Power for the 21st Century, Special Topic Conference and Exhibition, Convention Centre, Kassel, Germany, 25–27 September 2000.

[12] Landberg, L., Joensen, A., Giebel, G., Madsen, H., Nielsen, T. S. (2000a) 'Zephyr: The Short-term Prediction Models', presented at Wind Power for the 21st Century, Special Topic Conference and Exhibition, Convention Centre, Kassel, Germany, 25–27 September 2000.

[13] Landberg, L., Joensen, A., Giebel, G., Madsen, H., Nielsen, T. S. (2000b) 'Short Term Prediction Towards the 21st Century', in *Wind Forecasting Techniques: 33 Meeting of Experts, Technical Report from the International Energy Agency, R&D Wind*, Ed. S.-E. Thor, FFA, Sweden, pp. 77–85.

[14] Nielsen, T. S., Madsen, H., Tofting, T. (2000) 'WPPT: A Tool for On-line Wind Power Prediction', in *Wind Forecasting Techniques: 33 Meeting of Experts, Technical Report from the International Energy Agency, R&D Wind*, Ed. S.-E. Thor, FFA, Sweden, pp. 93–116.

[15] Sánchez, I. *et al.* (2002) 'SIPREÓLICO – A Wind Power Prediction System Based on Flexible Combination of Dynamic Models. Application to the Spanish Power System', in *First Joint Action Symposium on Wind Forecasting Techniques, International Energy Agency, R&D Wind*, Ed. S.-E. Thor, FFA, Sweden.

[16] Tande, J., Landberg, L. (1993) 'A 10 Sec. Forecast of Wind Turbine Output with Neural Networks', presented at European Wind Energy Conference 1993.

Useful websites

Anemos Project, http://anemos.cma.fr.
Danish Meteorological Institute, http://www.dmi.dk; see also http://www.dmi.dk/eng/ftu/index.html.
Deutscher Wetterdienst (German Weather Service), http://www.dwd.de.
HIRLAM, http://www.knmi.nl/hirlam; see also Danish Meteorological Institute.
HONEYMOON, http://www.honeymoon-windpower.net.
Institut für Solare Energieversorgungstechnik, http://www.iset.uni-kassel.de.
Prediktor, http://www.prediktor.dk; see also Risø National Laboratory.
Previento, http://www.previento.de.
Risø National Laboratory, http://www.risoe.dk.
Zephyr: see Risø National Laboratory.

18

Economic Aspects of Wind Power in Power Systems

Thomas Ackermann and Poul Erik Morthorst

18.1 Introduction

Wind power can reach a significant penetration level in power systems, as the examples in this part of the book have shown. To define the optimum penetration level of wind power is hardly a technical problem; it is, rather, an economic issue. Considering that the power system in most countries was designed around conventional generation sources such as coal, gas, nuclear or hydro generation (see Chapter 3), the integration of wind power may require a redesign of the power system and/or a change in power system operation. Both will affect the costs of the overall power supply.

Many of the conflicts between network owners or network operators and wind project developers worldwide are therefore often only indirectly related to technical issues being concerned rather with the question of what costs wind power causes and how those costs should be split between wind farm operator, network owner or network operator and other network customers.

In this chapter we discuss some of the relevant aspects – namely, network connection, network upgrade and system operation costs. The chapter includes examples from different countries of how costs for network connection and network upgrading can be divided. We also use empirical data to illustrate the effect of wind power on power prices and balancing costs.

Wind Power in Power Systems   Edited by T. Ackermann
© 2005 John Wiley & Sons, Ltd   ISBN: 0-470-85508-8 (HB)

18.2 Costs for Network Connection and Network Upgrading

Different countries use different approaches regarding the determination of connection costs for new wind turbines and wind farms. In some countries, the regulator sets the connection charges, independent of the actual connection costs. In most re-regulated market environments the connection costs have to represent the actual costs for the connection. However, the approaches differ when determining what should be included in the specific network connection costs. Three different approaches can be distinguished: deep, shallowish and shallow connection charges. Table 18.1 shows the network connection charging principals for renewable energy projects in the EU. The differences between the three approaches will be explained next.

18.2.1 Shallow connection charges

The shallow connection cost approach includes only the direct connection costs, that is, the costs for new service lines to an existing network point, and partially also the costs for the transformer that is needed to raise the voltage from the wind farm to the voltage in the distribution or transmission network. The details of the shallow connection approach, however, can vary significantly between countries, as can be seen from the following examples.

Example: Denmark

In the Danish approach, public authorities assign an area to a wind farm. The wind farm owner pays all network costs within this area (e.g. for the interconnection of the wind turbines). On the border of this area, the distribution network company defines a connection point (CP; see Figure 18.1). All costs from this connection point, including the costs for the connection point (transformer station), are paid by the network company, which can 'socialise' all these costs.

According to Danish regulations, network integration costs can be socialised (i.e. the costs are equally divided between all network customers). For wind power the network integration costs that can be socialised include those from the CP up to the 20 kV network. Today, new rules even allow the socialisation of network integration costs associated with wind power for a network voltage level of up to 100 kV.

In Denmark, the interconnection of a wind farm consists therefore typically of a newly built connection, usually an underground cable, between the CP and a strong

Table 18.1 Network connection charging principals in the EU

Approach	Country
Deep	Greece, Ireland, Luxembourg, Netherlands, Spain, UK
Shallowish	UK[a]
Shallow	France, Belgium, Denmark, Germany, Netherlands, Italy, Portugal, Sweden

[a] Proposed by the regulator Ofgem and expected to start 1 April 2005 (Ofgem, 2003).

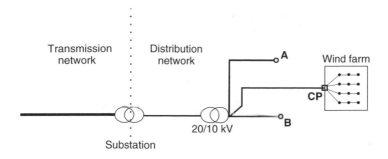

Figure   18.1 Danish connection approach. *Note*: CP = connection point

point within the distribution network, most likely an existing substation. This has the advantage that it solves a number of technical issues:

• As no other customers are connected to the new line, larger voltage variations can be allowed; flicker as well as harmonics are a minor concern.
• The comparatively long underground cable will result in a decrease of voltage variations at the strong connection point in the distribution network.
• In cases where local generation exceeds local demand, the protection system needs only to be redesigned between the CP in the distribution network and the transmission system.
• Power flow management will be required only if local generation exceeds the rating of the equipment between the CP in the distribution network and the transmission system.

Table 18.2 provides an overview of the annual network integration costs for combined heat and power (CHP) projects and wind power in the Western Danish system. The costs include the costs for building a network connection between the CP and the existing network as well as possible network upgrades within the Danish power system. Table 18.2 illustrates the following:

• Network Integration Costs per megawatt can vary significantly. For example, for wind power they can vary between DKr1.1 million per megawatt and DKr0.18 million per megawatt (€15 000–€24 000 per megawatt). This shows that network integration costs are very case-specific.
• Also, the average costs for CHP – DKr0.14 million per megawatt (€18 860 per megawatt) – are almost 50 % lower than the average costs for wind power – DKr0.26 million per megawatt (€35 000 per megawatt). The reason is that CHP is usually built close to populated areas and hence are closer to strong CPs than are wind farms, which are usually built at a larger distance from any population.

Example: Germany

In Germany, the distribution network company will define the capacity that can be connected at point A, B or at the substation (see Figure 18.2). For the calculation of the available capacity, the distribution network company takes into account voltage limits, increases in short-circuit levels and power flow issues, for instance.

Table 18.2 Annual integration cost for combined heat and power (CHP) and wind power in the Western part of Denmark (exchange rate: 1 € = 7.45 DKr)

Year	CHP			Wind power		
	MW installed	total cost per year (DKr millions)	cost per MW (DKr millions)	MW installed	total cost per year (DKr millions)	cost per MW (DKr millions)
1993	30	26	0.86	0	0	—
1994	180	58	0.32	15	2	0.13
1995	230	25	0.10	10	11	1.1
1996	250	19	0.07	70	60	0.8
1997	230	20	0.08	125	65	0.52
1998	120	3	0.02	280	52	0.18
1999	80	7	0.08	210	68	0.32
2000	40	11	0.27	210	131	0.62
2001	30	2	0.06	420	51	0.12
2002	20	19	0.95	220	76	0.34
Total	1300	190	0.14	1950	516	0.26

— Not applicable
Source: Eltra, http://www.eltra.dk.

The wind farm owner has to pay for the costs for the connection between the wind farm and the network CP, including the costs for the CP (i.e. the substation). Hence a German wind project will incur higher costs for being connected to the network than a similar Danish project. The German project will have to include these costs when calculating the overall financial feasibility of the project.

The distribution network company will mainly pay for the redesign of the protection system and the voltage control system. Only if a larger number of wind farms are connected within the same area may the distribution network company incur costs for network upgrades (e.g. between the distribution system and the transmission system). In Germany, all these costs can in principal be socialised (i.e. all wind power related

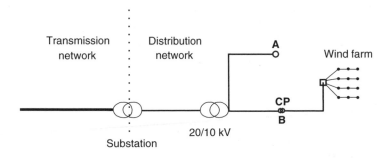

Figure 18.2 German network connection approach. *Note*: CP = connection point

network upgrade costs are shared by all network customers). As wind power in Germany is developed mainly along the coastline see Plate 1, in Chapter 2, wind power related network upgrade costs are higher in coastal areas than in other areas. Hence the German approach assures that the network related wind power costs are divided equally between all network users in Germany.

Example: Sweden

In Sweden, distribution network companies (DNCs) are required to have a license from the regulator in order to be able to operate a network within a certain geographical area. Only one license for a geographical area is available, hence if a renewable distributed generation (DG) project (e.g. example a wind farm) is under development the distribution company that has the license for this geographical area must build a network connection to this new power plant. The distribution network company can charge the new power plant with the costs for the construction of this new line (connection costs). In addition, the wind farm has to pay for using this distribution network (use-of-system charges). However, if the wind farm increases capacity-related costs for the DNC (e.g. capacity costs for the use of the transmission system) the DNC is not allowed to charge the wind farm for this. It is not uncommon for wind farms to increase capacity-related costs, because wind power is often developed in areas with low population density (i.e. in areas with good wind speeds). Hence, the locally generated power cannot be used entirely in the distribution network. It is therefore exported to the transmission network, which may cause additional costs. The disadvantage of the Swedish approach is that those additional costs must be shared by all distribution network customers of this particular distribution network company and are not covered by the wind farm operator or all Swedish network customers. Hence, DNCs have comparatively low incentives to connect larger wind farms.

18.2.2 Deep connection charges

Connection costs include the connection assets (transformer) as well as all or parts of the costs for necessary network reinforcements (i.e. network reinforcements at the transmission and distribution level).

The approach may cause the following problems:

- It is not always possible to determine exactly which costs are related to the connection of an additional generation project (i.e. a wind power project) or which costs might be caused by load increase or by ageing equipment.
- The network is usually upgraded in steps. The voltage level is, for example, increased from 20 kV to 60 kV. Hence a wind farm project might have to pay for a network upgrade that exceeds the actual requirements. Hence, any generation project that is connected later and benefits from this network reinforcement will not be charged for it; hence it uses the network for 'free' (the free-rider problem). In some cases, different wind power projects have voluntarily agreed to share the costs of a network upgrade, even though the actual rules do not require this.

In practice, deep connection charges caused significant conflicts between network owners or network operators, on the one hand, and wind farm developers, on the other hand, about the appropriate costs for network reinforcements.

18.2.3 Shallowish connection charges

Shallowish connection charges are a combination of deep charges and shallow charges, as the connection charges include a contribution to reinforcement costs based upon the proportion of increased capacity required by the connectee.

Example: UK

The English and Welsh regulator Ofgem concluded, after long investigations, that deep connection charges for wind power projects or other DG projects cannot be considered fair as new larger electricity customers that require a network upgrade when being connected to the distribution network usually pay only a shallow connection charge. (Ofgem, 2003). Hence demand and generation is treated differently, which can be considered unfair. In addition, it can also be argued that distributed and centralised generation are generally treated differently, because usually centralised generation pays only shallow connection fees (i.e. a large fossil fuel power station will not pay for network upgrade costs).

As a result of the findings, Ofgem has decided to introduce the following changes in England and Wales from 1 April 2005:

- Wind power or other DG requiring connection to a distribution network will be treated in a similar way to demand (i.e. it will no longer face deep connection charges).
- Connection charges will include a contribution to reinforcement costs based upon the proportion of increased capacity required by the connectee. This policy will apply to the connection of both demand and generation.

18.2.4 Discussion of technical network limits

Network upgrades are required because certain limits (e.g. capacity or stability limits) of the existing network are reached. It is often assumed that network limitations are of a purely technical nature. However, technical limitations are often closely related to economic aspects. For instance, distribution network operators usually have to take into account the loadability of the installed equipment when defining the possible wind power capacity that can be connected. Overloaded components can lead to voltage stability problems or to an interruption of the supply caused by the tripping of the overloaded equipment. Voltage stability problems are hardly acceptable; however, overloading of equipment may not automatically result in equipment failure. Some network components may tolerate short-term overloading, but the lifetime of the equipment will most likely be reduced when short-term overloading is acceptable.

The impact that an overloading of components has on the remaining equipment lifetime is influenced by numerous factors and is statistical in character. The exact effect of overloading on the lifetime of equipment is difficult to determine. That is the reason why most network companies do not allow any overloading during normal operation. Economically, a short overloading of some equipment may be economically efficient for the distribution company as it may be able to generate an additional income (i.e. via network tariffs) without causing any upgrade costs.

In relation to wind power, this discussion can be very important. As wind power generation as well as local demand varies, certain extreme events that define the technical limits of a network occur comparatively seldom. Typical extreme events in Europe are nights with very low local loads and extreme high wind speeds, for instance. Such events usually occur for only a few hours per year; hence the most economic solution to avoid network upgrades would either be

- to accept short-term overloading of network equipment or
- to reduce wind power production during such extreme events.

In the first case, the network operator would accept the economic risk associated with overloading equipment over a short period, because the operator would receive a higher income via network tariffs for the remaining time of the year.[1] In the second case, the wind power producer would spill some wind generation; however, in return the network operator would accept the connection of more wind power capacity to the network than it would in a situation without spilling in extreme events.[2]

In general, it seems difficult to include the above aspects into the definition of network interconnection charges. On the one hand, shallow connection charges provide little incentives for wind farm operators to consider voluntary spilling but may provide incentives for network companies to accept short-term overloading. On the other hand, shallowish charges may provide incentives for wind farm operators to consider voluntary spilling if this results in avoiding contributions to reinforcement costs.

18.2.5 Summary of network interconnection and upgrade costs

The above examples reflect very different general principals towards wind power and the related interconnection and upgrade costs. The German and Danish approach aims at simplicity and allows the socialisation of all wind power related upgrade costs. The principal idea behind the socialisation of wind power related upgrade costs is that wind power is a measure to create a sustainable energy supply for the benefit of all market participants. Hence, all market participants pay a share of the relevant costs (i.e. the network upgrade costs). The development in Germany and Denmark also shows that

---

[1] A few network operators in Europe therefore accept short-term overloading of network equipment caused by wind power. The equipment in question (i.e. transformers and overhead lines) is usually located in open terrain (i.e. the equipment is exposed to wind). Overloading is accepted because overloading is assumed to occur only during situations with high wind speeds (i.e. the overloaded equipment will be additionally cooled by the wind).
[2] Such cases are found in Europe. However, wind farm operators may encounter problems in financing such wind farms, because investors may not accept the uncertainty regarding the likelihood of spilling and hence the financial impact of such an agreement.

dividing the relevant costs between all market participants reduces interconnection barriers, which emerge when only some market participants have to cover the additional costs (as is currently the case in Sweden). The fast development of wind power in Germany and Denmark over the past decade is partly related to the fact that shallow interconnection charges have reduced interconnection barriers significantly.

The disadvantage of shallow connection charges is that there are no incentives for developing wind power projects in areas where network upgrade costs are low. Hence, with a shallow connection approach, network upgrade costs are not part of the overall economic decision of where to develop wind power projects, and that may lead to an overall inefficient economic solution.[3] The shallowish approach may help to find a more overall efficient economic solution. This will, however, depend very much on how the contribution to the reinforcement costs is actually calculated. Deep connection charges, however, are hardly justified because neither conventional power plants nor customers pay deep connection charges.

The final decision regarding which approach to implement depends on the overall policy aim (e.g. promoting the fast development of an emerging sustainable energy resource or providing incentives for the development of a sustainable energy resource under consideration of network upgrade costs). If the second approach is applied, all network customers should be treated in a similar way (i.e. conventional generation sources as well as customers should pay a contribution to the network reinforcement costs they cause).

18.3 System Operation Costs in a Deregulated Market

The intermittency of wind power affects system operation. It causes costs that are associated with balancing wind power in order to provide a continuous balance of demand and supply in a power system. The balancing task can be divided into primary and secondary control. Primary control refers to the short-term, minute-by-minute control of the power system frequency. Secondary control assumes the capacity tasks of the primary control 10 to 30 minutes later and thereby frees up capacity that is used for primary control. In the following, the relevant economic issues related to wind power and primary and secondary control are discussed (for an illustration, see also Figure 8.3, page 148).

Other system operation costs related to quality of supply other than frequency (e.g. voltage control, system restoration or restart costs following a blackout) are not discussed because it is usually very difficult to determine clearly what causes these costs. If a wind farm directly requires additional voltage control equipment, this could be considered network upgrade costs. The actual operating costs of the voltage control equipment, however, are difficult to determine and to allocate to a wind farm, for instance.

---

[3] It can, of course, be argued that a sustainable energy supply system takes into account external costs. A sustainable energy supply system therefore also aims at an overall efficient economic solution. However, shallow connection charges still provide no incentives for developing wind power projects in areas where network upgrade costs are low.

18.3.1 Primary control issues

As explained in Chapter 3, the impact of wind power on short-term (i.e. minute-by-minute) frequency variations is usually considered to be low. The sometimes significant power output variations of a single wind turbine are smoothed significantly if a large number of wind turbines are aggregated (see Figure 3.7, page 37). Hence, it is usually very difficult to allocate precisely primary control costs directly to wind power or any other network customer, as production or demand is not measured on a minute-by-minute basis.

Primary control is usually performed by frequency-sensitive equipment, mainly synchronous generators in large power plants. They will increase production when frequency decreases or will reduce production when frequency increases. If the frequency drops below a certain threshold, a special capacity that is reserved for primary control, known as primary control capacity or primary reserves, will be used.

A commonly asked question is whether wind power increases the need for primary control capacity and thus causes additional costs. Experience shows that under normal operating conditions wind power does not increase the need for primary control capacity. The maximum wind gradient per 1000 MW installed wind capacity observed in Denmark and Germany is 4.1 MW per minute increasing and 6.6 MW per minute decreasing (see also Table 9.1, page 175).[4] In the power system in Western Denmark, the primary control capacity is activated in the case of a frequency deviation of ±200 mHz, corresponding to a significant loss of production (around 3000 MW) within the UCTE (Union pour la Coordination du Transport d'Electricité) system. Based on the UCTE definition, Western Denmark requires a primary control capacity of 35 MW. This primary control capacity has not been increased over past years, despite a significant increase in the average wind power penetration over the past 10 years from almost 0 % to more than 18 % (see Chapter 10).

A need for additional primary control capacity can, however, arise if a larger number of wind turbines disconnect from the network within a comparatively short time. We can imagine two situations that can cause such an effect. The first case occurs if a storm with very high wind speeds approaches a large number of wind turbines with similar cutoff behaviour. If, in this case, the wind speed exceeds the cutoff wind speed the affected wind turbines will all shut down almost simultaneously. In Denmark and Germany, such high wind speeds occur only a few times a year and, owing to the large number of different turbines, no 'instant' shutdown of a large number of wind turbines that may require additional primary control capacity has been observed.[5] The impact might be different in other countries, for instance in locations with very high wind speeds, for example, New Zealand, where the wind speed exceeds the cutoff wind speed approximately every three to four days, and/or in areas with very large wind farms with similar cutoff behaviour. In such situations, additional primary control capacity might be required. A possible solution would be the installation of wind turbines with different

---

[4] In Germany and Denmark wind power is geographically well distributed. In the case of a large wind power project at a single location, the wind gradient can be significantly higher.
[5] The largest wind power decrease observed in the E.ON network in Germany as a result of high wind speeds was approximately 2000 MW. The disconnection, however, was not instantaneous, but occurred over approximately 5 hours.

cutoff behaviour, for instance, using a linear power reduction instead of a sudden shutdown of the wind turbine (for details, see Chapter 3). This way, the costs for possible additional primary control capacity can be avoided.

The second case occurs when a network fault causes frequency or voltage variations which lead to the disconnection of a very large number of wind turbines during times with high wind speeds (i.e. during times with high wind power production). New grid codes, however, require that wind turbines are able to stay connected to the network during certain faults (fault ride-through capability; see Chapter 7). Hence the likelihood of such situations is considerably reduced as a result of the new grid codes.

In summary, it is hardly possible clearly to determine the costs that wind power might cause regarding primary control. To reduce the likelihood of additional costs for primary control capacity caused by wind power, a certain shutoff behaviour for wind turbines can be required and/or new grid codes can be implemented. Whether this leads to the best overall economic solution depends very much on the specific case. As mentioned in Chapter 7, fault ride-through capability may increase the investment costs for wind turbines by up to 5 %. In power systems with sufficient primary control capacity, such additional investment costs may not be economically justified.

18.3.2 Treatment of system operation costs

Here, system operation costs are defined as the costs that are independent of network usage but that are essential for reliable system operation. Such costs are related to primary control and primary control capacity, black-start costs and investments in system robustness. The primary control costs are included because they can usually not be linked to any market participant. System operation costs should in principal be divided between all power system users based on their network usage, e.g. transmitted energy (kWh).

In some countries (e.g. England and Wales, and Australia) wind power is currently not required to contribute to these costs. Similar to the case of network upgrade costs, it can be argued that wind power is a measure to create a sustainable energy supply from which all market participants gain. Hence, these costs should be covered by market participants, excluding wind power generation.

The English and Welsh regulator Ofgem, however, concluded that such treatment may be considered unfair. Hence Ofgem is currently considering the implementation of a new tariff system that requires wind farms to contribute to the system operation costs (Ofgem 2003).

18.3.3 Secondary control issues

If there is an imbalance between generation and consumption, the primary control will jump into action to reduce the imbalance. This means that primary control reserves will be partly used up after the imbalance is restored and/or that the primary control will reduce the imbalance but the restored frequency will deviate from 50 Hz or 60 Hz. Secondary control is then used to solve this problem by freeing up capacity used for primary control and moving this required capacity under a secondary control regime (for an illustration, see Figure 8.3, page 148). Secondary control capacity should usually be available within 15 minutes notice. Hence, any difference between forecasted wind power production and actual wind power production is essentially allocated to secondary control capacity.

The figures in Table 9.1 (page 175) may be used to illustrate this. The maximum wind gradient per 1000 MW installed wind capacity observed in Denmark and Germany was 6.6 MW per minute decreasing. In such an extreme event the primary control must provide 6.6 MW per minute of additional capacity per 1000 MW of wind power to keep the defined frequency. This is usually not a problem and often amounts to less than load variations. However, if the wind power production decreases with this maximum gradient for 15 minutes, the requirements for primary control add up to almost 100 MW (15 min ×6.6 MW per min) per 1000 MW of installed wind power capacity. Secondary control will then start to free up the capacity used for primary control, which might be required if wind power production continues to decrease.

The demand on secondary control depends on the wind power production forecast error (i.e. the difference between the forecasted wind power production and the actual wind power production). The wind power production forecast error depends partly on the time span between the wind power forecast and actual production. That means that a forecast for wind power production 5 minutes prior to production will be more precise than a forecast 1 hour prior to production. In deregulated markets, the closing time of the power exchange defines the time of the last wind power production forecast that is possible (i.e. the closing time influences the forecast error and therefore the wind power related costs for secondary control).

Table 18.3 provides an overview of market closing times in different electricity markets. It should be noted that the mainly hydro based market power exchange (PX) Elspot in Scandinavia has a market closing time of 12 to 36 hours before actual delivery, despite the fact that a significant share of power generation is based on hydro powers which does not need such a long planning horizon or startup time. The Australian market, with a power generation mix that is based mainly on coal power, which can be considered rather inflexible in terms of startup times, allows rebidding until 5 minutes prior to actual delivery.

Table 18.3 Market closing times in various electricity markets

Market	Closing time
England and Wales	1 hour before the half-hour in question
Nord Pool Elspot (power exchange)	12:00 p.m. before the day in question; no changes possible after 12:00 p.m.
Nord Pool Elbas	1 hour before the hour in question; no changes possible after this
Australia power exchange	Rebidding possible until the resources are used for dispatch (i.e. up to 5 minutes before the time in question)
New Zealand power exchange	2 hours before the hour in question
PJM Market day-ahead market	12:00 p.m. before the day in question; no changes possible after 12:00 p.m.

The Elbas market can be seen as a supplement to Nord Pool's Elspot. Owing to the lengthy time span of up to 36 hours between Elspot price fixing and delivery, participants can use Elbas in the intervening hours to improve their balance of physical contracts. Elbas offers continuous trading up to 1 hour before delivery; however, the Elbas market can be used only by Swedish and Finish market participants and not by participants from Denmark.

The problem of long market closing times became particularly evident with the introduction of the New Electricity Trading Arrangement (NETA) in England and Wales in March 2001. Within NETA, wind power generators are forced to pay the imbalance costs between predicted wind power production and actual production. NETA first had a market closing time of 3.5 hours before the time of delivery. Hence, wind power generators had to pay for any imbalance between the wind power production that was predicted 3.5 hours prior to delivery and actual delivery. As a result, revenue for wind power generators fell by around 33 % (Massy, 2004). The market closing time was then reduced from 3.5 hours to 1 hour, which resulted in a significantly lower exposure to additional imbalance costs for wind power generators. Discussions with wind farm operators that operate several wind farms in England and Wales revealed that, now, wind power generators usually bid the actual aggregated wind power production measured 1 hour before the time of delivery into the PX. For large wind power generators, with a number of wind farms distributed over the country, the aggregation of the wind farms significantly reduces the possible imbalance within one hour. Hence, wind farm operators with a large number of wind farms consider the imbalance costs to be acceptable.

For wind power generators with a single wind farm the issue is different, as the geographical smoothing effect between different locations is smaller and hence variations between forecasted production and actual production can still be significant. It can, of course, be argued that those who cause the imbalance should also pay for keeping the balance. However, this will not result in an overall economically optimum solution. As mentioned before, power fluctuations from wind generation depend on the number of wind turbines. For wind turbines distributed over a large geographic area, fluctuations are reduced by $1/\sqrt{n}$, where n is the number of wind turbines. If all wind turbines in one power system are owned by the same market participant, the sum of all imbalance costs would be approximately $1/\sqrt{n}$ lower than for a large number of wind turbines being owned by different, independent owners where each owner pays the imbalance costs for each turbine. Hence, if all wind turbine owners pay individually, they pay together for a significantly higher total imbalance than they cause (the details depending very much on how the imbalance costs are determined). Therefore, advocates of wind power argue that a power system was built to aggregate generation and demand whereas individual imbalance pricing is reversing this by disaggregating generation in the name of competition.

A possible solution to this problem is to consider all independent wind farms as one generation source and divide the total imbalance costs caused by this aggregated wind farm between the different wind farm owners.

In summary, the approach for wind power imbalance pricing should consider the flexibility in a power system (i.e. the time required to start up or adjust generation sources). This time will vary according to the power generation mix in each power system as well as on the wind power penetration. Hence, the wind power imbalance costs depend very much on the overall design of the electricity market and on the approach to power system operation.

18.3.4 Electricity market aspects

Power system integration costs are usually discussed in relation to integration costs in the power system but seldom in combination with integration costs and benefits to the electricity market.

An important instrument of the electricity market is the PX. The PX defines a single market clearing price for each trading period. This single market price is the intersection where the marginal costs of the producer equals the price the last consumer is willing to pay (the equilibrium price). If there is a perfect competitive market, every power generator would bid with its short-run production costs, also known as marginal costs. These marginal costs usually represent the fuel costs for keeping the power unit(s) in operation for another time period. Wind power, however, has no fuel costs. Hence, wind power will always bid into the market with very low or zero costs to be able to operate whenever the fuel source, 'wind', is available. This can have the following two implications for the PX:

- Independent wind power generators may reduce the incentives for large generation companies to exercise market power; hence wind power may increase the overall economic efficiency of power markets (this conjecture is substantiated in Ackermann, 2004).
- If wind power production dominates the electricity market, it may be difficult for the market to reach an equilibrium between demand and supply. Hence the system operator may not be able to match demand and supply, which is required for secure operation of the power system. This aspect is discussed in more detail in the next section.

18.4 Example: Nord Pool

In the following, the impact of wind power in Denmark on the PX in Scandinavia, Nord Pool's Elspot, and on the balancing market, which is a part of the secondary control, is discussed in more detail.

The Nord Pool power exchange is geographically bound to Norway, Sweden, Finland and Denmark. The market was established in 1991 and, until the end of 1995, the electricity exchange covered Norway only. In 1996, Sweden joined the exchange and the name was changed to Nord Pool. In 1998, Finland was included, and in 1999–2000 Denmark joined.[6] The market is at present dominated by Norwegian and Swedish hydro power, though power trade with neighbouring German markets increases and thereby reduces the dominance of hydro power. Because Denmark is situated between

---

[6] Within Nord Pool, Denmark is separated into two independent parts covering the Western Denmark area (the Jutland–Funen, or Eltra, area) and the Eastern Denmark area (the Zealand, or Elkraft, area), including the small neighbouring islands. These two parts do not have a direct electrical connection and therefore constitute two specific pricing areas in the market.

the large conventional fossil fuel based power systems of central Europe (especially Germany) and the hydro-dominated Nordic system, Denmark is a kind of buffer between these systems. This implies that the power price in Denmark relates partly to the Nordic market and partly to the German market, depending on the situation in these markets.

18.4.1 The Nord Pool power exchange

The Nord Pool PX Elspot is a daily market in which the price of power is determined by supply and demand. Power producers and consumers submit their bids to the market 12–36 hours in advance, stating quantities of electricity supplied or demanded and the corresponding price. Then, for each hour, the price that clears the market (equalises supply with demand) is determined at the Nord Pool PX. In principle, all power producers and consumers are allowed to trade at the exchange but, in reality, only big consumers (distribution and trading companies and large industries) and generators act on the market, with the small ones forming trading cooperatives (as is the case for wind turbines) or commissioning larger traders to act on their behalf. A minor part of total electricity production is actually traded at the spot market. The majority is sold on long-term contracts, but the spot prices have a considerable impact on prices in these contracts.

Figure 18.3 shows a typical example of an annual supply and demand curve for the Nordic power system. As shown, the bids from hydro and wind power enter the supply curve at the lowest level owing to their low marginal costs, followed by combined heat and power (CHP) plants; condensing plants have the highest marginal costs of power

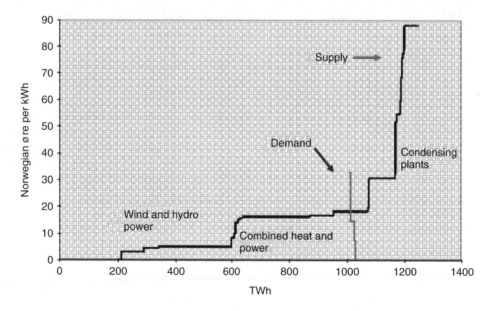

Figure   18.3 Supply and demand curve for the Nord Pool power exchange

production. In general, the demand for power is highly inelastic, with mainly Norwegian and Swedish electroboilers and power-intensive industry contributing to price elasticity in power demand.

If the trade of power can flow freely in the Nordic area (i.e. if there is no congestion of transmission lines between the areas) there will be only one price in the market, but if the required power trade cannot be handled physically because of transmission constraints, the market will be split into a number of submarkets. The pricing areas define these submarkets; for example, Denmark is split into two pricing areas (Jutland–Funen and Zealand). Thus, if the Jutland–Funen area produces more power than consumption and transmission capacity covers, this area would constitute a submarket, where supply and demand would be equalised at a lower price than that at the general Nord Pool market.

18.4.2 Elspot pricing

In a system dominated by hydropower, the spot price is heavily influenced by the precipitation in the area. This is shown in Figure 18.4, where the price ranges from a maximum of NKr350 per MWh (approximately €42 per MWh) to almost zero. Thus, there were very wet periods during 1992 and 1993 and, similarly, although to a lesser extent, in 1995 and 1998. Dry periods dominated in 1994 and at the beginning of 1995. In 1996, there was a long period with precipitation significantly below the expected level.

Looking more closely at the last two years, prices have been fairly stable (see Figure 18.5), with the exception of the end of 2002 and early 2003. The average price in 2001 and 2002 was approximately €25 per MWh, ranging from €15 per MWh to €100 per MWh. But in 2002 the autumn was very dry in both Norway and Sweden, and this heavily affected the prices in October, November and December 2002 as well as in January 2003. Thus, excluding the last three months of 2002 from the considered period,

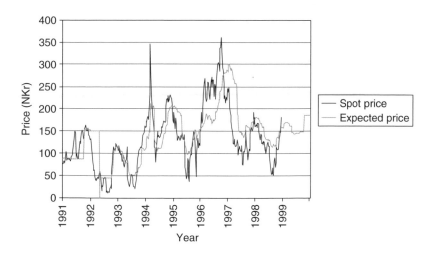

Figure **18.4** Price of power in the Nordic market for the period 1991–1999, in Norwegian krone (NKr). *Note*: the labels on the year axis represent January of the given year

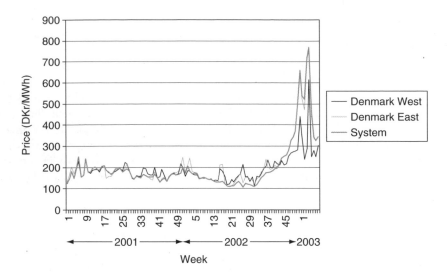

Figure 18.5 Power prices at the Nord Pool market in 2001, 2002 and early 2003: weekly average (exchange rate 1 € = DKr7.45)
Source: Nord Pool, 2003

the price ranged from €15 per MWh to €33 per MWh. But the draught in Norway and Sweden meant that prices rose to a high of more than €100 per MWh, and power prices in the Nordic market did not return to normality before Spring 2003 owing to the importance of hydro power. But high prices are only part of the abnormality in the present power market.

As shown in Figure 18.5, the system price and the two area prices for Denmark West and East, respectively, are normally closely related, implying that congestion in the transmission system seldom has a major impact upon price determination. But this close relationship seemed to have vanished by the end of 2002 and early 2003, when prices in the Western Danish area differed quite considerably from system prices, mainly owing to the high level of wind power penetration in this area. This phenomenon will be investigated in more detail below.

18.4.3 Wind power and the power exchange

In relation to the power system, wind power has two main characteristics that significantly influence the functioning of wind power in the system:

- Wind power is an intermittent energy source that is not easy to predict. The daily and weekly variations are significant, which introduces high uncertainty in the availability of wind-generated power even within relatively short time horizons.
- Wind power has high up-front costs (investment costs) and fairly low variable costs. Because part of the variable costs consists of annual fixed expenses, such as insurance and regular maintenance, the marginal running costs are seen to be even lower.

Bearing these two main characteristics in mind, a number of questions arise when wind power is introduced into a liberalised market:

- How much wind power can be introduced into a power exchange without jeopardising the overall functioning of the power exchange and the system operation?
- How will wind power influence the price at the power spot market in the short and long term?
- What is the need for secondary control from wind plants in relation to the time from gate closure to real-time dispatch?
- What is the cost of wind power when it fails to fulfil its bid to the market (i.e. the cost of regulating wind production in the system)?

We will try to answer these questions in what follows by using the experience gained in Denmark, especially in relation to the power system of Western Denmark. This area has been chosen because it has a number of specific characteristics, some of them related to wind power:

- It has a very large share of wind-produced energy – in 2002, around 18 % of total power consumption was covered by wind power (see also Chapter 10). Presently, most of wind-generated power is covered by prioritised dispatch, hence the system operator must accept wind power production at any time.
- It has a large share of decentralised CHP, which is paid according to a three-level tariff and is also covered by prioritised dispatch.
- It lies at the border between the fossil fuel based German power system and the hydro power dominated Nordic system and thus is heavily affected by both areas.

In 2002, there were around 2300 MW of wind power in the Western Denmark power system and thus wind power has a significant influence on power generation and prices.

18.4.3.1 The share of wind power in the power exchange

How much wind power can be introduced into a power exchange without jeopardising the overall functioning of the power exchange and the system operation? In the following, this question will be further investigated using a small case study carried out on the Western part of Denmark.

Figure 18.6 illustrates for December 2002 the share that wind-generated electricity had in relation to total power consumption in the Jutland–Funen area. In total, 33 % of domestic electricity consumption in this area was supplied by wind power in that month.

As shown in Figure 18.6, in December 2002 the share of wind-generated electricity in relation to total power consumption in the Jutland–Funen area reached almost 100 % at certain points in time. That means that in this area all power consumption at that time could be supplied by wind power. A large part of the power generated by wind turbines is still covered by priority dispatch in Denmark, and this is also the case for power produced by decentralised CHP plants. This implies that these producers do not react to the price signals from the spot market – wind producers under priority dispatch are paid

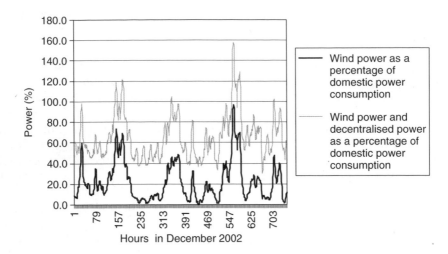

Figure 18.6 Wind-generated power and decentralised power as percentage of total power consumption on an hourly basis in December 2002 for the Jutland–Funen area of Denmark

the feed-in tariff for everything they produce, whereas decentralised CHP plants are paid according to a three-level tariff, highest during the day and lowest at night. Thus, CHP plants will only produce at the low tariff if there is a need to fill up the heat storage.

In December 2002, total prioritised production exceeded domestic power demand during several hours, requiring power to be exported and thus adding to the problem of congestion of transmission lines.

The consequences are clearly shown in Figure 18.7, which depicts deviations between the Nord Pool system price and the realised price in Western Denmark. The Western Denmark price is significantly lower than the system price for a large number of hours. Note that the power price in Western Denmark becomes lower than the system price if transmission lines from the area are utilised entirely for exporting power but supply still exceeds demand, forcing conventional power plants to reduce their load. Thus, during such situations, the Western Denmark power system is economically decoupled from the rest of the Nordic power market and constitutes a separate pricing area, where it has to manage on its own. The conventional power plants have to reduce their production until equilibrium between demand and supply is reached within the area. The consequence is that the price of power is also reduced. If power production cannot be reduced as much as required, a system failure or even breakdown might result. That the power price in the Western Denmark area becomes zero during a number of hours is a clear indication of the pressure put upon the power system.

Nevertheless, although the power system in Western Denmark sometimes has been under pressure, there have been no system failures due to too much wind power in the system. Thus it has been possible for the system operator to handle these large amounts of wind and decentralised CHP. In the future, the prioritised status of decentralised CHP plants is expected to be changed, moving these to act on the power market as other conventional power plants. In that case, an even higher share of wind power might be accepted in Western Denmark.

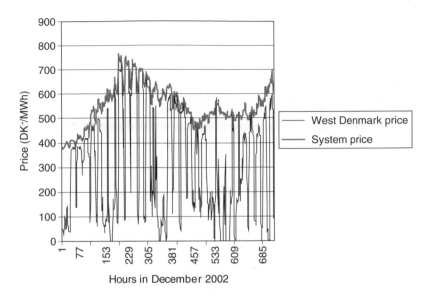

Figure   **18.7** Deviations of the price of power in Western Denmark from the system price of Nord Pool

18.4.3.2 The impact of large amounts of wind power on the power price

Wind power is expected to influence prices in the power market in two ways. First, wind power normally has a low marginal cost and therefore enters close to the bottom of the supply curve. This, in turn, shifts the supply curve to the right, resulting in a lower system power price, depending on the price elasticity of the power demand. If there is no congestion in the transmission of power, the system price of power is expected to be lower during periods with high winds compared with periods with low winds.

Second, as mentioned above, there may be congestion in power transmission, especially during periods with high wind power generation. Thus, if the available transmission capacity cannot cope with the required power export, the supply area is separated from the rest of the power market and constitutes its own pricing area. With an excess supply of power in this area, conventional power plants have to reduce their production, because wind power normally will not limit its power production. In most cases, this will lead to a lower power price in this submarket.

Figure 18.8 shows how the large capacity of wind power in the Western Denmark area affects the power system price. Five levels of wind power production and the corresponding system power prices are depicted for each hour of the day during periods without any congestion of transmission lines. As shown, the greater the wind power production, the lower the system power price. At very high levels of wind power production, the system price is reduced significantly during the day but increases at night. This phenomenon is difficult to explain. It could be a consequence of spot market bidders expecting high nocturnal levels of wind power. Nevertheless, there is a significant impact on the system price, which might increase in the long term if even larger shares of wind power are fed into the system.

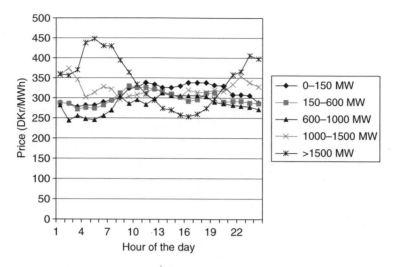

Figure   18.8 The impact of wind power on the spot power system price

The second hypothesis concerns power prices in cases where transmission line capacity is completely utilised. As shown in Figure 18.6, during December 2002 the share of wind-generated electricity in relation to total power consumption for the Jutland–Funen area was close to 100 % at certain points in time. This means that at that time all power consumption could be supplied by wind power in this area. If the prioritised production from decentralised CHP plants is added on top of wind power production, there are several periods with an excess supply of power. Part of this excess supply might be exported. However, when transmission lines are completely utilised, we have a congestion problem. In that case, equilibrium between demand and supply has to be reached within the specific power area, requiring conventional producers to reduce their production, if possible. The consequence to the market is illustrated in Figure 18.9. Again, five levels of wind power production and the corresponding power prices for the area are depicted for each hour of the day during periods with congestion of transmission lines to neighbouring power areas. As shown in Figure 18.9, there is a significant correlation between wind production and power price. Thus, the greater the wind power production, the lower the power price in the area.

The size of the impact of wind power on the power price at the spot market will depend heavily on the amount of wind power produced and the size and interconnections of the power market. Experience in Denmark shows the following:

- Even within the large Nordic power system, wind power has a small but clearly negative impact on the power price. The more wind power that is supplied the lower the power system price.
- When Western Denmark is economically separated from the rest of the power market because of congestions of transmission lines, wind power has a strong and clearly negative impact on power prices, both during the day and during the night.

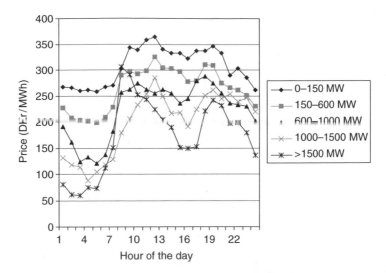

Figure 18.9 The impact of wind power on the spot power price of Western Denmark, when there are congestions in the power system between countries

18.4.4 Wind power and the balancing market

Currently, the transmission system operator in Western Denmark, Eltra, controls a secondary control capacity of 550 MW upwards and 300 MW downwards. The secondary control capacity is purchased via a tender process. According to Eltra, the capacity demand for secondary control as a result of wind power has increased significantly. An exact analysis of the additional secondary control requirements is difficult, because Eltra can also buy secondary control capacity on the balance market in Sweden and Norway.

The balance market has similar requirements on generation as secondary control (i.e. generation must be up and running 15 minutes after an order). In general, the cheapest alternative for manual secondary control is used; however, in the case of network congestion the local contracts must be used. The balance market in Sweden and Norway can be considered competitive, which is not always the case with similar balance markets.[7] The impact of wind power in Denmark on the Swedish–Norwegian balance market will therefore be discussed next in more detail.

18.4.4.1 The balancing market

If forecasted production and actual demand are not in balance, the regulating market has to step in. This is especially important for wind-based power producers. The

---

[7] If the power plants that are technically capable of participating in balance services are in the hands of a small number of market participants (i.e. former monopolists) these market participants may exercise market power to influence the prices for balancing services. Often, these higher prices are used to determine the costs of balancing wind power production, even though the determination of the costs is based on an inefficient market.

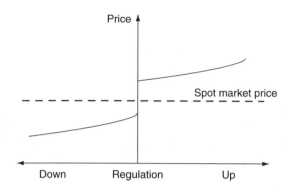

Figure   18.10 The functioning of the balancing market

producers that take part in the balancing market submit their bids to the balancing market 1–2 hours before the actual production hour, and power production from the bidding actors should be available within a notice of 15 minutes. For that reason, only fast-responding power capacity will normally be able to supply regulating power.

Usually, wind power production can be predicted 12–36 hours in advance to only a limited extent. Thus it will be necessary to pay a premium for the difference. Figure 18.10 shows how the regulating market works. If the power production from the wind turbines exceeds the initial production forecast, other producers will have to regulate down. In this case, the price that the wind power producer gets for the produced excess electricity is lower than the spot market price. If wind power production is lower than the initial forecast, other producers will have to regulate up to secure the power supply. These other producers will get a price above the spot market price for the extra electricity produced – an additional cost, that has to be borne by the wind power producer. The larger the difference between forecasted and actual wind power production, the higher the expected premium, as shown in Figure 18.10 by the difference between the regulatory curves and the stipulated spot market price.

Until the end of 2002, each country participating in the Nord Pool market had its own balancing market. Thus, the above approach is fairly close to the Norwegian way of operating the balancing market. In Denmark, balancing was handled by agreements with the largest power producers. In addition, the transmission system operator (TSO) had the option to purchase balancing power from abroad if the domestic producers were too expensive or not able to produce the required volume of regulatory power. A common Nordic regulatory market was established at the beginning of 2003.

When bids have to be submitted to the spot market 12–36 hours in advance it will not be possible for wind producers to generate the forecasted amount at all times. If other power producers are to regulate and compensate for the unfulfilled part, this causes costs, and with wind power gradually entering the spot market, in the future this cost will have to be borne by the owners of wind power plants. Until the end of 2002, the TSOs were obliged to handle the regulation in Denmark, wind power being prioritised dispatch. Since the beginning of 2003 the owners of wind turbines supplying the spot market have been financially responsible for balancing the power themselves. They can either continue to let the TSOs handle the balancing and pay the associated costs or they can contract private companies to do this work. In Denmark, some wind turbine owners have formed a cooperative for handling the spot market trading and balancing of their turbines. At the time of writing, the majority of owners in the spot market are in this cooperative.

18.4.4.2 The need for balancing of wind power

When wind power cannot produce according to the production forecasts submitted to the power market, other producers have to increase or reduce their power production in order to make sure that demand and supply of power are equal (balancing). But other actors on the spot market also might require balancing power, owing to changes in demand, shutdowns of power plants and so on. Now, the Danish TSOs have joined the common Nordic balancing market, but, until 2003, most, if not all, of the balancing was performed within the separate TSO areas.[8]

The capacities shown in Figure 18.11 are related to all regulation (i.e. not only to regulation in connection with wind power failing to produce at the forecasted production level).[9] Nevertheless, although not very significant, there is a clear tendency that the more wind power produced, the higher the need for down-regulation. Correspondingly, the less wind power produced, the higher the need for up-regulation. Note that Figure 18.11 shows that forecasts for wind power production tend to be too low when

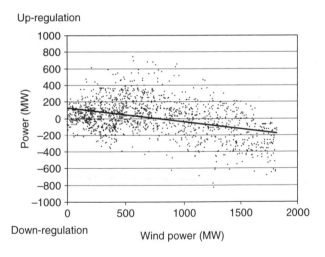

Figure   18.11 Regression analysis of down-regulation or up-regulation against the amount of wind power for the Jutland–Funen area: hourly basis for January–February 2002

---

[8] In the Nordic region, Norway, Sweden and Finland each have their own TSO. Because there is no interconnection over the Great Belt, Denmark is divided into two TSO areas, similar to the pricing areas in Denmark.
[9] From the available data, it is not possible to discern the specific cases when wind power failed to produce according to the forecasts.

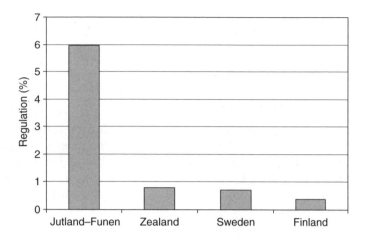

Figure   18.12 The need for regulation depending on the amount of wind power in the power system. *Note*: regulation is shown as a percentage of consumption

large amounts of wind power are produced and tend to be too high when only small amounts of wind-generated power are fed into the system.

Figure 18.12 compares power areas with different capacities of wind power and shows that wind power strongly increases the need for regulation. Note that at the Nordic power market the bidding for the spot market is carried out 12–36 hours in advance, which is one of the reasons why wind power often requires regulatory power.

Sweden and Finland comprise large areas and have very little wind power capacity. Zealand (the Eastern part of Denmark) has approximately 10 % of wind-generated power in relation to total domestic power consumption, whereas Jutland–Funen (the Western part of Denmark), as mentioned, has a coverage of more than 20 % of total power consumption. Figure 18.12 clearly illustrates the consequences for the regulation of power. In the Western Denmark area, regulation as a percentage of consumption is more than six times higher than in the other areas.

In general, more wind power in the power system should be expected to increase the need for regulation. However, the closer the time of gate closure to the actual time of dispatch, the smaller the divergence between actual wind power production and the submitted production bids should be.

18.4.4.3 The cost for balancing of wind power in the power market

In the Nordic power market, wind turbine owners who produce more than their initial production forecast will receive the spot price for all their production. However, they will have to pay a premium because other power plants have to regulate down as a result of the production exceeding the forecast. If the owners produce less than their bid, they will correspondingly have to pay a premium for the part that other generators have to produce in up-regulation. Figure 18.13 shows the costs of regulation in the Jutland–Funen area of Denmark on an hourly basis for January and February 2002.

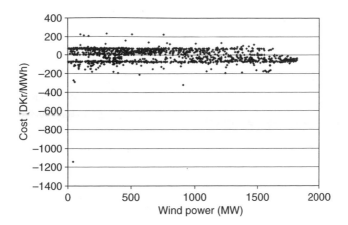

Figure 18.13 The cost of regulation in the Jutland–Funen area: hourly basis for January–February 2002

The picture shows a clear 'band' of costs, both for up-regulation and for down-regulation, almost independent of how much wind power is generated during the hour in question. Even though the need for regulation increases with higher quantities of wind power production, the regulating costs appear to be almost independent of the level of required regulation. During January–February 2002, the average cost of regulating up reached €8 per MWh regulated, and the corresponding cost of regulating down amounted to €6 per MWh.

Figure 18.14 shows the regulation costs for the whole of 2002, calculated as monthly averages. As the figure shows, the cost of up-regulation is constantly above the cost of down-regulation, probably because the marginal cost of up-regulation is higher than for power producers regulating down. Moreover, as expected, the cost of regulation – again, especially up-regulation – increases with the general level of the spot price, which

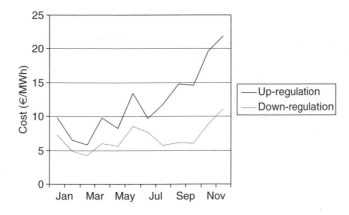

Figure 18.14 The cost of regulation, calculated as monthly averages for the year 2002 for the Jutland–Funen area

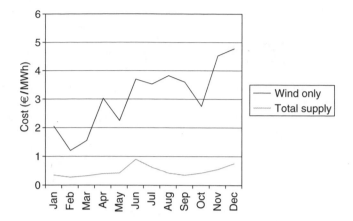

Figure   18.15 Regulation costs calculated as monthly averages for the Jutland–Funen area for 2002: costs incurred either by wind power only or in relation to the total power supply

increases substantially towards the end of 2002.[10] In 2002, the average up-regulation cost reached €12 per MWh regulated, and the cost of down-regulation amounted to €7 per MWh regulated.

As already mentioned, the regulated quantities relate not only to wind power but also to the total system, including nonfulfilment of bids from demand and conventional power producers as well. The estimate constitutes the upper limit. In Figure 18.15 we relate the 2002 monthly regulation costs for the Western part of Denmark to wind power only. Finally, Figure 18.15 gives the corresponding costs related to the total power supply.

Figure 18.15 shows that regulation costs per megawatt-hour borne by wind power only are lowest during periods with plenty of wind-generated power (i.e. during Winter and Spring 2002), and higher during the summer, when less wind power is produced. However, the high spot prices of Autumn and Winter 2002 are an exception. For the year 2002, the average regulation cost if borne by wind power amounts only to €3 per MWh.[11] As mentioned above, these estimates constitute an upper limit of the regulation costs for wind energy, because the regulated quantities do not relate purely to wind power. If the regulation costs are distributed across the total power supply, the cost per megawatt-hour is, of course, much lower and, if calculated as an average for 2002, the cost amounts to €0.5 per MWh.

18.5 Conclusions

Power systems in most countries were designed around conventional generation sources such as coal, gas, nuclear or hydro generation. Nevertheless, wind power can reach a

---

[10] At the end of 2002, there was a draught in Norway and Sweden and the power system prices reached extremely high levels.
[11] It is not known whether 2002 is a representative year for regulating costs.

significant penetration level in these power systems, although the integration of wind power may require a redesign of the power system and/or a change in power system operation, which might affect the cost of the overall power supply. In general, the optimum penetration level of wind power is hardly a technical problem; it is more of an economic issue.

The costs are related to interconnection and network upgrade costs as well as the increasing requirements for power system balancing (i.e. secondary control). Different approaches for the allocation of interconnection costs and upgrade costs can be used, depending on the overall goal. It is important to consider that certain interconnection charges (e.g. deep connection charges) may generate significant barriers for the deployment of wind energy.

In general, the costs of integrating wind energy into the power system depends on the amount of wind power in relation to the overall power market and the design of the power exchange, which can also significantly influence the requirements for secondary control.

Experience from the Western Denmark power system where wind power capacity corresponds to approximately 20 % of power consumption results in the following conclusions:

- Although wind-generated power exceeded domestic demand for electricity during several hours, there were no system failures as a result of too much wind power in the system. This means that the system operator was able to handle these large amounts of wind and decentralised CHP. If CHP plants were not treated as prioritised dispatch then even more wind power might be integrated into the system.
- Even within the large Nordic power system, wind power has a small but clearly negative impact on the power price. The more wind power that is supplied the lower the power system price.
- If Western Denmark is economically separated from the rest of the power market because of the congestion of transmission lines, wind power has a strong and significantly negative impact on power prices, both during the day and during the night.
- Although not very significant, there is a clear tendency for an increase in wind power production to result in a larger need for down-regulation. Correspondingly, the less wind power produced, the larger the need for up-regulation.
- For 2002, the average cost of up-regulation reached €12 per MWh regulated, and the cost of down-regulation amounted to €7 per MWh regulated. In general, the cost of up-regulation is higher than the cost of down-regulation. For the year 2002, the average cost of regulation if born by wind power amounts only to €3 per MWh for wind-generated power (upper limit). If the costs are distributed across the total power supply, the regulation costs are much lower, the average cost for 2002 amounting to €0.5 per MWh.

References

[1] Ackermann, T. (2004) *Distributed Resources in a Re-regulated Market*, PhD thesis, Department of Electrical Engineering, Royal Institute of Technology, Stockholm, Sweden.
[2] Massy, J. (2004) 'A Dire Start that Got Gradually Better', *Windpower Monthly* (February 2004) 46.
[3] Nord Pool (2003). http://www.nordpool.com.

[4] Ofgem (Office of Gas an Electricity Markets) (2003) Structure of Electricity Distribution Charges – Initial Decision Document, November 2003, available at http://www.ofgem.gov.uk/.

[5] Watson, J. (2002) 'The Regulation of UK Distribution Networks: Pathways to Reform', in *Proceedings of the Second International Symposium on Distributed Generation: Power System and Market Aspects*, Ed. T. Ackermann, Royal Institute of Technology, Stockholm, Sweden.

Part C
Future Concepts

19

Wind Power and Voltage Control

J. G. Slootweg, S. W. H. de Haan, H. Polinder and W. L. Kling

19.1 Introduction

The main function of an electrical power system is to transport electrical power from the generators to the loads. In order to function properly, it is essential that the voltage is kept close to the nominal value, in the entire power system. Traditionally, this is achieved differently for transmission networks and for distribution grids. In transmission networks, the large-scale centralised power plants keep the node voltages within the allowed deviation from their nominal value and the number of dedicated voltage control devices is limited. These power plants use the energy that is released by burning fossil fuels or by nuclear fission for generating electrical power. The power they generate makes up the largest part of the consumed electricity. Distribution grids, in contrast, incorporate dedicated equipment for voltage control and the generators connected to the distribution grid are hardly, if at all, involved in controlling the node voltages. The most frequently used voltage control devices in distribution grids are tap changers – transformers that can change their turns ratio – but switched capacitors or reactors are also applied. By using large-scale power plants to regulate voltages in the transmission network and by using dedicated devices in distribution grids to regulate the voltages at the distribution level, a well-designed, traditional power system can keep the voltage at all nodes within the allowed band width.

However, a number of recent developments challenge this traditional approach. One of these is the increased use of wind turbines for generating electricity. Until now, most wind turbines have been erected as single plants or in small groups and are connected to distribution grids. They affect the currents, or power flows, in the distribution grid to which they are connected, and, because node voltages are strongly related to power flows, they also change node voltages. This can lead to problems if the devices installed in the distribution

Wind Power in Power Systems Edited by T. Ackermann
© 2005 John Wiley & Sons, Ltd   ISBN: 0-470-85508-8 (HB)

grid cannot compensate the impact of the wind turbines on the node voltages. In this case, the voltages at some nodes within the distribution grid cannot be kept within the allowed deviation from their nominal value, and appropriate measures have to be taken.

A similar problem arises when large-scale wind farms are connected to the transmission network. Again, the wind farm affects the power flows and hence the node voltages, but this time in the transmission network. Here, voltages are controlled mainly by large-scale conventional power plants. If their capability to control voltages within the transmission network is not sufficient to compensate the impact of the wind farm on the node voltages, again, the voltage at some nodes can no longer be kept within the allowable deviation from its nominal value and appropriate measures have to be taken.

This chapter looks at the impact of wind power on voltage control. First, we will review the essential and basic principles of voltage control and comment on the impact of wind power on voltage control. Where appropriate, a distinction will be made between transmission networks and distribution grids. Then, we will briefly introduce the three wind turbine types that are most frequently used today and discuss their voltage control capabilities. We will illustrate voltage control with wind turbines and use steady-state and dynamic simulation results for this. To conclude, we will discuss the drawback of voltage control with wind turbines, namely, the increase in the rating of the power electronic converter.

19.2 Voltage Control

19.2.1 The need for voltage control

Voltage control is necessary because of the capacitance, resistance and inductance of transformers, lines and cables, which will hereafter be referred to as branches. Because branches have a capacitance, resistance and inductance (the first is often referred to as susceptance, the latter two as impedance), a current flowing through a branch causes a voltage difference between the ends of the branch (i.e. between the nodes being connected by the branch). However, even though there is a voltage difference between the two ends of the branch, the node voltage is not allowed to deviate from the nominal value of the voltage in excess of a certain value (normally 5 % to 10 %). Appropriate measures must be taken to prevent such deviation. Voltage control refers to the task of keeping the node voltages in the system within the required limits and of preventing any deviation from the nominal value to become larger than allowed.

It should be noted that if branches would not have an impedance and susceptance, the voltage anywhere in the system would be equal to that at the generators, and voltage control would not be necessary. This is a hypothetical case, though. It is also important to stress that node voltage is a local quantity, as opposed to system frequency, which is a global or systemwide quantity. It is therefore not possible to control the voltage at a certain node from any point in the system, as is the case with frequency (provided that no branch overloads are caused by the associated power flows). Instead, the voltage of a certain node can be controlled only at that particular node or in its direct vicinity. It is very important to keep this specific property of the voltage control problem in mind, because otherwise it is impossible to understand what impact the replacement of conventional generation by wind power has on voltage control.

There are various ways to affect node voltages. They differ fundamentally between transmission networks and distribution grids. This is because of the different characteristics of the branches in transmission networks and distribution grids and the divergent numbers and characteristics of the generators connected to both. Transmission networks consist of overhead lines with very low resistance. The voltage difference between two ends of a line with a high inductive reactance, X, when compared with its resistance, R (i.e. with a low R/X ratio) is strongly affected by what is called the reactive power flow through the line. Owing to the characteristics of transmission networks and the connected generators, node voltages are controlled mainly by changing the reactive power generation or consumption of large-scale centralised generators connected to the transmission network. These generators are easily accessible by network operators because of their relatively small number and the fact that they are continuously monitored and staffed. Further, they are very flexible in operation and allow a continuous control of reactive power generation over a wide range, as can be seen in Figure 19.1, which depicts a generator loading capability diagram.

Sometimes, dedicated equipment is used [e.g. capacitor banks or technologies referred to as flexible AC transmission systems (FACTS)]. These are, in principle, controllable reactive power sources. There are, however, differences with respect to how the reactive power is generated and with respect to the speed and accuracy of the control of the generated amount of reactive power. A detailed explanation of the

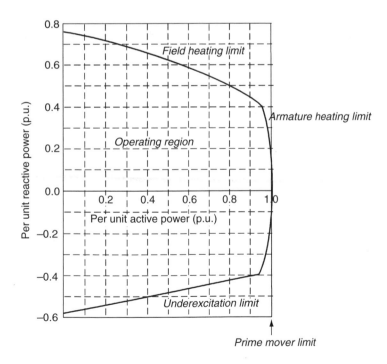

Figure 19.1 Example of a synchronous generator loading capability diagram

working principles and pros and cons of the various devices that are capable of generating controllable reactive power and that therefore can be used for voltage control purposes is beyond the scope of this chapter (for further reading, see, for example, Hingorani and Gyugyi, 1999).

In contrast, distribution grids consist of overhead lines or underground cables in which the resistance is not negligible when compared with the inductance (i.e. that have a much higher R/X ratio than transmission lines). Therefore, the impact of reactive power on node voltages is less pronounced than in the case of transmission networks. Further, the generators connected to distribution grids are not always capable of varying their reactive power output for contributing to voltage control. Node voltages in distribution grids are therefore controlled mainly by changing the turns ratio of the transformer that connects the distribution grid to the higher voltage level and sometimes also by devices that generate or consume reactive power, such as shunt reactors and capacitors. In general, distribution grids offer far fewer possibilities for node voltage control than do transmission networks.

19.2.2 Active and reactive power

In most cases, there is a phase difference between the sinusoidal current supplied to the grid by an alternating current (AC) generator and the voltage at the generator's terminals. The magnitude of this phase difference depends on the resistance, inductance and capacitance of the network to which the generator supplies its power and on the characteristics and the operating point of the generator itself. In the case of AC generation, the amount of generated power is, in general, not equal to the product of the root-mean-square (RMS) value of generator voltage and current, as in case of direct current (DC) generation. The amount of generated power depends not only on the amplitude of voltage and current but also on the magnitude of the phase angle between them.

The current of an AC generator can be divided into a component that is in phase with the terminal voltage and a component that is shifted by 90°. It can be shown that only the component of the current that is in phase with the voltage feeds net power into the grid (Grainger and Stevenson, 1994). Generated power is hence equal to the product of the RMS value of the in-phase component of the generator current and the terminal voltage. This quantity is named the active power or real power. It is measured in watts (W) or megawatts (MW) and represented by the symbol P.

The product of the RMS values of the source voltage and the out-of-phase (90° shifted) current component is called the reactive power. The unit of reactive power is voltamperes reactive (VAR). It is represented by the symbol Q and is often referred to as 'VARs'. The origin of reactive power is the electromagnetic energy stored in the network inductances and capacitances. The product of the RMS magnitudes of voltage and current is called the apparent power, which is represented by the symbol S. It is measured in volt amperes (VA). The apparent power S does not take into account the phase difference between voltage and current. It is therefore, in general, not equal to active power P unless there is no phase difference between the generator current and terminal voltage.

Another important quantity in this context is the power factor (PF). The PF is defined as follows:

$$PF = \cos \varphi = \frac{P}{S} = \frac{P}{\sqrt{P^2 + Q^2}},$$ (19.1)

where φ is the phase angle between the terminal voltage and terminal current, in degrees or radians. If φ equals 0, its cosine equals 1, reactive power Q equals 0 and apparent power S equals real power P. In that case, only active power is exchanged with the grid and there is no reactive power. This mode of operation is referred to as 'cos φ equal to one' or as 'unity power factor'.

19.2.3 Impact of wind power on voltage control

19.2.3.1 Impact of wind power on voltage control in transmission systems

When discussing the impact of wind power on voltage control, we have to distinguish between the impact of wind power on voltage control in transmission networks and on voltage control in distribution grids. As discussed above, in transmission networks, the task of voltage control was traditionally assigned to large-scale conventional power plants using synchronous generators, although some dedicated equipment, such as capacitor banks or FACTS, has been used as well. These traditional, vertically integrated, utilities have operated power generating units, on the one hand, and power transmission and distribution systems, on the other hand. They also have handled the voltage control issue, both short-term (day-to-day dispatch of units) and long-term (system planning).

Owing to recent developments, this is, however, changing. First, the liberalisation and restructuring of the electricity sector that is currently being implemented in many countries has resulted in the unbundling of power generation and grid operation. These activities are no longer combined in vertically integrated utilities as they used to be. As a consequence, voltage control is no longer a 'natural part' of the planning and dispatch of power plants. Now, independent generation companies carry out the planning and dispatch, and, in the long term, conventional power stations that are considered unprofitable will be closed down without considering their importance for grid voltage control. In addition, grid companies today often have to pay generation companies for generating reactive power. The grid companies also have to solve any voltage control problem that may result from the decisions taken by generation companies, either themselves or supported by the generation company. In the short term, this can be done by requiring the generation companies to redispatch. In the long term, additional equipment for controlling the voltage can be installed. Another recent development is that generation is shifted from the transmission network to the distribution grid.

As a result of these two developments (unbundling and decentralisation), the contribution of conventional power plants to voltage control in transmission networks is diminishing. It is becoming more difficult to control the voltage in the entire transmission network from conventional power stations only. Grid companies respond by installing dedicated voltage control equipment and by requiring generation equipment to have reactive power capabilities independent of the applied technology. This means that no exception is made for wind power or other renewables any longer, as has often been the case until now.

There is a third development that is specific to wind power. Wind farms that are large enough to be connected to the transmission system tend to be erected in remote areas or offshore because of their dimension and impact on the scenery. Given that the node voltage is a local quantity, it can be difficult to control the voltage at these distant places by use of conventional power stations elsewhere in the grid. Therefore, wind turbines have to have voltage control capabilities. The voltage control capabilities of various wind turbine types are expected to become an increasingly important consideration regarding grid connection and the turbines' market potential (Wind, 1999).

A final impact of wind power on voltage control that we would like to mention is that large-scale wind farms may make it necessary to install voltage control devices in the transmission network, irrespective of the voltage control capabilities of the wind turbines themselves. In other words: even if the wind turbines have exactly the same voltage control capabilities as the conventional synchronous generators whose output they replace, there will be no guarantee that they can fulfil the voltage control task of these generators. Therefore, it may be inevitable to take additional measures to control the grid voltage.

The local nature of grid voltage is the reason for this. In most cases, wind turbines will not be erected at or near the location of the conventional power plant whose output they replace. It is likely that the power generation of a conventional power plant in or near a load centre, such as a city or a large industrial site, will be replaced by a distant (offshore) wind farm. However, because voltage is a local quantity, this wind farm will not be able to control the grid voltage in the vicinity of the power plant that is replaced. Therefore, additional equipment must be installed near the location of this power plant in order to control the grid voltage. The reason for the necessary additional measures for voltage control is therefore not that conventional power generation is replaced by wind power generation as such, but that generation moves away from the vicinity of the load to a more distant location.

There are also, however, other developments that lead to a geographical displacement of generation, such as power transit between various countries as a result of market liberalisation. They can lead to voltage control problems, too, and additional equipment for voltage control in transmission networks may have to be installed. The problem is therefore not specific to the replacement of conventional generation by wind power. It can be induced by any development that leads to significant changes in generator locations.

19.2.3.2 Impact of wind power on voltage control in distribution systems

Traditionally, the two most important approaches towards voltage control in distribution grids are the use of tap-changing transformers (i.e. transformers in which the turns ratio can be changed) and devices that can generate or consume reactive power (i.e. shunt capacitors or reactors). Use of tap-changing transformers is a rather cumbersome way of controlling node voltages. Rather than affecting the voltage at one node and in its direct vicinity, the whole voltage profile of the distribution grid is shifted up or down, depending on whether the transformer turns ratio is decreased or increased. Capacitors and reactors perform better in this respect, because they affect mainly the voltage of the node to which they are connected. However, the sensitivity of the node voltage to changes in reactive power is rather limited and therefore relatively large capacitors and reactors are necessary. This disadvantage is due to the high R/X ratio of the branches in distribution grids when compared with that in transmission networks.

As was the case for transmission networks, recent developments are complicating the task of maintaining node voltages throughout distribution grids. More and more distributed generation (also referred to as embedded, decentralised or dispersed generation; see Ackermann, Andersson and Söder, 2001), such as wind turbines, solar photovoltaic (PV) systems and small combined heat and power (CHP) generation is being connected to distribution grids. These generators affect the power flows in distribution grids. In particular, if their output power does not correlate with the load, as is the case with generators using an uncontrollable prime mover, such as wind or sunlight, the variations in the current through the branches and therefore in the node voltages increase. The maximum and minimum value of the current through a certain branch used to depend on the load only, but, with the connection of distributed generation, the current limits have become dependent on the load as well as on the output of the distributed generator. The limits are now determined by a situation with minimum generation and maximum load, on the one hand, and maximum generation and minimum load on the other, rather than only by the difference between minimum and maximum load, as used to be the case.

One might argue that with an increasing number of generators connected to the distribution grid, the voltage control possibilities might increase as well. However, in many cases these small-scale generators are more difficult to use for voltage control than the generators in large-scale power plants connected to transmission systems because:

- they are not always able to vary reactive power generation (depending on the applied generator type and rating as well as on the rating of the power electronic converter, if existing);
- it may be (very) costly to equip them with voltage control capabilities;
- equipping them with voltage control capabilities could increase the risk of 'islanding' [i.e. a situation in which (part of) a distribution network remains energised after being disconnected from the rest of the system];
- there are many of them, which makes it very cumbersome to change controller parameters, such as the voltage set point or time constants, which may be necessary after a change in the network topology, for example.

There are many reasons for increasing the amount of distributed generation, such as increased environmental awareness or the desire to reduce investment risk. However, the impact of an increased amount of distributed generation on power flows and node voltages in distribution grids may lead to problems with maintaining node voltages in distribution grids. In that case, appropriate measures have to be taken. One approach would be, for example, to install additional tap changers (sometimes with a nominal turns ratio of 1 further down into the distribution grid; the device is then called a voltage regulator), reactors and/or capacitors. Another one would be to oblige distributed generators to contribute to voltage control, despite the above complications and disadvantages.

19.2.3.3 The importance of wind turbine voltage control capabilities

As argued above, the voltage control capabilities of wind turbines erected in large-scale wind farms and connected to the transmission network become increasingly important. The contribution of conventional synchronous generators to voltage control is decreasing as a result of unbundling and decentralisation. Further, wind farms are connected at distant locations at which it is difficult to control the voltage out of conventional power plants. It has also been shown that as a result of the impact of distributed generation, such as wind turbines, the range over which currents in distribution grids vary becomes larger. Because branch currents and node voltages are strongly correlated, this also applies to the range of variation in node voltages. Wind turbines with voltage control capabilities could counteract this effect.

It is becoming increasingly important, for wind turbines connected to the transmission system and for those connected to the distribution system, to be able to contribute to voltage control. Therefore, the remainder of this chapter will look at the voltage control capabilities of the various types of wind turbines.

19.3 Voltage Control Capabilities of Wind Turbines

19.3.1 Current wind turbine types

The vast majority of wind turbines that are currently being installed use one of the three main types of electromechanical conversion system. The first type is known as the Danish concept. In Section 4.2.3, the Danish concept is introduced as Type A. An (asynchronous) squirrel cage induction generator is used to convert the mechanical energy into electricity. Owing to the different operating speeds of the wind turbine rotor and the generator, a gearbox is necessary to match these speeds. The generator slip slightly varies with the amount of generated power and is therefore not entirely constant. However, because these speed variations are in the order of 1 %, this wind turbine type is normally referred to as constant-speed or fixed-speed. The Danish, or constant-speed, design is nowadays nearly always combined with stall control of the aerodynamic power, although pitch-controlled constant-speed wind turbine types have been built too (type A1; see Section 4.2.3).

The second type uses a doubly fed induction generator instead of a squirrel cage induction generator, and was introduced as Type C in Section 4.2.3. Similar to the previous type, it needs a gearbox. The stator winding of the generator is coupled to the grid, and the rotor winding to a power electronic converter, nowadays usually a back-to-back voltage source converter with current control loops. In this way, the electrical and mechanical rotor frequencies are decoupled, because the power electronic converter compensates the difference between mechanical and electrical frequency by injecting a rotor current with variable frequency. Variable-speed operation thus becomes possible. This means that the mechanical rotor speed can be controlled according to a certain goal function, such as energy yield maximisation or noise minimisation. The rotor speed is controlled by changing the generator power in such a way that it equals the value derived from the goal function. In this type of conversion system, the control of aerodynamic power is usually performed by pitch control.

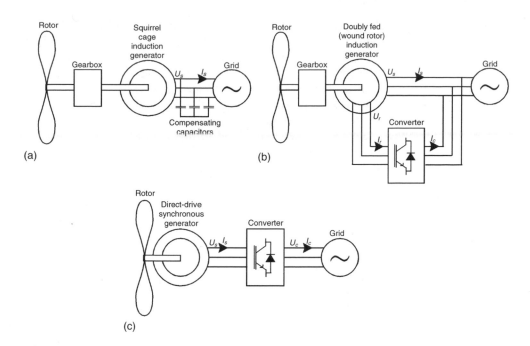

Figure 19.2 Widely used wind turbine types: (a) constant-speed wind turbine (Type A); (b) variable-speed wind turbine with doubly fed induction generator (Type C); and (c) direct-drive variable-speed wind turbine with multipole synchronous generator (Type D) *Note*: U = voltage; I = current; subscripts s, r, c, stand for stator, rotor and converter, respectively

The third type is called the 'direct-drive wind turbine' because it does not need a gearbox. It corresponds to Type D in Section 4.2.3. A low-speed multipole synchronous ring generator with the same rotational speed as the wind turbine rotor converts the mechanical energy into electricity. The generator can have a wound rotor or a rotor with permanent magnets. The stator is not coupled directly to the grid but to a power electronic converter. This may consist of a back-to-back voltage source converter or a diode rectifier with a single voltage source converter. The electronic converter makes it possible to operate the wind turbine at variable speed. Similar to Type C, pitch control limits the mechanical power input. Figure 19.2 presents the three main wind turbine types.

Apart from the wind turbine types depicted in Figure 19.2, other types have been developed and used. They include wind turbines with directly grid coupled synchronous generators and with conventional synchronous as well as squirrel cage induction generators combined with a gearbox and coupled to the grid with a full-scale power electronic converter. Currently, there are hardly any of these wind turbine types on the market, though, and we will not take them into consideration here.

19.3.2 Wind turbine voltage control capabilities

As mentioned above, node voltages are dependent on branch characteristics and on branch currents. In transmission networks and distribution grids, node voltage and reactive power are correlated and therefore node voltages can be controlled by changing the reactive power generation or consumption of generators. An analysis of the voltage control capabilities of the wind turbine types described in Section 19.3.1 has thus to show to what extent they can vary their reactive power output. In the following, we will discuss the voltage control capabilities of each of the wind turbine types described above.

Constant-speed wind turbines (Type A) have squirrel cage induction generators that always consume reactive power. The amount of the reactive power consumption depends on the terminal voltage, active power generation and rotor speed. Figure 19.3 illustrates the relation between terminal voltage, rotor speed, active power generation and reactive power consumption. This figure shows that a squirrel cage induction generator cannot be used for voltage control, because it can only consume and not generate reactive power and because the reactive power exchange with the grid cannot be controlled but is governed by rotor speed, active power generation and terminal voltage.

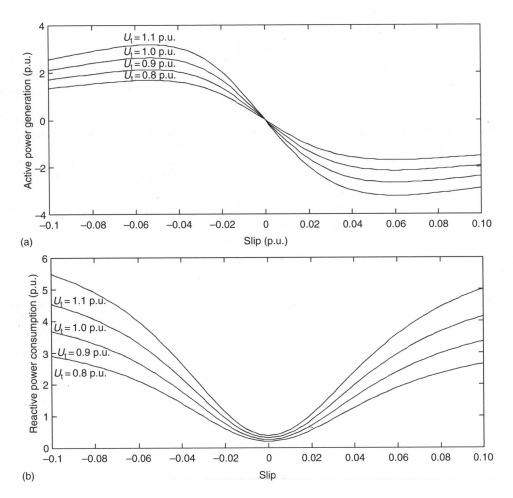

Figure 19.3 Dependence of (a) active and (b) reactive power of a squirrel cage induction generator on the rotor speed, with the terminal voltage, U_t, as a parameter

The fact that a squirrel cage induction generator consumes reactive power can be a disadvantage, particularly in the case of large wind turbines or wind farms and/or weak grids. In such cases, the reactive power consumption may cause severe node voltage drops. Therefore, the reactive power consumption of the generator is in most cases compensated by capacitors, as depicted in Figure 19.2(a). In this way, the reactive power exchange between the combination of the generator and the capacitors, on the one hand, and the grid, on the other, can be reduced thus improving the power factor of the system as a whole.

A conventional capacitor is an uncontrollable source of reactive power. By adding compensating capacitors, the impact of the wind turbine on the node voltages is reduced. But this is only a qualitative improvement. The voltage control capabilities as such are not enhanced, because there is still a unique relation between rotor speed, terminal voltage and active and reactive power generation. The voltage control capabilities of a constant-speed wind turbine can be enhanced only with use of more advanced solutions instead of conventional capacitors. Such advanced solutions include controllable sources of reactive power, such as switched capacitors or capacitor banks, a static condensor (Statcon) or a static VAR compensator (SVC). An example of this is the use of an American Superconductor D-VAR^{TM} system in combination with a wind farm with constant-speed wind turbines at Minot, ND.[1]

The reactive power generation of a doubly fed induction generator (Type C wind turbine) can be controlled by the rotor current, as will be discussed in more detail in Chapter 25. In this case, there is no unique relation between reactive power and other quantities, such as rotor speed and active power generation. Instead, at a particular rotor speed and the corresponding active power generation a widely varying amount of reactive power can be generated or consumed.

Figure 19.4 illustrates the operating range of a doubly fed induction generator at nominal terminal voltage. It shows that the amount of reactive power is, to a certain extent, affected by rotor speed and active power generation, as in the case of a squirrel cage induction generator, even though it does not directly depend on these quantities. The reason is that both generator torque and reactive power generation depend directly on the current that the power electronic converter feeds into the rotor. The part of the current that generates torque depends on the torque set point that the rotor speed controller derives from the actual rotor speed. The current that is needed to generate the desired torque determines, in turn, the converter capacity that is left to circulate current to generate or consume reactive power.

In the case of a direct-drive variable-speed wind turbine (Type D), the reactive power exchange with the grid is not determined by the properties of the generator but by the characteristics of the grid side of the power electronic converter. The generator is fully decoupled from the grid. Therefore, the reactive power exchange between the generator itself and the generator side of the converter as well as between the grid side of the converter and the grid are decoupled. This means that the power factor of the generator and the power factor of the grid side of the converter can be controlled independently.

---

[1] D-Var^{TM} is a trademark of American Superconductor.

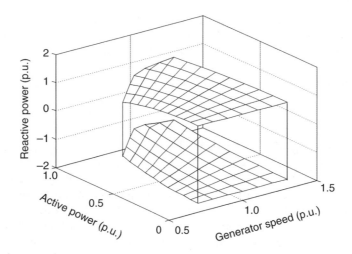

Figure   19.4 Operating range of a doubly fed induction generator

Figure 19.5 shows the operating range of a variable-speed wind turbine with a direct-drive synchronous generator, with the terminal voltage as a parameter. The rotor speed is not taken into account. As the generator and the grid are decoupled, the rotor speed hardly affects the grid interaction. It is assumed that at nominal voltage and power the wind turbine can operate with a power factor of between 0.9 leading and 0.9 lagging. Figure 19.5 illustrates that a variable-speed wind turbine with a direct-drive synchronous generator allows control of reactive power or terminal voltage, because many values of reactive power correspond to a single value of active power. It should be noted

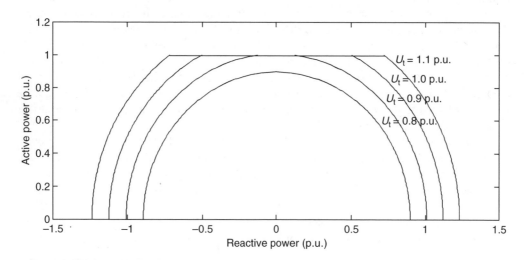

Figure   19.5 Operating range of a wind turbine with a direct-drive synchronous generator with the terminal voltage, U_t, as a parameter; the curves indicate the boundary of the operating range with U_t as a parameter

that in Figure 19.5 the curves indicate the boundary of the operating region, whereas in Figure 19.3, they indicate a unique relationship.

19.3.3 Factors affecting voltage control

As discussed above, a wind turbine with voltage control capability can control node voltages by changing the amount of reactive power that is generated or consumed. To govern the control actions, the voltage at a certain node is measured and fed into the voltage controller. The controller determines the amount of reactive power to be generated or consumed, according to the transfer function of the controller. It is easiest to let the wind turbine control its own terminal voltage but sometimes the voltage at a node somewhere else in the grid is controlled, although that node must be in the vicinity of the turbine because of the local nature of the grid voltage. When the measured voltage is too low, reactive power generation is increased; when it is too high, reactive power generation is decreased.

The following factors influence the relation between the amount of reactive power that is generated or consumed and the converter current that controls the reactive power generation:

- generator parameters (only in a wind turbine with a doubly fed induction generator);
- the set point of the voltage controller;
- values of the resistance, R, and the reactance, X, of the wind turbine grid connection;
- the amount of active power flowing through the grid connection.

This chapter will look only at the influence of the last factor (i.e. the amount of active power flowing through the grid connection). With respect to the other factors, the following can be said:

- Machine parameters are strongly related to machine size; therefore, all doubly fed induction generator based wind turbines of a given nominal power will have similar generator parameters. Therefore, generator parameters will hardly make a difference in this context (Heier, 1998).
- In practice, the set point of the terminal voltage will always be (nearly) equal to the nominal value of the voltage. Therefore, the terminal voltage controller set point will not play a major role.
- For a more detailed discussion on the influence of the values of R and X of the grid connection on the amount of reactive power that needs to be generated to keep the voltage at the reference value, see Svensson (1996), among others.

19.4 Simulation Results

19.4.1 Test system

We use a simple artificial test system in order to quantitatively investigate the voltage control capabilities of the various wind turbine concepts. The test system consists of a wind turbine that is connected through an impedance to a strong grid, represented by an

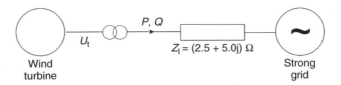

Figure   19.6 Test system for investigating the voltage control capabilities of wind turbine types. *Note*: U_t = terminal voltage; P = active power; Q = reactive power; Z_l = line impedance; $j = \sqrt{-1}$

infinite bus. The nominal power of the wind turbine equals 2 MW, and its nominal voltage is 700 V. The transformer is a three phase 700 V/10 kV transformer; the grid voltage is 10 kV. The nominal frequency is 50 Hz. The resistance of the transformer and the cable is 2.5 Ω and their inductance is 16 mH, which corresponds to a reactance of 5j Ω. Figure 19.6 shows the setup.

19.4.2 Steady-state analysis

First, the constant-speed wind turbine is analysed. For this type, voltage control is not possible when a compensating capacitor of fixed size is used. There are studies of systems with a reactive power source that can be controlled, such as capacitor banks or FACTS (e.g. see Abdin and Xu, 2000). However, the product portfolios of major wind turbine manufacturers show that this type of equipment is not part of standard products. Further, when this kind of additional equipment is used it is still not the wind turbine itself that controls the voltage but the extra device. As this chapter discusses the voltage control capabilities of wind turbines, rather than dedicated voltage control equipment, no further attention will be paid to such devices.

Chapter 25 includes the equations that can be used to calculate the amount of reactive power consumed by a squirrel cage induction generator at a certain terminal voltage and power output. The generator used here consumes 1.05 MVAR reactive power when generating nominal active power at nominal terminal voltage. It is assumed that this reactive power demand is fully compensated. To this end, a capacitor of approximately 6.8 mF has to be installed at each phase.

Figure 19.7 presents the results of the steady-state analysis of the constant-speed wind turbine. The horizontal axis shows the active power generated by the wind turbine. The vertical axis depicts the terminal voltage and the reactive power supplied to the grid. The reactive power supplied to the grid equals the reactive power generated by the capacitor minus the reactive power consumed by the squirrel cage induction generator.

Figure 19.7 shows that the reactive power decreases and the terminal voltage increases when active power generation increases. The decrease in reactive power generation can be explained by the fact that when the active power generation increases, the reactive power consumption of the squirrel cage induction generator increases as well. Therefore, a minor part of the reactive power generated by the capacitor will be supplied to the grid. The exact quantitative behaviour in a specific situation depends on the parameters of the generator and the grid connection (Svensson, 1996).

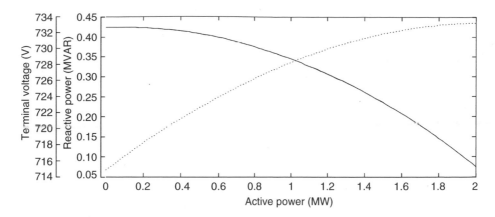

Figure 19.7 Results of the steady-state analysis of a constant-speed wind turbine using the test system of Figure 19.6. *Note*: reactive power generated = solid line; terminal voltage = dotted line

Although the two variable-speed wind turbine types studied here have a different generation system, they both allow full active and reactive power control, as described above. In the steady-state analysis, only the terminal voltage and the exchange of active and reactive power with the grid are studied. Therefore, it is unnecessary to analyse the steady-state behaviour of the two variable-speed wind turbine types separately.

Figure 19.8 shows the results of the steady-state analysis of variable-speed wind turbines. In both cases, the wind turbine voltage controller controls the voltage at the low-voltage (700 V) side of the transformer. The horizontal axis gives the active power generated by the wind turbine. The vertical axis depicts the terminal voltage and the generated reactive power. There is a marked difference between Figures 19.7 and 19.8. In Figure 19.8 the terminal voltage varies; in Figure 19.8 it is constant.

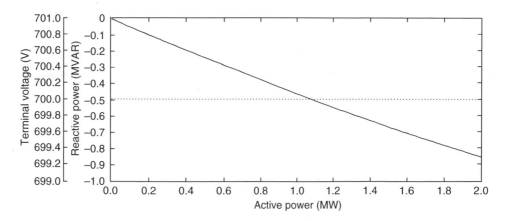

Figure 19.8 Results of the steady-state analysis of a variable-speed wind turbine with voltage controller, using the test system of Figure 19.6. *Note*: reactive power generated = solid line; terminal voltage = dotted line

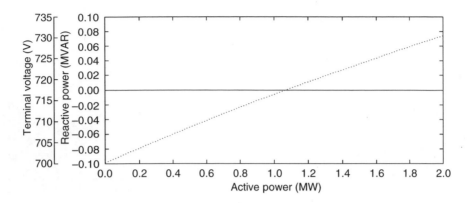

Figure 19.9 Results of the steady-state analysis of a variable-speed wind turbine with unity power factor using the test system of Figure 19.6. *Note*: reactive power generated = solid line; terminal voltage = dotted line

Presently, converters in variable-speed wind turbines are often controlled in such a way that no reactive power is exchanged with the grid. That means that a current with the same phase angle as the grid voltage is injected, and the power factor, PF, equals 1 (see Section 19.2.2). This mode of operation will be further referred to as unity power factor operation. Figure 19.9 presents the results of the steady-state analysis of this operating mode. Owing to the relatively high impedance of the grid connection in the test system, the variable-speed wind turbines did not perform very well at unity power factor operation.

19.4.3 Dynamic analysis

This section analyses the dynamic behaviour of the three wind turbine concepts. We use the same setup as for the steady-state analysis. When analysing the dynamic behaviour, the two different variable-speed wind turbine types must be treated separately, because they have different controllers for controlling the reactive power generation. In the wind turbine with the doubly fed induction generator, reactive power is generated by controlling the rotor current. In the wind turbine with a direct-drive synchronous generator, reactive power is changed by controlling the grid side of the converter. Because the transfer functions are different, the performance of the control systems may also differ and the two types of variable-speed wind turbines are therefore simulated separately. To illustrate the difference between wind turbines with and without a voltage controller, the grid voltage will be dropped after 30 seconds from 10 kV to 9.75 kV. In reality, this can happen when a large load is switched on or when a nearby synchronous generator trips.

Figure 19.10 shows the results of the simulation for a measured wind speed sequence. Figure 19.10(a) depicts the wind speed sequence. Then, for each of the wind turbine types, the generated active and reactive power and the terminal voltage are shown. For the variable speed turbines, both voltage control operating mode and unity power factor operating mode are studied. In the figure, the individual curves are labelled with the wind turbine type to which they refer.

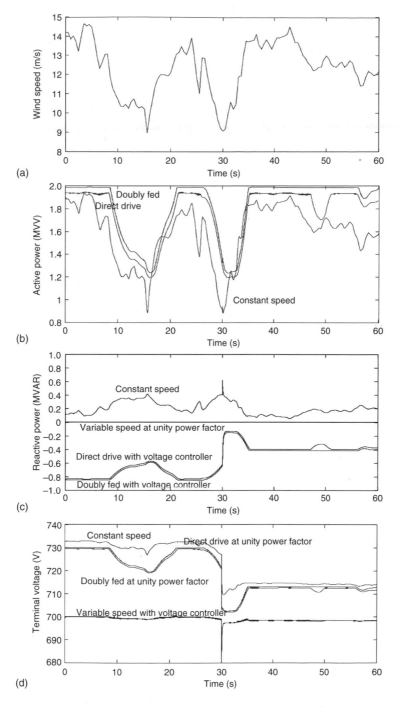

Figure 19.10 Results of dynamic simulation of a constant-speed wind turbine and a variable-speed wind turbine in either voltage control operating mode or unity power factor operating mode: (a) wind speed, (b) active power, (c) reactive power and (d) terminal voltage. The individual curves are labelled with the wind turbine type to which they refer

The following conclusions can be drawn from Figure 19.10:

- The voltage fluctuates most rapidly in the case of constant-speed wind turbines. The reason is that the rotor has a constant rotational speed, because the generator is directly coupled to the grid. Therefore, the rotating mass cannot be used as an energy buffer, as is the case in variable-speed wind turbines. Changes in wind speed, as well as the effect of the tower shadow, are directly reflected in changes in active power. The reactive power consumed by the generator depends on the active power, which in turn influences both the reactive power generated and the terminal voltage. This effect would be even more pronounced if the tower shadow effect were included in the models.
- The amplitude of the terminal voltage variations is high in the case of constant-speed wind turbines and variable-speed wind turbines in unity power factor mode. This is caused by the high impedance of the grid connection.
- The terminal voltage variation is smoothest in the case of variable-speed wind turbines with a voltage controller. As expected, a wind turbine with a terminal voltage controller performs better than one without a terminal voltage controller.
- Although the voltage controller works in a different way, the performance of both variable-speed wind turbines is similar. If the controllers are well designed, both variable speed-wind turbines perform equally well with respect to terminal voltage control.
- Finally, it is clear that wind turbines equipped with a voltage controller compensate the grid voltage drop and keep the voltage at its reference value. Only variable-speed wind turbines with a voltage controller are capable of controlling terminal voltage independently of the grid voltage, unless their operating limits are exceeded.

19.5 Voltage Control Capability and Converter Rating

The conclusion is that a variable-speed wind turbine with a voltage controller performs best with respect to voltage control. Unfortunately, this does not come for free. For both variable-speed wind turbines, voltage control requires power electronic converters with a rating that is higher than the rating for operation at unity power factor.

In order to illustrate this, the converter current was calculated when the wind turbines studied above generated nominal real power and:

- did not generate reactive power;
- generated nominal reactive power;
- consumed nominal reactive power.

We assumed that grid voltage and frequency equal their nominal values. The nominal reactive power exchange with the grid was assumed to equal 1 MVAR. This corresponds to a power factor of approximately 0.9. For reactive power consumption and generation, we also calculated the relative change in converter current when compared with the converter current in the unity power factor operation mode. The results are given in Table 19.1. Even though the quantitative values are specific to the wind turbines studied here, the qualitative conclusions can be applied more generally.

Table 19.1 Converter phase current for varying amounts of reactive power, Q, for wind turbine Types B and C when frequency, f, is 50 Hz, active power, P, is 2 MW and terminal voltage, U_t, is 700 V

Q	Converter current rating (A)	
	Doubly fed	Direct drive
0 MVAR	537	1650
1 MVAR generation	845	1845
percentage change[a]	+57.4	+11.8
1 MVAR consumption	618	1845
percentage change[a]	+15.0	+11.8

[a] Percentage change relative to $Q = 0$ MVAR case (unity power factor operating mode).

Table 19.1 shows that, compared with the unity power factor operating mode, the converter current rating must be higher for reactive power generation and consumption. In the case of doubly fed induction generators, for reactive power consumption the converter current rating can be lower than for reactive power generation. The reason is that the generator in this wind turbine type is grid coupled. The magnetising current can be drawn from the grid instead of being provided by the converter. In this mode of operation, reactive power is consumed and less converter current is needed than for the generation of the same amount of reactive power. Full voltage control capability requires that reactive power can be both generated and consumed. Therefore, reactive power generation is the determining factor in sizing the converter when equipping a doubly fed induction generator based wind turbine with a voltage controller.

Finally, Table 19.1 shows how much the converter current rating has to be increased in order to achieve voltage control capability. The relative increase of converter size is largest in the case of the doubly fed induction generator based turbine. It now could be hastily concluded that it is more expensive to equip a doubly fed induction generator based wind turbine with a voltage controller than a direct-drive wind turbine. However, this conclusion is *not* necessarily correct. The converter in a direct-drive wind turbine is larger and thus more expensive than in a doubly fed induction generator based wind turbine. This means that, although the relative increase in converter cost will be smaller in the case of the direct-drive wind turbine, the absolute cost increase may be substantially higher. Therefore, the costs for equipping a wind turbine with voltage control capability must be studied carefully for both concepts.

19.6 Conclusions

This chapter has discussed the impact of wind power on voltage control and the voltage control capabilities of various current wind turbine types. It was pointed out

- that voltage control is necessary because of the susceptance and impedance of the transformers, lines and cables that make up electrical networks;
- that node voltages are strongly related to branch currents, or power flows;
- that owing to recent developments the voltage control capabilities of wind turbines are becoming increasingly important, independent of whether they are connected to transmission networks or to distribution grids.

We have also looked at the most important current wind turbine concepts and their voltage control capabilities. It was concluded that the directly grid coupled asynchronous generator is not capable of terminal voltage control whereas turbines that are equipped with a doubly fed induction generator and a direct-drive synchronous generator can control the terminal voltage. However, in practice, this possibility is often not used. Therefore, the terminal voltage behaves similarly to a conventional directly grid coupled asynchronous generator.

Dynamic simulations were carried out and it was concluded that the terminal voltage variation is smoothest in the case of variable-speed wind turbines with voltage control. Furthermore, it was shown that only wind turbines with voltage control can compensate a drop in grid voltage. Finally, the drawback of terminal voltage control by wind turbines was discussed: that is, the need for an increased converter rating when compared with unity power factor operation.

References

[1] Abdin, E. S., Xu, W. (2000) 'Control Design and Dynamic Performance Analysis of a Wind Turbine–Induction Generator Unit', *IEEE Transactions on Energy Conversion* 15(1) 91–96.

[2] Ackermann, T., Andersson, G., Söder, L. (2001) 'Distributed Generation: A Definition', *Electric Power Systems Research* 57(3) 195–204.

[3] Grainger, J. J., Stevenson Jr., W. D. (1994) *Power System Analysis*, McGraw-Hill, New York.

[4] Heier, S. (1998) *Grid Integration of Wind Energy Conversion Systems*, John Wiley & Sons Ltd., Chichester, UK.

[5] Hingorani, N. G., Gyugyi, L. (1999) *Understanding FACTS*, Wiley–IEEE Press, New York.

[6] Svensson, J. (1996) 'Possibilities by Using a Self-commutated Voltage Source Inverter Connected to a Weak Grid in Wind Parks', in *1996 European Union Wind Energy Conference and Exhibition, Göteborg, Sweden, 20–24 May 1996*, pp. 492–495.

[7] Wind, T. (1999) 'Wind Turbines Offer New Voltage Control Feature', *Modern Power Systems* 15(11) 55.

20

Wind Power in Areas with Limited Transmission Capacity

Julija Matevosyan

20.1 Introduction

Historically, transmission systems are built together with power production installations in order to meet the expected electricity consumption. For economic reasons, they are usually not overdimensioned and therefore cannot guarantee power transmission capacity for new power plants for 100 % of the year.

Wind power plants have to be installed in the immediate proximity of the resource – wind. The best conditions for an installation of wind power can usually be found in remote, open areas with low population densities. The transmission system in such areas might not be dimensioned to accommodate additional large-scale power plants. The thermal limits of the conductors, voltage and transient stability considerations restrict transmission capability during extreme situations, such as during low local loads and high wind power production.

This chapter covers the issues concerning the integration of new generation (particularly from wind power) in areas with limited transmission capacity. We will start in Section 20.2 with an overview of different congestion types and some conventional measures to overcome them. In Section 20.3 we present different approaches used by transmission system operators (TSOs) to determine transmission capacity. In Section 20.4 we look at 'soft' and 'hard' grid reinforcement measures to increase transmission capacity. The problems associated with the integration of new wind generation in areas with limited transmission capacity are discussed in the Section 20.5. In Section 20.6 we present methods to estimate the amount of wind power that can be integrated in areas with limited transmission capacity before grid reinforcement becomes the more

Wind Power in Power Systems   Edited by T. Ackermann
© 2005 John Wiley & Sons, Ltd   ISBN: 0-470-85508-8 (HB)

economical option. The example of the Swedish power system is used to illustrate the results of the estimation methods. In Section 20.7 we briefly present conclusions.

20.2 Transmission Limits

Power transmission in a system may be subjected to the thermal limits of the conductors as well as to limits in relation to voltage and transient stability considerations. Thermal limits are assigned to each separate transmission line and the respective equipment. Limits arising from voltage and transient stability considerations are always studied by taking into account the operation of the whole power system or a part of it. In the following sections we provide details regarding transmission limits.

20.2.1 Thermal limit

Overhead transmission lines reach their thermal limit if the electric current heats the conductor material to a temperature above which the conductor material will start to soften. The maximum permissible continuous conductor temperature varies between 50 °C and 100 °C, depending on the material, its age, the geometry, the height of the towers, the security standards that limit the clearance from the ground and so on. (Haubrich *et al.*, 2001).

The thermal limit or current-carrying capacity of the conductors depends on the ambient temperature, the wind velocity, solar radiation, the surface conditions of the conductor and the altitude above sea level (House and Tuttle, 1958), (see Figure 20.1). The load on short

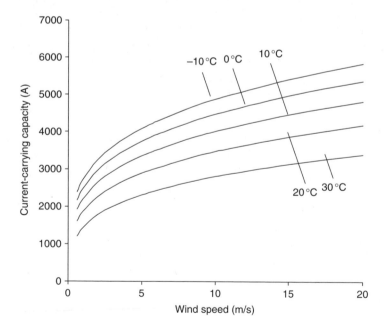

Figure   20.1 Dependence of the current-carrying capacity of the conductor on wind conditions and ambient temperature

transmission lines (less than 100 km long) is usually restricted by the heating of the conductors rather than by stability considerations. As Figure 20.1 shows, the current-carrying capacity at higher wind speeds is higher than at lower wind speeds. In places, where wind speeds are higher in winter, the low temperatures also contribute to an increase in the current-carrying capacity. The wind speed measurements from wind farms can be used for online estimations of the current-carrying capacity of short transmission lines. This will allow an increase in power transmission during higher wind speeds. This method is, for example, applied on the Swedish island of Gotland. Such online estimations of the current-carrying capacity are difficult to carry out for longer transmission lines. Wind speed and temperature changes considerably with distance. It would therefore be necessary to have measuring equipment in many places along the transmission line, which is expensive.

Other network elements such as breakers, voltage and current transformers and power transformers could further restrict the transmission capacity of some network branches. The thermal limit of the transmission line is then set by the lowest rating of the associated equipment.

The current-caring capacity of the conductor is related to the maximum allowed active power transfer by the following expression:

$$P_{max} = I_{max} U_{min} \cos \varphi_{min}, \tag{20.1}$$

where I_{max} is the current-caring capacity, U_{min} is the minimum voltage level that is expected during normal operating conditions and $\cos \varphi_{min}$ is the expected minimum power factor at full load (Wiik et al., 2000). Thus, by improving the load power factor and increasing the minimum voltage the allowed active power transfer can be increased.

20.2.2 Voltage stability limit

Voltage stability is the ability of the system to maintain steady acceptable voltages at all buses in the system under normal conditions and after being subjected to a disturbance. Instability occurs in the form of a progressive fall or rise of voltages in some buses. A possible result of voltage instability is a loss of load in an area, or outages. Furthermore, such outages or operation under field current limits may lead to a loss of synchronism (Van Cutsem and Vournas, 1998).

For a real power system, it may be difficult to separate voltage stability and angle stability. To give an overview of the pure voltage stability problem, a two-terminal network is analysed (see Figure 20.2). It consists of a load Z_{LD} with a constant power factor $\cos \varphi_{LD}$ supplied by a constant voltage source U_S via a transmission line. The transmission line is modelled as a series impedance $Z_L \angle \theta_L$.

The current magnitude, I, can be calculated by using Ohm's law as the magnitude of the sending-end voltage, U_S, divided by the magnitude of total impedance between sending and receiving ends:

$$I = \frac{U_S}{[(Z_L \cos \theta_L + Z_{LD} \cos \varphi_{LD})^2 + (Z_L \sin \theta_L + Z_{LD} \sin \varphi_{LD})^2]^{1/2}}, \tag{20.2}$$

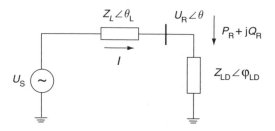

Figure 20.2 Two-terminal network. *Note*: U_R = receiving-end voltage; U_s = sending-end voltage of the constant voltage source; P_R = active power at receiving end; Q_R = reactive power at receiving end; Z_L = line impedance; Z_{LD} = load impedance; $\cos\varphi_{LD}$ = power factor; θ = angular difference between two terminal voltages; θ_L = arctan (X_L/R_L), where X_L is the line reactance, and R_L is the line resistance

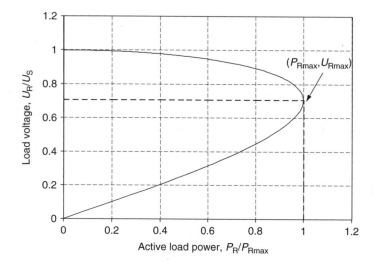

Figure 20.3 Nose curve for the system shown in Figure 20.2; $U_S = 1$ p.u. load with no reactive power demand $[Q_R = 0\ (\varphi_{LD} = 0°)]$; a purely reactive impedance of the transmission line is assumed $[Z_L = -jX_L = j0.2\,\text{p.u.}\ (\theta_L = 90°)]$. *Note*: U_S = sending-end voltage; U_R = receiving end voltage; P_R = active power at receiving end; P_{Rmax} = Maximum value of P_R; Q_R = reactive power at receiving end; φ_{LD} = load angle; Z_L = line impedance; X_L = reactance of line; R_L = resistance of line; θ_L = arctan (X_L/R_L)

The relationship between transferred power and the voltage can be illustrated by a so-called nose curve (see Figure 20.3). Voltage magnitude at the receiving end, U_R, is then equal to IZ_{LD}, and active power at the receiving end, P_R, can be calculated as $U_R I \cos\varphi_{LD}$, that is, $P_R = I^2 Z_{LD}\cos\varphi_{LD}$. If load impedance Z_{LD} is gradually decreased, the current increases. As long as Z_{LD} is larger than the line impedance Z_L the increase in current is more significant than the decrease in voltage at the receiving end, and the transferred power increases. This corresponds to the upper branch of the nose curve. When Z_{LD} becomes lower than Z_L the decrease in voltage is faster than the increase in current, and the transmitted power decreases. This corresponds to the lower branch of the nose curve and gives much higher losses in the system as well as unsatisfactory operating conditions for many devices that are not modelled in this simple case, such as onload tap changers, current limiters and so on.

It can be seen that the transmitted power P_R cannot exceed a value P_{Rmax}. Point (P_{Rmax}, U_{Rmax}) in the nose curve is called the *maximum loadability point*. The point marks the maximum possible power transfer for a particular power factor.

Figure 20.4 shows the nose curves for the two-terminal model in Figure 20.2 for different values of the load power factor. The locus of critical operation points is given by the broken line. As already mentioned, normally only the operating points above the critical point represent satisfactory operating conditions; a sudden change of the power factor can cause the system to go from stable operating conditions to unsatisfactory or unstable operating conditions.

Improving the power factor at the receiving end of the line by local reactive power compensation (to unity or leading load power factor) can enhance the voltage stability of the system. However, if large amounts of reactive power are transferred from the sending end (lagging load power factor) the voltage stability of the system will deteriorate (Kundur, 1993). Local reactive power compensation also results in a voltage increase at the receiving end. The amount of compensation should thus be chosen to keep the voltage at the receiving end within acceptable limits.

The line length also has a significant impact on the voltage stability as heavily loaded lines consume reactive power. Line reactance increases with line length, and reactive power consumption of the transmission line at heavy loading causes the maximum power

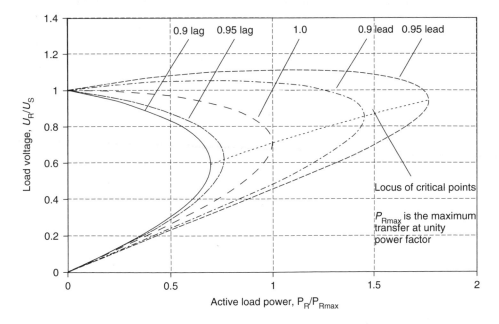

Figure 20.4 Nose curve for the system shown in Figure 20.2, with different load power factors. *Note*: U_R, U_S, voltage at the receiving and sending end, respectively; P_R, active power at receiving end

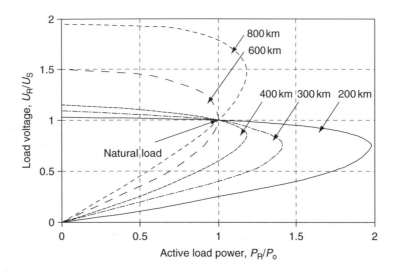

Figure 20.5 Relationship between receiving-end voltage, line length and load of lossless line. *Note*: U_R, U_S, voltage at receiving end and sending end, respectively; P_R, P_o; active power of receiving end and natural load, respectively

transfer to decrease. As the values in Figure 20.3 are normalised, it is difficult to see the impact of line length. In addition, it becomes important for lines that are longer than 100 km to take into consideration the shunt capacitances of the transmission line. Figure 20.5 illustrates the performance of lines of different length at unity load power factor.

In Figure 20.5, P_o is the so-called natural load; that is, the power delivered by the transmission line if it is terminated by its surge impedance Z_c (i.e. $Z_{LD} = Z_c$). Surge impedance, Z_c, is defined as $\sqrt{(L/C)}$, where L and C are respectively, the inductance and capacitance of the line per unit length. For a line exceeding 400 km the shunt capacitance of the transmission line partly compensates the reactive power consumption of the line but results in a high voltage at the receiving end, especially during light loading when the reactive power consumption of the line is lower. For lines longer than 600 km, at natural load the receiving voltage U_R is on the lower branch of the nose curve. This means that the voltage during such an operation is likely to be unstable (Kundur, 1993). For lines longer than 600 km, series compensation is used to reduce the impedance of the line and therefore the consumption of reactive power.

20.2.3 Power output of wind turbines

As the power output of wind turbines is very important in voltage stability studies and for the estimation of current-carrying capacity margins, it should be mentioned that it depends on the temperature and height above sea level. Figure 20.6 shows the power curve for a stall-regulated BONUS 600 kW turbine.

Figure 20.6 illustrates that for certain conditions (Curve 2), the power curve of the wind turbine is almost the same as for standard conditions (Curve 1). However, in some

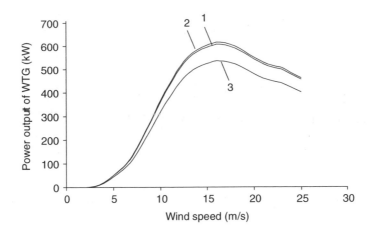

Figure   20.6 Power curves for a BONUS 600 kW wind turbine generator (WTG) for various heights above sea level, h_{sea}, pressure, P, and temperature, T. Curve 1: $h_{sea} = 0$ m; $P = 1013$ mbar; $T = 15°$ C.   Curve   2: $h_{sea} = 1500$ m; $P = 910$ mbar, $T = -18°$ C.   Curve   3: $h_{sea} = 1500$ m; $P = 910$ mbar; $T = 20°$ C

cases the difference is more significant (Curve 3). The effect of temperature and altitude varies with the type of turbine, though, and should be taken into account if appropriate.

20.2.4 Transient stability

Transient stability is the ability of the power system to maintain synchronism when subjected to severe transient disturbances. Stability depends on both the initial operating state and the severity of the disturbance (Kundur, 1993).

Looking at it from the point of view of thermal limits or steady-state voltage stability, wind power is like any other type of power generation. However, the behaviour of wind turbines during and after disturbances is different from that of conventional generators.

The disturbances, which are usually analysed in transient stability studies, are phase-to-ground, phase-to-phase-to-ground or three-phase short circuits. They are usually assumed to occur on the transmission lines. The fault is cleared by opening the appropriate breakers to isolate the faulted element.

Small wind farms usually do not contribute to the transient stability of the transmission system. The impact of a large wind farm on the transient stability of the system depends on the type of wind turbines used and is discussed in detail in Chapter 28.

20.2.5 Summary

In this section we discussed factors that determine the power transmission capability of power lines. Short transmission lines (shorter than 100 km) typically have thermal limits. The current-carrying capacity of the line can be calculated independent of the system configuration. However, the maximum active power transfer is influenced by the load power factor and therefore depends on the particular system. If power is transferred over long distances there are usually voltage stability problems before the thermal limits are reached. Voltage stability and transient stability limits depend on the system configuration and loading.

20.3 Transmission Capacity: Methods of Determination

The transmission capacity is determined by the TSO. Interconnected systems are usually operated by a number of different TSOs, and each TSO has access to data regarding network configuration, the associated equipment and operational statistics for its own system only. Therefore there are different methods for determining the transmission capacity within the area of the respective TSO and the cross-border transmission capacity. The transmission capacity that is available within each area depends mainly on the technical properties of the respective system, whereas the determination of the cross-border transmission capacity depends also on assumptions made by the respective TSOs.

20.3.1 Determination of cross-border transmission capacity

To provide consistent capacity values, European TSOs publish net transmission capacities (NTCs) twice a year. For each border or set of borders, the NTC is determined individually by all adjacent countries and, in the likely case of different results, the involved TSOs have to negotiate the NTC.

(Haubrich *et al*. 2001) studied the approaches that TSOs use to determine the transmission capacity between EU member states Norway and Switzerland. The results showed that the methods applied by the TSOs follow the same general pattern:

1. A base case network model reflecting typical load flow situations is prepared.
2. According to the transport direction for which the power capacity has to be determined, the generation in the exporting country is increased by a small amount and is decreased by the same amount in the importing country, thus simulating an incremental commercial power exchange ΔE.
3. The resulting simulated network state is checked as to whether it complies with the security criteria of the individual TSOs.
4. The highest feasible exchange denotes how much power can additionally be transmitted in the given base scenario. The base case commercial exchange is then added to ΔE in order to obtain the total transmission capacity.
5. There are numerous sources for uncertainty regarding the determination of the transmission capacity. Some of them are included in the security assessment (Step 3). Others are treated implicitly and lumped into the transmission reliability margin. The transmission reliability margin is then subtracted from the total transmission capacity in order to obtain the final NTC.

Although the general pattern is similar, different TSOs use different interpretations and definitions. Net transmission capacity, for example, depends on the assumptions of the base case (Step 1). The base scenario may change in the next calculation cycle and influence the NTC, even if the technical parameters remain constant. TSOs also model the generation increase or decrease, ΔE, in different ways.

When determining the limits of a feasible network operation, some TSOs include thermal limits that vary over the year and across their respective areas. Other TSOs apply probabilistic models based on meteorological statistics. Some TSOs assume constant ambient conditions throughout the year. The maximum allowed continuous conductor temperature differs largely from one TSO to the other, with values ranging from 50 °C to 100 °C. Some TSOs allow higher continuous current limits in $(n - 1)$ contingency situations (i.e. the outage of a single network element). However, the percentage of accepted overloads is also different for different TSOs. Many TSOs tolerate higher current limits in contingency situations only if the TSO itself is able to decrease the loading to a level below normal limits within a short period of time (10–30 min). Voltage stability is assessed for normal system operation as well as for the operation after $(n - 1)$ contingencies that are considered relevant for the security assessment. Some TSOs investigate not only single failures but also certain failure combinations. Nordic TSOs (Fingrid, Statnett, and Svenska Kraftnät) also include busbar failures, as possible consequences may be severe and endanger the security of the system. The TSOs also have diverging methods for assessing uncertainty as well as sources of uncertainty.

Apart from determining the NTC twice a year, the available transmission capacity is also determined on a daily or weekly basis for deay-ahead congestion forecasts. The method follows the pattern described above, but here the base case reflects a load flow forecast based on snapshots from the preceding day, and sometimes weather forecasts. The system models are updated according to changes in topology and switching status. Weather forecasts are used by some TSOs to allow for higher thermal transmission limits. As there are less uncertainties related to the short-term horizon, transmission reliability margins can be decreased. As a result of these factors, the actual allocation of transmission capacity can vary substantially from the NTC values that are calculated twice a year.

20.3.2 Determination of transmission capacity within the country

Each TSO applies to its area of responsibility complete system models for assessing the available transmission capacity. All TSOs use similar methods, but the security standards and base scenario assumptions again can vary considerably.

As mentioned in Section 20.2.1, thermal limits can be evaluated for each line depending on material, age, geometry, the height of the towers, the security standards that limit the clearance to the ground and so on. There can be different policies regarding the tolerance to overloads during contingencies, too.

The load flow is calculated for different scenarios. These calculations are then used for assessing the voltage stability during normal operation as well as after the most frequent and severe contingencies. The selection of failures to be assessed is based on the subjective distinction between 'frequent' and 'rare' failures and between 'severe' and 'minor' consequences (Haubrich et al., 2001). As was discussed in Section 20.3.1, the types of contigencies that are included in the assessment vary considerably.

Transient stability is analysed by performing dynamic simulations of the system during and after the fault. The transmission limits for the respective lines are defined by the most severe set of conditions.

The transmission limits within the country are taken into account:

- assessing the security for the cross-border capacity allocation;
- in order to adopt corrective measures in the case of congestions, such as redispatching generators, switching and load shedding;
- in order to prepare the technical prerequisites for new generation and so on.

20.3.3 Summary

Transmission capacity does not always reflect only the technical properties of the system. Determining the transmission capacity is a complex task and its results depend heavily on the assumptions made by each individual TSO. Although the general scheme for capacity allocation is similar for all European TSOs, differences in definitions and assumptions can lead to lower calculated transmission capacities. Thus a harmonised approach to determining the transmission capacity can increase transmission capacity without requiring substantial investment. This and other methods to increase transmission capacity will be discussed in the next section.

20.4 Measures to Increase Transmission Capacity

20.4.1 'Soft' measures

'Soft' measures may result in an increase of transmission capacity at low cost. They mostly concern cross-border transmission capacity; however, some of them may be applied to increase transmission capacity within a country:

- Determination methods between adjacent TSOs can be harmonised, with transparency regarding the base case assumptions (Haubrich et al., 2001).
- Ambient temperature and wind speed statistics can be considered when determining the transmission capacity.
- Temperature and wind speed forecasts for day-ahead capacity allocation can be used.
- On the one hand, the deterministic $(n-1)$ criterion leads under certain conditions to unnecessarily high congestion costs for power producers and for grid operators. On the other hand, during, for example, under unfavourable weather conditions, $(n-1)$ constraints may not provide sufficient security. A novel approach to online security control and operation of transmission systems is suggested by Uhlen et al. (2000). Their approach is based on a probabilistic criterion aiming at the online minimisation of total grid operating costs, which are defined as the sum of expected interruption costs and congestion costs during specified periods of operation. This approach is currently being studied by the Norwegian TSO and by the SINTEF (Stiftelsen for industriel og teknisk forskning ved Norges tekniske høgskole) research institute. But even the approach currently applied by the Norwegian system operator allows flexible

transmission limits by taking into account various criteria, such as the costs for redispatching, weather conditions and system protection. The value of the reduced congestion is then weighted against the expected interruption costs and/or possible congestion management costs. If the value of the reduced congestion is substantially higher than the increase in expected interruption costs, the system is operated at $(n - 0)$ reliability (Breidablik, Giæver and Glende, 2003).

- Temporary overloading after contingencies may be allowed. Depending on ambient temperature, a short-term loading of 30–50 % in excess of rated capacity is accepted for transformers, breakers and other components without a substantial loss of lifetime (Breidablik, Giæver and Glende, 2003).
- Unjustified, rare, fault considerations can be excluded from the determination of the transmission capacity.
- A probabilistic evaluation of operational uncertainties can be carried out. In this case, risk could be defined as the probability of having to adopt undesired measures during the operational phase multiplied by the costs caused by such measures (Haubrich *et al.*, 2001).

This list of measures concerns mainly improvements to the methods applied for determining transmission capacity. Extensive use of system protection schemes, especially of automatic control actions following critical line outages, may help to relax transmission congestions without substantial investments; measures include:

- the shedding of preselected generators;
- the shedding of preselected loads (industrial);
- network splitting to avoid cascading events.

These schemes are implemented primarily to allow increased transmission limits in addition to reducing the consequences of disturbances or interruptions (Breidablik, Giæver and Glende, 2003).

20.4.2 Possible reinforcement measures: thermal limit

By taking the thermal current limit as the critical factor, reinforcement measures (apart from the construction of new lines) can aim either at increasing the current limits of individual lines and/or the associated equipment, such as breakers, voltage and current transformers, or at optimising the distribution of load flows in order to decrease the loading of the critical branches. To increase current limits of individual lines, the following measures can be applied (Haubrich *et al.*, 2001):

- Determine the dynamic current-carrying capacity rating by means of real-time monitoring of line tension or sag as well as line current and weather conditions. This method reduces the probability of line overheating compared with the case where conservative weather conditions are assumed (wind speed 0.6 m/s; temperature 15 °C), (Douglas, 2003).
- Shorten insulators.
- Increase the tensile stress of conductors.

- Increase the height of the towers.
- Replace underdimensioned substation equipment.
- Install conductors with higher loadability, where old supporting structures can be reused, (Douglas, 2003). Recently developed high-temperature low-sag conductors may have the same diameter as the original conductors and thus the existing structure may not have to be reinforced.

If the associated line equipment limits the capacity it can be replaced by equipment with a higher rating.

The following measures can potentially influence the distribution of load flows:

- use of phase-shifting transformers;
- use of series capacitors or series reactors to adjust the impedances of the lines;
- inclusion of FACTS elements (including DC links) that use power electronics to control voltage and active – reactive power.

Phase-shifting transformers and series capacitors or series reactors are usually the preferred options. FACTS are extremly flexible, but also very expensive. FACTS devices may also negatively affect the reliability of the network by diverting power to weaker parts of the network (Douglas, 2003).

20.4.3 Possible reinforcement measures: voltage stability limit

If voltage limits or voltage stability are the determining factor for the transmission capacity, additional sources of reactive power (i.e. shunt capacitors, shunt reactors or FACTS elements) can be installed at critical locations in order to smoothen the steady-state voltage profile and to increase reserves against the loss of voltage stability. If voltage instability is caused by power transfer over long distances, series capacitors can be installed to decrease the impedance of the lines (see Section 20.2.2). The applicability of the suggested reinforcement measures depends on the individual network topology.

The obvious and most effective measure to increase transmission capacity is to build a new transmission line. However, this is a time-consuming solution as it takes about five years as well as being an expensive option. The cost of transmission line is approximately SEK4 million per kilometre (Arnborg, 2002).[1] Over the past few years, as a result of environmental concern it has become difficult to obtain permission to build new overhead transmission lines.

20.4.4 Converting AC transmission lines to DC for higher transmission ratings

One way to increase transmission capacity is to convert power lines from high-voltage AC (HVAC) to high-voltage DC (HVDC). This allows one to increase the power

---

[1] As at 7 September 2004, SEK1 = € 0.11.

transmission rating 2–3 times and reduce transmission losses (Häusler, Schlayer and Fitterer, 1997). The convertion includes changes to the conductor arrangement on the tower, the insulator assemblies and the configuration of the conductor bundle. The actual tower structure or the number of towers does not have to be changed, though. Depending on the condition of the existing conductors, it is possible to reuse them for the new HVDC link. In the case of multiple-system lines, some tower designs even allow a convertion in stages, and transmission can be continued.

An advantage of converting long HVAC lines (longer than 300 km) to HVDC is that the thermal limit rating can be fully used. This particulary increases the avalability of the double bipole systems. In the case of an outage of one HVDC line, the remaining bipole system can transmit double the power. As mentioned in Section 20.2.2, long HVAC transmission lines often cannot be loaded to their thermal rating because of voltage stability problems.

Depending on the condition of the existing system, the cost of converting the line from HVAC to HVDC can be about 30 % to 50 % lower compared with building a new transmission line. However, this figure does not include the construction of two converter stations. The disadvantage of a conversion to HVDC is that the power systems on both ends of the HVDC link become decoupled regarding frequency control. This may cause problems in power systems where the controllable units are concentrated at one end of the HVDC link.

20.5 Impact of Wind Generation on Transmission Capacity

With the integration of new generation comes an increased need for additional transmission capacity. The issuses discussed above are relevant to any type of new generation, not only wind power. However, wind power has some special features that have to be taken into account when assessing transmission capacity.

First, wind power production has to be evaluated taking into account its low utilisation time (2000–3000 hours per year), the spacial smoothing effect and the fact that the power output is a function of the ambient conditions (see Section 20.2.3). After that, wind power can be treated as any conventional generation when evaluating the thermal limits. Wind speed measurements from wind farms can even be used for the online estimation of the current-carrying capacity of short transmission lines (see Section 20.2.1).

The induction generators that are used in wind power applications consume reactive power. If there is no reactive power compensation, this results in a lagging power factor at the wind farm connection point. As shown in Figure 20.4, this may decrease the maximum power transfer from the wind farm to the network, if the limit is defined by voltage stability considerations. Reactive power compensation of wind turbines is usually provided by shunt capacitor banks, SVC or AC/DC/AC converters. Reactive power compensation provided by shunt capacitor banks depends on the voltage at the connection point and therefore may not be sufficient for lower voltage. However, if continuous reactive power compensation is used through AC/DC/AC converters, for example, wind power does not affect the maximum power transfer if the limit is defined by voltage stability considerations. Moreover, if at the wind farm connection point a leading power factor is provided, the maximum power transfer over the considered line could be increased, especially if it is acceptable to have a higher voltage at the wind farm connection point (see Section 20.2.2).

During and after faults in the system, the behaviour of wind turbines is different from that of conventional power plants. Conventional power plants mainly use synchronous generators that are able to continue to operate during severe voltage transients produced by transmission system faults. Variable-speed wind turbines are disconnected from the grid during a fault in order to protect the converter. If a large amount of wind generation is tripped because of a fault, the negative effects of that fault could be magnified (Gardner et al., 2003). This may, in turn, affect the transmission capacity in areas with significant amounts of wind power, as a sequence of contingencies would be considered in the security assessment instead of only one contingency. During a fault, fixed-speed wind turbines may draw large amounts of reactive power from the system. Thus, the system may recover much more slowly from the fault (see also Chapter 28). This could also affect transmission capacity.

There are several reasons why the integration of large-scale wind power may have a particular impact on the methods that are used for determining the available transmission capacity:

• The power output of wind farms depends on wind speed, therefore TSOs should include wind forecasts in the base case for determining the day-ahead transmission capacity and also use wind speed statistics in the base case that is used for determining the NTC twice a year. There may be higher uncertainties associated with prediction errors regarding the generation distribution and this may result in an increased transmission reliability margin, which in other word corresponds to a decrease in transmission capacity.
• Compared with conventional generation, for wind farms, less sophisticated models of generator characteristics are used. This could make simulation results less reliable (i.e. some TSOs may choose to increase transmission reliability margins to account for that).

Apart from the impact that wind power has on the methods for determining transmission capacity, its integration also requires greater investment regarding some of the measures for achieving an increased transmission capacity. It may, for instance, be significantly more expensive to provide sophisticated protection schemes for wind farms that are distributed over a certain area than for conventional generation of an equivalent capacity (Gardner et al., 2003). Wind farms are built in remote areas where the grid reinforcements are more urgent and more expensive than in areas close to industrial loads, where conventional generation is usually situated. Owing to the low utilisation rate of the wind turbines, the energy produced per megawatt of new transmission is low. The resulting question to what extent it is economic to increase transmission capacity for a particular wind farm will be addressed in the next section.

20.6 Alternatives to Grid Reinforcement for the Integration of Wind Power

Transmission capacity should not be increased at any cost. The optimal balance between extra income and the cost of additional transmission capacity has to be struck. Usually, it is considered not optimal completely to remove a bottleneck. Furthermore, the increase of transmission capacity is time-consuming. Therefore, other measures are necessary to handle congestion and the related problems and to allow a large-scale integration of wind power. Such measures include reducing the power production in conventional generation units, where fast regulation is possible, and/or curtailing excess wind energy, when the transmission system is congested.

20.6.1 Regulation using existing generation sources

One possibility to integrate wind power with only partial or without any grid reinforcement is to regulate power transmission with power storage devices or with conventional power sources, if fast regulation of the power output is possible. Examples of this approach are the use of gas turbines, or hydro power in power systems with large hydro reservoirs (e.g. as in Sweden and Norway). During high wind power production and simultaneous congestion problems in the transmission system, hydro power production at the same end of the bottleneck can be reduced and water can be saved in hydro reservoirs (see Figure 20.7). Thus, wind power production would not be affected by the congestion problem and it would not be necessary to curtail the production of wind energy. The saved water can then be optimally used by the hydro power plant (HPP) once the bottleneck is relieved (Matevosyan and Söder, 2003). Redispatching HPPs can, however, affect the efficiency of the hydro system. Furthermore, in areas where water inflow and wind speeds are correlated, the capacity of the hydro reservoirs will put additional limitations on the regulating capability.

The regulation of power transmission by hydro power can be relatively easily accomplished if wind farms and HPPs are owned by the same company. In this case, wind power production forecasts can be included in the production planning of HPPs (Vogstad, 2000). Forecast errors can be accounted for by frequent updates of the wind power production forecasts, and hydro power production can be changed accordingly so that transmission limits will not be exceeded. In the case of separate ownership, congestion problems caused by wind power production will be solved within the framework of the electricity market.

20.6.2 Wind energy spillage

Another option to introduce a certain amount of large-scale wind power without grid reinforcement is to 'spill' wind if the transmission system is highly loaded.

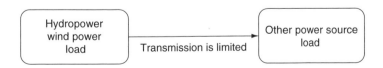

Figure 20.7 A congested system with hydro and wind power production on the same side of the bottleneck

Methods have been developed for estimating the amounts of spilled energy at various wind power penetration levels. These methods can be used at the early stage of a prefeasibility study for a project if wind speed measurements are already available. These estimation methods can also be used by TSOs for defining the technical prerequisites for the connection of a wind farm, or by the authorities for evaluating large-scale projects.

In the following subsections, we will describe two methods: the first method gives a simplified estimation based on several extreme assumptions, and the second estimation is similar to a probabilistic production cost simulation (Booth, 1972). The estimation methods are compared and the results are presented, using the Swedish transmission system as an example. The impact of the different assumptions on the results will be discussed at the end of this section.

20.6.2.1 Simplified method

A straightforward way to estimate the capacity of wind power that can be installed in an area with limited transmission capability is to compare the present transmission duration curve (TDC) for the bottleneck and the expected wind power production duration curve (WPDC), (see Figure 20.8). The WPDC is obtained by using wind speed measurements from the studied area. This is converted into power and scaled up to the respective installed capacity. The smoothing effect within a wind farm is not included in the considerations, and all wind turbines are assumed to be 100 % available.

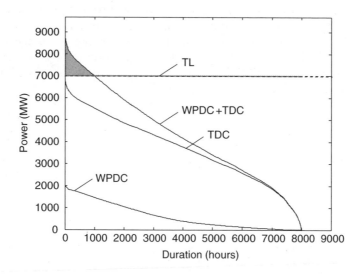

Figure   20.8 A transmission duration curve (TDC), a wind power production duration curve (WPDC), their sum (WPDC + TDC) and the transmission limit (TL); the area below WPDC + TDC and above TL corresponds to the amount of wind power that has to be spilled

This can be considered an initial estimation, but it will give an idea of the capacity of wind power that can be installed. The following assumptions are made:

- The transmission limits are given and are constant for the studied period.
- Only active power flows are considered.
- The produced wind power can be consumed on the other side of the bottleneck.
- Hourly variations of expected wind power production (calculated from wind speed measurements) and hourly variations of the actual power transmission from the studied area are assumed to be simultaneous. This means, for instance, that the maximum power transmission occurs at the same time as the maximum wind power production.

The last assumption may be unrealistic, but it represents the most extreme case that can occur.

When wind power production is high, it may be impossible to transfer all the generated power because of the transmission limits in the network. If the excess power cannot be stored in hydro reservoirs or by other means it has to be spilled. Given the assumptions listed above, the spilled energy, W_{spill}, can be calculated in the following way:

$$W_{\text{spill}} = \int_{0}^{t_{\text{TL}}} [P_{\text{TDC}}(t) + P_{\text{WPDC}}(t) - P_{\text{TL}}]\, dt, \tag{20.3}$$

where $P_{\text{TDC}}(t)$ and $P_{\text{WPDC}}(t)$ are the values of power, in megawatts, at time t on the TDC and WPDC, respectively; P_{TL} is the transmission limit in megawatts; t_{TL} is the number of hours during which the transmission limit is exceeded (i.e. when $P_{\text{TDC}}(t) + P_{\text{WPDC}}(t) \geq P_{\text{TL}}$).

The percentage expected spilled wind energy, k_{spill}, is calculated as follows:

$$k_{\text{spill}} = \frac{W_{\text{spill}}}{W_{\text{W}}} \times 100, \tag{20.4}$$

where W_{W} is the energy generated by the wind farm during the analysed period T_{h}:

$$W_{\text{W}} = \int_{0}^{T_{\text{h}}} P_{\text{WPDC}}(t)\, dt. \tag{20.5}$$

Figure 20.8 presents the results of this simplified estimation method. The results are based on the existing TDC, corresponding to the transmission from the northern to the southern and central parts of Sweden in 2001 (Arnborg, 2002). To obtain the WPDC, we have used wind speed measurements from Suorva in northern Sweden (Bergström, Källstrand, 2000) and the power curve for a pitch-regulated wind turbine. The calculated wind power output is scaled up to represent 2000 MW of wind power in northern Sweden. The effect of the fact that the wind farms are distributed over large area is not included. From Equation (20.3), the shaded area below WPDC + TDC and above TL in Figure 20.8 corresponds to the amount of wind energy that has to be spilled during a year.

20.6.2.2 Probabilistic estimation methods

Another approach to estimate the total installed capacity of wind power that can be integrated in areas with bottleneck problems is to calculate the probability of exceeding the transmission capacity limit. This approach is similar to production cost simulations (Booth 1972), and the estimation has to be based on statistical data. The assumptions are the same as those for the simplified method, except for the last assumption, which is changed as follows:

• Wind power output and transmitted power are independent variables. This means that it is equally probable to have high wind power production during times of high power transmission as during times of low transmission.

In the following subsections, X is the power flow in megawatts that is transmitted through the bottleneck in addition to the wind power generation, and Y corresponds to the generated wind power in megawatts.

Discrete probabilistic method

If the estimation method is based on statistical data from measurements, then X and Y are considered discrete variables.

The distribution function for the transmitted power and the corresponding probability mass function are:

$$F_X(x) = P(X \leq x),$$

$$f_X(x) = P(X = x),$$

where $P(X \leq x)$ is the probability of transmission X having a value less than or equal to a level x, and $P(X = x)$ is the probability of transmission X being exactly x. Distribution and probability mass functions can be obtained from long-term measurements of X.

Similarly, a distribution function and probability density function can be expressed for the wind power output Y:

$$F_Y(y) = P(Y \leq y),$$

$$f_Y(y) = P(Y = y).$$

Using long-term wind speed measurements, the power output Y of the planned wind farms can be obtained from the power curve of the wind turbines. Then, distribution and probability mass functions of Y are calculated.

Now we introduce the discrete variable Z, with $Z = X + Y$. Z is the desired power transmission after wind power is installed in the area with the bottleneck problems. Its probability mass function $f_Z(z)$ is obtained from a convolution, as follows (Grimmett and Strizaker, 1992):

$$f_Z(z) = \sum_x f_X(x) f_Y(z - x) = \sum_y f_X(z - y) f_Y(y). \qquad (20.6)$$

The distribution function of the discrete variable Z is

$$F_Z(z) = \sum_{i:\, z_i \leq Z} f_Z(z_i). \qquad (20.7)$$

It is more convenient to use a duration curve instead of a distribution function as it shows the probability of exceeding a certain level (e.g. $1 - F_Z(z) = P(Z \geq z)$). Figure 20.9 illustrates the results of the discrete probabilistic estimation for the same case as before. In this figure, the value $1 - F_Z(z = \text{TL})$ corresponds to the probability of exceeding the transmission limit. If the transmission limit is equal to 7000 MW, the probability that the transmission limit is exceeded is 0.05.

Continuous probabilistic method
The discrete probabilistic method can be generalised for applications where long-term measurements are not available. Variables X, Y and Z are then assumed to be continuous with a known distribution law. For variable X, we assume the Gaussian function with a known mean value m_G and a standard deviation σ:

$$f_X(x) = \frac{1}{\sigma(2\pi)^{1/2}} \exp\left[\frac{-(x - m_G)^2}{2\sigma^2}\right]. \qquad (20.8)$$

The corresponding distribution function is

$$F_X(x) = \int_0^x f_X(u)\,\mathrm{d}u = \frac{1}{\sigma(2\pi)^{1/2}} \int_{-\infty}^x \exp\left[\frac{-(u - m_G)^2}{2\sigma^2}\right]\mathrm{d}u, \qquad (20.9)$$

where u is the integration variable.

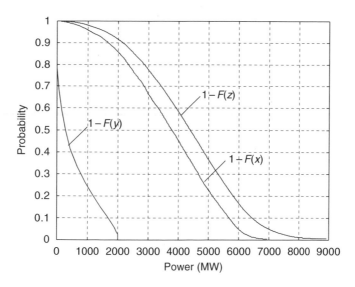

Figure 20.9 Wind power production, actual transmission and desired transmission duration curves – $1 - F_X(x)$, $1 - F_Y(y)$ and $1 - F_Z(z)$ respectively – obtained from a discrete probabilistic estimation (transmission limit = 7000 MW)

For wind power applications, the Reighly or, more generally, the Weibull functions are representative (Burton *et al.*, 2001). In this case, we will use the Weibull function. For this function, the shape parameter k can be adjusted to fit better the discrete data:

$$f_V(v) = \frac{k}{g}\left(\frac{v}{g}\right)^{k-1}\exp\left[-\left(\frac{v}{g}\right)^k\right], \tag{20.10}$$

$$F_V(v) = \frac{k}{g^k}\int_0^v u^{k-1}\exp\left[-\left(\frac{u}{g}\right)^k\right]du, \tag{20.11}$$

where v is the wind speed in m/s; and k and g are, respectively, the shape and scale parameters of the Weibull distribution.

Now it is necessary to arrive from wind speed v at power output y in megawatts. If a power curve,

$$y = \frac{1}{2}c_p A_R \rho V^3 \times 10^{-6},$$

of a pitch-regulated wind turbine is approximated by a smooth, increasing function, it can be inverted according to the following equation:

$$v(y) = \left[\frac{2y \times 10^6}{A_R c_p(y)\rho N_{WT}}\right]^{1/3}, \tag{20.12}$$

where A_R is the swept area in m^2, $c_p(y)$ is the performance coefficient of the turbine, ρ is the air density in kg/m^3, and N_{WT} is the number of wind turbines (WTs).

Substituting Equation (20.12) in Equation (20.10), we obtain the probability density and distribution functions of the wind power production:

$$f_Y(y) = \frac{k}{g^k}\left[\frac{2y \times 10^6}{A_R c_p(y)\rho N_{WT}}\right]^{(k-1)/3}\exp\left[\frac{2y \times 10^6}{A_R c_p(y)\rho N_{WT}g^3}\right]^{k/3}, \tag{20.13}$$

$$F_Y(y) = \int_0^y f_Y(u)\,du = \frac{k}{g^k}\int_0^y v(u)^{k-1}\exp\left[-\left(\frac{v(u)}{g}\right)^k\right]dv(u) \tag{20.14}$$

where v is expressed according to Equation (20.12).

The probability density function for the desired transmission $f_Z(z)$ is obtained from the convolution expression (Grimmett and Strizaker, 1992):

$$f_Z(z) = \int_o^{y_{max}} f_X(z-y)f_Y(y)\,dy. \tag{20.15}$$

Finally, the distribution function of the continuous variable Z is calculated:

$$F_Z(z) = \int_{-\infty}^{z} f_Z(u)\, du. \tag{20.16}$$

The quantity $1 - F_Z(z = \text{TL})$ corresponds to the probability of exceeding the transmission limit. The integral of $1 - F_Z(\text{TL} \leq z < \infty)$ is equal to the wind energy that has to be spilled. The details of this method are given in Sveca and Söder (2003).

As was mentioned above, the simplified method will give a rough estimation, but it will represent the worst case that can occur in the area with bottleneck problems. This method is also easily applied and does not require major computational resources.

If detailed data about transmission and wind speeds are available, simplified and discrete probabilistic estimations may be used. The continuous method should be applied if only the mean value and deviation are available. However, the continuous method has one limitation. We have assumed that all turbines are of the same type, although in reality this is not always true. This complication can be dealt with easily in the simplified or discrete probabilistic method, but for the continuous method the solution will be rather complex.

20.6.2.3 Application to the Swedish transmission system

In order to present a possible application of the suggested methods, we use them to study the large-scale integration of wind power in northern Sweden and to weigh the cost of the spilled wind energy against the cost of grid reinforcement. Some of the results have already been presented in the preceding sections to illustrate the suggested methods. In the following, we will provide some facts regarding the Swedish power system.

Swedish power system
The Swedish transmission system was built to utilise hydro power as efficiently as possible, to match the increased electricity consumption. The total installed capacity of the generating units is about 32 000 MW, of which 51.2 % is hydro power, 29.8 % is nuclear power and 18.1 % thermal power. Installed wind power capacity corresponds to about 1.1 % of total installed capacity. The largest hydropower plants are situated in the north, and eight long 400 kV transmission lines connect the northern part of the transmission system with the central and southern parts, where the main load is concentrated. During cold days, when power consumption is high, or during the spring flood, when power production is high, a lot of power is transferred from the north, and the power transmission capacity almost reaches its limits. There are also other transmission limits inside the Swedish power system and towards neighbouring countries (Nordel, 2000).

The development of wind power in Sweden is much slower compared with Denmark, Germany and Spain, for instance. The installed capacity of wind power amounts to 426 MW (Sept. 2004) divided among 706 wind turbines. The wind farms and wind turbines are mostly spread along the southern coast, onshore and near the shore of the islands of Gotland and Öland.

The Swedish Government proposed a bill (Governmental Bill 2001/02: 143) on planning objectives for wind power. According to this bill, the objective of 10 TWh of wind power production per year has to be achieved up to 2015. The bill was passed in 2002 and since then several studies have been conducted in order to draw up prerequisites for the integration of wind power in Sweden.

Energimyndigheten, the country's energy authority, has prepared a report pointing out areas that have suitable conditions for the integration of wind power. One alternative that was suggested was to develop large-scale wind power in the mountainous area in the north of the country. Favourable wind conditions and better economic conditions than for offshore wind farms make the area attractive for this purpose, even though the transmission capacity is limited.

In 2002, the Swedish government commissioned TSO Svenska Kraftnät to define general prerequisites for an integration of large-scale wind power (10 TWh) in the mountain area and offshore. The resulting report considers a case with 4000 MW of installed capacity with a utilisation time of 2500 h in northern Sweden (Arnborg, 2002). According to the report, five new 400 kV transmission lines at a total cost of SEK 20 billion would be required to guarantee electricity transmission available for 100 % of the year (for the exchange rate, see Footnote 1, page 444).

Assumptions and results
The following assumptions are made:

- Data from 2001 were used to arrive at the probability distributions.
- Only the inner transmission limit is considered. It is assumed that exports to neighbouring countries are not affected.
- The transmission limit is assumed to be constant.
- Power generated in northern Sweden can be consumed in the central and southern parts of the country.
- Only active power flows are considered, assuming that reactive power compensation can be controlled in hydro power plants.
- Variations of active power output at wind farms are not compensated by regulation at hydro power plants.

In addition, wind speed measurements available from the existing wind turbines in Sourva (northern Sweden) were used. The expected power output is obtained from a typical power curve for a pitch-regulated wind turbine. The calculated power output was then scaled up to represent different levels of installed capacity.

Figure 20.10 illustrates the percentage of wind energy that has to be spilled at various wind power penetration levels. The results were obtained by usind the probabilistic and the simplified method. For a comparison, the calculation is also perfomed to obtain hourly values in 2001. Here, wind power production is modelled using hourly wind speed measurements from Sourva and scaling up the power output in order to represent different levels of installed capacity. Actual hourly power transmission from 2001 is included in the analysis. The expected power transmission with wind power is calculated by summing these two variables for each hour. The resulting power transmission is compared with the transmission limit of 7000 MW and the annual energy spillage is

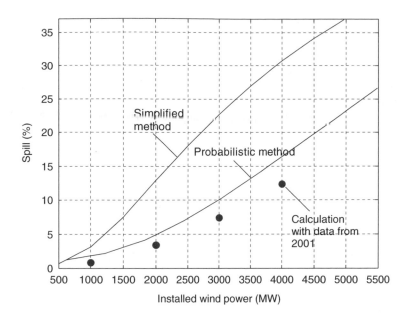

Figure 20.10 The percentage of wind energy that has to be spilled at various wind power penetration levels, estimated by the simplified method, the discrete probabilistic method and calculated with the actual correlation coefficient for the year 2001 (regulation with hydro power is not included)

calculated for each wind power penetration level. During 2001 the correlation between the wind speed measured in Sourva and the power transmission was −0.06. That means that during 2001, the highest wind speeds in northern Sweden were reached during periods when the transmission was slightly lower than the maximum. This result is closer to the probabilistic estimation methods, where transmission and wind power generation are considered to be independent variables. Therefore, we will mainly use the results of the discrete probabilistic method in the following analysis.

Economic evaluation
According to (Arnborg 2002), 4000 MW of additional transmission capacity will cost SEK 20 billion, which for a 10 % interest rate and 40-year lifetime will amount to SEK 2 billion per year. This is the cost of five new 400 kV transmission lines with a transmission capacity of approximately 800 MW per line. Therefore, if a new line is built, it will have to reduce wind energy spillage by at least SEK 400 million per year in order to be economically feasible.

In Figure 20.11(a) the spillage of wind energy illustrated in Figure 20.10 is expressed in millions of SEK per year assuming spillage costs of SEK 0.3 per kilowatt-hour. Using the discrete probabilistic method, the wind energy spillage at a penetration level of 4000 MW of wind power will cost approximately SEK 455 million per year. If a new transmission line is built, 800 MW of wind power can be transmitted without spillage and the residual 3200 MW will be subject to a spillage of 11.6 % of the annual wind energy output. The cost of the spillage is then SEK 260 million per year. This means that

the new transmission line will decrease costs for spillage to SEK195 million per year
(i.e. SEK(455 − 260) million per year). The cost of a line is SEK400 million per year.
Therefore, there is no economic incentive to build a new line, given the assumptions
made above.

If we assume that wind energy spillage costs SEK0.4 per kilowatt-hour and analyse
total costs for spillage versus grid reinforcement costs in the same way as above
(Figure 20.11(b)), the costs for spilled energy at 4000 MW wind power penetration are

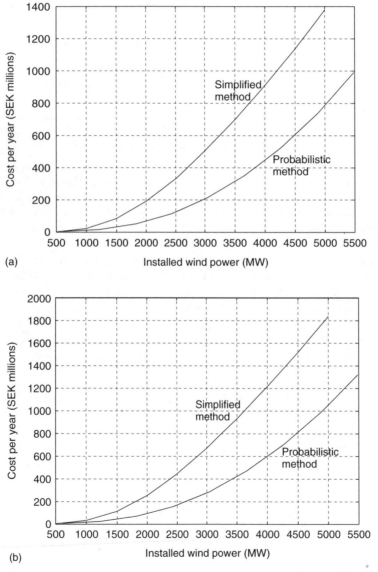

Figure   20.11 Total cost of spilled wind energy at (a) SEK0.3 per kWh and (b) SEK0.4 per kWh,
calculated by the simplified method and the discrete probabilistic method

SEK 600 million per year. For 3200 MW (i.e. 4000 MW − 800 MW), these costs are SEK 346 million per year. Consequently, a new transmission line decreases the cost of wind energy spillage to SEK 254 million per year (i.e. SEK (600 − 346) million per year). The conclusion is that even in this case there is no economic incentive to build a new transmission line at 4000 MW of wind power penetration in northern Sweden.

Discussion of assumptions

In this subsection, the impact of different assumptions on the results of the estimation is summarised:

- In the analysis, wind data from only one site and the power curve of only one turbine were used and scaled up to represent various wind power penetration levels. In reality, however, the wind turbines will be of various types and spread over a wide area. This would lead to a decrease in spilled energy because spillage becomes necessary at high wind speeds, which are not likely to occur simultaneously over a large area.
- It was assumed that there is no regulation in hydro power plants during wind power production peaks. In reality, such regulation could be profitable both for hydro power plants and for wind farm owners if an arrangement can be reached such as one where the water can be stored in reservoirs and a certain revenue received later (see also Section 20.6.1), as wind power cannot be stored (for large-scale wind farms, battery storage is an expensive option). If such an approach is used, wind energy spillage can be substantially reduced.
- If wind energy spillage is lower than determined in the analysis above, the grid reinforcement will appear even less economically attractive. However, grid reinforcement may be seen as attractive if a lower interest rate is used in the calculation. Another consideration may be that the grid reinforcement will enhance competition and increase transmission capacity, which reduces the risk of a capacity deficit.
- It was assumed in the analysis that all wind turbines are of the same type and that reactive power compensation is provided by hydro power plants. However, when it comes to large-scale wind power, the wind turbine type is very important in terms of reactive power regulation capabilities and possible impact on transmission limits.
- Transmission limits are defined by the TSO, based on $n − 1$ criteria and are not subject to substantial variation. However, it is possible to define the transmission limits as a function of weather conditions, power prices and other factors affecting the grid operating costs, for instance (Uhlen et al., 2000). In this case, the transmission limit would become variable, and that could have an impact on the spillage of wind energy.

The analysis above shows that the spilling of wind energy during periods of high hydro power production and full water reservoirs (e.g. during the spring flood) is a natural consequence because expansion of transmission capacity is a more expensive option.

20.6.3 Summary

In this section we have introduced alternative methods for the large-scale integration of wind power into areas with limited transmission capacity. There are two general methods: storage of excess wind power, or spillage. Even though spillage of energy may seem irrational it has been shown that in some cases it could be a more economic option than the construction of new transmission lines. The results of such an analysis, however, are highly dependent on input data and assumptions. Therefore, the estimation of wind energy spillage should be carefully considered for each individual project. The possibilities of energy storage, measures for grid reinforcement and even 'soft' measures to increase transmission capacity should be weighed against the estimated wind energy spillage.

20.7 Conclusions

The available transmission capacity is not a fixed value determined only by the technical properties of the system. It also reflects the policies and assumptions of the respective TSOs. Therefore, 'soft' measures to increase transmission capacity that concern methods of capacity determination and 'hard' measures that concern the installation of new equipment and the building of new transmission lines can be applied.

Such measures are relevant for any type of new generation. The integration of wind power may, however, put additional constraints on the determination of transmission capacity. However, because of its low utilisation time, there are alternative ways to increase the penetration of wind power in congested areas. For each particular project, there is an optimal combination of measures.

References

[1] Arnborg, S. (2002) *Overall Prerequisites for Expansion of Large-scale Wind Power Offshore and in Mountainous Areas*, Svenska Kraftnät, Stockholm.
[2] Bergström, H., Källstrand, B. (2000) 'Measuring and Modelling the Wind Climate in Mountain Valley in Northern Sweden', in *Proceedings of Wind Power Production in Cold Climate Conference, BOREAS V, Finland.*
[3] Booth, R. R. (1972) 'Power System Simulation Model Based on Probability Analysis', *IEEE Transactions on PAS*, 91, 62–69.
[4] Breidablik, Ø., Giæver, F., Glende, I. (2003) Innovative measures to increase the utilization of Norwegian transmission, in *Proceedings of IEEE Power Tech Conference, Bologna*, edited and published by Faculty of Engineering, University of Bologna, Bologna.
[5] Burton, T., Sharpe, D., Jenkins, N., Bossanyi, E. (2001) *Wind Energy Handbook*, John Wiley & Sons Ltd, Chichester, UK.
[6] Douglas, D. (2003) 'Can Utilities Squeeze more Capacity out of the Grid?' *Transmission and Distribution World* (November) 38–43; available at www.tdworld.com.
[7] Gardner, P. Snodin, H., Higgins, A., McGoldrick, S. (2003) *The Impacts of Increased Levels of Wind Penetration on the Electricity Systems of the Republic of Ireland and Northern Ireland: Final Report*, Garrad Hassan and Partners Ltd.; available from www.cer.ie/cerdocs/cer03024.pdf.
[8] Grimmett, G. R., Strizaker, D.R. (1992) *Probability and Random Processes*, Clarendon Press, Oxford, pp. 33–114.
[9] Haubrich, H.-J., Zimmer, C., von Sengbusch, K., Fritz, W., Kopp, S. (2001) 'Analysis of Electricity Network Capacities and Identification of Congestion', report commissioned by the European Commission Directorate-General Energy and Transport; available from www.iaew.rwth-aachen.de/publikationen/EC_congestion_final_report_main_pact.pdf.
[10] Häusler, M., Schlayer, G., Fitterer, G. (1997) 'Converting AC Power Lines to DC for Higher Transmission Ratings', *ABB Review* 3, 4–11.
[11] Heier, S. (1998) *Wind Energy Conversion Systems*, John Wiley & Sons Ltd, Chichester, UK.

[12] House, H. E., Tuttle, P. D. (1958) 'Current-carrying Capacity of ACSR', presented at the AIEE Winter General Meeting, New York.

[13] Kundur, P. (1993) *EPRI Power System Engineering Series: Power System Stability and Control*, McGraw-Hill, New York.

[14] Matevosyan, J., Söder, L (2003) 'Evaluation of Wind Energy Storage in Hydro Reservoirs in Areas with Limited Transmission Capacity', in *Proceedings of the Fourth International Workshop on Large-scale Integration of Wind Power and Transmission Networks for Offshore Wind Farms, October 2003, Billund, Denmark*, edited and published by the Department of Electrical Engineering, Royal Insitute of Technolgy, Stockholm, Sweden.

[15] Nordel (2000) 'Nordel Annual Report', http://www.nordel.org.

[16] Sveca, J., Söder, L. (2003) 'Wind Power in Power Systems with Bottleneck Problems', in *Proceedings of IEEE Power Tech Conference, Bologna*, edited and published by the Faculty of Engineering, University of Bologna, Bologna.

[17] Swedish Government (2001) 'Collaboration for Secure, Efficient and Environmentally Friendly Energy Supply', Government Bill 2001/02: 143, Swedish Government, Stockholm [in Swedish].

[18] Uhlen, K., Kjølle, G. H., Løvås G. G., Breidablik, Ø. (2000) 'A Probabilistic Security Criterion for Determination of Power Transfer Limits in a Deregulated Environment', presented at Cigre Session, Paris.

[19] Van Cutsem, T., Vournas, C. (1998) *Voltage Stability of Electric Power Systems*, Kluwer Academic, Dordrecht.

[20] Vogstad, K.-O. (2000) 'Utilising the Complementary Characteristics of Wind Power and Hydropower through Coordinated Hydro Production Scheduling using the EMPS Model', in *Proceedings from Nordic Wind Energy Conference, NWPC' 2000, Trondheim, Norway, 13–14 March 2000*, edited and published by the Norwegian Institute of Technology, Norway, pp. 107–111.

[21] Wiik, J., Gustafsson, M., Gjengedal, T., Gjerde, J. O. (2000) 'Capacity Limitations on Wind Farms in Weak Grids', in *Proceedings of the International Conference: Wind power for the 21st century, Kassel*, edited and published by WIP-Renewable Energies, Germany, www.wip-munich.de.

21

Benefits of Active Management of Distribution Systems

Goran Strbac, Predrag Djapić, Thomas Bopp and Nick Jenkins

21.1 Background

The penetration of distributed generation (DG) and wind power in particular is expected to increase significantly over the coming years, and a paradigm shift in control, operation and planning of distribution networks may be necessary if this generation is to be integrated in a cost-effective manner. The transition from passively to actively managed distribution networks is driven by (a) government environmental commitments to connect a large number of small-scale generation plants; (b) technological advances in energy generation and storage as well as in information and communication technologies; (c) regulatory reform and unbundling of the energy industry.

The historic function of 'passive' distribution networks is viewed primarily as the delivery of bulk power from the transmission network to the consumers at lower voltages. Traditionally, these were designed through deterministic (load flow) studies considering the critical cases so that distribution networks could operate with a minimum amount of control. This practise of passive operation can limit the capacity of distributed generation that can be connected to an existing system.

In contrast, active management techniques enable the distribution network operator to maximise the use of the existing circuits by taking full advantage of generator dispatch, control of transformer taps, voltage regulators, reactive power management and system reconfiguration in an integrated manner. Active management of distribution networks can contribute to the balancing of generation with load and ancillary services. In the future, distribution management systems could provide real-time network monitoring and control at key network nodes by communicating with generator controls,

Wind Power in Power Systems   Edited by T. Ackermann
© 2005 John Wiley & Sons, Ltd   ISBN: 0-470-85508-8 (HB)

loads and controllable network devices, such as reactive compensators, voltage regulators and on-load tap-changing (OLTC) transformers. State estimation and real-time modelling of power capability, load flow, voltage, fault levels and security could be used to make the right scheduling and constraining decisions across the network. These techniques will probably be applied gradually rather than fulfilling all the above listed attributes right from the beginning (Bopp *et al.*, 2003).

From a regulator's perspective, active management should enable open access to distribution networks. It has the ability to facilitate competition and the growth of small-scale generation. In addition, the use of the existing distribution assets should be maximised to minimise the costs for consumers.

Therefore, an integral understanding of the interrelated technical, economic and regulatory issues of active management and DG is important for the development of the future distribution systems (Jenkins *et al.*, 2000).

21.2 Active Management

In this section, the fundamental features of passive distribution networks are examined and their inability to accommodate increased amounts of DG is discussed. It is demonstrated how the voltage rise effect, the main limiting factor for connecting generators in rural areas, can be effectively controlled within an active network environment and, as a result, enable considerably higher levels of penetration of DG to be connected into existing systems.

In this chapter, local and area-based coordinated voltage control schemes are discussed. Qualitative analysis of these alternative schemes is then carried out on a simplistic distribution network model where the effect of these controls can be easily understood. It should be noted that the principles described are applicable to DG in general, not just wind farms.

Distribution network operators prefer to connect DG to higher voltage levels in order to minimise the impact on the voltage profiles. On the other hand, the developers of DG favour connections to lower voltage levels since connection costs increase notably with the voltage level. Active management enables the voltage rise to be mitigated effectively. Therefore it can have a significant impact on the commercial viability of DG projects and the level of DG penetration that can be connected to an existing system (Strbac *et al.*, 2002).

21.2.1 Voltage-rise effect

The voltage rise effect is a key factor that limits the amount of additional DG capacity that can be connected to rural distribution networks. The impact of DG on the voltage profile can be qualitatively modelled by means of a simple distribution system representation, as shown in Figure 21.1 (Liew and Strbac, 2002). The distribution system (DS) is connected to a 33/11 kV OLTC transformer. The 11 kV voltage side of the OLTC transformer is connected to busbar 1, which is interconnected to busbar 2 via the impedance (Z) of an overhead line. The load, the generation (Gen.) and the reactive compensation device (Q Comp) are connected to busbar 2. The active and reactive power of the load (P_L and

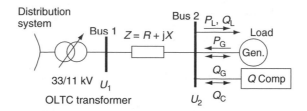

Figure   21.1 Simple system for voltage-rise modelling. *Note*: Gen. = generation; OLTC = on-load tap-changing; Q Comp = reactive compensation device; U_i = voltage at bus i; Z = impedance; R = resistance; X = reactance; P_L = active power of load; Q_L = reactive power of load; P_G = active power of generator; Q_G = reactive power of generator; Q_C = reactive power of compensation device; j= $\sqrt{-1}$

Q_L, respectively), the generator (P_G and Q_G, respectively) and the reactive power of the reactive compensation device (Q_C) are marked as arrows in Figure 21.1.

The voltage U_2 at busbar 2 can be approximated as follows:[1]

$$U_2 \approx U_1 + R(P_G - P_L) + (\pm Q_G - Q_L \pm Q_C)X, \qquad (21.1)$$

where X is the reactance.

The impact of active management control actions, the distributed generator, the load, the OLTC setting U_1 (voltage at busbar 1) and the reactive compensator on the voltage at busbar 2 (U_2) can be derived qualitatively from Equation (21.1).

For passively operated distribution networks, the maximum generation capacity that can be connected can be determined by analysing the worst case. Usually, the highest voltage rise is expected for the coincidence of maximum generation ($P_G = P_G^{max}$) and minimum load ($P_L = Q_L = 0$). For the sake of simplicity, a unity power factor is assumed ($\pm Q_G \pm Q_C = 0$). The voltage for these extreme conditions can be expressed as:

$$U_2 \approx U_1 + RP_G^{max}. \qquad (21.2)$$

The maximum capacity of generation that can be connected to busbar 2 is limited by the maximum allowed voltage at busbar 2, U_2^{max}. This can be seen from Equation (21.3):

$$P_G^{max} \le \frac{U_2^{max} - U_1}{R}, \qquad (21.3)$$

Another important parameter that determines the maximum permissible generation capacity is the resistance, R, of the network impedance, Z. The reactance, X, does not need to be considered in the formula as long as unity power factor operation of the generator is assumed. In this context, it should be mentioned that the voltage rise or drop can be mitigated by reducing the network impedance. This can be achieved by network reinforcement.

---

[1] The convention applied considers flows towards the busbar as positive.

21.2.2 Active management control strategies

21.2.2.1 Active power generation curtailment

Active power generation curtailment controls the voltage by constraining the active power of DG to limit the voltage rise (for a detailed description of active management, see Strbac *et al.*, 2002). Periods when constraints are placed on generation (constraint-on periods) tend to be relatively short because of the low probability of the coincidence of maximum demand and minimum load conditions. Generation curtailment can be profitable for a distributed generator if, in return, a larger plant capacity can be connected. The resulting higher output during periods when there are no constraints can outweigh the monetary impact of the generation curtailment. The price of electricity during the constraint-on periods is anticipated to be relatively low because these periods typically coincide with low load conditions. The impact of the curtailed active power P_G^{curt} on the resulting maximum capacity of generation that can be connected, P_G^{max}, can be seen by comparing Equation (21.3) and with Equation (21.4):

$$P_G^{max} \approx P_G^{curt} + \frac{U_2^{max} - U_1}{R}. \tag{21.4}$$

Given that active power generation curtailment is a local voltage control scheme, this scheme does not require communication systems.

21.2.2.2 Reactive power management

Particularly in weak overhead distribution networks, reactive compensation can effectively limit the voltage rise and can facilitate the connection of higher generation capacities. This is possible because of the relatively high reactance of overhead lines and hence the ability to influence voltage by manipulating reactive flows.

The reactive power absorbed by the network is represented by

$$Q_{import} = -(\pm Q_G - Q_L \pm Q_C).$$

The maximum amount of generation capacity that can be connected to the network under low-load conditions can be expressed as:

$$P_G^{max} \approx \frac{(U_2^{max} - U_1)}{R} + \frac{Q_{import} X}{R}. \tag{21.5}$$

The second term on the right-hand side of the Equation (21.5) quantifies the additional generation capacity that can be connected compared with that represented by Equation (21.3). Note that this scheme will not be very effective in urban systems, as the reactance of underground cables is typically four times smaller than that of overhead lines. Note also that a decreased power factor leads to an increase in network losses, and therefore losses have to be included in the assessment for reactive compensation schemes.

Reactive power compensation is a local voltage control scheme that does not need communication systems.

21.2.2.3 Coordinated voltage control

Coordinated voltage control with on-load tap changers enables the connection of an increased capacity of DG by actively changing the OLTC transformer setting and maintaining the voltages of a distribution network within defined limits.

The OLTC transformer can regulate the voltage U_1. The minimum possible value U_1^{min} allows the highest generation capacity P_G^{max} to be expressed as follows:

$$P_G^{max} \leq \frac{U_2^{max} - U_1^{min}}{R}. \tag{21.6}$$

However, in conditions of maximum load and maximum generation it can be beneficial to separate the voltage control of feeders, which mainly supply load from feeders with a significant generation. This can be achieved by taking advantage of voltage regulators on appropriate feeders. For a complex network configuration, the OLTC transformer and voltage regulator tap settings can be determined by optimisation techniques.

Coordinated voltage control with on-load tap changers and voltage regulators requires a number of measurements from key network points as well as communication systems.

21.3 Quantification of the Benefits of Active Management

21.3.1 Introduction

As mentioned in Section 21.2, the current operating policy based on passive operation of distribution networks limits the capacity of generation that can be connected to the existing networks. It has also been described how the voltage-rise effect could be effectively controlled within an active network environment and, as a result, enable higher levels of penetration of DG into existing systems. The ability of active networks to accommodate DG is now illustrated through use of a characteristic case study of connecting a wind farm to a weak distribution network. It is shown that the amount of DG that can be connected to the existing system can be increased significantly by changing the operating philosophy from active to passive. In this section, the benefits of following four main control strategies are quantified:

- **Active power generation curtailment**: this type of control may be particularly effective when the generator is connected to a weak network, with a high R/X ratio. The generator may find it profitable to curtail some of its output for a limited period if allowed to connect larger capacity and avoid network reinforcement. This may be particularly suitable for wind generation, as the generation curtailment is likely to be required during times when minimum demand coincides with high output, such as summer nights in the UK. However, during that time, it is expected that the energy price will be relatively low.
- **Reactive power management**: the absorption of reactive power can be beneficial in controlling the voltage-rise effect, especially in weak overhead networks with DG. In this respect, a reactive compensation facility, such as a STATCOM, at the connection point may be used (see Saad-Saoud *et al.*, 1998). However, reactive power management as a means of increasing the penetration of DG has not been previously

investigated. In this example, an optimal power flow (OPF) method is applied to illustrate the potential benefits of reactive power management.

- **Area based coordinated voltage control of OLTC transformers**: the introduction of a coordinated voltage control policy may be particularly beneficial for increasing the capacity of wind generation that can be connected to the existing distribution network. Present voltage control in distribution networks is primarily carried out by OLTC transformers. Voltage control is usually based on a simple constant voltage policy or a scheme that takes into account circuit loading while determining the voltage that should be maintained. It is important to bear in mind that this voltage control policy was designed for passive networks with strictly unidirectional power flows. In active distribution networks with multidirectional power flows the validity of this local control voltage practice becomes inherently inadequate. In fact, this practice limits the degree of openness and accessibility of distribution networks and therefore has a considerable adverse impact on the amount of generation that can be accommodated. Alternative voltage control practices that go beyond the present local voltage control, such as an area-based control of OLTC transformers, are considered in this study and the benefits of such policies are quantified. The aim of area-based coordinated voltage control is to maintain the voltage of the network area concerned within the statutory limits.

- **Application of voltage regulators**: in the context of the voltage-rise effect, conditions of minimum load and maximum generation are usually critical for the amount of generation that can be connected. However, it may also be necessary to consider conditions of maximum load and maximum generation. This is because the use of OLTC transformers to reduce the voltage on the feeder where the generator is connected may produce unacceptable voltage drops on adjacent feeders that supply load. In this case, it may be beneficial to separate the control of voltage on feeders that supply load from the control of voltage on feeders to which the generator is connected. This can be achieved by the application of voltage regulators.

The benefits of active management of a distribution network, exercised through the above alternative control strategies, can be quantified by the volume of annual energy and corresponding revenue that can be generated for various capacities of wind generation installed. An OPF tool that minimises the amount of generation curtailment necessary to maintain the voltage with the prescribed limits while optimising the settings of available controls (such as tap position of the OLTC transformer) is used in this study. This analysis is carried out on a characteristic 33 kV network that exhibits all phenomena of interest. Further benefits of active management of distribution networks are discussed qualitatively.

21.3.2 Case studies

21.3.2.1 Description of the system

The 33 kV distribution network on which the case study is carried out is shown in Figure 21.2. The network is fed from a 132 kV network (busbar 1) through an OLTC transformer. Loads are connected to busses 2, 3, 4 and 5. The load at busbar 2 represents the aggregated loads of the remaining part of the system. Distributed wind generation is connected at bus 6, where power factor correction is also connected. Branch parameters of this network are given in Table 21.1.

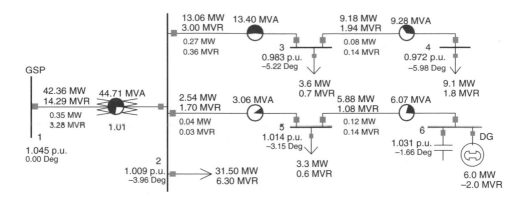

Figure 21.2 Distribution test system in Power World Simulator™, with values shown for the maximum loading condition. *Note*: The buses are numbered 1 to 6; DG = distributed generation; GSP = grid supply point

Table 21.1 Branch parameters of the distribution network considering a 100 MVA base; for an illustration of the system, see Figure 21.2

	From	To	Reactance (p.u.)	Resistance (p.u.)
Branch 1	Bus 1	Bus 2	0.01869	0.17726
Branch 2	Bus 2	Bus 3	0.15500	0.20174
Branch 3	Bus 3	Bus 4	0.09000	0.15770
Branch 4	Bus 2	Bus 5	0.40000	0.30000
Branch 5	Bus 5	Bus 6	0.35700	0.40000

A mixture of residential, industrial and commercial loads is allocated to each of the busbars. Hourly averages of active and reactive power are used to form *annual load profiles*. These take into account daily and seasonal variations of load and may be considered to be typical. Reactive demand profiles are also modelled following generally accepted rules characteristic for the majority of load (variations in active load are considerably greater than that in reactive load). Peak values of the loads are given in Figure 21.2, together with line flows and losses.

Modelling of the hourly output of wind generation is based on the stochastic Markov model presented in Masters *et al.* (2000). This model is used to create a *normalised* annual generation profile, which was the basis for constructing wind generator output for various installed capacities. Reactive demand of the generator is approximately modelled and, together with two sizes of power factor correction applied, average power factors of 0.95 and 0.98 are achieved (absorbing VARs).

21.3.2.2 Base case scenarios

Base case scenarios represent the present, passive approach to voltage control. Two characteristic conditions are considered: (1) minimum load and maximum generation, and (2) maximum load and maximum generation. In all studies, it was assumed that the voltage could vary within $\pm 3\%$ from the nominal value, and the tap-changer setting was adjusted to accommodate maximum DG capacity. Applying Condition (1) it was found that a 10 MW wind generator could be connected; Condition (2) reduces this to only 6 MW. This shows that Condition (1) is not always the worst-case scenario.

21.3.2.3 Tool for modelling the operation of an active distribution system

A distribution management system controller is simulated that applies the concept of optimal power flow to schedule the available controls optimally and to quantify the benefits of alternative voltage control strategies. This benefit is expressed in terms of the increased amount of DG that can be connected.

The main purpose of an OPF applied in the context of active management of distribution networks is to determine the optimal schedule of available controls (generation curtailment, VAR absorption, turns ratio of the OLTC transformer, load shedding, etc.) that minimise the total cost of taking these actions while satisfying voltage and thermal constraints.

In this particular case, the general optimisation task is reduced to a problem of minimising the amount of generation that needs to be curtailed in order to satisfy voltage constraints. Clearly, if the voltage rises above the maximum value as a result of high wind generation output the OPF would enforce constraints by curtailing the generation while keeping the power factor of the wind farm unchanged.

21.3.2.4 Generation curtailment

In this exercise, generation curtailment is used to manage the voltage-rise effect. The OLTC transformer maintains a constant voltage at its terminals. The effect of applying generation curtailment to larger wind farm schemes is investigated. Two case studies are performed: first, operation with an average power factor of 0.98, and, second, operation with an average power factor of 0.95.

Power factor of 0.98
Figure 21.3 shows the resultant *annual* energy produced and curtailed with installed capacity from 4 MW to 20 MW, in steps of 2 MW. Based on current practice, the capacity of DG allowed for connection is generally limited by the extreme situation of minimum or maximum loading and maximum generation output. This condition allows only 6 MW of generation to be connected. For this level of output no generation curtailment occurs and any increase in the level of penetration will lead to a violation of voltage limits at the connection point.

The lighter bars in Figure 21.3 represent the net energy generated in the course of one year, and the darker bars represent the curtailed energy. Net energy generated increases until penetration reaches 12 MW.

Net energy generated increased from 11 102 MWh (for a 6 MW installed capacity) to about 13 972 MWh (for an 8 MW installed capacity), with 898 MWh (or 6.04 %) being curtailed.

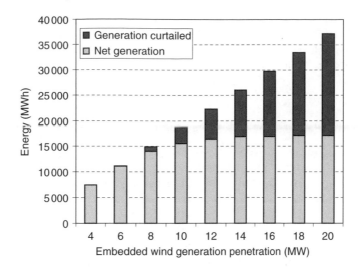

Figure　21.3 Produced and curtailed annual energy for a power factor of 0.98

Power factor of 0.95

A similar case study is performed for an average power factor of 0.95. The results are shown in Figure 21.4. In this case, the net energy generated increases beyond 12 MW, although the energy curtailed for installations larger than 10–12 MW is significant. Comparing this case with the above clearly shows the benefits of operating with lower power factors.

In this case, the net energy generated increased from 11 152 MWh (for a 6 MW installed capacity) to about 14 631 MWh (for an 8 MW installed capacity), where

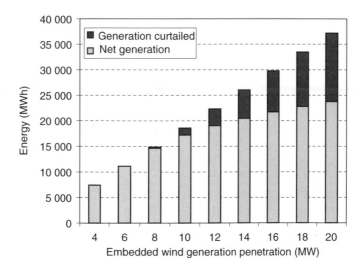

Figure　21.4 Produced and curtailed annual energy for a power factor of 0.95

239 MWh (or 1.61 %) is curtailed. For the 8 MW wind farm, the reduction in average power factor from 0.98 to 0.95 reduces the amount of energy curtailed from 898 MWh to 239 MWh. This clearly shows that the request to operate wind farms with unity power factor (used in some utilities) will severely limit the amount of generation that can be connected.

21.3.2.5 Reactive compensation and voltage control

The effect of applying reactive power compensation to manage the voltage-rise effect and hence allow an increase in penetration of wind generation is achieved by absorbing reactive power at the point of connection. In this case, active power generation would be curtailed only when the reactive power absorbed is insufficient for maintaining the voltage within permissible limits. The capacity of reactive power compensating plants may therefore be important in limiting generation curtailment. The effectiveness of reactive compensation will be system-specific.

In order to simulate this, a static VAR compensator (SVC), with various capacities, has been added to the point of connection. Figure 21.5 shows the amount of energy generated for various penetration levels and various reactive compensation capacities. Figure 21.5 shows that the effect of reactive compensation in this particular case is not very significant. In other words, the reduction of the power factor below 0.95 will not be very beneficial (in this particular system, this power factor appears to be optimal). It is important to bear in mind that in other systems (Liew and Strbac, 2002), the effect of reactive compensation may be considerable. This will be determined primarily by the X/R ratio (driven by the type and the capacity of the circuit).

21.3.2.6 Area-based voltage control by on-load tap-changing transformer

In the previous cases, the OLTC transformer was used to maintain voltage magnitude at busbar 2 to a constant value of 1.00 p.u. In this case, the tap position of the OLTC transformer is optimised in order to minimise generation curtailment. The results are shown in Figure 21.6.

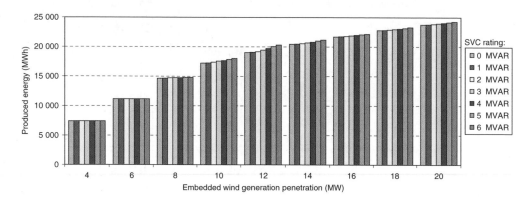

Figure   21.5 Energy generated by the wind farm, by static VAR compensator (SVC) rating

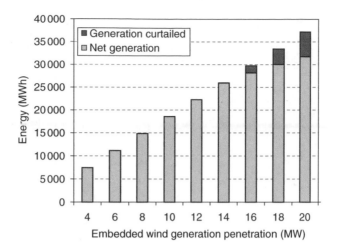

Figure 21.6 Energy generated and curtailed with the application of on-load tap-changing (OLTC) transformer area-based voltage control

Year-round analysis using the OPF tool shows that penetration levels up to 14 MW can be achieved with virtually no energy curtailed (for a 14 MW wind farm, only 87 MWh is expected to be curtailed in one year). For a wind farm of 16 MW, the amount of energy curtailed is 1578 MWh, or 5.3 %. Clearly, area-based voltage control by OLTC transformers allows a considerable increase in penetration of DG. It is, however, important to remember that this technique of voltage regulation will need the implementation of a distribution management system (with appropriate communication systems).

21.3.2.7 Area-based voltage control by on-load tap-changing transformer and voltage regulator

As indicated in Section 21.3.1, the use of OLTC voltage control in order to reduce the voltage on the feeder where the generator is connected may produce unacceptable voltage drops on adjacent feeders that supply loads. The separation of the control of voltage on feeders that supply load, from the control of voltage on feeders to which the generators are connected, is achieved by the application of voltage regulators, and this is investigated in this section.

In order to examine the benefits of this option, a voltage regulator is inserted at the beginning of the feeder, which accommodates the wind farm. This allows independent voltage regulation on feeders with loads by the OLTC transformer, while the voltage regulator controls the voltage on the feeder with the wind farm. In the OPF, the voltage regulator is modelled in a similar way as an OLTC transformer with the ability to change the turns ratio from 0.9 p.u. to 1.1 p.u. continuously. The results of this case are shown in Figure 21.7. It is clear that this voltage control policy increases the amount of generation that can be connected (up to 20 MW) with almost no curtailed energy.

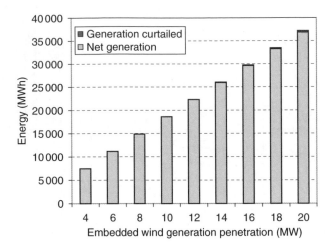

Figure 21.7 Energy generated and curtailed energy from a wind farm over a year, for different values of wind farm penetration

21.3.2.8 Impact of voltage controls on losses

Connection of distributed generators alters the loss performance of distribution networks. A low level of penetration of DG tends to reduce network power flows and thus network losses. However, when penetration is high then DG will export power to the grid and may cause increased losses. In this particular system, with no generation being connected, network losses are 2860 MWh per year. Figure 21.8 shows the change in network losses for various levels of penetration and various voltage control strategies. For a low level of penetration, losses are reduced regardless of voltage control. For the

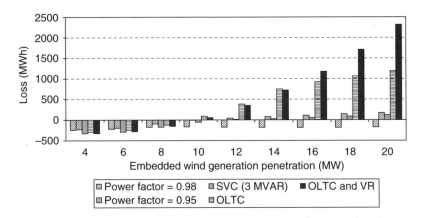

Figure 21.8 The losses for different voltage controls and distributed wind generation penetration. *Note*: use of static VAR compensator; OLTC = use of on-load tap-changing transformer; VR = use of voltage regulator

higher level of penetration, made possible by the application of OLTC transformers and voltage regulators, losses are significant because of the increased amount of renewable energy generated.

21.3.2.9 Economic evaluation

The development of DG will depend on governments' environmental actions, deregulation, novel technologies and economics. Amongst these factors, economics are likely to be decisive for the selection of future technologies and policies. Therefore, an economic evaluation of DG is indispensable. A cost–benefit analysis (CBA) enables an evaluation and comparison of the cost-effectiveness of projects. Monetary values are placed on costs and benefits to find out whether the benefits outweigh the costs. The CBA used for the studies here takes into account: capital costs; operation, maintenance and repair (OM&R) costs; costs of active management schemes; savings from economies of scale; and revenue from energy sales and environmental incentives. Based on an assumed discount rate and a useful plant life time, an evaluation and comparison of net present values (NPVs) of alternative projects can be carried out. A positive NPV indicates that a project tends to be profitable.

The overall benefits of the studied active management schemes are summarised in Figure 21.9 for various levels of DG penetration. The studies performed clearly show

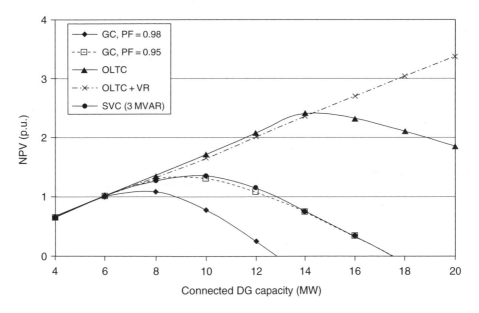

Figure   21.9 Comparison of net present values (NPVs) of active management schemes. *Note*: NPV of 6 MW care is taken to represent 1 p.u.; DG = distributed generation; GC = generation curtailment; PF = power factor; SVC = use of static VAR compensator; OLTC = use of on-load tap-changing transformer; VR = use of voltage regulator

significant benefits of the use of active control in the distribution network. The most beneficial are schemes with area-based voltage control by OLTC transformers and voltage regulators, achieving a threefold increase in the capacity of DG that can be connected.

The aim of the economic evaluation was to investigate the optimum combination of plant capacity and the applied active management scheme. The export capacity of the plant was varied from 4 MW to 20 MW. Five active management schemes were studied. These included generation curtailment (GC) with power factors (PFs) 0.98 and 0.95, 3 MVAR reactive compensation, and OLTC transformer voltage control schemes without voltage regulators (labelled OLTC) and with voltage regulators (labelled OLTC + VR). The NPVs of all cases studied were calculated. Positive NPVs represent overall benefit. The NPV of 6 MW DG represents the maximum DG export capacity that does not require network reinforcement or active management. It is therefore taken as the reference, 1 p.u. NPVs greater than 1 p.u. result from increased income streams that outweigh the costs of active management and additional plant capacity. Maximum NPVs, as depicted in Figure 21.9, indicate the optimum DG plant capacity and active management scheme combinations. In this study, all five active management schemes enable increased energy exports but have different technical and economic limits:

- Generation curtailment with a power factor of 0.98: this has its highest NPV at around 8 MW, but if 13 MW capacity is exceeded it becomes unprofitable.
- Generation curtailment with a power factor of 0.95: this reaches its highest NPV at around 10 MW and seems to be profitable up to around 17.5 MW. Note therefore that higher energy exports are possible at the expense of a decrease in power factor. The difference between both curves quantifies the financial benefit of a decreased power factor. The resulting costs such as cost of network losses and charges for reactive power are not considered in the NPV calculation.
- Use of SVC (3 MVAR): this shows almost the same characteristic as generation curtailment with a power factor of 0.95. The benefit of SVC application in networks with a higher X/R ratio is expected to be higher than in this example with a relative low X/R ratio.
- Use of an OLTC transformer: this scheme has its highest NPV at around 14 MW, but higher DG capacities, up to 20 MW, are still economically feasible.
- Use of an OLTC transformer and voltage regulators: benefit of voltage regulator application can be perceived in particular for DG capacities exceeding the point of best performance (14 MW) of the OLTC scheme, being quantified as the difference between the OLTC and OLTC + VR curves in Figure 21.9.

To access many of these benefits new commercial arrangements have to be developed, and techniques must be established that allow one to determine the positive or negative contribution of all network participants.

The paradigm shift in energy generation, transmission and distribution will also result in costs, but it is expected that active management has the potential to reduce these costs compared with conventional system reinforcement. The cost of active management equipment seems often to be lower than the cost for additional transformers, cables and overhead lines. That is why active management control schemes can be a competitive alternative to conventional system reinforcement with regard to mitigating the voltage-rise effect.

21.3.2.10 The benefits of active management

In this section the benefits of active distribution management have been quantified in terms of the amount of the additional capacity of DG that can be connected to the existing distribution system. The benefits of the studied active management control strategies have been quantified by the volume of annual energy produced and the corresponding revenue that can be generated for various capacities of wind generation installed.

However, the added value of the active management of distribution networks is multifaceted. The benefits of active management are not limited to the facilitation of higher penetration of DG in existing networks or to the reduction of connection costs through the maximisation of the use of existing networks. For instance, energy generated close to demand is considered to have a higher value because of reduced network losses and avoided utilisation of upstream network assets. Another potential benefit of active management is the provision of services. It can enable DG to provide ancillary services (e.g. reactive power management) and provide network support (contributing to security of supply) to the distribution network. However, the distribution network operator could provide the service of active management, to yield higher profits for DG as a result of increased energy output. The prices and the commercial arrangements for such services are yet to be established and therefore they are difficult to value at present. Other benefits of active management lie in its potential to defer or to reduce investment in distribution plants (e.g. in voltage control and peak demand shaving). A benefit associated with reduced investment is a lower risk with respect to stranded assets as a result of potential load or generation migration. In future, active distribution networks may enable islanded operation, and this could increase the quality of service delivered by distribution networks. At present, it is difficult to quantify such benefits of active management and distributed generation since a considerable amount of judgement is required.

Value assessment methodologies must be capable of reflecting technical and economic impacts. One possible approach to identify and quantify the benefit of DG islanding in a portion of the distribution network normally connected to the grid is a cost–benefit assessment approach based on the customer outage cost methodology (Kariuki and Allan, 1996a, 1996b). This approach would assess the value of DG islanding to consumers. The benefit of DG islanding in future active distribution networks can be seen as offering an improved quality of service to customers. The monetary value of the improved quality of service can be determined as a reduction in customer outage costs in a service area. Figure 21.10 shows how reduced system outage durations as a result of DG islanding results in reduced customer outage costs. In order for islanding to be viable, the identified value or benefit must outweigh the costs of making islanded operation possible.

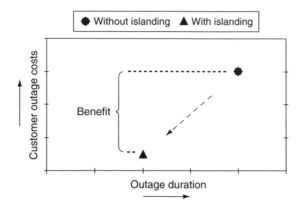

Figure 21.10 Reduced customer outage cost resulting from islanding of distributed generation

21.4 Conclusions

This chapter described the principles of active management of distribution networks. The main benefits of active management were identified and illustrated in worked examples. Four active management schemes were studied: (A) use of active power generation curtailment, (B) use of reactive power management, (C) use of area-based coordinated voltage control of OLTC transformers and (D) the application of voltage regulators. The studies performed clearly showed significant benefits regarding the active control of distribution networks. It was shown that a considerable increase in penetration levels can be achieved when distributed generators are allowed to absorb reactive power. The use of OLTC transformer voltage control appears to be most beneficial as it allows for a large increase in penetration at a low cost, from the primary plant perspective, but would require more complex control of distribution network operation. An increase in DG based on these controls would be conditioned by the operation of an adequate ancillary services market. Further benefits of active management were discussed, and one methodology to assess the value of DG islanding was presented. It can be concluded that the exploitation and assessment of the benefits of active management often depend on the development of appropriate technical information tools and commercial arrangements. These arrangements and frameworks must consider the requirements of an unbundled electricity industry and must allow the marketing of such benefits.

References

[1] Bopp, T., Shafiu, A., Chilvers, I., Cobelo, I., Jenkins, N., Strbac, G., Haiyu, L., Crossley, P. (2003) 'Commercial and Technical Integration of Distributed Generation into Distribution Networks', presented at CIRED, 17th International Conference on Electricity Distribution.
[2] Jenkins, N., Allan, R., Crossley, P., Kirschen, D., Strbac, G. (2000) *IEE Power and Energy Series: 31. 'Embedded Generation'*, Institute of Electrical Engineers (IEE), London.
[3] Kariuki, K. K., Allan, R. N. (1996a) 'Evaluation of Reliability Worth and Value of Lost Load', *IEE Proc. Gener. Transm. Distrib.* 143(2) 171–180.

[4] Kariuki, K. K., Allan, R. N. (1996b) 'Application of Customer Outage Costs in System Planning, Design and Operation', *IEE Proc. Gener. Transm. Distrib.* 143(4) 305–312.

[5] Liew, S. N., Strbac, G. (2002) 'Maximising Penetration of Wind Generation in Existing Distribution Networks', *IEE Proc. Gener. Transm. Distrb.* 149(3) 256–262.

[6] Masters, L., Mutale, J., Strbac, G., Curcic, S., Jenkins, N. (2000) 'Statistical Evaluation of Voltages in Distribution Systems with Wind Generation', *IEE Proc. Gen. Trans. Dist* 147(4) 207–212.

[7] Saad-Saoud, Z., Lisboa, M., Ekanayake, J., Jenkins, N., Strbac, G. (1998) 'Application of STATCOM to Wind Farms', *IEE Proc. C, Gen. Trans. Dist.* 145(5) 511–516.

[8] Strbac, G., Jenkins, N., Hird, M., Djapic, P., Nicholson, G. (2002) *Integration of Operation of Embedded Generation and Distributed Networks*, UMIST, Manchester, and Econnect, Future Energy Solutions, publication K/EL/0262/00/00.

22

Transmission Systems for Offshore Wind Farms

Thomas Ackermann

22.1 Introduction

The interest in the utilisation of offshore wind power is increasing significantly world-wide. The reason is that the wind speed offshore is potentially higher than onshore, which leads to a much higher power production. A 10 % increase in wind speed results, theoretically, in a 30 % increase in power production. The offshore wind power potential can be considered to be significant. For Europe, the theoretical potential is estimated to lie between 2.8 and 3.2 PWh (Germanischer Loyd, 1998) and up to 8.5 PWh (Leutz *et al.*, 2001). In addition, countries in central Europe, particularly Germany, are running out of suitable onshore sites. Some countries promote offshore wind power because they believe that offshore wind power can play an important role in achieving national targets for the reduction of emissions of carbon dioxide (CO_2). However, investment costs for offshore wind power are much higher than those for onshore installations. Fortunately, in many central European waters, water depth increases only slowly with distance from the shore, which is an important economic advantage for the application of bottom-mounted offshore wind turbines.

Until now, practical experience from offshore wind energy projects is limited (for an overview of existing projects, see Table 22.1). The installation and supporting structure of offshore wind turbines is much more expensive than that of onshore turbines. Therefore offshore wind farms use wind turbines with a high rated capacity (≥ 1.5 MW), which are particularly designed for high wind speeds. The first offshore projects materialised in Denmark, the Netherlands, Sweden, the United Kingdom and Ireland. Most existing offshore wind farms tend to be demonstration projects close to the shore.

Wind Power in Power Systems   Edited by T. Ackermann
© 2005 John Wiley & Sons, Ltd   ISBN: 0-470-85508-8 (HB)

Table 22.1 Offshore wind energy projects, as of 31 January 2003

Name and Location	Year	No. turbine	Rated capacity (MW)	Total capacity (MW)	Hub height (m)	Minimum distance from shore (km)	Water depth (m)
Nogersund, Sweden[a]	1991	1	0.22	0.22	37.5	0.25	6
Vindeby, Baltic Sea, Denmark	1991	11	0.45	4.95	37.5	1.5	2–5
Lely, Ijsselmeer, The Netherlands	1994	4	0.5	2	41.5	0.80	5–10
Tunø Knob, Baltic Sea, Denmark	1995	10	0.5	5	43	6.0	3–5
Irene Vorrink, Ijsselmeer, The Netherlands	1996	28	0.6	16.8	50	0.02	1–2
Bockstigen, Baltic Sea, Sweden	1997	5	0.55	2.75	41.5	4.0	5–6
Lumijoki/Oulu, Baltic Sea, Finland	1999	1	0.66	0.66	50	0.5	2–3
Utgrunden, Baltic Sea, Sweden	2000	7	1.425	9.975	65	8.0	7–10
Blyth, North Sea, UK	2000	2	2.0	4	58	1.0	6–9
Middelgrunden Baltic Sea, Denmark	2001	20	2.0	40	60	2–3	3–6
Yttre Stengrund, Baltic Sea, Sweden	2001	5	2.0	10	60	5.0	6–10
Horns Rev, Denmark	2002	80	2.0	160	80	14.0	6–12
Frederikshaven, Baltic Sea, Denmark	2002	2 1	2.3 3	7.6	80	0.5	1
Rønland, Denmark	2002	4 4	2 2.3	17.2	70	0.5	2–3
Samso, Baltic Sea, Denmark	2003	10	2.3	23	61	3.5	20
Nysted, Baltic Sea, Denmark	2003	72	2.3	165.6	68	6–10	6–9.5
North Hoyle, UK	2003	30	2.0	60	80	7–8	12
Arklow Bank, Republic of Ireland	2003	7	3.6	25.2	73.5	10	5–15

[a] Decommissioned in 1998.

However, a shift towards large-scale commercial projects has already begun. At the time of writing, the largest existing projects are the two Danish wind farms: Horns Rev, with 80 × 2 MW (160 MW) and Nysted, with 72 × 2.3 MW (165.6 MW).

Further offshore projects are in the planning stage, particularly in Denmark, Germany (with proposals for a total of 60 GW!), the Netherlands, France, the United

Kingdom and Sweden. Also, in the USA the first offshore wind energy projects are currently being discussed (e.g. off Cape Cod and Long Island).

Future projects not only will include wind turbines with a higher rated capacity (≥ 3 MW) but also will be significantly larger in total size and further away from the shore, hence most likely also to be in deeper waters, of up to 40 m. Typical future project proposals for offshore wind farms are in the range of 250–1000 MW. This will require new concepts for the overall transmission system, including concepts for transmission within the offshore wind farm as well as to shore and for network integration into the onshore power system.

This chapter will focus mainly on the different options for the transmission system between a large ($\gg 100$ MW) offshore wind farm and the onshore power network.

22.2 General Electrical Aspects

The internal electric system of an offshore wind farm and its connection to the main power system pose new challenges. Onshore, the standard solution is an AC network within the wind farm, which collects the power production of each wind turbine. The voltage level within a wind farm is usually the same as the medium voltage level of the distribution network point. As most wind turbine generators operate at a generator voltage level of 690 V, transformers that are installed directly in or close to the basement of each wind turbine are used to increase the generator voltage level to the voltage level in the wind farm network. The highest medium voltage level used within an onshore wind farm lies usually in the range of 33–36 kV, as the market offers competitive standardised equipment for this voltage range.

However, offshore wind farms tend to be bigger, and the distance between wind turbines is usually greater than that in onshore wind farms because of the larger wake effect offshore. Often, the distance to the shore or to the next (offshore) transformer station is also significantly longer than for onshore wind farms. For large offshore wind farms with an AC network, higher voltage levels will certainly be useful in order to minimise power losses, but higher voltage levels may result in bigger transformers and higher costs for these transformers. The transformers will be placed in the nacelle, the tower or in a container next to the wind turbine; hence the size of the transformer might become an issue. In addition, the cost and size of switchgears will also increase with voltage levels. Today, a collection voltage of up to 36 kV is considered the standard solution. Research by Lundberg (2003), however, indicates that for a small offshore wind farm (around 60 MW) that uses Type C turbines and has a maximum distance to shore of 80 km the most cost-effective voltage level is 45 kV.[1] In the future, there may be other more cost-effective solutions, once wind turbine manufacturers start to offer wind turbines with generator voltage levels higher than 690 V (e.g. 4000 V or more). Higher voltage levels, however, will require better trained maintenance staff and enhanced safety, which will increase costs.

Existing smaller offshore wind farms that are reasonably close to the shore have opted for a comparatively low voltage level. The reason is that the reduction in (load) losses is not sufficient to justify the additional costs for the equipment required for higher

---

[1] For definitions of Type A to D wind turbines, see Section 4.2.3.

voltage levels. The same applies to smaller offshore wind farms that are currently being planned. The seven turbines of the Swedish Utgrunden offshore wind farm (9.975 MW), for instance, are interconnected at a voltage level of 20 kV. The cable to shore has a length of 8 km and is operated at 20 kV. Onshore, the voltage is raised to 50 kV.

For the Danish Middelgrunden offshore wind farm, which is somewhat larger, at 40 MW, a voltage level of 30 kV was chosen for the connection between the 20 turbines and the approximately 3 km long cable to shore (Middelgrunden, 2004). The Danish Horns Rev wind farm, which was constructed during 2002, has a capacity of 160 MW and uses a voltage level of 36 kV within the wind farm. Horns Rev is the first offshore wind farm that uses an offshore transformer platform. At the offshore transformer station, the voltage level is raised to 150 kV to feed the approximately 15 km long AC cable to shore (Horns Rev, 2004). The Danish Nysted offshore wind farm, with a capacity of 165.6 MW, uses 33 kV as the collection voltage. It also includes an offshore transformer station where the voltage is raised to 132 kV for the approximately 10 km AC connection to shore (Steen Beck, 2002).

The network design of an offshore wind farm does not necessarily correspond to the most energy-efficient network layout (i.e. with lowest losses). The reason is that offshore transformer stations are rather complex and include large support structures. Hence, offshore transformer stations are very expensive and there is hardly any experience yet regarding the reliability of offshore transformer stations.

22.2.1 Offshore substations

The Horns Rev offshore substation is designed as a tripod construction, in contrast to the Nysted substation, which is based on a monopile construction. The Horns Rev substation includes a steel building with a surface area of approximately 20×28 m which is placed about 14 m above mean sea level. The platform accommodates, among others things, the following technical installations (Horns Rev, 2004):

- 36 kV switch gear;
- 36/150 kV transformer;
- 150 kV switch gear;
- a control and instrumentation system, as well as a communication unit;
- an emergency diesel generator, including 2×50 tonnes of fuel;
- sea-water-based fire-extinguishing equipment;
- staff and service facilities;
- a helipad;
- a crawler crane;
- an MOB (man overboard) boat.

The Horns Rev and Nysted offshore substations are unique. Nothing of this kind had been built in the world before, not even for the offshore gas and oil industry, which usually only operates at voltage levels of 13.8 kV. The two existing offshore substations will provide important information about the reliability of offshore substations for future projects.

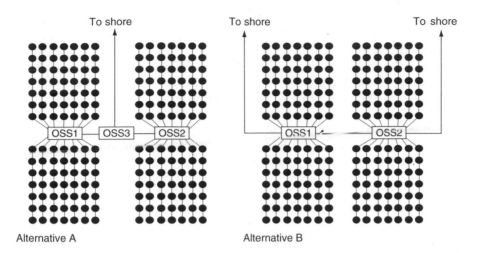

Figure   22.1 Possible layouts for a 980 MW (196 × 5 MW) offshore wind farm. *Note*: OSS = offshore substation

However, offshore project developers are likely to adopt the simplest approach possible, which might be also the most economic rather than the most energy-efficient solution. They tend to avoid the installation of offshore transformer platforms whenever possible. The planned 100 MW NoordZeeWind project in the Netherlands, for instance, will not include an offshore substation. The internal voltage level of the wind farm and of the 8 km connection to shore will most likely be 33 kV or 36 kV.

Very large wind farms ($\gg$ 250 MW) that are located far off the shore may require a number of offshore substations. The first step will be to connect each wind farm module with a maximum capacity of 250 MW to a substation where the voltage is stepped up [or converted to high-voltage direct current (HVDC)] for the transmission link to shore. Another option is to connect each substation to shore (see also Figure 22.1).

Large offshore wind farms can certainly have different layout configurations. Looking at Figure 22.1, for instance, in Alternative B it might be possible to increase the voltage level within each wind farm block and to have only one offshore substation per block. This might reduce total investment costs yet increase the risk that the whole wind farm may be lost in the case where there are technical problems with the single network connection. Hence, designers of general wind farm layout have to consider not only the main wind direction but also the technical transmission solution, the investment and operating costs and the overall reliability as well as issues related to the location, such as water depth throughout the proposed area (for a more detailed discussion of possible layouts for offshore wind farms, see Lundberg, 2003; Martander, 2002).

22.2.2 Redundancy

Most electric networks of offshore wind farms that are currently under discussion have no or very little redundancy. The turbines of a wind farm may be connected to a 'radial' network, as for example in Horns Rev. A radial network design means that a number of turbines are connected to one feeder. The maximum number of turbines on each feeder is determined by the maximum rating of the cable (Brakelmann, 2003). If the cable is damaged the whole feeder will be disconnected until the fault is repaired. Redundancy might be achieved by enabling connections between feeders. The last wind turbine in one feeder could be connected to the last turbine in the next feeder, for instance. During normal operation, this connection would be open. If there is a fault in one of these two feeders, the connection would be closed.

For this, additional equipment would be required to isolate the fault. The goal is that as many wind turbines as possible remain connected to the grid. It has also to be taken into account that the number of wind turbines that may be connected to one feeder will be lower, as the maximum cable rating must be able to cope with twice as many turbines in case the fault is close to the first turbine within one feeder. Today, the likelihood of a fault and the associated costs are assumed to be lower than the costs for the additional equipment – therefore, redundancy is not taken into consideration. In order to reduce the likelihood of damage, the cables are usually buried 1–2 m into the seabed to protect them from ships' anchors and strong sea currents.

Usually, the cable from the offshore wind farm to the shore does not include any redundancy either (see also Figure 22.1, Alternative A). A fault on this cable, however, will result in a loss that is equivalent to the entire wind farm. The economic consequences from such a fault may be huge. The repair might take months, depending on the availability and current position of cable repair ships. It is, however, very difficult to protect the cable from damage, particularly if the cable crosses major shipping routes. At the First International Workshop on Transmission Networks for Offshore Wind Farms in Stockholm, Sweden, it was reported that the anchors of large ships could dig down up to 13 m into the seabed. It is not practically viable to bury a cable at such a depth.

Redundancy could be achieved by using another, backup, cable following another route to shore (see Figure 22.1, Alternative B). Besides the significant costs of such a second connection to shore, it might be also very difficult to carry this out in practice. At many locations, environmental restrictions would make it very difficult to find a second cabling route. A second connection point to the onshore network might not always be available either. Currently, no developers of offshore wind farms seriously consider any redundancy for the onshore cabling. Experience from the operation of the first large offshore wind farms will show whether redundancy for onshore connections will become necessary in future.

22.3 Transmission System to Shore

For the cabling to shore, either high-voltage alternating-current (HVAC) or high-voltage direct-current (HVDC) connections could be used. For HVDC connections, there are two technical options: line commutated converter (LCC) based HVDC and voltage source converter (VSC) based HVDC technology.

All offshore wind farms that are currently operating (as of December 2003) have adopted an AC alternative, and all those planned to be installed within the near future (1–2 years) will also use an AC solution. This is because of the comparatively small size and/or the short distance between the shore and the existing wind farms. As the size of future wind farms and the distance to shore is likely to increase, this might change.

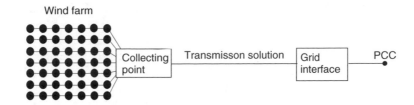

Figure   22.2 Schematic layout of an offshore wind farm; the collecting point can be an offshore substation. *Note*: PCC = point of common coupling

In the following, three different options for the connection of offshore wind farms to shore are discussed in more detail. [The general layout of an offshore wind farm, which may relate to either HVAC or HVDC transmission, is shown in Figure 22.2.]

22.3.1 High-voltage alternating-current transmission

An HVAC transmission system consists of the following main components: an AC based collector system within the wind farm; perhaps an offshore transformer station including offshore reactive power compensation; three-core polyethylene insulation (XLPE) HVAC cable(s) to shore, and, onshore, possibly a static VAR compensator (SVC; Eriksson *et al.*, 2003, Häusler and Owman, 2002; see also Figure 22.3).

The HVAC solution was implemented at the Horns Rev 160 MW wind farm in Denmark. As this wind farm is only 21 km offshore, no compensation was needed at the offshore transformer station. The cable used was a 170 kV three core XLPE cable with 630 mm^2 copper conductor.

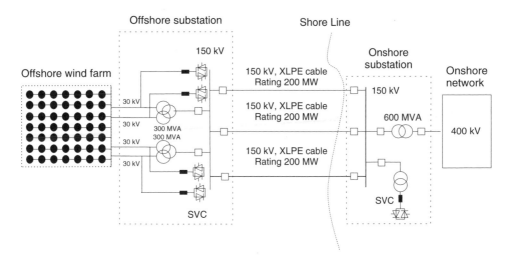

Figure   22.3 The basic configuration of a 600 MW wind farm with a high-voltage alternating-current (HVAC) solution. *Note*: SVC = static VAR compensator; XLPE = polyethylene insulation
Sources: based on Eriksson *et al.*, 2003; Häusler and Owman, 2002.

With an increasing distance to shore, reactive power compensation will be required at both ends of the cable (i.e. at the offshore substation, too). A 400 MW offshore wind farm with two 150 kV, 120 km, cable systems, for instance, will require 150 MVAR compensation offshore as well as onshore (Eriksson *et al.*, 2003). Furthermore, the maximum rating of AC cables is currently limited to about 200 MW per three-phase cable, based on a voltage rating of 150–170 kV, compensation at both ends of the cable and a maximum cable length of 200 km. That means that a 1000 MW wind farm would need five cables for the connection to shore, and that does not even take possible requirements for redundancy into account.

For shorter distances, higher voltage ratings of up to 245 kV might be possible, which would increase the maximum rating to 350–400 MW (Rudolfsen, 2001; Rudolfsen, 2002). A three-core, 1000 mm^2, conductor cable operated at a system voltage of 230 kV with a maximum of $\pm 10\%$ voltage variation between no load and full load can, for instance, transmit 350 MW over 100 km with losses of 4.3%, or 300 MW over 200 km with losses of 7.3%, if the cable is compensated equally at both ends (Rudolfsen, 2001). Higher voltage levels of up to 400 kV are under development, which may allow a maximum transmission of 1200 MVA over a maximum distance of 100 km. It is, however, uncertain when such solutions will become commercially available.

Finally, the main disadvantage of HVAC solutions must be mentioned: with increasing wind farm size and distance to shore, load losses will increase significantly. Also, an increase in the transmission voltage level will lead to larger and more expensive equipment (e.g. transformers) as well as more expensive submarine cables. Hence, an increase in the voltage level is often only justifiable if an increase in capacity is required.[2]

22.3.2 Line-commutated converter based high-voltage direct-current transmission

The advantage of LCC based HVDC connections is certainly their proven track record. The first commercial LCC HVDC link was installed in 1954 between the island of Gotland and the Swedish mainland. The link was 96 km long, 20 MW rated and used a 100 kV submarine cable. Since then, LCC based HVDC technology has been installed in many locations in the world, primarily for bulk power transmission over long geographical distances and for interconnecting power systems, such as for the different island systems in Japan and New Zealand (Hammons *et al.*, 2000). Other well-known examples for conventional HVDC technology are:

- the 1354 km Pacific Interie DC link, with a rating of 3100 MW at a DC voltage of ± 500 kV;
- the Itaipu link between Brazil and Paraguay, rated at 6300 MW at a DC voltage of ± 600 kV (two bipoles of 3150 MW).

---

[2] The capacity charging current of long HVAC cable increases linearly with the voltage level as well as linearly with the lengths of the cable. Hence, for a given wind farm capacity (e.g. 200 MW) the losses related to the capacity charging current will be higher for a 245 kV solution than for a 150 kV system design.

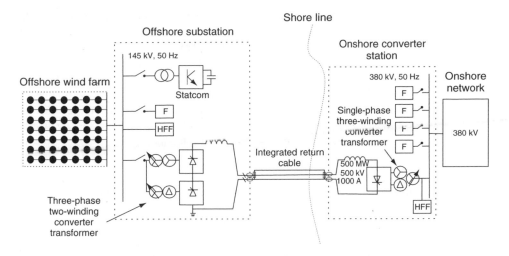

Figure 22.4 Basic configuration of a 500 MW wind farm using a line-commutated converter (LCC) high-voltage direct-current (HVDC) system with a Statcom (for a configuration for a 1100 MW wind farm using an LCC HVDC system with diesel generators on the offshore substation, see Kirby *et al.*, 2002). *Note*: F = filter; HFF = high-frequency filter; the statcom can be replaced with a diesel generator
Source: based on Cartwright, Xu and Saase, 2004.

However, there is no experience regarding LCC based HVDC offshore substations or onshore in combination with wind power.

A thyristor based LCC HVDC transmission system consists of the following main components: an AC based collector system within the wind farm; an offshore substation with two three-phase two-winding converter transformers as well as filters and either a Statcom or a diesel generator that supply the necessary short-circuit capacity; DC cable(s); and an onshore converter station with a single-phase three-winding converter transformer as well as the relevant filters (Cartwright, Xu and Saase, 2004; Kirby *et al.*, 2002; see also Figure 22.4).

This technology requires comparatively large converter stations both on- and off-shore, as well as auxiliary service at the offshore converter station. The auxiliary service has to keep up a rather strong AC system at the offshore converter in order to enable the operation of the line-commutated converters even during periods with no or very little wind. This auxiliary offshore service will be most likely supplied by diesel generator(s) (Kirby *et al.*, 2002). Another solution, also referred to as hybrid HVDC, comprises a LCC based HVDC converter and a Statcom, see Cartwright *et al.*, 2004. The Statcom provides the necessary commutation voltage to the HVDC converter and reactive power compensation to the offshore network during steady state, dynamic and transient conditions.

The total conversion efficiency from AC to DC and back to AC using the two converters (offshore and onshore) lies in the range of 97–98 % and depends on the design details of the converter stations. A system design with 98 % efficiency will have higher investment costs compared with a design with lower efficiency. Hence, the advantage of an LCC HVDC solution are comparatively low losses (i.e. 2–3 % for a

500 MW transmission over 100 km, including the losses in both converters but excluding consideration of the requirements for the auxiliary service at the offshore converter station). In addition, the higher transmission capacity of a single cable compared with HVAC transmission or VSC based transmission can be an advantage for very large offshore wind farms.

The requirement for a strong network offshore (and onshore) as well as the need for comparatively large offshore substation converter stations, however, reduce the likelihood of that LCC based HVDC solutions will be considered for small to medium-sized wind farms (i.e. those that are rated at less than about 300 MW).

22.3.3 Voltage source converter based high-voltage direct-current transmission

VSC based HVDC technology is gaining more and more attention. It is marketed by ABB under the name HVDC Light and by Siemens under the name HVDC Plus. This comparatively new technology has only become possible as a result of important advances in high-power electronics, namely, in the development of insulated gate bipolar transistors (IGBTs). In this way, pulse-width modulation (PWM) can be used for the VSCs as opposed to thyristor based LCCs used in the conventional HVDC technology.

The first commercial VSC based HVDC link was installed by ABB on the Swedish island of Gotland in 1999. It is 70 km long, with 60 MVA at ±80 kV. The link was built mainly in order to provide voltage support for the large amount of wind power installed in the South of Gotland (see also Chapter 13). Between 1999 and 2000, a small demonstration project with 8 MVA at ±9 kV was built in Tjæeborg, Denmark. The project is unique, as the link is used to connect a wind farm (three wind turbines with a total capacity of 4.5 MW) to the Danish power system (Skytt, Holmberg and Juhlin, 2001). In 2000, a link – 65 km long, with 3 × 60 MVA at ±80 kV – was built in Australia between the power grids of Queensland and New South Wales. A second link was installed in Australia in 2002. With a length of 180 km, this connection is the longest VSC based HVDC link in the world. It has a capacity of 200 MVA at a DC voltage of ±150 kV. In the USA, a 40 km submarine HVDC link was installed between Connecticut and Long Island in 2002. The link operates at a DC voltage of ±150 kV and has a capacity of 330 MVA. For 2005, ABB plans the first installation of a VSC offshore converter station for the offshore Troll A platform in Norway. The link will be 67 km long with a rating of 2 × 41 MVA at ±60 kV (Eriksson et al., 2003).

A VSC based HVDC transmission system consists of the following main components: an AC based collector system within the wind farm; an offshore substation with the relevant converter(s); DC cable pair(s); and an onshore converter station [Häusler and Owman, 2002; Eriksson et al., 2002; see also Figure 22.5].

A VSC based HVDC link does not require a strong offshore or onshore AC network; it can even start up against a nonload network. This is possible because in a VSC the current can be switched off, which means that there is no need for an active commutation voltage. Furthermore, the active and reactive power supply can be controlled independently. In addition, it is important to mention that the VSC based HVDC connection is usually not connected to ground. Therefore, the VSC based HVDC always needs two conductors (cables).

VSCs use IGBT semiconductor elements with a switching frequency of approximately 2 kHz. This design results in comparatively high converter losses of up to 2 % per converter station. Research currently focuses on how to reduce these losses. The advantage of the high switching frequency are low harmonic levels and, hence a reduced need for filters. Also, the rating per converter is presently limited 300–350 MW, while the cable rating at ±150 kV is 600 MW. Hence, if more power than 300 MW is to be transferred, the number of converter stations has to be increased [see Figure 22.5(a)]. It is, however, expected that the available converter rating will be increased to approximately 500 MW by the market leader in the near future. Figure 22.5(b) shows the possible design for a 500 MW offshore wind farm based on converter stations with a 500 MW rating.

The total efficiency of the two converter stations of a VSC based HVDC system is less than that of a LCC HVDC system. There is hardly any information on the total system efficiency published, but at the International Workshops on Transmission Networks for

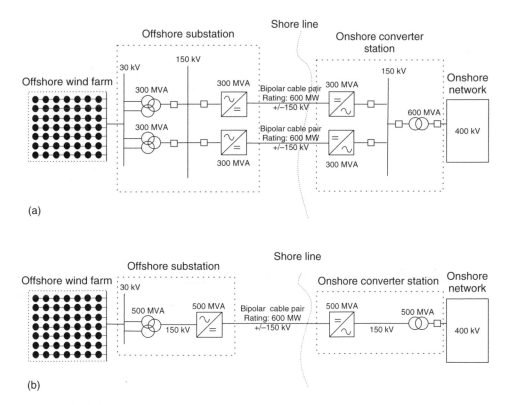

(a)

(b)

Figure   22.5 (a) A 600 MW wind farm using two voltage source converter (VSC); high-voltage direct-current (HVDC) systems, each converter station with a 300 MW rating. (b) A 500 MW wind farm using one VSC HVDC system based on a converter station with a 500 MW rating
Source: for part (a) based on Eriksson *et al.*, 2003.

Offshore Wind Farms of 2000 to 2003 the total efficiency, including both converter stations, was reported to lie in the range of 90–95 %.

Other often-emphasised advantages of a VSC based HVDC solution is the capability of four-quadrant operation, the reduced number of filters required, black-start capability and the possibility of controlling a number of variables such as reactive power, apparent power, harmonics and flicker when feeding the power system from a VSC (Burges *et al.*, 2001; König, Luther and Winter, 2003).

22.3.4 Comparison

The following comparison of the three different technical transmission solutions is divided into technical, economic and environmental issues.

22.3.4.1 Technical issues

The relevant technical issues are: rating, losses, size of offshore installation, grid impact and implementation issues.

Rating

Presently, AC cables have a maximum rating of about 200 MW per three-phase cable. This rating is based on a voltage level of 150–170 kV, compensation at both ends of the cable and a maximum cable length of around 200 km. For shorter distances, voltage ratings may increase to 245 kV, which would raise the maximum rating to 350 MW over a maximum of 100 km, or 300 MW over 150–200 km (Rudolfsen, 2001). Bipolar cable pairs for VSC based HVDC, in comparison, can have a maximum rating of 600 MW for a voltage level of ±150 kV, independent of the cable length. Currently, only converter stations with a maximum rating of 300–350 MW are in operation, which means that two converter stations would be needed for the full utilisation of the maximum cable rating. Converter stations with a maximum rating of 500 MW, however, are announced to be becoming available in the near future, which will reduce the number of cables required for VSC based HVDC solutions (see Table 22.2).

Table 22.2 Number of cables needed for different wind farms and different technical solutions

Wind farm capacity (MW), 100 km transmission distance	HVAC (150 kV)	LCC HVDC		VSC HVDC (150 kV)	
		150 kV, bipolar cable	450 kV, monopolar cable	300 MW CS rating, bipolar cable	500 MW CS rating, bipolar cable
300	2	1 + 1	1	1 + 1	1 + 1
500	3	2 + 2	1	2 + 2	1 + 1
900	5	4 + 4	2	3 + 3	2 + 2
1200	6	5 + 5	2	4 + 4	3 + 3

Note: CS = converter station; HVAC = high-voltage alternating-current; HVDC = high-voltage direct-current; LCC = line-commutated converter; VSC = voltage source converter.

For LCC based HVDC, the cable and converter ratings are not limiting factors regarding the maximum capacity ($< 1000\,\text{MW}$) of the offshore wind farms that are presently under discussion.

Table 22.2 shows the number of cables required for wind farms with different total capacity. It can be seen that a 1200 MW offshore wind farm requires up to six cables for a HVAC link, but only two for an LCC HVDC solution. The VSC HVDC solution would require 3–4 cables.

The number of required cables will influence the total investment costs. It should be taken into account, though, that overall system reliability may increases if two or more cables are used. In this respect, it should be noted that the different cables should follow different cable routes to maximise reliability benefits. This may not always be feasible, though.

Losses

For HVAC transmission, the power loss depends to a large extent on the length and characteristics of the AC cable (i.e. losses increase significantly with distance; see also Santjer, Sobeck and Gerdfes, 2001). The losses of HVDC connections show only a very limited correlation with the length of the cable, depending on the efficiency of the converter stations. As explained above, the efficiency of LCC stations is usually higher than that of VSCs. This means that for short distances the losses from a HVAC link are lower than those from a HVDC connection, owing to the comparatively high converter losses. There is, however, a distance X where the distance-related HVAC losses reach similar levels to those of HVDC links (see Figure 22.6). For distances larger than X, losses in the HVDC solution are lower than those for HVAC links. The distance X depends on the system configuration (e.g. on cable type and voltage levels) but is, however, usually longer for VSC HVDC than for LCC HVDC technology.

The critical distance X depends very much on the individual case. For medium-sized wind farms of around 200 MW, the critical distance X for HVAC compared with VSC HVDC is around 100 km, for instance. This means that an HVAC link shows lower losses for distances less than 100 km, but if the distance exceeds 100 km a VSC HVDC solution will result in lower losses.

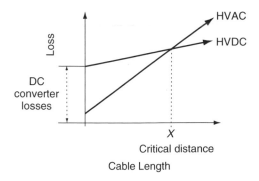

Figure   22.6 Comparison of losses for high-voltage alternating current (HVAC) and high-voltage direct current (HVDC)

Size of offshore substation

Different technical transmission solutions have widely divergent requirements regarding the size of the offshore substation (Kirby, Xu and Siepman, 2002). In general, the size of an AC offshore substation will be only about a third of the size of the corresponding HVDC solution, owing to the significant space required by the converter stations. For onshore HVDC converter stations, LCC based converter stations need considerably more space than do VSC based systems. Eriksson *et al.* (2003) argue that a 300 MW VSC offshore converter station requires a space of approximately $30 \times 40 \times 20$ m (width $\times$ length $\times$ height). Regarding VSC converter stations, it is important to remember that the maximum possible rating at present is 300 MVA; hence a larger capacity demand will require multiple VSC converter stations. For very large capacity requirements ($>> 300$ MW), the possible advantages of VSC based solutions regarding space requirements compared with LCC solutions may be significantly reduced.

Grid impact

Owing to the considerable rating of offshore wind farms, the impact of the entire offshore wind farm system on the onshore power system has to be taken into account (i.e. the type of wind turbines, the transmission technology and the grid interface solution). It is also important to consider that some of the countries that expect a significant development of offshore wind farms already have an onshore network with a significant amount of onshore wind power (e.g. on Germany, see Chapter 11; on Denmark, see Chapter 10).

Transmission network operators in Denmark and Germany, for instance, have therefore already defined new grid connection requirements for connecting wind farms to the transmission system. These regulations are also binding for offshore wind farms (for details, see Chapter 7). Other transmission network operators currently prepare similar requirements. The new regulations will try to help the onshore network to remain stable during faults. An example of such requirements is that a wind farm will have to be able to reduce the power output to 20 % below rated capacity within 2 s of the onset of a fault. After the fault, the wind farm output has to return to the prefault level within 30 s.

During the past few years, a number of studies have been conducted on the impact of the different transmission solutions on the grid and their capability to comply with the new grid connection requirements (see Bryan *et al.*, 2003; Cartwright, Xu and Saase, 2004; Eriksson *et al.*, 2003; Grünbaum *et al.*, 2002; Häusler and Owman 2002; Henschel *et al.*, 2002; Kirby, Xu and Siepman, 2002; Kirby *et al.*, 2002; König, Luther and Winter, 2003; Martander 2002; Schettler, Huang and Christl, 2000; Søbrink *et al.*, 2003). It can be concluded from these studies that the grid impact depends very much on the individual case (i.e. the grid impact depends on the detailed design of the various solutions). The manufacturers of the various technical transmission solutions currently develop appropriate system designs to minimise the grid impact and to comply with the new grid connection requirements. Many of the above-cited studies are performed by or in cooperation with manufacturers of the various technical transmission solutions in order to demonstrate their technical capabilities. In other words, possible drawbacks of certain technical solutions regarding grid integration are minimised with additional equipment. Hence the main decision criteria for or against a certain technical solution will be based mainly on the overall system economics, which should include the cost for the additional equipment. There will certainly be more in-depth research in this area over the next years, such as the EU project, DOWNVIND (http://www.downvind.com).

It must, however, be mentioned that both HVDC technologies have a significant advantage over an HVAC solution: HVDC technologies significantly reduce the fault contribution to the onshore power network. Expensive upgrades of existing onshore equipment such as transformers and switchgears may thus become unnecessary. In addition, VSC based HVDC technology has the capability of providing ancillary services to the onshore network (e.g. providing active as well as reactive power supply and voltage control). This capability could also be used within the offshore AC networks for controlling the reactive power in the network, for instance. However, HVAC or an LCC based HVDC solution in combination with additional equipment (e.g. an SVC or a Stacom) might be able to provide similar benefits.

Implementation
Many wind farms will be built in two steps: at first, there will be a small number of wind turbines and then a larger second phase. Therefore, it is important to point out that XLPE cables can be used for AC as well as for HVDC links. During the first step, an AC solution may be applied as a VSC based HVDC system will not be economic because of the small size of this first phase of the project. In the second phase, the AC system will be converted to an HVDC system, which requires a converter station onshore and offshore, and possibly more cables to shore. The existing cable, however, may be incorporated into the HVDC link. This approach might be seen in some German offshore wind farms. The first phase with the AC solution will comprise a wind farm with a capacity of 50–100 MW. The distance to the onshore grid may be around 60–100 km. During the second phase, up to 1000 MW may be added. At that point, the link from the offshore wind farm has to be extended to the 380 kV onshore network (an extra 35 km), and, most likely, HVDC technology will be used, incorporating the cable that was already installed during phase one.

Summary
In Table 22.3 the three standard transmission solutions are briefly compared. The technical capabilities of each system can probably be improved by adding additional equipment to the overall system solution.

22.3.4.2 Economic issues

When comparing HVAC and HVDC links, the total system cost for equivalent energy transmission over a similar distance should be considered. The total system cost comprises investment costs and operating costs, including transmission losses and converter losses. Investment costs change with rating and operating costs (i.e. losses) and with the distance from a strong network connection point onshore. Therefore the economic analysis has to be carried out based on specific cases. Over the past years, a number of studies have been conducted (e.g. Burges *et al.*, 2001; CA-OWEA, 2001; Häusler and Owman, 2002; Holdsworth, Jenkins and Strbac, 2001; Lundberg, 2003; Martander, 2002). As it is rather difficult to obtain good input data, in particular regarding the various costs but also regarding the converter losses of HVDC solutions, significant

Table 22.3 Summary of transmission solutions: high-voltage alternating-current (HVAC) transmission, line-commutated converter (LCC) based high-voltage direct-current (HVDC) transmission and voltage source converter (VSC) based HVDC transmission

	Transmission solution		
	HVAC	LCC based HVDC	VSC based HVDC
Maximum available capacity per system	200 MW at 150 kV 350 MW at 245 kV	~1200 MW	350 MW 500 MW announced
Voltage level	Up to 245 kV	Up to ±500 kV	Up to ±150 kV
Does transmission capacity depend on distance?	Yes	No	No
Total system losses	Depends on distance	2–3 % (plus requirements for ancillary services offshore)	4–6 %
Does it have Black-start capability?	Yes	No	Yes
Level of faults	High compared with HVDC solutions	Low compared with HVAC	Low compared with HVAC
Technical capability for network support	Limited	Limited	Wide range of possibilities
Are offshore substations in operation?	Yes	No	Planned (2005)
Space requirements for offshore substations	Small	Depends on capacity; converter is larger than VSC	Depends on capacity; converter is smaller than LCC but larger than HVAC substation

differences between the different solutions can be found. In the following, we will present some general economic conclusions (for a summary, see Figure 22.7). It should be emphasised, though, that the results are very specific to the individual cases. Also, the economic impact of a possible 500 MW converter station rating for the VSC based HVDC solution is not considered, because the relevant data are currently not available. More detailed research in this area will certainly be performed in the near future (e.g. within the proposed EU project, DOWNVIND).

Offshore wind farms of up to 200 MW

In general, the investment costs for a bipole DC cable and a single three-core 150 kV XLPE AC cable with a maximum length of 200 km are very similar, with the DC cable probably having a slight cost advantage over the AC cable. However, the investment

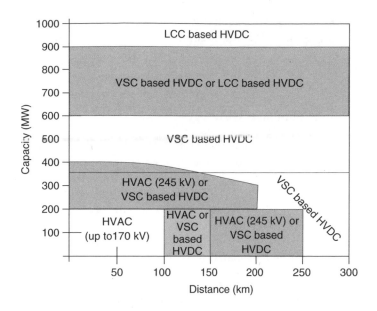

Figure 22.7 Choice of transmission technology for different wind farm capacities and distances to onshore grid connection point based on overall system economics (approximation): economics of high-voltage alternating-current (HVAC) links, line-commutated converter (LCC) based high-voltage direct-current (HVDC) links and voltage source converter (VSC) based HVDC link

cost of VSC converters is up to 10 times higher than that of an HVAC infrastructure (e.g. a transformer station). Hence, for maximum distances of approximately 100 km and a maximum rating of 200 MW, an HVAC link operated at a maximum voltage level of 170 kV is usually considered the most economic solution. With a larger distance, the increasing losses in the HVAC link may justify the investment in a VSC based HVDC solution. For distances between 150 and 250 km, VSC based HVDC and HVAC links operated at a maximum voltage level of 245 kV are rather close as far as their economics is concerned. Once a distance of 250 km is exceeded, theoretically only VSC based HVDC links are technically feasible. HVAC solutions may be technically viable if compensation is installed along the cables, which may require an offshore platform for the compensation equipment (Eriksson *et al.*, 2003). A distance of 150 km or more is not unlikely because strong grid connection points onshore might be at some distance inland and the distance onshore has to be included in the total transmission distance.

Offshore wind farms between 200 and 350 MW
For wind farms between 200 and 350 MW, either two 150 kV three-core XLPE AC cables are required or one 245 kV cable. That means that the cost of an HVAC link increases and a VSC based HVDC solution may become economically competitive. However, for distances exceeding 100 km, the technical feasibility of current HVAC solutions based on maximum voltage levels drops to about 300 MW at 200 km. Hence, VSC based HVDC connections are most likely to be more economic than a second AC cable.

Offshore wind farms between 350 and 600 MW

For a maximum size of 400 MW and a comparatively short distance to a strong grid interconnection point, HVAC operated at 245 KV might be a very competitive solution. For larger capacities, HVAC links will need at least two three-core XLPE AC cables operated at 245 kV or even three cables operated at 150 kV. VSC based HVDC links, on the other hand, will still only require one bipole DC cable. Hence, VSC HVDC seems to be the most economic solution.

Offshore wind farms between 600 and 900 MW

For a wind farm rating of 600 MW or more, VSC HVDC links will also require two bipole DC cables as well as three converter stations onshore as well as offshore. An LCC based HVDC link requiress only one DC cable and only one converter station onshore and one offshore and probably will therefore economically lie very close to a VSC HVDC solution. However, reliability issues may result in a solution where two cables to shore are preferred because the risk of losing one cable and consequently the whole offshore wind farm might be considered too high.

Offshore wind farms larger than 900 MW

For wind farms larger than 900 MW, LCC based HVDC links are probably the most economic solution. However, as mentioned above, reliability issues may lead to two independent cable systems to shore. In that case, a VSC based HVDC link will most likely be the more economic solution.

Summary

It remains to be seen how technical development will affect the economics of the different solutions in the future. Advocates of VSC based HVDC solutions argue that cost reduction in power electronics will make this technology cheaper in the near future, whereas HVAC advocates hope that a future increase in transmission voltage will provide similar benefits.

22.3.4.3 Environmental issues

From an environmental perspective, two main issues are of interest: the magnetic field of the submarine cables as well as the number of submarine cables buried in the seabed.

Submarine cables installed and operated from offshore installations to the shore often pass through very sensitive areas environmentally. The impact of submarine cables on these areas is therefore often a very important part of the environmental permitting process. The permit granting authorities will most likely favour the technical solution with the lowest impact on these sensitive areas, which means a solution with a minimum number of cables as well as low magnetic fields for the submarine cables. In general, three-core AC submarine cables have a lower magnetic field than DC submarine cables; however, AC solutions may require more cables than DC solutions. Hence, it is not directly obvious which solution will have the lowest environmental impact and is therefore very much case-dependent.

In addition, it should be mentioned that diesel generators on offshore platforms combined with a significant diesel storage capacity might cause environmental concerns.

22.4 System Solutions for Offshore Wind Farms

The installation of a large offshore wind farms combined with a transmission link built for the sole purpose of transmitting the power of an offshore wind farm is different from typical onshore solutions. Onshore, a wind farm typically is connected to a transmission or distribution system that is at least partly already in place and services a number of customers. The design of an onshore installation must take this into account.

There are usually no customers connected to offshore wind farms or to the transmission system to shore. Only at the point of common coupling (PCC) onshore (see Figure 22.2) do grid codes have to be fulfilled. Furthermore, HVDC transmission solutions decouple the offshore wind farm from the onshore power system. This condition may allow the application of different wind turbine design concepts (e.g. based on different generator technologies or different control approaches). The ultimate goal of new concepts would then be to find the optimal economic solution for the overall system of wind turbines and transmission system rather than to focus either on the wind turbines or on the transmission system individually.

In the following, we will present the most interesting system solutions that are currently under discussion.

22.4.1 Use of low frequency

Schütte, Gustavsson and Ström (2001) suggest the use of a lower AC frequency within the offshore wind farm. Frequencies lower than 50 or 60 Hz are currently used mainly in electrified railway systems. The railway systems in Germany, Switzerland, Austria, Sweden and Norway, for instance, use 16 2/3 Hz at 15 kV, Costa Rica uses 20 Hz and the USA mainly 25 Hz.

Now, if an HVDC transmission link is chosen for an offshore wind farm, the low AC frequency would be applied only within the collector system of the offshore wind farm. If an HVAC transmission solution is used, the low AC frequency can be applied in the internal wind farm network and for the transmission system to shore. Onshore, a frequency converter station would be required to convert the low frequency of the offshore network to the frequency of the onshore network (see also Figure 22.8).

The advantages of a low AC frequency approach lie in two areas. First of all, a low network frequency would allow a simpler design in the offshore wind turbines. This is

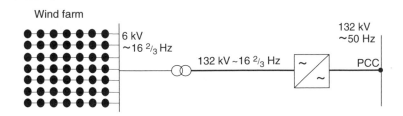

Figure   22.8 Connection of an offshore wind farm using a low AC frequency. *Note*: PCC = point of common coupling
Source: based on Schütte, Gustavsson and Ström, 2001.

mainly because of the fact that the aerodynamic rotor of a large wind turbine operates rather slowly (i.e. the rotor of a 3–5 MW turbine has a maximum revolution of 15 to 20 rpm). A lower AC frequency would therefore allow a smaller gear ratio for wind turbines with a gearbox, or a reduction of pole numbers for wind turbines with direct-driven generators, both consequences resulting in lighter turbines that are thus likely to be cheaper. Second, a low AC frequency will increase the transmission capacity of HVAC transmission links or the possible maximum transmission distance, as a capacity charging current is significantly reduced at lower frequency. The disadvantage of this concept is that the transformer size will increase significantly and therefore transformers will be more expensive.

As far as we know, the idea of low AC frequency for offshore wind farms is currently not being pursued further by the industry.

22.4.2 DC solutions based on wind turbines with AC generators

When using wind turbines equipped with a back-to-back (AC/DC/AC) converter it is theoretically possible to separate the converter into an AC/DC converter installed at the wind turbine, followed by a DC transmission to shore and a DC/AC converter close to the PCC. In other words, the DC bridge in the converter is replaced with a (VSC based) HVDC transmission system. As the AC generator usually operates at 690 V and the (VSC based) HVDC transmission at around 150 kV, an additional DC/DC transformer (DC/DC switch mode converter/buck booster) is usually required to reach the required HVDC voltage. The disadvantage of this approach is that if all wind turbines are connected to the same DC/DC transformer, they will all work at the same operational speed. This operational speed can vary over time. Large offshore wind farms, however, will cover such large areas that only a few turbines will be exposed to the same wind speed at any given time. The operational speed of most wind turbines will not lead to optimal aerodynamic efficiency. Therefore wind turbines are connected in clusters of approximately five turbines to the DC/DC transformer (see Figure 22.9). The five turbines of a cluster will operate at the same speed, which can vary over time. As the wind speed can also vary between those five turbines, the overall aerodynamic efficiency of this solution will still be lower than in the case of individual variable speeds at each turbine. The idea, however, is that the cost benefits of using clusters are larger than the drawbacks of the lower aerodynamic efficiency.

Variations of the principal design concept are possible. They are discussed in more detail elsewhere (Courault, 2001; Lundberg, 2003; Macken, Driesen and Belmans, 2001; Martander, 2002; Pierik *et al.*, 2001, 2004; Weixing and Boon-Teck, 2002, 2003). The studies cited also include detailed economic analyses of this concept, but with partly different conclusions. Some companies find this approach interesting and promising enough to investigate it further.

22.4.3 DC solutions based on wind turbines with DC generators

Finally, the AC generator can be replaced with a DC generator or an AC generator with an AC/AC–AC/DC converter (another option would be a DC generator with a gearbox that allows variable-speed operation, as proposed by Voith Turbo GmbH, Crailsheim,

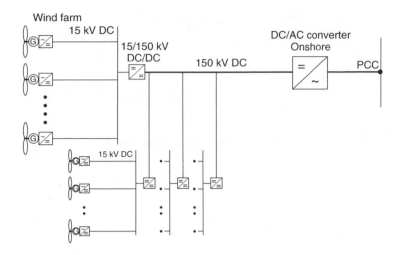

Figure 22.9 DC wind farm design based on wind turbines with AC generators. *Note*: PCC = point of common coupling
Source: based on Martander, 2002.

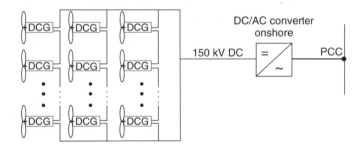

Figure 22.10 DC wind farm design based on wind turbines with DC generators (DCGs). *Note*: PCC = point of common coupling
Source: based on Lundberg, 2003.

Germany). One design option for a wind farm would then be similar to the design shown in Figure 22.7, but without AC/DC converters close to the wind turbines. Another option is to connect all wind turbines in series in order to obtain a voltage suitable for transmission (see Figure 22.10). This option, however, would require a DC/AC–AC/DC converter for a DC generator to allow each turbine to have an individual variable speed. The advantage of a series connection of DC wind turbines is that it does not require offshore substations (for a more detailed discussion of this concept, see Lundberg, 2003).

22.5 Offshore Grid Systems

There is a wide range of ideas under discussion in the area of transmission grids for offshore wind farms. One idea is that of a large offshore grid, often referred to as the

'DC Supergrid', for instance. It assumes that an offshore LCC or VSC based HVDC transmission network can be built. It would range from Scandinavia in the North of Europe down to France in the South of Europe, with connections to all countries that lie inbetween, including the United Kingdom and Ireland. All offshore wind farms in the area would be connected to this supergrid. Such a system is assumed to be able to handle redundancy, and it might better solve possible network integration issues as it aggregates wind power production distributed over a large geographical area.

The cost of such an offshore network would be enormous. First studies suggest that it would be more cost-effective to upgrade the existing onshore networks in order to incorporate the additional offshore wind power than to build an offshore transmission network (see also PB Power, 2002). The conversion of the existing onshore AC transmission lines to LCC or VSC based HVDC could be a very interesting and useful approach to upgrading the onshore network. In this way, existing transmission rights-of-way and infrastructure could be used. This constitutes a major advantage as it is very difficult to obtain permission to build new transmission infrastructure projects. Switching to HVDC could at least double the capacity in comparison with existing AC high-voltage transmission lines.

Initial studies on local offshore grid structures found that interconnecting wind farms within smaller geographical areas such as between the British Isles (Watson 2002) or off the shores of Denmark (Svenson and Olsen, 1999) or the Netherlands seems to be economically more justified. More detailed studies, however, are certainly necessary for further evaluation.

22.6 Alternative Transmission Solutions

Finally, what alternatives are there for transporting the energy produced by an offshore wind farm? Some studies focus on the offshore production of hydrogen. This hydrogen could then be transported to shore via pipeline or even on large ships. The German government has already indicated that it might not tax any hydrogen produced by offshore wind farms. In this way, hydrogen would have to compete with petrol (gasoline, 95–98 octane); that is, at a price of around € 1.1 per litre, which is the typical price in Europe. First studies imply that hydrogen produced by offshore wind farms could be competitive at this price level. However, at the Third International Workshop on Transmission Networks for Offshore Wind Farms, Stockholm (2004), Steinberger-Wilckens emphasised that it is actually more economic to transmit the energy to shore by electric transmission and to produce the hydrogen onshore (see Chapter 23).

22.7 Conclusions

In summary, it can be said that there are many alternatives for the design of the internal electric system of a wind farm and of the connection to shore. The technically and economically appropriate solutions depend very much on the specific case. For operational and economic reasons, though, the long-standing principle held by offshore engineers should not be discarded lightly: keep offshore installations as simple as possible. Many of the commonly discussed solutions for the electric system of offshore wind farms, however, can hardly be called simple. Further research, in particular regarding system solutions that focus on integrating wind turbine design and transmission solutions, is certainly recommended to create simpler electric design concepts for offshore wind farms.

Acknowledgement

I would like to thank Per-Anders Löf for valuable comments as well as all the presenters and participants in the Stockholm and Billund workshops for their contribution.

References

[1] Brakelmann, H. (2003) 'Aspects of Cabling in Offshore Windfarms', in *Proceedings of the Fourth International Workshop on Large-scale Integration of Wind Power and Transmission Networks for Offshore Wind Farms*, Eds J. Matevosyan and T. Ackermann; Royal Institute of Technology, Stockholm, Sweden.

[2] Bryan, C., Smith J., Taylor, J., Zavadil, B. (2003) 'Engineering Design and Integration Experience from Cape Wind 420 MW Offshore Wind Farm', in *Proceedings of the Fourth International Workshop on Large-scale Integration of Wind Power and Transmission Networks for Offshore Wind Farms*, Eds J. Matevosyan and T. Ackermann; Royal Institute of Technology, Stockholm, Sweden.

[3] Burges, K., van Zuylen, E. J., Morren, J., de Haan, S. W. H. (2001) 'DC Transmission for Offshore Wind Farms – Concepts and Components', in *Proceedings of the Second International Workshop on Transmission Networks for Offshore Wind Farms*, Ed. T. Ackermann; Royal Institute of Technology, Stockholm, Sweden.

[4] CA-OWEA (2001) 'Grid Integration, Energy Supply and Finance, a CA-OWEA' (Concerted Action on Offshore Wind Energy in Europe) Report, Sponsored by the EU', available at http://www.offshore-wind.de/ media/article000320/CA-OWEE_Grid_Finance.pdf.

[5] Cartwright, P., Xu, L., Saase, C. (2004) 'Grid Integration of Large Offshore Wind Farms Using Hybrid HVDC Transmission', in *Proceedings of the Nordic Wind Power Conference, Held at Chalmers University of Technology, Sweden, March 2004*, CD produced by Chalmers University of Technology, Sweden.

[6] Courault, J. (2001) 'Energy Collection on Large Offshore Wind Farms – DC Applications', in *Proceedings of the Second International Workshop on Transmission Networks for Offshore Wind Farms*, Ed. T. Ackermann; Royal Institute of Technology, Stockholm, Sweden.

[7] Eriksson, E., Halvarsson, P., Wensky, D., Hausler, M. (2003) 'System Approach on Designing an Offshore Windpower Grid Connection', in *Proceedings of Fourth International Workshop on Large-scale Integration of Wind Power and Transmission Networks for Offshore Wind Farms*, Eds J. Matevosyan and T. Ackermann; Royal Institute of Technology, Stockholm, Sweden.

[8] Germanischer Lloyd (1998) 'Offshore Wind Turbines', Report Executive Summary, Germanischer Lloyd, Wind Energy Department, Hamburg, Germany.

[9] Grünbaum, R., Halvarsson, P., Larsson, D., Ängquest L. (2002), 'Transmission Networks Serving Offshore Wind Farms Based on Induction Generators', in *Proceedings of the Third International Workshop on Transmission Networks for Offshore Wind Farms*, Ed. T. Ackermann; Royal Institute of Technology, Stockholm, Sweden.

[10] Hammons, T. J., Woodford, D., Loughtan, J, Chamia, M., Donahoe, J., Povh, D., Bisewski, B., Long, W. (2000) 'Role of HVDC Transmission in Future Energy Development', *IEEE Power Engineering Review* (February 2000) 10–25.

[11] Häusler, M., Owman, F. (2002) 'AC or DC for Connecting Offshore Wind Farms to the Transmission Grid?' in *Proceedings of the Third International Workshop on Transmission Networks for Offshore Wind Farms*, Ed. T. Ackermann; Royal Institute of Technology, Stockholm, Sweden.

[12] Henschel, M., Hartkopf, T., Schneider, H., Troester, E. (2002) 'A Reliable and Efficient New Generation System for Offshore Wind Farms with DC Farm Grid', in *Proceedings of the 33rd Annual Power Electronics Specialists Conference: Volume 1*, IEEE, New York, pp. 111–116.

[13] Holdsworth, L., Jenkins, N., Strbac, G. (2001) 'Electrical Stability of Large Offshore Wind Farms', in *Proceedings of the Seventh International Conference on AC–DC Power Transmission, November 2001*, IEEE, New York, pp. 156–161.

[14] Horns Rev (2004), http://www.hornsrev.dk/.

[15] Kirby, N. M., Xu, L., Siepman, W. (2002) 'HVDC Transmission Options for Large Offshore Windfarms', in *Proceedings of Third International Workshop on Transmission Networks for Offshore Wind Farms*, Ed. T. Ackermann; Royal Institute of Technology, Stockholm, Sweden.

[16] Kirby, N. M., Xu, L., Luckett, M., Siepman, W. (2002) 'HVDC Transmission for Large Offshore Wind Farms', *IEE Power Engineering Journal* 16(3) 135–141.

[17] König, M., Luther, M., Winter, W. (2003) 'Offshore Wind Power in German Transmission Networks', in *Proceedings of the Fourth International Workshop on Large-scale Integration of Wind Power and Transmission Networks for Offshore Wind Farms*, Eds J. Matevosyan and T. Ackermann; Royal Institute of Technology, Stockholm, Sweden.

[18] Leutz, R., Ackermann, T. Suzuki, A., Akisawa, A., Kashiwagi, T. (2001) 'Technical Offshore Wind Energy Potentials around the Globe', in *Proceedings of the European Wind Energy Conference 2001, Copenhagen, Denmark, July 2001*, WIP-Renewasle Energies, Munich, Germany, pp. 789–792.

[19] Lundberg, S. (2003) 'Configuration Study of Large Wind Parks', licentiate thesis, Technical Report 474 L, School of Electrical and Computer Engineering, Chalmers University of Technology, Göteborg, Sweden.

[20] Macken, K. J. P., Driesen, L. J., Belmans, R. J. M. (2001) 'A DC Bus System for Connecting Offshore Wind Turbines with the Utility System', in *Proceedings of the European Wind Energy Conference 2001, Copenhagen, Denmark, July 2001*, WIP-Renewable Energies, Munich, Germany, pp. 1030–1035.

[21] Martander, O. (2002) 'DC Grids for Wind Farms', licentiate thesis, Technical Report 443L, School of Electrical and Computer Engineering, Chalmers University of Technology, Göteborg, Sweden.

[22] Middelgrunden (2004), Homepage http://www.middelgrunden.dk/.

[23] PB Power (2002) 'Concept Study – Western Offshore Transmission Grid', report by PB Power for the UK Department of Trade and Industry (DTI); available at http://www.dti.gov.uk/energy/renewables/publications/pdfs/KEL00294.pdf.

[24] Pierik, J., Damen, M. E. C., Bauer, P., de Haan, S. W. H. (2001) 'ERAO Project Report: Electrical and Control Aspects of Offshore Wind Farms, Phase 1: Steady State Electrical Design, Power Performance and Economic Modeling, Volume 1: Project Results', Technical Report ECN-CX-01-083, June 2001, ECN Wind Energy, Petter, The Netherlands.

[25] Pierik, J., Morren, J. de Haan, S. W. H., van Engelen, T., Wiggelinkhuizen, E., Bozelie, J. (2004) 'Dynamic Models of Wind Farms for Grid-integration Studies', in *Proceedings of the Nordic Wind Power Conference, Held at Chalmers University of Technology, Sweden, March 2004*, CD produced by Chalmers University of Technology, Sweden.

[26] Rudolfsen, F. (2001) 'Strømforsyning til offshoreplattformer over lange vekselstrøm sjøkabelforbindelser'. Prosjektoppgave Høsten 2001, NTNU, Institute for Electrical Power Technology, Norway.

[27] Rudolfsen, F. (2002) 'Power Transmission over Long Three Core Submarine AC Cables', in *Proceedings of the Third International Workshop on Transmission Networks for Offshore Wind Farms*. Ed. T. Ackermann; Royal Institute of Technology, Stockholm, Sweden.

[28] Santjer, F., Sobeck, L.-H., Gerdfes, G. J. (2001) 'Influence of the Electrical Design of Offshore Wind Farms and of Transmission Lines on Efficiency', in *Proceedings of the Second International Workshop on Transmission Networks for Offshore Wind Farms*, Ed. T. Ackermann; Royal Institute of Technology, Stockholm, Sweden.

[29] Schettler, F., Huang, H., Christl, N. (2000) 'HVDC Transmission Systems using Voltage Sourced Converters Design and Applications', in *Power Engineering Society Summer Meeting, 2000, IEEE, Volume 2*, Institute of Electrical and Electronic Engineers (IEEE), New York, pp. 715–720.

[30] Schütte, T., Gustavsson, B., Ström, M. (2001) 'The Use of Low Frequency AC for Offshore Wind Power', in *Proceedings of the Second International Workshop on Transmission Networks for Offshore Wind Farms*, Ed. T. Ackermann; Royal Institute of Technology, Stockholm, Sweden.

[31] Skytt, A.-K., Holmberg, P., Juhlin, L.-E. (2001) 'HVDC Light for Connection of Wind Farms', in *Proceedings of the Second International Workshop on Transmission Networks for Offshore Wind Farms*, Ed. T. Ackermann; Royal Institute of Technology, Stockholm, Sweden.

[32] Søbrink, K., Woodford, D., Belhomme, R., Joncquel, E. (2003) 'AC Cable versus DC Cable Transmission for Offshore Wind Farms – A Study Case', in *Proceedings of the Fourth International Workshop on*

Large-scale Integration of Wind Power and Transmission Networks for Offshore Wind Farms, Eds J. Matevosyan and T. Ackermann; Royal Institute of Technology, Stockholm, Sweden.

[33] Steen Beck, N. (2002) 'Danish Offshore Wind Farm in the Baltic Sea', in *Proceedings of the Third International Workshop on Transmission Networks for Offshore Wind Farms*, Ed. T. Ackermann; Royal Institute of Technology, Stockholm, Sweden.

[34] Svenson, J., Olsen, F. (1999) 'Cost Optimising of Large-scale Offshore Wind Farms in the Danish Waters', in *Proceedings of the European Wind Energy Conference, Nice France, March 1999*, James & James, London, pp. 294–299.

[35] Watson, R. (2002) 'An Undersea Transmission Grid to Offload Offshore Wind Farms in the Irish Sea', in *Proceedings of the Third International Workshop on Transmission Networks for Offshore Wind Farms*, Ed. T. Ackermann; Royal Institute of Technology, Stockholm, Sweden.

[36] Weixing, L., Boon-Teck, O. (2002) 'Multiterminal LVDC System for Optimal Acquisition of Power in Wind-farm using Induction Generators', *IEEE Transactions on Power Electronics* 17(4) 558–563.

[37] Weixing, L., Boon-Teck, O. (2003) 'Optimal Acquisition and Aggregation of Offshore Wind Power by Multiterminal Voltage-source HVDC', *IEEE Transactions on Power Delivery* 18(1) 201–206.

23

Hydrogen as a Means of Transporting and Balancing Wind Power Production

Robert Steinberger-Wilckens

23.1 Introduction

Today's electricity power generation is based on a complex system of frequency and voltage control and electricity exchange between subgrids. Power production is ruled by a time schedule with about 24-hour ahead prediction. The schedule takes the forecast of load and basic meteorological parameters into account, which are both today fairly well understood. The introduction of a fluctuating energy source as constituted by solar or wind electricity introduces an additional stochastic component to power system scheduling. This may lead to power mismatch in the case of conflicts with the flexibility of conventional (backup) power production. As a result, additional control power (and energy production) is required from conventional, fast-responding electricity generation; otherwise, renewable energy production is lost. Such losses also occur when renewable energy potential is 'stranded' as a result of limitations in transmission capacity, the impossibility of installing transmission lines and so on. Questions relating to these issues are receiving growing interest in the context of offshore wind projects.

To avoid the spilling of renewable energy production, an energy storage medium needs to be incorporated into the generation system in order to allow flexible usage of the power generated. Hydrogen offers several interesting characteristics in this context:

- It can be reconverted to electricity with a reasonably high efficiency if it is used as fuel cells.
- It enables peak power production and load following, either from central installations or from virtual power stations (i.e. it offers decentralised generation capacity).

Wind Power in Power Systems   Edited by T. Ackermann
© 2005 John Wiley & Sons, Ltd   ISBN: 0-470-85508-8 (HB)

- It can constitute an alternative means of energy transport (e.g. using pipelines where electricity cables are undesirable) while offering high energy density and low transport losses.
- It can be sold as industrial gas outside the electricity market; thus, on the one hand, it reduces market pressure and, on the other hand, develops alternative markets for renewable energies (e.g. in transport fuels).

23.2 A Brief Introduction to Hydrogen

Hydrogen is the element with the lowest atomic weight. Compared with the next lightest gaseous fuel relevant in power production – natural gas (methane) – it has a heating value of roughly one third with respect to volume and more than double with respect to weight. Table 23.1 presents some physical and chemical properties of hydrogen in comparison with methane (natural gas) and propane. The aspects that are most important in the context of power production are energy density (defining the storage density and volume of storage, and the transportation energy effort) as well as ignition limits, diffusion coefficient and explosion energy (safety concerns).

The important role hydrogen plays in the context of renewable energy arises from the fact that it can be produced easily from water and electricity by the process of electrolysis. It can then either be burned as a fuel as a substitute for gaseous fossil fuels or be converted to electricity in fuel cells in an electrochemical process that exceeds the efficiency of conventional electricity generation. Within certain limitations, hydrogen can thus be used for storing renewable electricity and can either be sold off as a product in its own right or be reconverted to electricity.

Table 23.1 The physical and chemical properties of hydrogen as compared with methane and propane

Properties	Gas		
	Hydrogen (H_2)	Methane (CH_4)	Propane (C_3H_8)
Density (kg/Nm^3)[a]	0.0838	0.6512	1.870
Lower heating value:			
kWh/kg	33.31	13.90	12.88
kWh/Nm^3	2.80	9.05	24.08
Upper heating value:			
kWh/Nm^3	3.30	10.04	26.19
Diffusivity in air at NTP (cm^3/s)	0.61	0.16	0.12
Ignition energy (mJ)	0.02	0.29	0.26
Explosion limits in air (vol %)	4–75	5.3–15.0	2.1–9.5
Explosion energy [(kg TNT)/m^3]	2.02	7.03	20.5
Autoignition temperature (°C)[b]	585	540	487

[a] Density for a normal cubic metre at 293.15 K, 0.101 MPa.
[b] For comparison, the temperature range for gasoline is 228–471 °C.
Note: NTP = normal temperature and pressure; Nm^3 = norm cubic metres.
Sources: Alcock, 2001; Fischer, 1986.

An inspection of Table 23.1 shows that hydrogen has a rather low volumetric energy density (0.0838 kg/Nm^3), even though the energy density per unit weight (33.31 kWh/kg) is the highest of all energy vectors. Owing to its low weight, hydrogen disperses very quickly in the atmosphere (0.61 cm^3/s). This offsets the low ignition energy (0.02 mJ) and the unfortunately wide limit of the range of explosive mixtures with air (4–75 vol %) – much wider than that for the other fuel gases listed. The use of hydrogen in closed spaces (laboratories, garages and so on) therefore calls for appropriate safety measures (e.g. gas sensors). The explosion energy (2.02 kg TNT/m^3), though, is considerably lower than that for methane (7.03 kg TNT/m^3) for instance. In addition, hydrogen causes problems with ferric materials, giving rise to embrittlement.

The specific security precautions for hydrogen are not too great. The hazards associated with accidents involving hydrogen are not greater than those involving conventional fuel (in some aspects, though, the hazards are different) and are manageable. The widespread use of hydrogen in industry and the existing transport pipelines and networks that have been operated over many decades have proved this (see Section 23.3.3).

23.3 Technology and Efficiency

23.3.1 Hydrogen production

Hydrogen can be generated directly from electricity by using electrolysis. In an electrolytic solution, water is split into its components – hydrogen and oxygen – at two electrodes. The gases are produced separately at the electrodes and have a high purity. Electrolysers can be operated at pressures between ambient pressure and 200 bar. The high-pressure processes are more efficient but also lead to higher equipment costs and more complicated systems. However, they can feed a pipeline system directly without any additional compression. Electrolysis requires an input of desalinated and demineralised water.

Electrode lifetime, mechanical stability and process efficiency suffer when input power varies. For this reason, electrolysers conventionally are operated at constant rated power. During shutdown, standard electrodes will corrode unless a protective voltage is applied. This requires an electricity backup. Nowadays, new improved electrode materials (e.g. catalytic laminated nickel plates) have been developed (Steinberger-Wilckens and Stolzenburg, 2000). With these electrodes, no protective voltage is required for stabilisation.

The efficiency of electrolytic hydrogen production is about 65 %, resulting in 4.2–4.8 kWh of energy cost for the electrolysis of 1 Nm^3 of hydrogen at atmospheric pressure [referring to the lower heating value (LHV)]. Using high-temperature and pressurised electrolysis as well as advanced materials, efficiency can be raised considerably, by up to 85–90 % (Pletcher and Walsh, 1990; Wendt and Bauer, 1988).

Electrolysers are available at various sizes and operating pressures. The largest installations realised to date were rated at around 150 MW of electric power input and consisted of 2 MW units. Large-scale electrolysis is used predominantly in the synthesis of ammonia for the manufacture of fertiliser, the two largest installations being in India and Egypt, both fed by hydro power (Wendt and Bauer, 1988). Depending on process pressure and system requirements, the hydrogen is either stored directly or is first pressurised.

Electrolysers are generally made of high-quality materials that have to be corrosion proof if applied in a marine environment. Today, they are already used on drilling platforms and produce hydrogen for welding and cutting.

As well as direct production of hydrogen by electrolysis, other processes of synthesising hydrogen-rich products are potentially possible. These products would ideally display a higher volumetric energy content, thus reducing the amount of storage volume and/or transport energy. Options include the synthesis of products such as ammonia, methanol, hydrazin and borates (Peschka, 1988). Their production will require the supply of additional base materials, such as nitrogen or carbon dioxide, which may both be obtained from the air (with an additional energy cost). In the case of borates, closed cycles can be established (Mohring, 2002), which implies that the 'used' hydrogen carrier will be returned to the site of the hydrogen source.

23.3.2 Hydrogen storage

Once gaseous hydrogen is produced, the question arises over what type of storage to use. Hydrogen can be stored in a variety of ways, each with its specific advantages and disadvantages. The main criteria are safety, efficiency, energy density and volume. In the following, the different storage methods that are commercially available today are described.

23.3.2.1 Compressed hydrogen

Compressed gaseous hydrogen (CGH_2) is stored in high-pressure tanks. This requires compression energy corresponding to 4–15 % of the hydrogen energy content stored, depending on the intake and outlet pressure of compression (Feck, 2001). Storage volume per unit of energy is generally high because of the low energy density of compressed hydrogen compared with that of, for instance, gasoline. Today, high-pressure tanks are being developed to achieve 700–1000 bar and thus reduce geometric volume. This raises safety concerns which, in turn, require enhanced security measures. Results from the design, development and testing of composite-material vessels, though, have shown the technical feasibility and the manageability of the risk of such an approach (Chaineaux *et al.*, 2000).

23.3.2.2 Liquid hydrogen

Liquid hydrogen typically has to be stored at 20 K ($-253\,^{\circ}$C), leading to a considerable energy cost for compressing and chilling the hydrogen into its liquid state. The cooling and compressing process results in a net loss of about 30 % of energy (Wagner, Angloher and Dreier, 2000). The storage tanks are insulated, to maintain the temperature, and are reinforced to store the liquid hydrogen under (low) pressure. Given the energy required for liquefaction and the cost of maintaining storage pressure and temperature, liquid hydrogen storage is expensive in comparison with other aggregate forms.

23.3.2.3 Metal hydrides

Metal hydrides are created from specific metallic alloys that can incorporate hydrogen into their metal lattice, emitting heat in the process. The hydrogen can be released again through heating the metal hydride vessel. The total amount of hydrogen absorbed is generally 1–2 % of the total weight of the storage medium. This surprisingly low figure is attributable to the disparity between the very low weight of hydrogen and the high weight of the metal alloys. In terms of volumetric storage capacity, metal hydride tanks perform similarly to liquid hydrogen (storing 60 and 70 kg H_2/m^3, respectively). Some metal hydrides are capable of storing 5–7 % of their own weight, but require unloading temperatures of 250 °C or higher (Carpetis, 1988).

The main disadvantage is that the inert mass of metal has to be installed or even moved around (in the case of vehicle tanks or transport vessels). Still, hydrides offer an interesting solution for the safe storage of hydrogen as they have the advantage of delivering hydrogen at constant pressure (30 to 60 bar) over a broad range of discharging levels. The lifetime of a metal hydride storage tank, though, is related directly to the purity of the hydrogen, since the hydriding process will suffer from impurities that are deposited in the crystal lattice.

23.3.2.4 Storage systems

Generally, storage containers for gaseous hydrogen are divided into several compartments that are shaped in spheres or, most commonly, cylinders ('bottles'). These are then connected into 'banks' of storage vessels. A storage container may include several 'banks' in order to maintain as constant a pressure as possible during unloading (e.g. in filling stations). This is the most cost-effective way of pressure storage since the cylinders are standard industry equipment manufactured in large numbers. More rarely, single-vessel pressure tanks are found.

Liquid hydrogen, in contrast, is stored in cryotanks, which are built as compact as possible in order to save costs and minimise losses from evaporation. They do not usually include a refrigeration unit. The slow warming of the tanks results in an over-pressure from the formation of a gaseous phase, which regularly has to be relieved by 'blowing off' the gas. For large tanks, the rate of evaporative loss amounts to about 0.1 % per day. For smaller, road transport, tanks it still remains below 1 %.

Tanks and all interconnection equipment (such as valves, piping, washers and so on) are available in materials that are suitable for a marine installation.

23.3.3 Hydrogen transport

Hydrogen is generally transported in the compressed gaseous (CGH_2) and liquefied (LH_2) aggregate state. In the case of small-scale deliveries for industrial purposes, compressed hydrogen is typically transported by vehicle. Bulk delivery generally stems from hydrogen production at customer-site production plants – so called 'captive hydrogen' – [e.g. via electrolysis or steam reforming of hydrocarbons at refineries, chemical plants and in the pharmaceutical industry; see Zittel and Niebauer, 1998]. Gaseous hydrogen is also transported via gas pipelines. Today, there are hydrogen distribution systems in the North of France, in the Benelux region (Belgium, The Netherlands and Luxemburg) and in Germany, for instance. It might be also an option to adjust the natural gas pipeline network in Europe for the transport of hydrogen.

Modifications to, for example, seals and connections because of the lower viscosity of hydrogen will be necessary in order to prevent losses and to secure a safe operation. Hydrogen losses during pipeline transport are very low, at about 0.001 % per 100 km (Wagner, Angloher and Dreier, 2000). However, the required compression energy to transfer hydrogen in order to deliver a given energy amount at a given time is ten times higher than that for natural gas (Feck, 2001).

For the transport of liquid hydrogen by sea, marine vessels will have to be used – so-called 'barge carriers' – which in principle consist of large fibreglass vacuum-insulated tanks. The storage capacity is about 15 000 m³. The energy consumption of such vessels is about 400 kWh/km (Feck, 2001). The losses can be estimated at 0.1 % per day (Gretz, 2001) or 0.013 % per 100 km (Feck, 2001) of the transported energy. The subsequent onshore transport of LH_2 by road will consume 4 kWh per km (Fritsche et al., 2001), with an effective mass per vehicle of 3.5 t. Liquid hydrogen losses from 'boil-off' (i.e. evaporation in the steady state) is estimated to average 0.3 % per day (Wagner, Angloher and Dreier, 2000).

This comparison shows that transport via pipeline would be by far the most energy-efficient means of delivery. If the hydrogen were transported in a pipeline over a distance of 100 km offshore followed by 150 km transport on land, pipeline delivery would 'cost' 6.5 % of the energy content of the hydrogen gas flow. Transport by barge in pressure vessels at 350 bar or in cryogenic tanks for LH_2 followed by rail transport (preferably using the same containers) would amount to losses of around 17.5 % and 33 %, respectively, of the hydrogen energy content if finally delivered at 20 mbar pressure (as required for most end-use appliances). If delivery at a pressure of 350 bar is necessary (e.g. for hydrogen vehicle-filling stations) the figures change to 16.5 % for transport by pipeline, 17.5 % for use of pressure vessels and 48 % for use of cryogenic tanks, because of the final compression for achieving the required end-pressure (all values here have been calculated with use of data from Feck, 2001; Wagner, Angloher and Dreier, 2000).

Finally, hydrogen does not need be transported in a 'pure' state or in an infrastructure that is separate from the conventional energy supply system. Similar to town gas that was used until the middle of the last century in many parts of Europe (and still is used, for instance, in Stockholm), hydrogen can be mixed with natural gas and thus reduce the greenhouse effect of natural gas. We will not discuss the implications of this here, but will refer anybody interested to the NaturalHy project led by GasUnie in the Netherlands, a European level integrated project starting in the year 2004.

23.4 Reconversion to Electricity: Fuel Cells

Hydrogen can be reconverted to electricity by means of conventional gas-fired equipment, such as gas motors [e.g. in combined heat and power (CHP) plants] or gas turbines. These processes have a rather limited electrical efficiency, of 28–35 %. The total efficiency of converting electricity to hydrogen (electrolyser) and back to electricity will lie in the range of 18–24 %, neglecting transport losses. Heat recovery in CHP increases the overall efficiency but the original energy content of the electricity generated from wind power will still be wasted to a large extent.

Table 23.2 Comparison of fuel cell types: polymer electrolyte fuel cells (PEFCs), phosphoric acid fuel cells (PAFCs), alkaline fuel cells (AFCs), solid oxide fuel cells (SOFCs) and molten carbonate fuel cells (MCFCs)

	Low-temperature fuel cells			High-temperature fuel cells	
	PEFC	PAFC	AFC	SOFC	MCFC
Electrolyte	Polymer	Phosphoric acid	Alkali	Ceramic	Molten carbonate salt
Operating temperature (°C)	80	220	150	850	650
Fuel	H_2	H_2	H_2	H_2, CO and CH_4	H_2, CO, CO_2 and CH_4
Oxidant	O_2 or air	O_2 or air	O_2 (!)	O_2 or air	O_2 or air
Efficiency[a] (%)	35–45	35–45	50–70	50–55	40–50

[a] Higher heating value.
Source: Larminie and Dichs, 2000.

Fuel cells offer an alternative, as they convert hydrogen (and oxygen) into electrical energy by means of an electrochemical process. In contrast to the case for thermal power plants, the conversion in fuel cells is not limited by the Carnot process efficiency. Basically, the process consists of the catalytic ionisation of hydrogen and oxygen. The ions pass an electrolyte membrane, separating the fuel and the oxidant gas stream. The recombination of hydrogen and oxygen ions then yields pure water and electricity.

Fuel cells are categorised by their operating temperature. Low-temperature cells are operated at 80–100 °C, high-temperature cells at 650–1000 °C (see Table 23.2). The higher the temperature the more efficient the electrochemical processes. A further distinction is made according to the type of electrolyte. Low-temperature cells work with liquid electrolytes [i.e. alkaline fuel cells (AFCs) or phosphoric acid fuel cells (PAFCs)] or with polymer membranes [i.e. polymer electrolyte fuel cells (PEFCs)]. High-temperature cells use zirconium oxides [i.e. solid oxide fuel cells (SOFCs)] or molten carbonates [i.e. molten carbonate fuel cells (MCFCs)].

Three aspects influence the choice of a fuel cell for reconversion to electricity:

- high (electrical) efficiency in order to minimise overall conversion losses;
- suitability for pure hydrogen obtained from electrolysis;
- fast response to load variations required for load-following capability.

The first two criteria are self-explanatory, but the third provides potential for increasing the controllability in connection with fluctuating electricity sources, such as wind energy. Fuel cells can be used to reduce the demands of conventional voltage and frequency control. MCFCs are not suitable for pure hydrogen feed gas, and the efficiency and cost effectiveness of PAFCs and AFCs are insufficient. PEFCs show a high efficiency in operation with pure oxygen, which means that PEFCs, and SOFCs, seem to be the most suitable for being used in equipment for converting hydrogen to electricity.

SOFCs are difficult to cycle through low part load and cannot easily be shut down or refired, but they have the highest electrical efficiency. SOFCs integrated into a gas-and-steam cycle power plant can contribute to a total electrical efficiency of up to 70 % (Palsson 2002). PEFCs, in contrast, are flexible in operation and fast in response to load changes. They are suitable for customer site installations of CHP at various scales. Also, these fuel cells will constitute the major part of fuel cells in vehicle applications. Therefore, prices for PEFC are expected to fall drastically after market introduction of fuel cell vehicles, maybe from 2010 onwards. A combination of the two types seems ideal for covering both base and peak electricity loads.

23.5 Hydrogen and Wind Energy

Today, wind and solar electricity constitute only a small part of the European electricity production. This situation may change in the future when the use of renewable electricity generation gradually approaches the technically exploitable potential. If renewable power reaches high penetration levels in the electricity distribution networks, the fluctuating power output will at times even exceed the load requirements and thus cause excess power and surplus energy production, which is equivalent to a temporary shutdown of renewable power plants.

The continental European grids are interconnected in the network of the Union pour la Coordination du Transport d'Electricité (UCTE, 2003). This network maintains the stability of the electricity grid through distributing control responsibilities among its members. The network allows the aggregation of loads and generation over a very large area. Therefore, the integration of high wind power penetration levels will be easier than in isolated power systems (see also Chapter 3). Power trading between individual regional grids, though, is limited to contingents agreed upon in advance (24 hours and more) and is governed by business agreements, even though the physical electricity flow itself cannot be easily controlled within the grid. The members may even risk penalties if they do not comply with the prearranged power balance.

The amount of surplus energy caused by fluctuating sources in electricity networks depends on the amount of renewable power capacity installed, the characteristics of the renewable sources utilised and the characteristics of load and conventional power generation (Steinberger-Wilckens, 1993). In rigid grids with a large contribution from base load and/or with slow or limited response in the power generation to fast load gradients, a surplus situation will occur more often than it will in flexible grids. Depending on the type of the predominant renewable energy source, the ratio of peak power to average power varies. This ratio is reasonably low for wind energy (i.e. between 3:1 and 4:1, depending on the siting) and is even lower for offshore wind power.

Table 23.3 shows the percentage of surplus (wind) energy production normalised to the load as a function of wind power penetration. We define wind power penetration as the ratio of wind energy production to total load requirement. The electricity grid assumed corresponds to the German system in 1990, with a contribution of about 30 % nuclear energy and 4 % hydro energy, with the majority of contribution from coal-fired generation. Surplus wind energy production starts at a penetration level of about 25 % and reaches a value of 7 % percent at a penetration of about 50 %. Taking

Table   23.3 Model calculation for surplus wind energy (as a percentage of total consumption or 'load') that cannot be absorbed by the electricity grid as a function of renewable penetration in the grid, where renewable penetration is defined as the ratio of total renewable energy production and total electricity consumption (in the case depicted, wind energy is fed into the German grid at the national scale (Steinberger-Wilckens, 1993)

Wind power penetration (% of load)	Surplus energy (% of load)	Surplus wind energy (% of wind energy)	Surplus wind energy (TWh/year)
25	0.0	0.00	0.0
30	0.5	0.15	0.8
35	1.2	0.42	2.3
40	2.8	1.12	6.2
45	4.7	2.12	11.7
50	6.8	3.40	18.8
55	9.2	5.06	28.0
60	12.0	7.20	39.8

into account the normalisation, that means that 14 % of the wind energy generation is discarded. The calculations include the effect of the geographical distribution in a nationwide grid.

Hydrogen as an intermediate storage vector can 'absorb' this surplus energy and 'release' it at times of low (or lower) renewable power production. The conversion of electricity to hydrogen and the storage and reconversion to electricity were discussed in Sections 23.3 and 23.4. Here, we will look at the influence of storage size on total system performance.

Table 23.4 shows the percentage of conventional (fossil fuel) energy in a grid with high wind and/or solar energy production in relation to penetration and storage size. The storage system in this example is assumed to have an efficiency of 70 %. If there is no storage (storage capacity $= 0$ h), at a penetration rate of 100 %, about 50 % of the renewable energy input has to be discarded, mostly because of the characteristics of the solar power plants. The share of renewable energy that can be utilised increases with storage size. Owing to storage losses, however, it does not reach a 100 % load match. It should be noted that the increase in renewable contribution rises sharply with the first increment in storage capacity but then approaches the limit of infinite storage very slowly. This indicates that short-term and daily patterns govern the characteristics of renewable (wind and solar) energies. Long-term temporal shifting of renewable power requires very large storage sizes.

Table 23.5 illustrates the potential wind power production for 2010 in comparison with the net electricity generation in 2000. The wind production was derived from the total onshore potential estimated by van Wijk and Coelingh (1993) plus the rated power of offshore projects in the planning phase around the year 1993. This will give an estimate of the possible level of wind power production in the area between 2006 and 2015. Percentages vary, but many countries may be faced with a wind power generation that substantially exceeds the surplus energy generation limits mentioned above. Data are presented only for European countries, since here the ratio between offshore wind potential and load requirements is high. This is because of the ratio of length of coastline

Table 23.4 Percentage of conventional electric energy necessary for load following (columns 2 through 5) as a function of renewable energy penetration (column 1; fixed mix of 90 % solar, 10 % wind energy; for a comparison with a 100 % wind system see Table 23.3) and storage size

Renewable power penetration (% of load)	Conventional energy (% of load), by storage capacity			
	0 h	3 h	6 h	infinite
0	100	100	100	100
10	90	90	90	90
20	81	81	81	81
30	73	72	72	72
40	67.5	63	62	62
50	63.5	58	55	54
60	60	52.5	49	46
70	57.5	50	44.5	37.5
80	55.5	47.5	40	30
90	54	45	37.5	22
100	53.5	42	35	16

Note: the storage capacity is normalised to the hourly average load (i.e. a storage capacity of 1 h denotes a storage that can deliver the average load for 1 hour). Total storage system efficiency in this example is set to 70 % (battery storage; Steinberger-Wilckens, 1993). Short-term storage reduces the contribution of conventional (fossil fuel) generation considerably. Mid- and long-term storage (above 12 hours) has an effect only at very high penetration rates, (of above 50%) as indicated in the last column ('infinite') which also gives an indication of the maximum achievable reduction in conventional energy use.

to mainland area. Coastal States in the USA may face similar problems in the future if the electric power flow is restricted for any reason (see above).

Apart from controlling wind resources, the more immediate role of hydrogen in offshore wind exploitation appears to be the transport of energy from offshore wind farms to the shore. The installation of a hydrogen pipeline is no more difficult than that of a sea cable. A hydrogen pipeline is likely to take up less space, which may be an important aspect given the massive wind capacity that may have to be transferred to the shore. Transport losses are lower for hydrogen, and the required investment costs for the production of hydrogen and its reconversion to electricity are similar to those for high-voltage transmission. Still, losses in conversion to and from hydrogen are high. Offshore hydrogen generation will be feasible only if the hydrogen is used for the controlling purposes described above and for establishing hydrogen power production. It does not make any sense to use the hydrogen only as a means of transport to the shore.

23.6 Upgrading Surplus Wind Energy

Use of hydrogen as an intermediate storage medium results in a new controllability in the wind power resource and in the possibility of optimising power supply. It may be

Table 23.5 Net electricity production in the countries of the EU-15 in 2000 (Bassan, 2002) and extrapolated wind energy production in about 2010 (as the sum of onshore potential plus offshore planning figures from 2003)

Country	Total production (TWh/year)	Wind production in 2010	
		TWh/year	%[a]
Belgium	80.1	6.20	7.74
Denmark	34.5	28.32	82.09
Germany	533.5	100.09	18.76
Greece	49.8	44.48	89.32
Spain	215.2	87.20	40.52
France	516.7	85.17	16.48
Ireland	22.7	45.56	200.70
Italy	263.6	69.00	26.18
Luxemburg	1.1	0.00	0.00
Netherlands	85.8	7.70	8.97
Austria	60.3	3.00	4.98
Portugal	42.2	15.00	35.55
Finland	67.3	7.00	10.40
Sweden	142.1	58.39	41.09
UK	358.6	118.04	32.92
Total EU-15	2473.5	675.15	27.30

[a] Wind energy as a percentage of total electricity supply (net production).
Note: '2010' denotes the period 2006–15; figures include all grid losses. (e.g. electricity consumption in Germany amounts to approximately 490 TWh/year).
Sources: hypothetical onshore potential, van Wijk and Coelingh, 1993; offshore operational and planning figures, IWR, 2003; Paul, 2002; Paul and Lehmann, 2003; NEA, 2001; WSH, 2003.

attractive, for instance, to enter electricity spot markets for peak load. Although in the past peak power was said to have a high value (Steinberger-Wilckens, 1993) this was not necessarily reflected in the price indexes (*E&M*, 2002).

It is still under debate whether or not these indexes constitute 'real' market prices since the effective price of control power supplied and peaking power station operation are difficult to separate from overall system operation. Theoretically, the electricity delivered by a marginal peaking power station may be 'infinite' if this generating capacity is not put into operation in a particular year, but this full-cost accounting is not reflected in the actual spot pricing since this cost is attributed to safety-of-supply risk management (and thus is included in the 'overhead' costs of electricity generation). Suppliers of 'green' electricity, who might be required to certify load-following capabilities, would be especially interested in 'green' peak load.

Hydrogen could offer an opportunity to store electricity over a considerable span of time [e.g. on a daily basis, from weekend to week day and, less likely, over longer periods (see Table 23.4)]. In the case of surplus electricity production, the energy cost of generating the hydrogen would be reduced to zero.

A cost analysis, though, would need to show that the extra cost of hydrogen production and of re-electrification installations can be compensated by extra revenue. Natural gas reforming could be used for the production of hydrogen as a backup in situations of a low supply of wind-derived hydrogen. This could enhance the flexibility of this system and help to guarantee power production under all circumstances.

23.6.1 Hydrogen products

Apart from being used in electricity production, hydrogen can also be marketed as an independent product. On the one hand, hydrogen has various industrial applications. In various places there is already a regional hydrogen infrastructure that connects the industry with a hydrogen surplus (derived from various chemical processes, such as the chlorine chemistry) with facilities that use hydrogen in their production processes (e.g. as a reducing agent). On the other hand, hydrogen is expected to be one of the most important future energy vectors in the transport market. Until now, it has commonly been believed that the first hydrogen vehicles will run on fuel that is derived decentrally (i.e. at the filling station) from natural gas by steam reforming. This process produces carbon dioxide emissions, though. Hydrogen produced from wind energy can constitute a more ecological alternative, since vehicles running on this fuel are practically emission-free even when analysed on a global scale (i.e. taking into account the balance of all well-to-wheel contributions; Feck, 2001).

If surplus wind-derived electricity is converted to hydrogen it is also branched out of the electricity market and the price pressure exerted by surplus production will be reduced. The spot market would react to an excess of electricity supply by an erosion of revenues which, in turn, would drastically cut the return on investment for offshore wind farms.

In addition to pure hydrogen, there are several ways of producing chemicals from hydrogen. Today, the possible synthesis of methanol from atmospheric carbon dioxide and wind-derived hydrogen is of major interest. Methanol can be transported by tankers more easily than can hydrogen, and just as easily by pipeline. Whether methanol or hydrogen will be first to be used as vehicle fuel is still under discussion. An analysis of the efficiencies definitely points towards the direct use of hydrogen (Feck, 2001).

Other possible hydrogen-derived products include ammonia, synthesised from atmospheric nitrogen and hydrogen. This technology has been discussed in the past as a vector for hydrogen fixation, storage and transport. However, there are no immediate advantages to an application in the context of offshore wind farms.

23.7 A Blueprint for a Hydrogen Distribution System

Wind-derived hydrogen can constitute the basis for an environmentally safe, emission-free energy supply. Figure 23.1 integrates the elements of hydrogen production, transport, storage and distribution into a complete system. Here, electricity and heating, transport fuel and industrial gas demands are all supplied by wind energy sources.

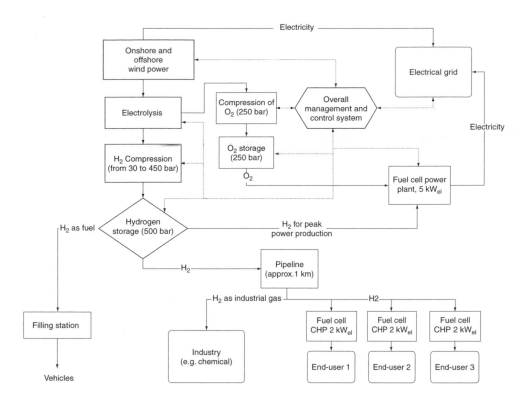

Figure 23.1 Flow diagram of the wind–hydrogen production and distribution system HyWind-Farm, which consists of electrolytic production of hydrogen from wind energy, diffusion of hydrogen by pipeline and use for small-scale combined heat and power (CHP) in residential applications, a small-scale fuel cell power plant (without heat recovery) and hydrogen fuel delivery; hydrogen can also be supplied to industrial customers. *Note*: an electrolyser, compressor, storage and filling station setup are included in the EUHYFIS module; the system is intended for a medium-sized prototype installation in Northern Germany (see EUHYFIS, 2000)

The HyWindFarm system is the result of a joint development by engineering companies, industrial gas suppliers, energy traders, control equipment manufacturers, and control and system management software engineers. It is expected to be built in Northern Germany next to an onshore wind farm. The EUHYFIS (European Hydrogen Filling Station; EUHYFIS, 2000) forms the 'core' of the system, thus simplifying the installation. The filling station is a compact, preassembled module that comprises electrolyser, compressor and storage. In this case, it would also be the interface with the distribution system.

The system shown uses low-temperature fuel cells (PEFCs) to supply heat and electricity to residential, administrative or industrial buildings and high-efficiency cells for the reconversion to electricity. These could either be high-temperature cells (SOFCs) or oxygen-driven PEFCs. The oxygen would have to be supplied by molecular sieves (used for separating air into oxygen and nitrogen) or by installing a separate supply of oxygen that is derived as a byproduct from the electrolytic production of hydrogen. The engineering of an autonomous supply system for an island or isolated community has resulted in a similar system layout (see Sørensen and Bugge, 2002).

23.7.1 Initial cost estimates

At this point, cost estimates are still rough and are likely to contain large error margins. Many of the technologies involved (e.g. offshore platforms, gas pipelines, and electrolysers) are widely used. Their application in the context of bulk hydrogen production, though, still poses substantial technical problems that need to be solved. The extrapolation of costs to large systems is thus rather speculative.

In addition, fuel cell technology is quickly developing and it is still difficult to come up with realistic cost estimates of electricity production. Projections offered by the industry are basically guidelines that need to be verified in practice. We therefore do not include costs for the production of electricity from hydrogen. Table 23.6 gives an overview of component costs for a 1000 MW (1 GW) hydrogen production plant.

It must be pointed out that the equipment costs also have to reflect the marine environment of the installations and the respective additional corrosion of the components. Also, cost statements that are available today often refer to components in an early stage of development. Projections for equipment at a mature stage of technical development manufactured in large quantities are thus neither linear nor very reliable. The figures in Table 23.6, however, correspond roughly to qualitative data published by Altmann and Richert (2001) for offshore applications and to an analysis of long-distance onshore transport of wind energy via hydrogen in the USA by Keith and Leighty (2002).

Table 23.7 shows estimates of the total cost of hydrogen (i.e. energy) including for production, distribution and storage systems. Providing that wind energy sells at approximately €0.06 per kilowatt-hour at the time in question (i.e. after 2010, when hydrogen systems of the size presented here can be actually built) the total cost of hydrogen produced by the HyWindFarm system will be €0.568 per norm cubic metre (Nm^3), or

Table 23.6 Cost estimate for a 1 GW hydrogen production plant on an offshore platform; the hydrogen is transported by pipeline to the shore and the transport is driven by the pressurised electrolyser

Component	Qualification	Specific costs (€/kW)	Total costs (€ millions)
Electrolyser, pressurised, 70 % efficiency	1 GW	800	800
Pipeline length	120 km	—	36
Ancillaries, platform, landing station, contingencies	N.A.	—	173
Total			1009

N.A. Not applicable.
— Not calculated.

Table 23.7 Estimated production costs of hydrogen of 1 GW (including pressurisation and storage)

Contribution	Specifics	Total costs (€ millions)	Specific costs (€/Nm³)
Hours of full operation	4000 h p.a.	—	—
Hydrogen production	933 million Nm³ p.a.	—	—
Annual capital costs	N.A.	185	
Annual operation and management	N.A.	62	
Cost of hydrogen production plant and storage	N.A.	—	0.266
Cost of energy for hydrogen production, including wind electricity costs	N.A.	—	0.302
Total cost	N.A.	—	0.568

N.A. Not applicable.
—Not Calculated.
Note: Nm³ = norm cubic metres.

€0.189 per kilowatt-hour. Considering that the reconversion to electricity is not yet included in these figures, delivery of stored energy to the electricity grid will be necessarily restricted to peak load hours and high-quality, fail-safe electricity production because of the associated costs. However, industrial hydrogen is sold at prices that are similar to those mentioned above. Taking vehicle fuels as a standard, the figures quoted are in the order of magnitude of taxed petrol and even diesel, taking into account that a fuel cell vehicle would be of superior efficiency (roughly factor 2 above a diesel engine). Even though the figures stated here are of limited reliability, this result is encouraging.

23.8 Conclusions

Hydrogen production from offshore wind energy is technically feasible and builds mainly on technology that is already available today. Further technological development will be necessary in order to scale up the components, improve efficiencies and provide equipment that endures marine environments.

The role of hydrogen will be to transport wind energy to shore and to supply a means of controlling wind energy output. The reconversion to electricity will be very sensitive to cost factors and will probably be restricted to high-quality power applications.

The use of hydrogen as fuel or industrial gas and its distribution to stationary fuel cells appear to offer a promising revenue stream for electricity generation.

References

[1] Alcock, J. L. (2001) 'Compilation of Existing Safety Data on Hydrogen and Comparative Fuels', EIHP2 Report, May 2001, EU contract ENK6-CT2000-00442.
[2] Altmann, M., Richert, F. (2001) 'Hydrogen Production at Offshore Wind Farms', presented at Offshore Wind Energy Special Topic Conference, Brussels, December 2001.

[3] Bassan, M. (2002) 'Electricity Statistics 2001', Statistics in Focus, Environment and Energy, Theme 8, March 2002; EUROSTAT, Luxemburg.

[4] Carpetis, C. (1988) 'Storage, Transport and Distribution of Hydrogen', in J. Winter and J. Nitsch (Eds), *Hydrogen as an Energy Carrier*, Springer, New York and Berlin pp. 249–290.

[5] Chaineaux, J., Devillers, C., Serre-Combe, P. (2000) 'Behaviour of a Highly Pressurised Tank of GH$_2$, Submitted to a Thermal or Mechanical Impact', in *Forum für Zukunftsenergien, HYFORUM 2000, Proceedings (Volume II), Bonn, 2000, 55–64*.

[6] E&M (*Energie & Management*) (2002) 'SWEP, EEX and APX Indices', *Energie & Management*, (1 November 2002), p. 3.

[7] EUHYFIS (2000) 'European Hydrogen Filling Station EUHYFIS – Infrastructure for Fuel Cell Vehicles Based on Renewable Energies', report, contract JOE-CT98-7043, July 2000.

[8] Feck, T (2001) *Ökobilanzierung unterschiedlicher Kraftstofflebenszyklen für Wasserstofffahrzeuge*, Carl-von-Ossietzky University Oldenburg, Oldenburg, Germany.

[9] Fischer, M. (1986) 'Safety Aspects of Hydrogen Combustion in Hydrogen Energy Systems', *Int. J. Hydrogen Energy* 11, 593–601.

[10] Fritsche, U. R., Schmidt, K. (2003) 'Handbuch zu Globales Emissions-Modell Integrierter System (GEMIS)' Version 4.1, Öko-Institut (Institut für angewandte Ökologie eV) Freiburg, Germany.

[11] Gretz, J. (2001) 'Note', Wasserstoffgesellschaft Hamburg e.V., Hamburg, Germany.

[12] Grube, Th., Höhlein, B., Menzer, R., Stolten, D. (2001) 'Vergleich von Energiewandlungsketten: Optionen and Herausforderungen von Brennstoffzellen-fahrzeugen' 13, Intl. AVL Meeting, Gvaz, September 6–7.

[13] IWR (2003) 'Offshore Windenergie' www.iwr.de/wind/offshore; accessed 10 March 2003.

[14] Keith, G., Leighty, W. (2002) 'Transmitting 4,000 MW of New Windpower from North Dakota to Chicago: New HVDC Electric Lines or Hydrogen Pipeline', prepared for the Environmental Law and Policy Centre, Chicago, IL.

[15] Larminie, J., Dicks, A. (2000) *Fuel Cells Explained*, John Wiley & Sons Ltd, Chichester, UK.

[16] Mohring, R. M. (2002) 'Hydrogen Generation via Sodium Borohydride', presented at Hydrogen Workshop, Jefferson Laboratory, November 2002.

[17] NEA, *et al* (Nieder. Energieagentur/Deutsches Windenergie-Institut/Niedersächsisches Institut für Wirtschaftsforschung) (2001) 'Untersuchung der wirtschaftlichen und energiewirtschaftlichen Effekte von Bau und Betrieb von Offshore-Windparks in der Nordsee auf das Land Niedersachsen', study for the Ministry of the Environment of Lower Saxony, Germany.

[18] Palsson, J. (2002) *Thermodynamic Modelling and Performance of Combined Solid Oxide Fuel Cell and Gas Turbine Systems*, doctoral thesis, Department of Heat and Power Engineering Lund University, Sweden.

[19] Paul, N. (2002) 'Offshore-Projekt in Nord- und Ostsee', *Sonne Wind & Wärme* 7, S64–66.

[20] Paul, N., Lehmann, K.-P. (2003) 'Offshore-Projekt in Europa', *Sonne Wind & Wärme* 2, S58–63.

[21] Peschka, W. (1998) 'Hydrogen Energy Applications Engineering', in J. Winter and J. Nitsch (Eds), *Hydrogen as an Energy Carrier*, Springer, New York and Berlin pp. 30–55.

[22] Pletcher, D., Walsh, F. C. (1990) *Industrial Electrochemistry*, Blackie Academic and Professional, Glasgow.

[23] Sørensen, J., Bugge, O. (2002) 'Wind–Hydrogen System on the Island of Røst', Norway' presented at 14th World Hydrogen Energy Conference (WHEC 2002), Montreal, Canada.

[24] Steinberger-Wilckens, R. (1993) *Untersuchung der Fluktuationen der Leistungsabgabe von räumlich ausgedehnten Wind- und Solarenergie-Konvertersystemen in Hinblick auf deren Einbindung in elektrische Verbundnetze*, Verlag Shaker, Aachen.

[25] Steinberger-Wilckens, R., Stolzenburg, K. (2000) 'EUHYFIS – European Hydrogen Filling Station, Renewable Energies for Zero Emission Transport', in *Proceedings of HYFORUM, September 2000, München* forum für Zukunftsenersien, Bonn, pp. 513–522.

[26] UCTE (Union pour la Coordination du Transport d'Electricité) (2003) 'Our World-Wug UCTE', http://www.ucte.org/ourworld/ucte.

[27] van Wijk, A. J. M., Coelingh, J. P. (1993) *Wind Potential in the OECD Countries*, University of Utrecht, Utrecht, The Netherland.

[28] Wagner, U., Angloher, J., Dreier, T. (2000) *Techniken und Systeme zur Wasserstoff-verbreitung*, WIBA, Koordinationsstelle der Wasserstoffinitiative Bayern, Munich, Germany.

[29] Wendt, H., Bauer, G. H. (1998) 'Water Splitting Methods', in J. Winter and J. Nitsch (Eds), *Hydrogen as an Energy Carrier*, Springer, New York and Berlin pp. 166–208.

[30] WSH (Wind Service Holland) (2003), 'WSH-Offshore wind energy', http://home.wxs.nl/~windsh/ offshore. html, accessed 10th November 2004.

[31] Zittel, W., Niebauer, P. (1988) 'Identification of Hydrogen By-product Sources in the European Union', Ludwig-Bölkow-Systemtechnik GmbH, Ottobrunn, funded by the European Union, contract 5076-92-11 EO ISP D Amendment 1.

Part D

Dynamic Modelling of Wind Turbines for Power System Studies

24

Introduction to the Modelling of Wind Turbines

Hans Knudsen and Jørgen Nygård Nielsen

Dedicated to our friend Vladislav Akhmatov, whose never-failing endeavour, zeal and perseverance during his PhD project brought the world of wind turbine aerodynamics to us.

24.1 Introduction

This chapter presents a number of basic considerations regarding simulations for wind turbines in electrical power systems. Though we focus on the modelling of wind turbines, the general objective is also to look at the wind turbine as one electrotechnical component among many others in the entire electrical power system.

The chapter starts with a brief overview of the concept of modelling and simulation aspects. This is followed by an introduction to aerodynamic modelling of wind turbines. We will present general elements of a generic wind turbine model and some basic considerations associated with per unit systems, which experience shows are often troublesome. Mechanical data will be discussed together with a set of typical mechanical data for a contemporary sized wind turbine. We will give an example of how to convert these physical data into per unit data. These per unit data are representative for a wide range of sizes of wind turbine and are therefore suitable for user applications in a number of electrical simulation programs. Finally, various types of simulation phenomena in the electrical power system are discussed, with special emphasis on what to consider in the different types of simulation.

Wind Power in Power Systems   Edited by T. Ackermann
© 2005 John Wiley & Sons, Ltd   ISBN: 0-470-85508-8 (HB)

We will not address the issues related to model implementation in various simulation programs except for some specific properties related to dynamic stability compared with full transient generator models.

24.2 Basic Considerations regarding Modelling and Simulations

Computer simulation is a very valuable tool in many different contexts. It makes it possible to investigate a multitude of properties in the design and construction phase as well as in the application phase. For wind turbines (as well as other kinds of complex technical constructions), the time and costs of development can be reduced considerably, and prototype wind turbines can be tested without exposing physical prototype wind turbines to the influence of destructive full-scale tests, for instance. Thus, computer simulations are a very cost-effective way to perform very thorough investigations before a prototype is exposed to real, full-scale, tests.

However, the quality of a computer simulation can only be as good as the quality of the built-in models and of the applied data. Therefore it is strongly recommended to define clearly the purpose of the computer simulations in order to make sure that the model and data quality are sufficiently high for the problem in question and that the simulations will provide adequate results. Otherwise, the results may be insufficient and unreliable.

Unless these questions are carefully considered there will be an inherent risk that possible insufficiencies and inaccuracies will not be discovered – or will be discovered too late – and subsequent important decisions may be made on a faulty or insufficient basis. Hence, computer simulations require a very responsible approach.

Computer simulations can be used to study many different phenomena. The requirements that a specific simulation program has to meet, the necessary level of modelling detail and requirements regarding the model data may differ significantly and depend on the objective of the investigation. Different parts of the system may be of varying importance, depending on the objective of the investigation, too. All this should be taken into consideration before starting the actual computer simulation work.

However, in order to take into account all this, it is necessary to have a general understanding of the wind turbine (or of any other system on which one wants to perform a computer simulation) and how the various parts of the wind turbine can be represented in a computer model. This basic overview will be provided in the subsequent sections of this chapter. After that, the different types of simulations and various requirements regarding accuracy will be discussed in more detail.

24.3 Overview of Aerodynamic Modelling

The modelling of different types of generators, converters, mechanical shaft systems and control systems is all well-documented in the literature. In the case of wind turbines it is therefore primarily the aerodynamic system that may be unfamiliar to those who work with electrotechnical simulation programs. For that reason, an introduction to the basic physics of the turbine rotor and the various ways in which the turbine rotor is commonly represented will be outlined. For more details, see, for example, Akhmatov (2003), where the subject is treated in great detail and where a comprehensive list of references regarding wind power in general is included. Some of the most recommendable references in this context are Aagaard Madsen (1991), Freris (1990), Hansen (2000), Heier (1996), Hinrichsen (1984), Johansen (1999), Øye (1986), Snel and Lindenburg (1990), Snel and Schepers (1995), Sørensen and Koch (1995), and Walker and Jenkins (1997).

24.3.1 Basic description of the turbine rotor

From a physical point of view, the *static* characteristics of a wind turbine rotor can be described by the relationships between the total power in the wind and the mechanical power of the wind turbine. These relationships are readily described starting with the incoming wind in the rotor swept area. It can be shown that the kinetic energy of a cylinder of air of radius R travelling with wind speed V_{WIND} corresponds to a total wind power P_{WIND} within the rotor swept area of the wind turbine. This power, P_{WIND}, can be expressed by:

$$P_{\text{WIND}} = \frac{1}{2}\,\rho_{\text{AIR}}\,\pi\,R^2\,V_{\text{WIND}}{}^3, \tag{24.1}$$

where ρ_{AIR} is the air density ($= 1.225\,\text{kg/m}^3$), R is the rotor radius and V_{WIND} is the wind speed.

It is not possible to extract all the kinetic energy of the wind, since this would mean that the air would stand still directly behind the wind turbine. This would not allow the air to flow away from the wind turbine, and clearly this cannot represent a physical steady-state condition. The wind speed is only *reduced* by the wind turbine, which thus extracts a fraction of the power in the wind. This fraction is denominated the power efficiency coefficient, C_P, of the wind turbine. The mechanical power, P_{MECH}, of the wind turbine is therefore – by the definition of C_P – given by the total power in the wind P_{WIND} using the following equation:

$$P_{\text{MECH}} = C_P\,P_{\text{WIND}}. \tag{24.2}$$

It can be shown that the theoretical static upper limit of C_P is 16/27 (approximately 0.593); that is, it is theoretically possible to extract approximately 59% of the kinetic energy of the wind. This is known as Betz's limit. For a comparison, modern three-bladed wind turbines have an optimal C_P value in the range of 0.52–0.55 when measured at the hub of the turbine.

In some cases, C_P is specified with respect to the electrical power at the generator terminals rather than regarding the mechanical power at the turbine hub; that is, the losses in the gear and the generator are deducted from the C_P value. When specified in this way, modern three-bladed wind turbines have an optimal C_P value in the range of 0.46–0.48. It is therefore necessary to understand whether C_P values are specified as a mechanical or as an electrical power efficiency coefficient.

If the torque T_{MECH} is to be applied instead of the power P_{MECH}, it is conveniently calculated from the power P_{MECH} by using the turbine rotational speed ω_{turb}:

$$T_{\text{MECH}} = \frac{P_{\text{MECH}}}{\omega_{\text{turb}}}. \tag{24.3}$$

It is clear from a physical point of view that the power, P_{MECH}, that is extracted from the wind will depend on rotational speed, wind speed and blade angle, β. Therefore, P_{MECH} and hence also C_P must be expected to be functions of these quantities.

$$P_{\text{MECH}} = f_{P_{\text{MECH}}}(\omega_{\text{turb}}, V_{\text{wind}}, \beta). \tag{24.4}$$

Now, the forces of the wind on a blade section – and thereby the *possible* energy extraction – will depend on the angle of incidence φ between the plane of the moving rotor blades and the relative wind speed V_{rel} (see Figure 24.1) as seen from the moving blades.

Simple geometrical considerations, which ignore the wind turbulence created by the blade tip (i.e. a so-called two-dimensional aerodynamic representation) show that the angle of incidence φ is determined by the incoming wind speed V_{WIND} and the speed of the blade. The blade tip is moving with speed V_{tip}, equal to $\omega_{\text{turb}} R$. This is illustrated in Figure 24.1. Another commonly used term in the aerodynamics of wind turbines is the tip-speed ratio, λ, which is defined by:

$$\lambda = \frac{\omega_{\text{turb}} R}{V_{\text{WIND}}}. \tag{24.5}$$

The highest values of C_P are typically obtained for λ values in the range around 8 to 9 (i.e. when the tip of the blades moves 8 to 9 times faster than the incoming wind). This means that the angle between the relative air speed – as seen from the blade tip – and the rotor plane is rather a sharp angle. Therefore, the angle of incidence φ is most conveniently calculated as:

$$\varphi = \arctan\left(\frac{1}{\lambda}\right) = \arctan\left(\frac{V_{\text{WIND}}}{\omega_{\text{turb}} R}\right). \tag{24.6}$$

It may be noted that the angle of incidence φ is defined at the tip of the blades, and that the local angle will vary along the length of the blade, from the hub (r = 0) to the blade tip (r = R) and, therefore, the local value of φ will depend on the position along the length of the blade.

On modern wind turbines, it is possible to adjust the pitch angle β of the entire blade through a servo mechanism. If the blade is turned, the angle of attack α between the blade and the relative wind V_{rel} will be changed accordingly. Again, it is clear from a

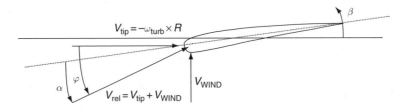

Figure 24.1 Illustration of wind conditions around the moving blade. *Note:* V_{tip} = tip speed; ω_{turb} = turbine rotational speed; R = rotor radius; V_{rel} = relative wind speed; V_{WIND} = wind speed; α = angle of attach; φ = angle of incidence between the plane of the rotor and V_{rel}; β = blade angle

physical perspective that the forces of the relative wind on the blade, and thereby also the energy extraction, will depend on the angle of attack α between the moving rotor blades and the relative wind speed V_{rel} as seen from the moving blades.

It follows from this that C_P can be expressed as a function of λ and β:

$$C_P = f_{C_p}(\lambda, \beta). \tag{24.7}$$

C_P is a highly nonlinear power function of λ and β. It should be noted that one main advantage of an approach including C_P, λ and β is that these quantities are normalised and thus comparable, no matter what the size of the wind turbine.

On older and simpler wind turbines, the blades have a fixed angular position on the hub of the wind turbine, which means the blade angle β is constant (β_{const}). This is called stall (or passive stall) control, because the turbine blades will stall at high wind speeds and thus automatically reduce the lift on the turbine blades. With a fixed pitch angle of the blades, the relation between the power efficiency coefficient $C_P(\lambda, \beta_{\text{const}})$ and the tip-speed ratio, λ, will give a curve similar in shape to the one shown in Figure 24.2(a).

Assuming a constant wind speed, V_{WIND}, the tip-speed ratio, λ, will vary proportionally to the rotational speed of the wind turbine rotor. Now, if the $C_P - \lambda$ curve is known for a specific wind turbine with a turbine rotor radius R it is easy to construct the curve of C_P against rotational speed for any wind speed, V_{WIND}. The curves of C_P against rotational speed will be of identical shape for different wind speeds but will vary in terms of the 'stretch' along the rotational speed axis, as illustrated in Figure 24.2(b). Therefore, the optimal operational point of the wind turbine at a given wind speed V_{WIND} is, as indicated in Figure 24.2(a), determined by tracking the rotor speed to the point λ_{opt}. The optimal turbine rotor speed $\omega_{\text{turb, opt}}$ is then found by rewriting Equation (24.5) as follows:

$$\omega_{\text{turb, opt}} = \frac{\lambda_{\text{opt}} V_{\text{WIND}}}{R}. \tag{24.8}$$

The optimal rotor speed at a given wind speed can also be found from Figure 24.2(b). Observe that the optimal rotational speed for a specific wind speed also depends on the turbine radius, R, which increases with the rated power of the turbine. Therefore, the larger the rated power of the wind turbine the lower the optimal rotational speed.

These basic aerodynamic equations of wind turbines provide an understanding that fixed-speed wind turbines have to be designed in order for the rotational speed to match the most likely wind speed in the area of installation. At all other wind speeds, it will not be possible for a fixed-speed wind turbine to maintain operation with optimised power efficiency.

In the case of variable-speed wind turbines, the rotational speed of the wind turbine is adjusted over a wide range of wind speeds so that the tip-speed ratio λ is maintained at λ_{opt}. Thereby, the power efficiency coefficient C_P reaches its maximum and, consequently, the mechanical power output of a variable-speed wind turbine will be higher than that of a similar fixed-speed wind turbine over a wider range of wind speeds. At higher wind speeds, the mechanical power is kept at the rated level of the wind turbine by pitching the turbine blades.

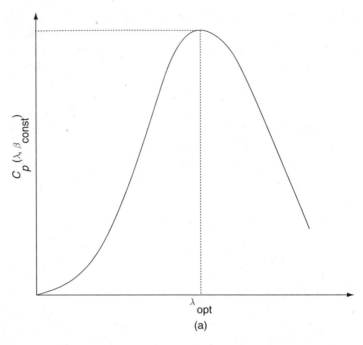

(a)

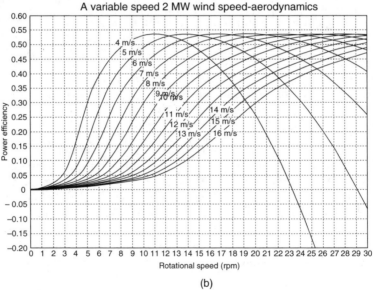

(b)

Figure   24.2 Power efficiency coefficient, C_P, for a fixed blade angle (a) C_P as a function of tip-speed ratio, λ, and (b) C_P as a function of rotational speed for various wind speeds (4–16 M/S). Reproduced from *Wind Engineering*, volume 26, issue 2, V. Akhmatov, 'Variable Speed Wind Turbines with Doubly-fed Induction generators, Part I: Modelling in Dynamic Simulation Tools', pp. 85–107, 2002, by permission of Multi-Science Publishing Co. Ltd

Figure 24.3 gives an example that illustrates the mechanical power from a fixed-speed and a full variable-speed wind turbine. It shows that the mechanical power output is higher for the variable-speed wind turbine at all wind speeds. Only at a wind speed of 7 m/s, is the mechanical power output the same.

Other things being equal, variable-speed wind turbines will yield greater annual power production compared with similar fixed-speed wind turbines. This improvement in efficiency is, however, obtained at the cost of greater complexity in the construction of the unit and also some additional losses in the power electronic converters, which enable the variable-speed operation (for more details, see Chapter 4). If the wind turbine is erected in an environment with high winds (e.g. offshore) the gain in annual energy production may be less significant, because this gain is achieved primarily in low-wind situations. Also, if the speed controllability is achieved at the cost of additional losses (e.g. in frequency converters) the net result might even be negative. Some fixed-speed wind turbines can, in a way, even be characterised as variable-speed wind turbines, or at least as two-speed wind turbines. Some manufacturers either include two generators – a high-power and a low-power generator – with a different number of pole pairs, into the wind turbine, or they apply a special generator, which is able to change the number of

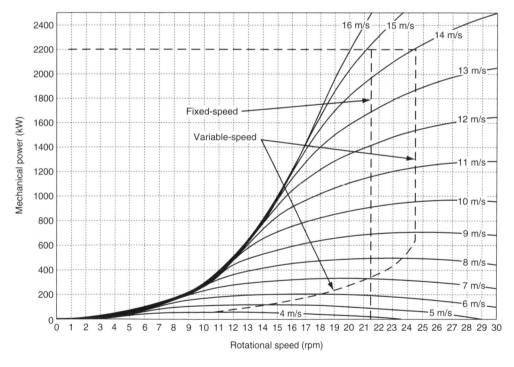

Figure 24.3 Illustrative mechanical power curves of fixed-speed and variable-speed wind turbines for various wind speeds (4–16 m/s). Reproduced from *Wind Engineering*, volume 26, issue 2, V. Akhmatov, 'Variable Speed Wind Turbines with Doubly-fed Induction generators, Part I: Modelling in Dynamic Simulation Tools', pp. 85–107, 2002, by permission of Multi – Science Publishing Co. Ltd

pole pairs (e.g. from 2 to 3) by changing the connections of the stator windings. In this way, a certain fraction of the increase in annual power production from a variable-speed turbine is obtained through a very simple and cost-effective measure.

At high wind speeds, the mechanical power with the optimal C_P value will exceed the nominal power for which the wind turbine is designed. It is therefore necessary to reduce the mechanical power. This is achieved by turning the blades away from the optimal pitch angle. There are two possibilities for doing this – either out of the wind or up against the wind:

- If the blades are turned out of the wind, the lift on the blades is gradually reduced. This is called pitch control and requires a relatively large change in pitch angle to reduce power significantly.
- If the blades are turned up against the wind, the turbine blades will stall and thus automatically reduce the lift on the turbine blades. This effect is obtained with a relatively small change in pitch angle. This is called active stall control and requires a more accurate control of the pitch angle because of the high angular sensitivity.

24.3.2 Different representations of the turbine rotor

After having described the basic aerodynamic properties of the wind turbine, we can now present the most commonly applied different ways of representing a wind turbine in simulation programs. The various model representations are most conveniently approached by first stating whether the representation is based on power, P, torque, T, or the power efficiency coefficient, C_P. It is beyond the scope of this general overview to go into the details of the various modelling approaches. In the following, we will only outline the major aspects in order to provide a general overview.

24.3.2.1 Constant power

The simplest possible representation of a wind turbine is to assume a constant mechanical input. The mechanical input can be chosen as either the mechanical power or the mechanical torque, and then the other quantity can be calculated by using Equation (24.3).

Even though both ways are possible, we *strongly* recommend the constant power representation. If constant torque is applied, the mechanical power in the model will vary proportionally with the rotational speed. In certain cases, this may result in a numerically unstable model system. In contrast, in a constant power representation the torque is an inverse function of the rotational speed and thus introduces an intrinsically stabilising term into the mechanical system. More importantly, a constant torque model will in most cases reflect the physical behaviour of the wind turbine less accurately than a constant power model.

24.3.2.2 Functions and polynomial approximations

Functions and polynomial approximations are a way of obtaining a relatively accurate representation of a wind turbine, using only a few parameters as input data to the turbine model. The different mathematical models may be more or less complex, and they may involve very different mathematical approaches, but they all must generate curves with the same fundamental shape as those of the physical wind turbine.

As previously mentioned, the advantage of a $C_P - \lambda - \beta$ representation is that it is a normalised representation. For that reason, $C_P - \lambda - \beta$ representations are convenient for use in connection with functions and polynomial approximations.

As a simple example of such a function, consider the following polynomial approximation:

$$C_P = C_{P,\max}\{1 - k_\lambda(\lambda - \lambda_{\text{opt}})^2\}\{1 - k_\beta(\beta - \beta_{\text{opt}})^2\}. \qquad (24.9)$$

This function is characterised by only five parameters: k_λ, k_β, λ_{opt}, β_{opt} and $C_{P,\max}$, where k_λ is the tip speed ratio coefficient, and k_β is the pitch angle coefficient. The example has never been applied in actual studies and is probably too simple for most wind turbine simulations. However, the example does fulfil a number of necessary conditions for a polynomial approximation (such as that a number of selected significant points match exactly and that there is a continuity between these points) and thereby also for achieving a reasonably high accuracy in the vicinity of the desired working point.

There are other alternatives, of course, which are based on more complicated approaches, such as higher-order approximations or completely different types of functions. The Fourier expansion with trigonometric functions, for instance, could be a reasonable way to represent the angular dependency of the turbine blades.

For the most simple wind turbines [i.e. passive-stall, constant-speed wind turbines (Type A turbines; Section 4.2.3)], the problem is only two-dimensional, since the blade angle is constant. Therefore, a simple power – wind speed curve will provide the information that is necessary to determine a $C_P - \lambda$ curve. This has been demonstrated in Akhmatov (1999).

For more advanced wind turbines with blade angle control (fixed-speed or variable-speed wind turbines) the problem becomes three-dimensional as the blade angle controllability has to be taken into account. Examples of actually applied functional approximations are included in Section 25.5.3 and in Slootweg, Polinder and Kling (2002).

24.3.2.3 Table representation

Instead of applying a functional or polynomial approximation to C_P it is also possible to apply a more cumbersome but direct approach by simply using a $C_P - \lambda - \beta$ table. If the value of C_P is specified for a number of combinations of λ and β values, the C_P values can be organised in a $\lambda - \beta$ matrix. A suitable interpolation method must then be applied between the $\lambda - \beta$ nodes in this matrix. The advantage of the table representation is that it is simple to understand and explain and that the necessary accuracy can be achieved simply by selecting a suitable resolution of the matrix. The disadvantage is equally obvious: the tables – and thereby the amount of necessary data – may be rather substantial.

24.3.2.4 Blade element momentum method and aeroelastic code

From a physical perspective, the wind pressure yields a force on each blade, which is turned into a torque on the shaft of the turbine rotor. A torque representation is therefore, from a physical perspective, the most natural way to model a turbine rotor. This model representation is known as the blade element momentum (BEM) method. This method is used to calculate the C_P values used in some of the previously mentioned approaches to represent the turbine rotor; e.g. the table representation in 24.3.2.3.

In short, this method is based on a separation of the blades into a number of sections along the length of each blade. Each blade section is then characterised by the blade geometry, and the aerodynamic properties are given for each section from the hub ($r = 0$) to the blade tip ($r = R$) as functions of the local radius r. We can calculate the *static* forces on the blade element, and consequently the corresponding shaft torque, for a given wind speed, V_{WIND}, a given rotational speed, ω_{turb}, of the turbine rotor and a given blade angle β.

Like all the previous representations, this is a *static* aerodynamic representation. The BEM method is based on the assumption that the turbine blades at all times are in a steady-state condition – or at least in a quasi steady-state condition. However, it is possible to represent the aerodynamic transition process during changes of wind speed, V_{WIND}, rotational speed, ω_{turb}, and/or blade angle, β, through some characteristic time constants. Øye (1986) suggested an engineering model that describes this modification of the BEM method. The verification of the aerodynamic rotor model in Section 27.2.3 shows the effect of this aerodynamic model representation. The same section also discusses the use of the static aerodynamic representation and the aerodynamic representation including the aerodynamic transition processes.

It is possible to extend the level of detail in the aerodynamic model by taking into account the flexibility of the blades. In that case, traditional beam theory is used to model the blades. This method is commonly referred to as an aeroelastic code (AEC).

24.4 Basic Modelling Block Description of Wind Turbines

Modern wind turbines are complex and technically advanced constructions. The wind turbine models in various simulation programs reflect this complexity. Further, the wind turbine models in various simulation programs – as well as the models of all other components – are normally designed to accommodate specific purposes for which each simulation program is intended. Therefore, wind turbine models in different simulation programs may differ substantially and may require very different data, often with widely varying levels of detail in the various parts of the construction.

However, considering the modelling, wind turbines can in most cases be represented by a generic model with six basic block elements and their interconnections (see Figure 24.4) or by something similar, depending on the specific wind turbine in question. The six model elements, which will be described in the following six subsections, are the model representation of the:

- aerodynamic system;
- mechanical system (turbine rotor, shafts, gearbox and the generator rotor);

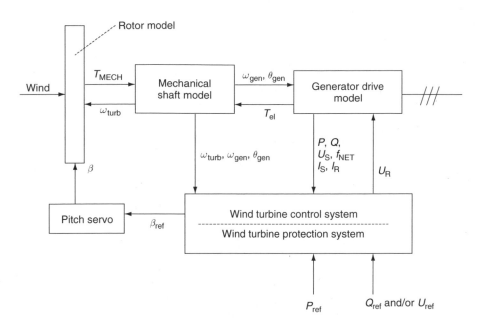

Figure 24.4 Block diagram of a generic wind turbine model. *Note:* $f_{\mathrm{NET}}=$ grid electrical frequency; $I_{\mathrm{s}}=$ stator current; $I_{\mathrm{R}}=$ rotor current; $P=$ active power; $P_{\mathrm{ref}}=$ active power reference; $Q=$ reactive power; $Q_{\mathrm{ref}}=$ reactive power reference; $U_{\mathrm{S}}=$ stator voltage; $U_{\mathrm{R}}=$ rotor voltage; $U_{\mathrm{ref}}=$ stator voltage reference $T_{\mathrm{el}}=$ electrical torque; $T_{\mathrm{MECH}}=$ Mechanical torque; $\omega_{\mathrm{turb}}=$ rotational speed of turbine; $\omega_{\mathrm{gen}}=$ rotational speed of generator rotor; $\theta_{\mathrm{gen}}=$ generator rotor angle; $\beta=$ pitch angle; $\beta_{\mathrm{ref}}=$ blade reference angle

- generator drive (generator and power electronic converters, if any);
- pitch control system;
- wind turbine control system;
- protection system of the wind turbine.

For any specific wind turbine model in any simulation program, it should be possible to identify all model parts with something in the generic model. Likewise, it should also be possible to associate all data in the model with a physical meaning in this generic model.

24.4.1 Aerodynamic system

The aerodynamic system of a wind turbine is the turbine rotor (i.e. the blades of the wind turbine). The turbine rotor reduces the air speed and at the same time transforms the absorbed kinetic energy of the air into mechanical power, P_{MECH}. A specific wind turbine rotor is represented by data that describe the constructional design of the wind turbine. In addition to constructional design data, the mechanical power output of the turbine depends on the wind speed, V_{WIND}, the blade angle, β, of the turbine blades, and the rotational speed, ω_{turb}, of the turbine rotor. This was described in more detail in Section 24.3. In this context – that is, in order to provide an overview of the generic model – it is sufficient to express the mechanical power output of a wind turbine with specific constructional data as a function of rotational speed, wind speed and blade angle, as in Equation (24.4).

Depending on the modelling environment and on the choices made by the model developer, the link between the aerodynamic system and the mechanical system will be either mechanical power, P_{MECH}, or mechanical torque, T_{MECH}. As previously stated, they are related to each other by the rotational speed, ω_{turb}, as shown in Equation (24.3).

24.4.2 Mechanical system

The mechanical system of a wind turbine is the drive train, which consists of the rotating masses and the connecting shafts, including a possible gear system. The major sources of inertia in this system lie in the turbine and in the generator rotors. The tooth wheels of the gearbox contribute only a relatively small fraction. For that reason, the inertia of the gear is often neglected and only the transformation ratio of the gear system is included. Thus, the resulting mechanical system model is a two-mass model with a connecting shaft, and with all inertia and shaft elements referred to the same side of the gearbox, as indicated in Figure 24.5. If necessary, it is also possible to include a representation of the gear system with inertias together with both the low-speed and the high-speed system. This will result in a system with three rotating masses and two shafts.

The significance of applying a two-mass model of the mechanical shaft system, and not a lumped mass model is illustrated in the partial verification of the shaft system model in Section 27.2.2 and in the full-scale model verification in Section 27.3. In wind turbines, the shaft representation is in general more important, because of the relatively soft shaft, in comparison with the shaft systems of traditional power plants. Shafts can be considered soft when the shaft stiffness is below 1 p.u., when measured in per unit (see Section 24.5). They are considered stiff when the stiffness exceeds 3 p.u.

24.4.3 Generator drive concepts

In this context, the term *generator drive* is a broad term covering everything from the shaft and the main terminals to the power grid. For Type C and Type D wind turbines, the converters are considered to be an integral part of the generator drive.[1]

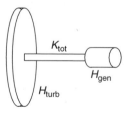

Figure 24.5 Two-mass shaft system model. *Note:* H_{gen} = inertia of generator; H_{turb} = inertia of the turbine; k_{tot} = total shaft stiffness

---

[1] For definitions of wind turbine Types A–D, see Section 4.2.3.

24.4.3.1 Fixed-speed wind turbines: Types A and B

The generator drive in a fixed-speed wind turbine is only the induction generator itself. In the case of a conventional induction generator with a short-circuited rotor (Type A wind turbine) the rotational speed is limited to a very narrow range, which is determined by the slip of the induction generator. Induction machine models are readily available in most power system simulation programs.

In the case of an induction generator with a variable rotor resistance (Type B wind turbine) it is possible to vary the speed over a somewhat wider range. However, the speed range is still rather limited and the speed cannot be controlled directly. Hence, looking at it from a control system perspective, this type of wind turbine must essentially be considered as a fixed-speed wind turbine. In addition to the model of the induction generator itself, a model of a generator drive with an induction generator with variable rotor resistance has to include a representation of the control system, which determines the instantaneous value of the rotor resistance. The vast majority of simulation programs do not contain such models as standard.

24.4.3.2 Variable-speed wind turbines: Types C and D

As the name indicates, variable-speed generator drives enable the wind turbine control system to adapt the rotational speed of the wind turbine rotor to the instantaneous wind speed over a relatively wide speed range. The electrical system has a fixed frequency, though. A generator drive connecting a variable-speed mechanical system with a fixed-frequency electrical system must therefore contain some kind of a slip or decoupling mechanism between the two systems.

In wind turbine technology, the doubly-fed induction generator drive and the full-load converter connected generator drive are the two most frequently applied variable-speed generator drive concepts. There are also other variable-speed generator drive types, but they are currently not generally applied in wind turbines. It would probably also be possible to use written pole synchronous generators, for instance, as a way of obtaining variable-speed capability in a generator drive. In short, all types of variable-speed high-power drives – electrical, mechanical or hydraulic with an electrical generator somewhere in the drive system – can, theoretically, be applied in wind turbines.

All variable-speed generator drives have one thing in common: they must be able to control the instantaneous active power output, otherwise it would be impossible to maintain a power balance in the rotating mechanical system and thus it would also be impossible to maintain the desired constant rotor speed of the turbine. For most variable-speed generator drives, such as the presently prevailing Types C and D, for instance, it will at the same time be possible to control the reactive power output. This implies that the drive will need externally defined reference values for active and reactive power.

In general, variable-speed generator drives will consist of a more or less traditional generator combined with power electronics to provide the slip or decoupling mechanism. As yet, most simulation programs do not include such models as standard. There may be standard models of the individual components (i.e. the synchronous and induction generators, and the frequency converters) but they are not integrated into a unified model with the necessary, additional, internal control systems needed in a variable-speed generator drive. Hence, at the moment it is up to the users to implement appropriate models of variable-speed generator drives. However, the increasing number of wind turbines using these types of generator drive creates a strong demand for such models. It is therefore probably only a matter of time before such standard models will be available in most commercial power system simulation programs.

24.4.3.3 Model implementation: dynamic stability vs. transient generator models

Once a model is developed, the implementation of the model in any simulation program is, in principle, straightforward, albeit often cumbersome. However, some types of simulation program contain limitations that make it impossible to implement complete models in a physically correct way. In dynamic stability programs (see Section 24.6.2.2) the basic algorithms of the program are traditionally based on the assumption that all electromagnetic transients have been extinguished and that only the electromechanical transients and the control system transients are present in the network. Therefore, in order to incorporate a model into a dynamic stability program, it is necessary to make a number of assumptions (e.g. see Kundur, 1994, pages 169–179 and 300–305).

If the electromagnetic transients are ignored in the model implementation, DC offsets in the machine stator currents will be neglected. This implies that the time derivatives of the fluxes in the stator windings are neglected and the stator fluxes are eliminated as state variables and instead calculated as algebraic variables. Model implementations of this type are referred to as dynamic stability models, as opposed to model implementations including the stator fluxes as state variables, which may be referred to as transient (or full transient) models. These model implementations are often referred to as third-order and fifth-order models, respectively, where the order denotes the number of state variables in the generator model. In the third-order model, the state variables are the rotor speed, and the rotor d-axis and q-axis fluxes. In the fifth-order models, these state variables are supplemented with the stator d-axis and q-axis fluxes.

It is, however, possible to incorporate a transient – or at least a semitransient – model into dynamic stability programs using appropriate ways of getting around the constraints in the simulation program. Kundur (1967) describes one way of doing this. It may be of significance to all generator drives, that include an electrical generator that is directly grid-connected (i.e. wind turbine Types A–C).

The partial verification of the induction generator model in Section 27.2.1 and the full-scale model verification in Section 27.3 show the significance of applying a full transient model as opposed to a dynamic stability model in the case of fixed-speed wind turbines with induction generators. The significance of the stator fluxes lies in that they cause a brief braking torque in the case of an external disturbance. In general, a small and brief braking torque would be of no importance, but in wind turbines the inertia of the generator rotor is quite small and, at the same time, the shaft system is relatively soft. The combination of braking torque, low-inertia rotor and soft shaft causes the generator rotor to enter the subsequent transient with a slightly slower rotor speed. Again, a small speed deviation would be insignificant for most other types of generators, but for induction generators the slip-dependence means that the generator will absorb more reactive power and generate less active power. Consequently, a small initial speed difference may lead to a significantly different end result. This phenomenon is illustrated in the speed curves in Figure 27.3.b (page 609) and is also described in Knudsen and Akhmatov (1999) and in Akhmatov and Knudsen (1999b).

Such a verification has not been performed for variable-speed wind turbines, but there is no doubt that the generator implementation will affect the simulated stator currents and, consequently, the possible actions of the overcurrent protection systems. The stator transient itself is probably irrelevant to the overall course of events in a variable-speed generator drive. However, if the transient itself can cause a protective action (see Section 24.4.6) the representation of the transients will be significant anyhow.

24.4.4 Pitch servo

In variable-pitch wind turbines, the blade angle is controlled by a pitch servo. The main control system produces a blade reference angle and the pitch servo is the actuator, which actually turns the turbine blades to the ordered angle.

The pitch servo is subject to constructional limitations, such as angular limits β_{min} and β_{max}. That means that the blades can only be turned within certain physical limits. For active-stall-controlled wind turbines, the permissible range will be between $-90°$ and $0°$ (or even a few degrees to the positive side), whereas for pitch-controlled wind turbines the permissible range will lie between $0°$ and $+90°$ (or even a few degrees to the negative side). The control system may impose other, normally narrower, limits on the reference angle, though.

Likewise, there are limitations on the pitch speed, $d\beta/dt$. The pitch speed limit is likely to be higher for pitch-controlled wind turbines than for active-stall-controlled wind turbines, which have a higher angular sensitivity (see Section 24.3.1). The pitch speed limit may differ significantly for a positive $(d\beta/dt_{pos, max})$ and negative $(d\beta/dt_{neg, max})$ turning of the blade. The pitch speed is normally less than $5°$ per second, although the pitch speed may exceed $10°$ per second during emergencies.

24.4.5 Main control system

The exact structure of the main control system is unique for each type of wind turbine, and even for the same type of wind turbine it may vary according to the individual manufacturer. However, the basic tasks of the control system are the same, namely, to control the power and speed of the wind turbine. The most significant difference is whether the control system is used in a fixed-speed or in a variable-speed wind turbine.

24.4.5.1 Fixed-speed wind turbines: Types A and B

For fixed-speed wind turbines, the generator can be considered to be a passive power-producing component. In effect, the turbine blade angle is the only controllable quantity in the entire wind turbine. Through measurements of a number of quantities such as wind speed, turbine rotor speed and active electrical power, the control system optimises the blade angle in relation to the incoming wind. In high winds, the control system can reduce the power from the wind turbine, thus keeping the power at the rated maximum power of the wind turbine. In emergency situations, the blade angle control can also be used for preventive rapid power reduction. It can be activated through an external signal from the grid control centre, for instance, or some other external source, or by a locally generated signal in the case of high rotor speed and/or very low AC terminal voltage (see Section 29.3).

24.4.5.2 Variable-speed wind turbines: Types C and D

For variable-speed wind turbines, the generator is a much more controllable element. In addition to the turbine blade angle, the instantaneous active and reactive power output of the generator can be controlled.

As mentioned in Section 24.3.1, the variable-speed feature makes it possible to adjust the turbine speed to the optimal speed given in Equation (24.8), thus optimising the power efficiency coefficient C_P. This means that the control system must contain some kind of speed control system and a way to determine a speed reference.

The optimal speed reference is provided by Equation (24.8) or by any other approach arriving at the same value, such as the speed reference derived from Figure 24.2(b). Additionally, for dimensioning reasons, the optimal reference speed is normally cut off at a minimum and a maximum permissible rotational speed corresponding to low and high wind speed situations.

The speed control system can be designed in many different ways. However, they all share the common feature that the mechanical power input from the rotating system can be controlled (at least up to an upper limit determined by the incoming wind) and that the electrical power emitted through the generator also can be controlled (at least in normal grid situations with nominal voltage). This means it is always possible to control the power balance in the rotating system and thereby also the speed. Consequently, the actions of the electrical power control system and the blade angle control system must be coordinated in some way. This coordination is incorporated in the design process of the control systems.

Control strategies may differ depending on the choices made by the manufacturer. It would be possible to construct a single control system that could work for all wind conditions. Another option is to let the blade angle control system control the speed in high winds, thus leaving it to the power control system to maintain constant, rated maximum power; in low and medium wind situations, the power control system can control the speed, thus leaving it to the blade angle control system to optimise the blade angle to the incoming wind, thereby optimising the power production.

Similar to fixed-speed wind turbines, the blade angle control can also be used for a preventive rapid power reduction in emergency situations. However, if a short-term overspeeding in the rotating system is permissible, the power control system will be able to reduce the generator terminal power more quickly.

Reactive power can also be controlled (see also Chapter 19). This makes it possible to use wind turbines for voltage control. Currently, this possibility is not made use of very often. It would seem to be only a matter of time, though, before the grid code will require large wind farms to supply such ancillary services.

24.4.6 Protection systems and relays

The protection scheme of wind turbines is based on measurements of various quantities, such as voltage, current and rotor speed, including possible measuring delays and the various relay limits. If these limits are exceeded more than a permissible period of times they will cause the relay to initiate protective action. Examples of such protective action are disconnection or preventive rapid power reduction, the second action being similar to the so-called fast valving process in thermal power plants. Obviously, such protection system action may have a significant impact on the outcome of any simulation.

24.5 Per Unit Systems and Data for the Mechanical System

Per unit systems (p.u. systems) are commonly and traditionally used in many power system simulation programs. Therefore, one is bound to encounter p.u. systems sooner or later when working with simulation programs for electrical power systems.

Experience shows that the questions relating to p.u. systems and data interpretation, especially for the mechanical part of the system, are a source of many serious misunderstandings, problems and errors [e.g. when an (electrical) engineer must convert manufacturer (mechanical) information into valid data in a simulation program]. For that reason we will describe this in more detail here, even though it, theoretically, seems to be basic knowledge. This way, we hope to raise awareness of this issue in order to avoid or at least reduce p.u. related problems.

At first glance, the p.u. concept may be rather confusing. However, it is nothing more than a definition of a new set of – conveniently and carefully chosen – basic measuring units for the physical quantities that are under consideration. That means that, for example, instead of measuring the power from a wind turbine in watts, kilowatts or megawatts the power may be measured as the percentage of the rated power from the wind turbine. In fact, the power can be measured as the percentage of any other constant power that is suitable to be used for a comparison in a given context. Likewise, all voltages at the same voltage level as the wind turbine generator can be measured as the percentage of the nominal voltage of the wind turbine. These basic 'measuring units' are the base values of the p.u. system.

Once the most fundamental base values for electrical quantities such as power and voltage have been chosen, it is possible to derive base values for other electrical quantities such as current, resistance and reactance on *each* voltage level in the system. Power is an invariant quantity at all voltage levels so it is only possible to choose one base value for power. For voltage, it is possible and necessary to define a base value for each voltage level in the system.

It is assumed that the reader is sufficiently familiar with the p.u. concept in electrical systems that there is no need to go into further detail. For more information, see standard text books that provide an introduction to the p.u. concept (e.g. Kundur, 1994).

Now, if the p.u. system is to be extended to the rotating mechanical system, it becomes necessary to define additional base values for angular speed and angle. The definition of these base values is in many cases not stated clearly, but they are nevertheless defined implicitly, otherwise the p.u. system would not be consistent. Using these two additional base values it is possible to derive consistent base values for other relevant quantities, such as torque (T_{base}) and shaft stiffness (k_{base}), which can be defined as:

$$T_{base} = \frac{P_{base}}{\omega_{base}}, \tag{24.10}$$

$$k_{base} = \frac{T_{base}}{\theta_{base}} = \frac{P_{base}}{\omega_{base}\theta_{base}}, \tag{24.11}$$

Where the subscript 'base' indicates the base value of the quantity concerned. Note that the base values for angular speed (ω_{base}) and angle (θ_{base}), and consequently all other derived base values, so far must be assumed to be in *electrical* radians (i.e. they are referred to the electrical system). However, if the mechanical system rotates at a different (typically lower) speed as a result of generator pole pairs and/or a mechanical gear system, supplementary base values must be defined accordingly. This definition of new base values for *each* 'speed level' corresponds to the fact that for each voltage level a separate voltage base value is chosen.

The most commonly used definition of base *electrical* angular rotational speed refers to synchronous operation; that is, 314.16 rad/s in 50 Hz systems, and 376.99 rad/s in 60 Hz systems.

The base angle, however, is not traditionally defined as a base value. However, we have found that it makes the p.u. concept easier to understand and more consistent. The introduction of a separate base angle also makes it possible to extend the p.u. system to all parts (or 'speed levels') of the mechanical system in a logical way that is consistent with the definition of separate base voltages at separate voltage levels. Since base angles are traditionally not explicitly defined, base angles have instead been defined implicitly (i.e. because of a lack of actual choice rather than as a deliberate choice). Consequently, shaft stiffness, for instance, is often specified in p.u. per electrical radian (el. rad) instead of in 'true' p.u. – and one electrical radian is therefore the most commonly used (implicit) definition of base angle. It follows from this that the resulting p.u. value for shaft stiffness, for example, will be exactly the same whether it is specified as p.u./el.rad or as 'true' p.u. with a base angle of 1 el. rad. It would be just as possible to choose another base angle value, such as 2π el. rads (corresponding to one electrical revolution). From various perspectives, this base value might be even more justified.

We have not had the opportunity to work in detail with models of Type D wind turbines. In this type of wind turbines, the generator electrical system is decoupled from the rest of the electrical system through full-load converters. Therefore the p.u. system presented above may not be suitable for that type of wind turbine. It will at least be necessary to consider carefully the definitions of base angular speed and base angle in the generator system, which – from an electrical perspective – is an island system. A suitable choice of base electrical angular speed would probably refer to the nominal operation of the wind turbine. That means that the suitable value of base electrical angular speed would depend on the nominal turbine rotational speed [and thereby on the turbine radius; see Equation (24.8)] and on the number of pole pairs, n_{PP}, of the generator. If there is a gearbox, the gear ratio, n_{gear}, has also to be included. Usually, Type D wind turbines do not have a gearbox, though.

As for inertia of the rotating parts (i.e. of the wind turbine rotor and generator rotor), these are typically specified in SI units such as J, GD^2 or similar traditionally used definitions. In p.u. systems, the inertia time constant, H, is generally used. H is defined as:

$$H = \frac{J\omega_{base}^2}{2P_{base}}. \tag{24.12}$$

Strictly speaking, H is not a p.u. value. Unlike real p.u. values, which by the very definition of p.u. systems are without physical unit, the inertia time constant is measured in seconds. However, it is a convenient and, for historical reasons, commonly acknowledged quantity.

Because of the various ways to specify inertia, it is very important to be absolutely clear about which definition the manufacturers are applying; otherwise, data may be misinterpreted, which would lead to erroneous simulation results.

Table 24.1 lists the typical range of physical values for the necessary mechanical data of a wind turbine in the 2 MW range with a gearbox in the shaft system and with a grid-connected generator (i.e. of Type A, B or C). It also includes an example of a dataset, which will be used to illustrate the application of the p.u. system.

In order to convert these values into p.u., the necessary base values must be defined. The power base value is:

$$P_{base} = 2 \text{ MW}.$$

In the following, the indices el., HS and LS denote, respectively; the electrical system, the system at the high-speed side of the gearbox (i.e. with the generator rotor) and the system at the low-speed side of the gearbox (i.e. with the turbine rotor).

Assuming a 50 Hz system, the electrical base angular speed and base angle, $\omega_{base, el.}$ and $\theta_{base, el.}$, respectively, become:

$$\omega_{base, el.} = 314.16 \text{ rad/s (corresponding to 50 Hz)};$$

$$\theta_{base, el.} = 1.0 \text{ rad (traditional choice)};$$

$$RPM_{nom, el.} = 3000.$$

Table 24.1 Typical mechanical data: physical values

Quantity	Typical range	Example value
Generator rotor intertia, J_{gen} (kg m^2)	65–130	121.5
Wind turbine inertia, J_{turb} (10^6 kg m^2)	3–9	6.41
High-speed shaft stiffness, K_{HS} (kNm/rad)	50–100	92.2
Low-speed shaft stiffness, K_{LS} (MNm/rad)	80–160	145.5
Number of pole pairs, n_{pp}	2	2
Gear ratio, n_{gear}	90–95	93.75

The corresponding base values in the HS and the LS system become:

$$\omega_{\text{base, HS}} = \frac{\omega_{\text{base, el.}}}{n_{\text{pp}}} = \frac{314.16}{2} \text{ rad/s} = 157.1 \text{ rad/s};$$

$$\theta_{\text{base, HS}} = \frac{\theta_{\text{base, el.}}}{n_{\text{pp}}} = \frac{1.0}{2} \text{ rad} = 0.5 \text{ rad};$$

$$\text{RPM}_{\text{nom, HS}} = \frac{\text{RPM}_{\text{nom, el.}}}{n_{\text{pp}}} = \frac{3000}{2} = 1500.$$

and

$$\omega_{\text{base, LS}} = \frac{\omega_{\text{base, el.}}}{n_{\text{pp}} n_{\text{gear}}} = \frac{314.16 \text{ rad/s}}{2 \times 93.75} = 1.676 \text{ rad/s};$$

$$\theta_{\text{base, LS}} = \frac{\theta_{\text{base, el.}}}{n_{\text{pp}} n_{\text{gear}}} = \frac{1.0 \text{ rad}}{2 \times 93.75} = 0.00533 \text{ rad};$$

$$\text{RPM}_{\text{nom, LS}} = \frac{\text{RPM}_{\text{nom, el.}}}{n_{\text{pp}} n_{\text{gear}}} = \frac{3000}{2 \times 93.75} = 16.$$

Using the angular speed base values, the inertia time constants can be calculated as follows:

$$H_{\text{gen}} = \frac{J_{\text{gen}} (\omega_{\text{base, HS}})^2}{2 P_{\text{base}}} = 0.75 \text{ s};$$

$$H_{\text{turb}} = \frac{J_{\text{turb}} (\omega_{\text{base, LS}})^2}{2 P_{\text{base}}} = 4.5 \text{ s};$$

Where J_{gen} and J_{turb} are the generator rotor intertia and wind turbine inertia, respectively. The base values for shaft stiffness are calculated equally easily:

$$k_{\text{base, el.}} = \frac{P_{\text{base}}}{\omega_{\text{base,el.}} \theta_{\text{base,el.}}} = 6366 \text{ kNm/rad},$$

$$k_{\text{base, HS}} = \frac{P_{\text{base}}}{\omega_{\text{base, HS}} \theta_{\text{base, HS}}} = 25.46 \text{ kNm/rad},$$

$$k_{\text{base, LS}} = \frac{P_{\text{base}}}{\omega_{\text{base, LS}} \theta_{\text{base, LS}}} = 223.8 \text{ MNm/rad},$$

thus making it possible to calculate the p.u. shaft stiffnesses:

$$k_{\text{HS}} = \frac{k_{\text{HS}}}{k_{\text{base, HS}}} = 3.62,$$

$$k_{\text{LS}} = \frac{k_{\text{LS}}}{k_{\text{base, LS}}} = 0.65.$$

The use of base angle and base angular speed in this p.u. system automatically ensures that all quantities are referred to the electrical side with n_{pp} and n_{gear} raised to the correct power exponent.

As indicated in Section 24.4.2, the inertia of the tooth wheels in the gearbox is often ignored. In that case, the total p.u. shaft stiffness k_{tot} can be calculated as a 'parallel connection' of the two shaft stiffnesses:

$$k_{tot} = k_{HS}||k_{LS} = \left(\frac{1}{k_{HS}} + \frac{1}{k_{LS}}\right)^{-1} - 0.55.$$

It is recognised that the stiffness of the low-speed shaft – when measured in p.u. – is so much smaller than that of the high-speed shaft that the low-speed shaft effectively determines the total shaft stiffness. This is the case in spite of the fact that the low-speed shaft stiffness – when measured in physical units (Nm/rad) – is more than a factor 1000 higher than the value of the high-speed shaft. Because of the gearbox, the physical values of the shaft stiffnesses are simply not comparable, but the use of the p.u. system neutralise this. Hence the p.u. shaft stiffnesses become directly comparable.

The calculated p.u. data are organised in Table 24.2. Although the typical physical values of the mechanical data in Table 24.1 are representative only for wind turbines in the 2 MW range, the p.u. values of the mechanical data will be representative for wind turbines of a much wider power range, reaching from below 1 MW up to 4 or 5 MW and maybe even wider (see also the chapter on 'scaling wind turbines and rules for similarity' in Gasch and Twele, 2002).

The range of typical values is, of course, quite wide. However, there is a tendency that data for active-stall-controlled wind turbines are at the upper end of the range, whereas those for pitch-controlled wind turbines are at the lower end of the range.

In some cases, the manufacturer supplies the torsional eigenfrequencies of the shaft instead of shaft stiffnesses. If the inertias are known, it is possible to calculate the shaft stiffness from these data, provided it is absolutely clear which eigenfrequencies have been specified.

From Equations (25.8) in Section 25.5.4, and setting the external electrical and mechanical torques, T_e and T_{wr}, respectively, in these equations to zero, the eigenoscillations of the drive train can be determined. The eigenfrequency of a so-called free–free shaft system, $f_{free-free}$, is then straightforwardly given by:

$$2\,\pi\,f_{free-free} = \left[\omega_{base,\,el.}\,\frac{k_{tot}}{2}\left(\frac{1}{H_{turb}} + \frac{1}{H_{gen}}\right)\right]^{1/2}. \tag{24.13}$$

Table   24.2 Typical mechanical data – p.u. values

Quantity	Typical range	Example value
Generator rotor inertia constant, H_{gen} (s)	0.4–0.8	0.75
Wind turbine inertia constant, H_{turb} (s)	2.0–6.0	4.5
High-speed shaft stiffness, k_{HS} (p.u.)	2.0–4.0	3.62
Low-speed shaft stiffness, k_{LS} (p.u.)	0.35–0.7	0.65
Total shaft stiffness, k_{tot} (p.u.)	0.3–0.6	0.55

Similarly, it can be shown that the eigenfrequency of a free–fixed shaft system is:

$$2 \, \pi \, f_{\text{free–free}} = \left(\omega_{\text{base, el.}} \frac{k_{\text{LS}}}{2} \frac{1}{H_{\text{turb}}} \right)^{1/2}. \tag{24.14}$$

In this case, the use of the LS shaft stiffness, k_{LS}, indicates that the system is assumed to be fixed at the gearbox, as opposed to the use of k_{tot} in Equation (24.13), which indicates that the entire drive train is assumed to take part in the free–free eigenoscillation.

Knowing the eigenfrequencies and the inertia time constants and using Equations (24.14) and (24.15), it is trivial to determine the p.u. shaft stiffnesses. Note that both the free–free and the free–fixed eigenfrequencies often are referred to merely as eigenfrequency. Therefore, such frequencies have to be interpreted correctly, making sure exactly which parts of the mechanical construction are actually included in the eigenoscillation under consideration.

24.6 Different Types of Simulation and Requirements for Accuracy

When performing computer simulations the available simulation software, the inherent models available in the software and the available data for these models may for obvious reasons often be used without any special considerations by some users. However, in all simulation work it is a given fact that the simulation type, the model accuracy and the data accuracy all may have a significant impact on the outcome of the investigations.

24.6.1 Simulation work and required modelling accuracy

In addition to the purpose of the simulations, it is also necessary to be aware of the individual parts of the total system model and to identify the parts that are most important with respect to the investigations in question. There must be a reasonable balance between the accuracy of all models in the simulated system, that is, of individual component models as well as a possible external system model. There has to be a level of proportionality, which means that the minimum acceptable accuracy of a model must increase with the significance that the specific model has to the phenomena under consideration.

In this context, it is important to keep in mind that accuracy of a model representation of a component is achieved through a combination of the accuracy of the component model itself and the accuracy of the model data. The accuracy of the component model refers to the validity of the model and the physical level of detail in the model, whereas the accuracy of the model data refers to the quality of the parameter measuring or estimation methods. Likewise, the accuracy of the total system model depends on the accuracy of the models of the individual components in the system.

Ultimately, the total quality of any simulation result depends on the least accurate model representation of the most significant components in the total system model. The consequence of all this is that the development of very accurate models can be justified only if it is also possible to obtain data with a corresponding accuracy. Likewise, it is only justifiable to apply very accurate model representations (i.e. model and data) of components if the model representation of other components that have a similar or higher significance to the simulation results has an equally high accuracy.

However, high-accuracy models and/or high-accuracy data for any model representations of components can be used whenever it is convenient and justified for other reasons, for instance in order to avoid constructing and verifying equivalents, or to produce models that are more realistic and therefore more trustworthy and intuitively understandable for the users.

24.6.2 Different types of simulation

As previously stated, computer simulations can be used to investigate many different phenomena. Therefore, requirements regarding simulation programs, the necessary level of detail in the models and the model data may differ substantially depending on what is to be investigated. Depending on the objective of the simulation, there is a large number of different software programs available that are each specially tailored for specific types of investigation.

In the following, the overall purpose of different types of simulations is briefly discussed, and for each different type of simulation the required level of accuracy of the various parts of the wind turbine model is briefly evaluated. We will also mention typical programs used for these types of simulations.

We would like to stress that the programs mentioned here are those programs we have become familiar with during our work. That does not mean in any way that other programs that are not mentioned here will be of inferior value.

24.6.2.1 Electromagnetic transients

Electromagnetic transients are simulated using special electromagnetic transients programs (EMTPs) such as Alternative Transient Program (ATP), the DCG/EPRI EMTP, EMTDC^{TM}, and Simpow^{TM}.[2] These programs include an exact phase representation of all electrical components, often with the possibility to include a complex representation of saturation, travelling wave propagation and short—circuit arcs, for instance. Simulations are, in general, performed in the time domain, and the immediate simulation output are the instantaneous values of voltages, currents and quantities derived therefrom.

EMTP simulations are used to determine fault currents for all types of faults, symmetrical as well as nonsymmetrical. Overvoltages in connection with switching surges, lightning surges as well as ferro-resonance in larger networks, for example, can also be determined, which is useful for coordinating the insulation. The exact functions of power electronics [e.g. high-voltage direct-current (HVDC) links, static VAR compensators (SVCs), STATCOMs, and voltage source converters (VSCs)] can also be simulated with use of such programs.

---

[2]ATP and DCG/EPRI EMTDC^{TM} is a trademark of Manitoba HVDC Research Centre Inc.; Simpow^{TM} is a registered trademark of STRI AB.

The general level of detail in EMTP programs requires a reasonable wind turbine model in such programs to include the significant electrical components [i.e. the generator, possible power electronics (including basic controls), possible surge arresters and possible static reactive compensation]. Other parts of the construction may be neglected or considered to be constant. This could be, for instance, the incoming wind, the incoming mechanical power, the secondary parts of the control system and, in some cases, the mechanical shaft system.

24.6.2.2 Dynamic stability and dynamic stability simulations

Transient and voltage stability are normally evaluated with use of special transient stability programs (TSPs), such as PSS/E[TM] (Power System Simulator for Engineers), Simpow[TM], CYMSTAB, PowerFactory, and Netomac[TM].[3] In general, these programs have a phasor representation of all electrical components. In some of these TSPs, only the positive sequence is represented, whereas other TSPs also include a representation of the negative and zero sequences. Simulations are, in general, performed in the time domain, and the immediate simulation output are the root mean square (RMS) values of voltages, currents and quantities derived therefrom.

TSP simulations are used to evaluate the dynamic stability of larger networks. The terms 'dynamic stability' and 'transient stability' are often used interchangeably for the same aspect of power system stability phenomena. The conceptual definition of 'transient stability' is outlined in Kundur (1994, pages 17–27), among others, and associated with 'the ability of the power system to maintain synchronism when subjected to a severe transient disturbance'. The term 'voltage stability' is also in accordance with the definition in Kundur (1994, page 27), and others, as 'the ability of a power system to maintain steady acceptable voltages at all buses in the system under normal operating conditions and after being subjected to a disturbance'. Neither of these stability terms should be confused with actual electromagnetic 'transient phenomena' such as lightning and switching transients. Such transients are characterised by significantly lower time constants (microseconds) and, consequently, EMTP simulations should be used to analyse these phenomena (see Section 24.6.2.1). Therefore, the terms 'dynamic analysis' and 'dynamic stability analysis' are commonly associated with the ability of the power system to maintain both the transient and the voltage stability. Bruntt, Havsager and Knudsen (1999) and Noroozian, Knudsen and Bruntt (2000), among others, give examples of wind power related dynamic stability investigations.

The general level of detail in such programs requires a reasonable wind turbine model in such programs to include the major electrical components: that is the generator, possible power electronics (including basic controls), possible static reactive compensation, main control systems and protection systems that may be activated and come into operation during simulated events, 'soft' mechanical shaft systems and the incoming mechanical power from the turbine rotor. Only very few other parts, (e.g. the incoming wind) may be neglected or considered constant. Chapter 25 and Slootweg, Polinder and

---

Kling (2002) give examples of wind turbine models, that have been implemented in TSPs. The examples also include the most significant typical data for the generator drive and the turbine rotor.

As shown in Section 24.3.2, the incoming mechanical power from the turbine rotor can be represented in many different ways with more or less detailed representations. However, a constant power representation is not adequate if the model must be able to represent a preventive rapid power reduction or similar actions resulting from adherence to the grid code, for instance. It remains to be determined whether a simple or an accurate static C_P approximation is sufficient or whether the exact aerodynamic model with representations of aerodynamic transition processes is required. Strictly speaking, the effects demonstrated in the verification of the aerodynamic rotor model in Section 27.2.3 show only that the more exact representation does include a short-lasting torque pulse. However, it is not known whether this torque pulse may significantly alter the final results of any overall transient stability study or whether it can be neglected without any significant consequences for the final results.

24.6.2.3 Small signal stability

Small signal stability is associated with the ability of a system – usually a large system, such as an entire interconnected AC power system – to return to a stable point of operation after a smaller disturbance. In a textbook example of a small signal stability analysis, the eigenvalues of the system, which are usually complex, and the corresponding eigenvectors are determined. However, since a large system may easily include many hundreds or even thousands of state variables – not to mention the exact same number of complex eigenvectors of exactly the same dimension – it is by no means a simple task to do this with adequate numerical accuracy.

Some dynamic stability programs include in their very nature the necessary model data for the physical system and the relevant control systems, having built-in modules to perform eigenvalue analyses. Examples of such models are PSS/E, with the module LSYSAN (Linear System Analysis), and Simpow. These two programs are compared in Persson *et al.* (2003). We also assume that other dynamic stability programs will include similar linear system analysis modules. For additional references, see, for example, Paserba (1996), which gives an overview of a number of linear analysis software packages. There are also special programs for eigenvalue analysis, such as AESOPS and PacDyn. Bachmann Erlich and Grebe (1999) provide an example of eigenvalue analysis on the UCTE (Union pour la Coordination du Transport d'Electricite) system with use of such a specialised eigenvalue analysis software.

If convenient, it is also possible to perform an indirect small signal stability analysis directly by using a time domain dynamic stability analysis program. Applying a small disturbance will in most cases excite the predominant eigenoscillations in the system, and the eigenoscillations will then be visible in many output variables such as voltages, currents, rotor speeds, and so on. A postprocessing, where the frequency and the damping of the various swing modes in a selected section of the postdisturbance output curves are calculated, will then give a good indication of the eigenvalues that were excited by the disturbance. This is an easy way to get an initial idea of a system or to confirm results obtained through eigenvalue analyses. This is addressed in Bachmann Erlich and Grebe (1999).

Hitherto, wind turbines have not been taken into account in connection with small signal stability investigations because of the nature of wind power production, which traditionally has been dispersed in the power system, and because of the size of the individual wind turbines, which are several orders of magnitude smaller than centralised power plants and even more orders of magnitude smaller than the total interconnected power system. However, the size of individual wind turbines and entire wind farms on land as well as offshore is increasing, resulting in a higher wind power penetration. In addition, the controllability of modern wind turbines is being improved. This feature is becoming increasingly important because of the increasingly demanding grid codes. Therefore, wind power will increasingly be included in small signal stability investigations.

The general level of detail in small signal stability programs requires a reasonable wind turbine model in such programs to include the components that may contribute to and have an impact on the eigenoscillations of the system (i.e. linearised representations of the shaft system, the control system, the generator and possible power electronics). Other parts of the construction may be neglected or considered constant, such as, the incoming wind and all discrete actions performed by the control and protection system.

24.6.2.4 Aerodynamic modelling and mechanical dimensioning

The design of a wind turbine involves many decisions regarding mechanical construction and aerodynamic design. In this process, the wind turbine itself is at the very centre of investigations. The strength and shaping of the blades, the dimensioning of the shafts and gear, the strength of the tower and even the tower foundation – especially for offshore wind turbines – have to be considered carefully when designing a wind turbine. As we have not been working with these aspects of wind turbine development, the following are rather general statements.

There are many CAD-like simulation tools that can handle mechanical constructions, on the one hand, and other tools that can deal with aerodynamic properties, on the other hand. However, in the design phase of a wind turbine, the mechanical and the aerodynamic properties must be regarded as a whole. Consequently, there is a need for specialised simulation tools that can deal with both aspects at the same time. One such tool is FLEX4, which can represent blade deflection, as well as the torsional twist of the shaft system and the tower. A newer version, FLEX5, can also cope with the deflections of the tower and the foundation (Christensen et al., 2001).

A reasonable wind turbine model in such programs has to include the major mechanical and aerodynamic components of the wind turbine [i.e. the blades, the pitch servo system (if pitching changes are to be evaluated), the shaft system (including gearbox), the generator (as a minimum representation, there must be a decelerating torque), the tower and, in some cases, the foundation]. Also, the emergency disc brake can be included if the mechanical impact of an emergency shutdown on the turbine is to be evaluated. Finally, as the most significant external factor influencing a wind turbine is the wind itself, it must be represented in a realistic way such as by measured wind series, preferably with frequent wind gusts.

So far, the external AC power system, and in many cases even the generator itself, has traditionally not been represented very accurately, probably because of the fact that the mechanical and aerodynamic properties have been much more decisive for the design. For traditional fixed-speed wind turbines with induction generators it has been an acceptable approximation to represent the generator and, consequently, the external AC power system by a simple speed-dependent generator air gap torque. However, with the controllability of variable-speed wind turbines and the available choices of generator control strategies (power or speed control; see Section 24.4.5.2), the option of choosing a control strategy will have to be taken into consideration as well.

24.6.2.5 Flicker investigations

In the everyday operation of all electric power systems there are all kinds of small disturbances, whenever loads are switched on or off, or whenever the operating point of any equipment is altered. All such small disturbances result in small changes in the voltage amplitude. If the short-circuit capacity of the power system is sufficiently high compared with the disturbance, or if the frequencies of the disturbances are outside the human visual susceptibility range, the disturbances are mostly without significance. However, in many cases they are not, and these disturbances are referred to as flicker. There are standardised methods of measuring flicker (IEC 61000-4-15; see IEC, 1997) as well as recommendations for maximum allowed flicker levels (IEC 61000-3-7; see IEC, 1996).

Most flicker is caused by discrete events, such as equipment being switched on or off (e.g. when motors start and capacitor banks switch in or out). But flicker can also be caused by periodic disturbances of any kind, and the tower shadow (i.e. the brief reduction in the shaft torque every time a blade is in the lee in front of the tower) is exactly such a periodic disturbance. This is often denoted as the 3p effect, because it is three times higher than the rotational speed of the wind turbine. If many wind turbines are grid-connected close to each other the cumulative effect may become large enough to cause flicker problems. Section 25.5.3 describes how the tower shadow can be included in a wind turbine model, and an empirical model of this periodic nature of the wind turbine output is described in Akhmatov and Knudsen (1999a), in which it is also outlined how dynamic stability simulations with a customised wind turbine flicker model can be used to evaluate the expected flicker contribution from wind turbines in areas with high wind power penetration.

It may be added that as wind turbines grow larger and, consequently, have a lower rotational speed [see Equation (24.8)] the flicker contribution from the 3p effect will occur at lower and lower frequencies, where the recommended flicker limits are higher. This fact will in itself mitigate possible flicker problems caused by the tower shadow.

Wind power also contributes to flicker through wind gusts, turbulence and random variations in incoming wind. These changes in the wind cause changes in the instantaneous power production of a wind turbine, which, in turn, will alter the instantaneous voltage and thereby contribute to the flicker level. A possible way of representing this is shown in Section 25.5.2.

24.6.2.6 Load flow and short-circuit calculations

Load-flow (LF) calculations require knowledge of the basic electrical properties of the power system, such as line and transformer impedances and transformer tap-changer data, in addition to operational data of the power system, such as location and size of momentary power consumption and production, voltage (or reactive power) set points of voltage controlling devices, and transformer settings. Short-circuit (SC) calculations require basically the same information but must also include generator impedances. All these data are comparatively simple to obtain. However, the amount of data required in the case of larger networks may be substantial.

LF and SC calculations are usually performed with use of special programs, that are customised for the task. There is a large number of programs available for that purpose (e.g. NEPLAN, Integral, PSS/E^{TM}, Simpow^{TM}, PowerFactory and CYMFLOW and CYMFAULT, to mention just a few). LF and SC calculations both assume a steady state (i.e. all time derivatives are considered zero). LF calculations are used to calculate the voltages in the nodes and the flow of active and reactive power in the lines between the nodes of the power system in various points of operation. SC calculations are used to calculate the level of fault currents in the case of short circuits anywhere in the power system.

In the case of LF calculations, it is necessary only to represent the input of active and reactive power, P and Q. Wind turbines with induction generators (Type A) must be represented with an induction machine, with the active power P set to the momentary value. Based on the machine impedance, the LF program will then calculate the reactive power consumption Q corresponding to the terminal voltage. (A few LF programs do not have an inbuilt induction generator model, in which case it is necessary to apply a PQ representation with the Q value adjusted to an appropriate value. A user-defined macro may be very convenient for that, if such an option is available in the program.) Variable-speed wind turbines, however, have reactive power controllability. Therefore, the most suitable representation is a fixed PQ representation or, if the wind turbine is set to voltage control, a PV representation.

In the case of SC calculations, the necessary wind turbine representation will depend on the specific generator technology applied. In general, the SC representation must include the parts that will provide support to the SC current, if there is a short circuit somewhere in the vicinity of the wind turbine. For fixed-speed wind turbines with induction generators (Types A and B) the machine itself is an appropriate representation. For variable-speed wind turbines with doubly-fed induction generators (Type C) the machine itself must be represented, whereas the smaller partial load converter often can be neglected. For variable-speed wind turbines with full-load converters (Type D) the generator is decoupled from the AC network and only the converter is directly connected to the grid; consequently, the appropriate SC representation must be a converter of suitable type and rating.

24.7 Conclusions

In this chapter we provided a general overview of aspects related to wind turbine modelling and computer simulations of electrical systems with wind turbines.

We provided an overview of wind turbine aerodynamic modelling, together with basic descriptions of how to represent the physical properties of a wind turbine in different mathematical models.

In order to arrive at a better understanding of wind turbine models in general, a generic wind turbine model was shown, where the various independent elements of a wind turbine and their interaction with each other were explained.

We presented a general overview of p.u. systems, with special emphasis on p.u. systems for the mechanical system of a wind turbine. Typical mechanical data for a contemporary sized wind turbine were given, and the conversion of these data from physical units to p.u. was presented in detail.

Finally, we gave an overview of various types of simulation and discussed what to include in a wind turbine model for each specific type of simulation.

References

[1] Aagaard Madsen, H. (1991) 'Aerodynamics and Structural Dynamics of a Horizontal Axis Wind Turbine: Raw Data Overview', Risø National Laboratory, Roskilde, Denmark.

[2] Akhmatov, V. (1999) 'Development of Dynamic Wind Turbine Models in Electric Power Supply', MSc thesis, Department of Electric Power Engineering, Technical University of Denmark, [in Danish].

[3] Akhmatov, V. (2002) 'Variable Speed Wind Turbines with Doubly-fed Induction Generators, Part I: Modelling in Dynamic Simulation Tools', *Wind Engineering* 26(2) 85–107.

[4] Akhmatov, V. (2003) *Analysis of Dynamic Behaviour of Electric Power Systems with Large Amount of Wind Power*, PhD thesis, Electric Power Engineering, Ørsted-DTU, Technical University of Denmark, Denmark.

[5] Akhmatov, V., Knudsen, H. (1999a) 'Dynamic modelling of windmills', conference paper, IPST'99 – International Power System Transients, Budapest, Hungary.

[6] Akhmatov, V., Knudsen, H. (1999b) 'Modelling of Windmill Induction Generators in Dynamic Simulation Programs', conference paper BPT99-243, IEEE Budapest Power Tech '99, Budapest, Hungary.

[7] Bachmann, U., Erlich, I., Grebe, E. (1999) 'Analysis of Interarea Oscillations in the European Electric Power System in Synchronous Parallel Operation with the Central-European Networks', conference paper BPT99-070, IEEE Budapest Power Tech '99, Budapest, Hungary.

[8] Bruntt, M., Havsager, J., Knudsen, H. (1999) 'Incorporation of Wind Power in the East Danish Power System', conference paper BPT99-202, IEEE Budapest Power Tech '99, Budapest, Hungary.

[9] Christensen, T., Pedersen, J., Plougmand, L. B., Mogensen, S., Sterndorf, H. (2001) 'FLEX5 for Offshore Environments', paper PG 5.8, EWEC European Wind Energy Conference 2001, Copenhagen, Denmark.

[10] Freris, L. (1990) *Wind Energy Conversion Systems*, Prentice Hall, New York.

[11] Gasch, R., Twele, J. (2002) *Wind Power Plants*, James & James, London, UK.

[12] Hansen, M. O. L. (2000) *Aerodynamics of Wind Turbines: Rotors, Loads and Structure*, James & James, London, UK.

[13] Heier, S. (1996) *Windkraftanlagen im Netzbetrieb*, 2nd edn, Teubner, Stuttgart, Germany.

[14] Hinrichsen, E. N. (1984) 'Controls for Variable Pitch Wind Turbine Generators', *IEEE Transactions on Power Apparatus and Systems* PAS-103(4) 886–892.

[15] IEC (International Electrotechnical Commission) (1996) 'EMC-Part 3: Limits – Section 7: Assessment of Emission Limits for Fluctuating Loads in MV and HV Power Systems – Basic EMC Publication. (Technical Report)', (IEC 61000-3-7, IEC), Geneva, Switzerland.

[16] IEC (International Electrotechnical Commission) (1997) 'EMC, Part 4: Testing and Measurement Techniques – Section 15: Flickermeter – Functional and Design Specifications', IEC 61000-4-15, IEC, Geneva, Switzerland.

[17] Johansen, J. (1999) 'Unsteady Airfoil Flows with Application to Aeroelastic Stability', Risø National Laboratory, Roskilde, Denmark.

[18] Knudsen, H., Akhmatov, V. (1999) 'Induction Generator Models in Dynamic Simulation Tools', conference paper, IPST'99, International Power System Transients, Budapest, Hungary.

[19] Kundur, P. (1967) *Digital Simulation and Analysis of Power System Dynamic Performance*, PhD thesis, Department of Electrical Engineering, University of Toronto, Canada.

[20] Kundur, P. (1994) *Power System Stability and Control*, EPRI, McGraw-Hill, New York.

[21] Noroozian, M., Knudsen, H., Bruntt, M. (2000) 'Improving a Wind Farm Performance by Reactive Power Compensation', in *Proceedings of the IASTED International Conference on Power and Energy Systems, September 2000, Marbella, Spain*, pp. 437–443.

[22] Øye, S. (1986) 'Unsteady Wake Effects Caused by Pitch-angle Changes', IEA R&D WECS Joint Action on Aerodynamics of Wind Turbines, 1st symposium, London, UK, pp. 58–79.

[23] Paserba, J. (1996) 'Analysis and Control of Power System Oscillations', technical brochure, July 1996, Cigŕe, Paris, France.

[24] Persson, J., Slootweg, J. G., Rouco, L., Söder, L., Kling, W. L. (2003) 'A Comparison of Eigenvalues Obtained with Two Dynamic Simulation Software Packages', conference paper BPT03-254, IEEE Bologna Power Tech, Bologna, Italy, 23–26 June 2003.

[25] Slootweg J. G., Polinder H., Kling W. L. (2002) 'Reduced Order Models of Actual Wind Turbine Concepts', presented at IEEE Young Researchers Symposium, 7–8 February 2002, Leuven, Belgium.

[26] Snel, H., Lindenburg, C. (1990) 'Aeroelastic Rotor System Code for Horizontal Axis Wind Turbines: Phatas II, European Community Wind Energy Conference, Madrid, Spain, pp. 284–290.

[27] Snel, H., Schepers, J. G. (1995) 'Joint Investigation of Dynamic Inflow Effects and Implementation of an Engineering Method', Netherlands Energy Research Foundation, ECN, Petten, The Netherlands.

[28] Sørensen, J. N., Kock, C. W. (1995) 'A Model for Unsteady Rotor Aerodynamics', *Wind Engineering and Industrial Aerodynamics* Vol. 58, 259–275.

[29] Walker, J. F., Jenkins, N. (1997) *Wind Energy Technology*, John Wiley & Sons Ltd, London, UK.

25

Reduced-order Modelling of Wind Turbines

J. G. Slootweg, H. Polinder and W. L. Kling

25.1 Introduction

In most countries and power systems, wind power generation covers only a small part of the total power system load. There is, however, a tendency towards increasing the amount of electricity generated from wind turbines. Therefore, wind turbines will start to replace the output of conventional synchronous generators and thus the penetration level of wind power in electrical power systems will increase. As a result, it may begin to affect the overall behaviour of the power system. The impact of wind power on the dynamics of power systems should therefore be investigated thoroughly in order to identify potential problems and to develop measures to mitigate such problems.

In this chapter a specific simulation approach is used to study the dynamics of large-scale power systems. This approach will hereafter be referred to as power system dynamics simulation (PSDS). It was discussed in detail in Chapter 24; therefore this chapter will include only a short introduction. It is necessary to incorporate models of wind turbine generating systems into the software packages that are used for PSDS in order to analyse the impact of high wind power penetration on electrical power systems. These models have to match the assumptions and simplifications applied in this type of simulation.

This chapter presents models that fulfil these requirements and can be used to represent wind turbines in PSDSs. After briefly introducing PSDS, the three main wind turbine types will be described. Then, the modelling assumptions are given. The models were required to be applicable in PSDSs. Many of the presented assumptions are a result of this requirement. Then, models of the various subsystems of each of the most

Wind Power in Power Systems   Edited by T. Ackermann
© 2005 John Wiley & Sons, Ltd   ISBN: 0-470-85508-8 (HB)

important current wind turbine types are given. Finally, the model's response to a measured wind speed sequence will be compared with actual measurements.

25.2 Power System Dynamics Simulation

PSDS software is used to investigate the dynamic behaviour and small signal stability of power systems. A large power system can easily have hundreds or even thousands of state variables. These are associated with branches, with generators and their controllers and, if dynamic rather than static load models are used, with loads. If the relevant aspect of power system behaviour is characterised by rather low frequencies or long time constants, long simulation runs are required. A simulation run of sufficient length would, however, take a substantial amount of computation time if all fast, high-frequency, phenomena were included in the applied models. If the investigated power system is large it will be difficult and time consuming to study different scenarios and setups.

To solve this problem, only the fundamental frequency component of voltages and currents is taken into account when studying low-frequency phenomena. In this chapter, this approach will hereafter be referred to as PSDS. It is also known as fundamental frequency or electromechanical transient simulation. First, this approach makes it possible to represent the network by a constant impedance or admittance matrix, similar to load-flow calculations. The equations associated with the network can thus be solved by using load-flow solution methods, the optimisation of which has been studied extensively and which can hence be implemented very efficiently, resulting in short computation times. Second, this approach reduces the computation time by cancelling a number of differential equations, because no differential equations are associated with the network and only a few are associated with generating equipment. This approach also makes it possible to use a larger simulation time step (Kundur, 1994). This way, reduced-order models of the components of a power system are developed.

Examples of PSDS packages are PSS/E^{TM} (Power System Simulator for Engineers) and Eurostag.[1] This type of software can be used when the phenomena of interest have a frequency of about 0.1–10 Hz. The typical problems that are studied when using these programs are voltage and angle stability. If the relevant frequencies are higher, instantaneous value simulation (IVS) software, such as the Alternative Transient Program (ATP), an electromagnetic transient program (EMTP) or SimPowerSystems from Matlab must be used.[2] These packages contain more detailed and higher-order equipment models than do PSDS software. However, their time step is much smaller and it is hardly feasible to simulate a large-scale power system in these programs.

Some advanced software packages, such as PowerFactory, Netomac^{TM} and Simpow^{TM}, offer both dynamic and instantaneous value modes of simulation.[3] They may even be able automatically to switch between the dynamic and instantaneous value domains, each with their own models, depending on the simulated event and/or the user's preferences. IVSs are also referred to as electromagnetic transient simulations.

---

[1] PSS/E^{TM} is a trademark of Shaw Power Technologies, Inc.
[2] Matlab^{TM} is a registered trademark.
[3] Netomac^{TM} is a trademark of Siemens; Simpow^{TM} is a trademark of STRI AB.

For more information on the different types of power system simulation and the type of problems that can be studied with them, see Chapter 24.

25.3 Current Wind Turbine Types

The vast majority of wind turbines that are currently installed uses one of the three main types of electromechanical conversion system. The first type uses an (asynchronous) squirrel cage induction generator to convert the mechanical energy into electricity. Owing to the different operating speeds of the wind turbine rotor and the generator, a gearbox is necessary to match these speeds (wind turbine Type A0).[4] The generator slip varies slightly with the amount of generated power and is therefore not entirely constant. However, because these speed variations are in the order of 1 %, this wind turbine type is normally referred to as a constant-speed or fixed-speed turbine. Today, the Danish, or constant-speed, design is nearly always combined with a stall control of the aerodynamic power, although in the past pitch-controlled constant-speed wind turbine types were also built.

The second type uses a doubly fed induction generator instead of a squirrel cage induction generator (Type C). Similar to the first type, it needs a gearbox. The stator winding of the generator is coupled to the grid, and the rotor winding to a power electronics converter, which today usually is a back-to-back voltage source converter (VSC) with current control loops. This way, the electrical and mechanical rotor frequencies are decoupled, because the power electronics converter compensates the difference between mechanical and electrical frequency by injecting a rotor current with a variable frequency. Thus, variable-speed operation becomes possible (i.e. control of the mechanical rotor speed according to a certain goal function, such as energy yield maximisation or noise minimisation is possible). The rotor speed is controlled by changing the generator power in such a way that it equals the value as derived from the goal function. In this type of conversion system, the required control of the aerodynamic power is normally performed by pitch control.

The third type is called a direct-drive wind turbine as it works without a gearbox (Type D). A low-speed multipole synchronous ring generator with the same rotational speed as the wind turbine rotor is used to convert the mechanical energy into electricity. The generator can have either a wound rotor or a rotor with permanent magnets. The stator is not coupled directly to the grid but to a power electronics converter. This may consist of a back-to-back VSC or a diode rectifier with a single VSC. By using the electronic converter, variable-speed operation becomes possible. Power limitation is again achieved by pitch control, as in the previously mentioned type. The three main wind turbine types are depicted in Figure 19.2 (page 421) and are discussed in more detail in Chapter 4.

25.4 Modelling Assumptions

This section discusses the assumptions on which the models in this chapter are based. First, in order to put a reasonable limit on data requirements and computation time, a

---

[4] For definitions of wind turbine Types A–D, see Section 4.2.3.

quasistatic approach is used to describe the rotor of the wind turbine. An algebraic relation between the wind speed at hub height and the mechanical power extracted from the wind is assumed. More advanced methods, such as the blade element impulse method, require a detailed knowledge of aerodynamics and of the turbine's blade profile characteristics (Heier, 1998; Patel, 2000). In many cases, these data will not be available or, in preliminary studies, a decision regarding the wind turbine type may not have been taken and, consequently, the blade profile will of course not be known. However, this is not considered a problem, because the grid interaction is the main topic of interest in power system dynamics studies, and the impact of the blade profile on the grid interaction can be assumed to be rather limited.

This chapter presents models for representing wind turbines in PSDSs. The models have to comply with the requirements that result from the principles on which PSDS software is based, as described in Section 25.2. We first make the following assumptions that apply to the model of each wind turbine type:

- Assumption 1: magnetic saturation is neglected.
- Assumption 2: flux distribution is sinusoidal.
- Assumption 3: any losses, apart from copper losses, are neglected.
- Assumption 4: stator voltages and currents are sinusoidal at the fundamental frequency.

Furthermore, the following assumptions apply to some of the analysed systems:

- Assumption (a): in both variable-speed systems (Types C and D), all rotating mass is represented by one element, which means that a so-called 'lumped-mass' representation is used.
- Assumption (b): in both variable-speed systems, the VSCs with current control loops are modelled as current sources.
- Assumption (c): in the system based on the doubly fed induction generator (Type C), rotor voltages and currents are sinusoidal at the slip frequency.
- Assumption (d): in the direct drive wind turbine (Type D), the synchronous generator has no damper windings.
- Assumption (e): when a diode rectifier is used in the direct drive wind turbine, commutation is neglected.

Assumption (a) is made because the mechanical and electrical properties of variable-speed wind turbines are decoupled by the power electronic converters. Therefore, the shaft properties are hardly reflected in the grid interaction, which is the main point of interest in power system studies (Krüger and Andresen, 2001; Petru and Thiringer, 2000).

Assumptions (b) and (c) are made in order to model power electronics in PSDS simulation software and are routinely applied in power system dynamics simulations (Fujimitsu et al., 2000; Hatziargyriou, 2001; Kundur, 1994). A full converter model would require a substantial reduction of the simulation time step and the incorporation of higher harmonics in the network equations. This is not in agreement with the intended use of the models described here.

Assumption (d) is made because the damper windings in the synchronous generator do not have to be taken into account. When a back-to-back VSC is used, generator speed is controlled by the power electronic converter, which will prevent oscillations. When a diode rectifier is used, damper windings are essential for commutation. However, commutation is neglected according to Assumption (e), and therefore the damper windings will be neglected as well.

25.5 Model of a Constant-speed Wind Turbine

25.5.1 Model structure

Figure 25.1 depicts the general structure of a model of a constant-speed wind turbine. This general structure consists of models of the most important subsystems of this wind turbine type, namely, the rotor, the drive train and the generator, combined with a wind speed model. Each of the blocks in Figure 25.1 will now be discussed separately, except for the grid model. The grid model is a conventional load-flow model of a network. The topic of load-flow models and calculations is outside the scope of this chapter and the grid model is therefore not treated any further. Instead, the reader is referred to textbooks on this topic (e.g. Grainger and Stevenson, 1994).

25.5.2 Wind speed model

The output of the first block in Figure 25.1 is a wind speed sequence. One approach to model a wind speed sequence is to use measurements. The advantage of this is that a 'real' wind speed is used to simulate the performance of the wind turbine. The disadvantage is, however, that only wind speed sequences that have already been measured can be simulated. If a wind speed sequence with a certain wind speed range or turbulence intensity is to be simulated, but no measurements that meet the required characteristics are available, it is not possible to carry out the simulation.

A more flexible approach is to use a wind speed model that can generate wind speed sequences with characteristics to be chosen by the user. This makes it possible to simulate a wind speed sequence with the desired characteristics, by setting the value of the corresponding parameters to an appropriate value. In the literature concerning the simulation of wind power in electrical power systems, it is often assumed that the wind

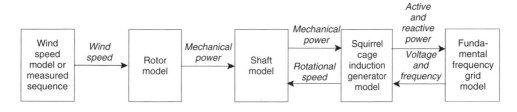

Figure   25.1 General structure of constant-speed wind turbine model

speed is made up by the sum of the following four components (Anderson and Bose, 1983; Wasynczuk, Man and Sullivan, 1981):

- the average value;
- a ramp component, representing a steady increase in wind speed;
- a gust component, representing a gust;
- a component representing turbulence.

This leads to the following equation:

$$\nu_w(t) = \nu_{wa} + \nu_{wr}(t) + \nu_{wg}(t) + \nu_{wt}(t), \tag{25.1}$$

in which $\nu_w(t)$ is the wind speed at time t; ν_{wa} is the average value of the wind speed; $\nu_{wr}(t)$ is the ramp component; $\nu_{wg}(t)$ is the gust component; and $\nu_{wt}(t)$ is the turbulence component. The wind speed components are all in 'metres per second', and time t is in seconds. We will now discuss the equations used in this chapter to describe the various components of Equation (25.1).

The average value of the wind speed, ν_{wa}, is calculated from the power generated in the load-flow case combined with the nominal power of the turbine. Hence, the user does not need to specify this value (Slootweg, Polinder and Kling, 2001). An exception is a variable-speed wind turbine with pitch control at nominal power. In this case, there is no unique relation between the generated power, indicated in the load-flow case, and the wind speed. Therefore, the user must give an initial value either of the wind speed or of the pitch angle. The equations describing the rotor, which will be discussed below, can then be used to calculate the pitch angle or the average wind speed, respectively.

The wind speed ramp is characterised by three parameters – the amplitude of the wind speed ramp, $\hat{A}_r$ (in m/s), the starting time of the wind speed ramp T_{sr} (in seconds), and the end time of the wind speed ramp, T_{er} (in seconds). The wind speed ramp is described by the following equation:

$$\left. \begin{array}{l} t < T_{sr}, \quad \text{for } \nu_{wr} = 0; \\[2mm] T_{sr} \leq t \leq T_{er}, \quad \text{for } \nu_{wr} = \hat{A}_r \dfrac{(t - T_{sr})}{(T_{er} - T_{sr})}; \\[2mm] T_{eg} < t, \quad \text{for } \nu_{wg} = \hat{A}_r. \end{array} \right\} \tag{25.2}$$

The wind speed gust is characterised by three parameters – the amplitude of the wind speed gust, $\hat{A}_g$ (in m/s), the starting time of the wind speed gust, T_{sg} (in seconds), and the end time of the wind speed gust T_{eg} (in seconds). The wind gust is modelled using the following equation (Anderson and Bose, 1983; Wasynczuk, Man and Sullivan, 1981):

$$\left. \begin{array}{l} t < T_{sg}, \quad \text{for } \nu_{wg} = 0; \\[2mm] T_{sg} \leq t \leq T_{eg}, \quad \text{for } \nu_{wg} = \hat{A}_g \left\{ 1 - \cos\left[2\pi \left(\dfrac{t - T_{sg}}{T_{eg} - T_{sg}} \right) \right] \right\}; \\[2mm] T_{eg} < t, \quad \text{for } \nu_{wg} = 0. \end{array} \right\} \tag{25.3}$$

Finally, we will describe the modelling of the turbulence. The turbulence component of the wind speed is characterised by a power spectral density. Here, the following power spectral density is used:

$$P_{Dt}(f) = l_{v_{wa}} \left[\ln\left(\frac{h}{z_0}\right)^2 \right]^{-1} \left[1 + 1.5\frac{fl}{v_{wa}} \right]^{-5/3}, \tag{25.4}$$

in which P_{Dt} is the power density of the turbulence for a certain frequency (W/Hz); f is the frequency (Hz); h is the height at which the wind speed signal is of interest (m), which normally equals the height of the wind turbine shaft; v_{wa} is the mean wind speed (m/s); l is the turbulence length scale (m), which equals $20h$ if h is less than $30\,m$, and equals 600 if h is more than $30\,m$; and z_0 is the roughness length (m). Through the parameter z_0, we take into account the dependence of the turbulence intensity on the landscape where the wind turbine is located. The roughness length depends on the structure of the landscape surrounding the wind turbine. Table 25.1 gives the values of z_0 for various landscape types (Panofsky and Dutton, 1984; Simiu and Scanlan, 1986).

The fact that PSDSs are carried out in the time domain, whereas the turbulence of the wind is described by a power spectral density given in the frequency domain, raises a specific problem: the translation of a power spectral density into a time sequence of values with the given power spectral density. To solve this problem, we use a method that is described in Shinozuka and Jan (1972). This method works as follows. A power spectral density can be used to derive information about the amplitude of a signal's component with a given frequency. Then, a large number of sines with a random initial phase angle and an amplitude calculated from the power spectral density are added for each time step. Thus we can generate a time domain signal with a power spectral density that is a sampled equivalent of the original power spectral density. The smaller the frequency difference between the components, the better the power spectral density of the artificially generated signal resembles the original power spectral density. The required computation time will increase, though. For more information on the method that we used for generating time domain signals with a given power spectral density, the reader is referred to Shinozuka and Jan (1972). Figure 25.2 shows an example of a wind speed sequence generated using this approach.

Table 25.1 Value of roughness length, z_0, for various landscape types

Landscape type	Range of z_0 (m)
Open sea or sand	0.0001–0.001
Snow surface	0.001–0.005
Mown grass or steppe	0.001–0.01
Long grass or rocky ground	0.04–0.1
Forests, cities or hilly areas	1–5

Sources: Panofsky and Dutton, 1984; Simiu and Scanlan, 1986.

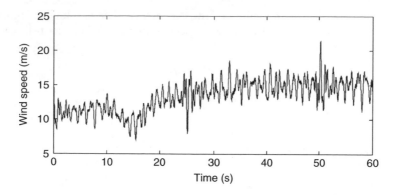

Figure   25.2 Example of a simulated wind speed sequence, with the following input values: average wind speed, v_{wa}, 11.5 m/s; start time of wind speed ramp, T_{sr}, 5 s; end time of wind speed ramp, T_{er}, 35 s; amplitude of wind speed ramp, $\hat{A}_r$, 4 m/s; start time of wind speed gust, T_{sg}, 5 s; end time of wind speed gust, T_{eg}, 15 s; amplitude of wind speed gust, $\hat{A}_g = -3$ m/s; and roughness length, z_0, 0.01 m

25.5.3 Rotor model

The following well-known algebraic equation gives the relation between wind speed and mechanical power extracted from the wind (Heier, 1998; Patel, 2000):

$$P_{wt} = \frac{\rho}{2} A_{wt} c_p(\lambda, \theta) v_w^3, \tag{25.5}$$

where P_{wt} is the power extracted from the wind in watts; ρ is the air density (kg/m^3); c_p is the performance coefficient or power coefficient; λ is the tip speed ratio v_t/v_w, the ratio between blade tip speed, v_t (m/s), and wind speed at hub height upstream of the rotor, v_w (m/s); θ is the pitch angle (in degrees); and A_{wt} is the area covered by the wind turbine rotor (m^2). Most constant-speed wind turbines are stall controlled. In that case, θ is left out and c_p is a function of λ only.

Manufacturer documentation shows that the power curves of individual wind turbines are very similar. We therefore do not consider it necessary to use different approximations for the $c_p(\lambda)$ curve for different constant-speed wind turbines in PSDSs. Instead, a general approximation can be used. We would like to stress that this does not necessarily apply to other types of calculations, such as energy yield calculations for financing purposes. Here, we use the following general equation to describe the rotor of constant-speed and variable-speed wind turbines:

$$c_p(\lambda, \theta) = c_1 \left(\frac{c_2}{\lambda_i} - c_3\theta - c_4\theta^{c_5} - c_6 \right) \exp\left(\frac{-c_7}{\lambda_i} \right), \tag{25.6}$$

where

$$\left[\left(\frac{1}{\lambda + c_8\theta} \right) - \left(\frac{c_9}{\theta^3 + 1} \right) \right]^{-1}. \tag{25.7}$$

The structure of this equation originates from Heier (1998). However, the values of the constants c_1 to c_9 have been changed slightly in order to match the manufacturer data better. To minimise the error between the curve in the manufacturer documentation and the curve we obtained by using Equations (25.6) and (25.7), we applied multidimensional optimisation. Table 25.2 includes both the original parameters and the parameters used here. Figure 25.3 depicts the power curves of two commercial constant-speed wind turbines, together with the generic numerical approximation from Table 25.2.

High-frequency wind speed variations are very local and therefore even out over the rotor surface, particularly when wind turbines become larger. To approximate this effect, a low-pass filter is included in the rotor model. Figure 25.4 illustrates the low-pass

Table 25.2 Approximation of power curves

	c_1	c_2	c_3	c_4	c_5	c_6	c_7	c_8	c_9
Heier (1998)	0.5	116	0.4	0	—	5	21	0.08	0.035
Constant-speed wind turbine	0.44	125	0	0	0	6.94	16.5	0	−0.002
Variable-speed wind turbine	0.73	151	0.58	0.002	2.14	13.2	18.4	−0.02	−0.003

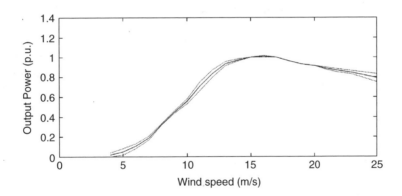

Figure 25.3 Comparison of the numerical approximation of the power curve of a stall-controlled wind turbine (solid curve) with the power curves of two commercial stall-controlled wind turbines (dotted lines)

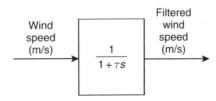

Figure 25.4 Low-pass filter for representing the evening out of high-frequency wind speed components over the rotor surface

filter. The value of the time constant τ depends on the rotor diameter as well as on the turbulence intensity of the wind and the average wind speed (Petru and Thiringer, 2000). For the wind turbine analysed here, τ was set to 4.0 s.

Finally, we would like to include the tower shadow in the rotor model. This can be done by adding a periodic pulsation to the mechanical power that is the output of the rotor model, as calculated with Equation (25.5). The frequency of this pulsation depends on the number of blades (normally three) and the rotational speed of the wind turbine rotor. The amplitude of the pulsation is in the order of a few percent. The tower shadow is particularly important in investigations concerning power quality and the mutual interaction between wind turbines that are situated close to each other, electrically.

25.5.4 Shaft model

It has been repeatedly argued in the literature that the incorporation of a shaft representation in models of constant-speed wind turbines (Type A) is very important for a correct representation of their behaviour during and after voltage drops and short circuits (e.g. see Akhmatov, Knudsen and Nielsen, 2000). This is because of the fact that the low-speed shaft of wind turbines is relatively soft (Hinrichsen and Nolan, 1982). The models presented here are to be used for PSDSs. These are, among others, used for analysing a power system's response to the mentioned disturbances. It is therefore essential to incorporate a shaft representation into the constant-speed wind turbine model. However, only the low-speed shaft is included. The gearbox and the high-speed shaft are assumed to be infinitely stiff. The resonance frequencies associated with gearboxes and high-speed shafts usually lie outside the frequency bandwidth that we deal with in PSDSs (Akhmatov, Knudsen and Nielsen, 2000; Papathanassiou and Papadopoulos, 1999). Therefore, we use a two-mass representation of the drive train. The two-mass representation is described by the following equations:

$$\left.\begin{aligned}
\frac{d\omega_{wr}}{dt} &= \frac{T_{wr} - K_s\gamma}{2H_{wr}}, \\
\frac{d\omega_m}{dt} &= \frac{K_s\gamma - T_e}{2H_m}, \\
\frac{d\gamma}{dt} &= 2\pi f(\omega_{wr} - \omega_m),
\end{aligned}\right\} \tag{25.8}$$

in which f is the nominal grid frequency; T is the torque; γ is the angular displacement between the two ends of the shaft; ω is frequency; H is the inertia constant; and K_s is the shaft stiffness. The subscripts wr, m and e stand for wind turbine rotor, generator mechanical and generator electrical, respectively. All values are in per unit, apart from K_s, γ and f, which are in p.u./el. rad., degrees, and hertz, respectively.

The resonance frequency of the shaft's torsional mode was experimentally determined as 1.7 Hz (Pedersen et al., 2000). When resonance frequency and the inertia constants of the generator and the turbine rotor are known, K_s can be calculated by using the general equation describing a two-mass system (Anderson, Agrawal and Van Ness, 1990). The shaft is depicted schematically in Figure 25.5, which includes some of the quantities from Equation (25.8).

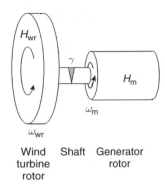

Wind   Shaft   Generator
turbine   rotor
rotor

Figure 25.5 Schematic representation of the shaft, including some of the quantities from Equation (25.8): H_{wr} and H_m = inertia constant for the wind turbine rotor and generator rotor, respectively; ω_{wr} and ω_m = angular frequency of wind turbinerotor and generator rotor, respectively; γ = angular displacement between shaft ends

25.5.5 Generator model

The voltage equations of a squirrel cage induction generator in the d – q (direct – quadrative) reference frame, using the generator convention, can be found in the literature (Kundur, 1994) and are as follows

$$\left.\begin{aligned}
u_{ds} &= -R_s i_{ds} - \omega_s \psi_{qs} + \frac{d\psi_{ds}}{dt}, \\[4pt]
u_{qs} &= -R_s i_{qs} + \omega_s \psi_{ds} + \frac{d\psi_{qs}}{dt}, \\[4pt]
u_{dr} &= 0 = -R_r i_{dr} - s\omega_s \psi_{qr} + \frac{d\psi_{dr}}{dt}, \\[4pt]
u_{qr} &= 0 = -R_r i_{qr} + s\omega_s \psi_{dr} + \frac{d\psi_{qr}}{dt},
\end{aligned}\right\} \tag{25.9}$$

in which s is the slip, u is the voltage, i is the current, R is the resistance, and ψ is the flux. All quantities are in per unit. The subscripts d and q stand for direct and quadrature component, respectively, and the subscripts r and s for rotor and stator, respectively. The generator convention is used in this equation, which means that a current leaving the machine is positive, whereas a current entering the machine is negative. The opposite of the generator convention is the motor convention, where a current entering the machine is positive whereas a current leaving the machine is negative.

The slip is defined as follows:

$$s = 1 - \frac{p}{2}\frac{\omega_m}{\omega_s}, \tag{25.10}$$

in which p is the number of poles. With respect to per unit (p.u.) quantities, it should be noted at this point that the goal of using them is to make impedances independent of voltage level and generator rating by expressing them as a percentage of a common base value. For an elaborate treatment of p.u. calculation and the correlation between physical and p.u. values, see general textbooks on power systems, such as Grainger and Stevenson (1994). The flux linkages in Equation (25.9) can be calculated by using the following equations, in which, again, the generator convention is used:

$$
\left.\begin{aligned}
\psi_{ds} &= -(L_{s\sigma} + L_m)i_{ds} - L_m i_{dr}, \\
\psi_{qs} &= -(L_{s\sigma} + L_m)i_{qs} - L_m i_{qr}, \\
\psi_{dr} &= -(L_{r\sigma} + L_m)i_{dr} - L_m i_{ds}, \\
\psi_{qr} &= -(L_{r\sigma} + L_m)i_{qr} - L_m i_{qs}.
\end{aligned}\right\}
\qquad (25.11)
$$

In these equations, ψ is flux linkage and L is the inductance. The indices m, r and σ stand for mutual, rotor and leakage, respectively. By inserting Equations (25.11) in Equations (25.9), while neglecting the stator transients, in agreement with the assumptions discussed above, the voltage current relationships become:

$$
\left.\begin{aligned}
u_{ds} &= -R_s i_{ds} + \omega_s[(L_{s\sigma} + L_m)i_{qs} + L_m i_{qr}], \\
u_{qs} &= -R_s i_{qs} - \omega_s[(L_{s\sigma} + L_m)i_{ds} + L_m i_{dr}], \\
u_{dr} &= 0 = -R_r i_{dr} + s\omega_s[(L_{r\sigma} + L_m)i_{qr} + L_m i_{qs}] + \frac{d\psi_{dr}}{dt}, \\
u_{qr} &= 0 = -R_r i_{qr} - s\omega_s[(L_{r\sigma} + L_m)i_{dr} + L_m i_{ds}] + \frac{d\psi_{qr}}{dt}.
\end{aligned}\right\}
\qquad (25.12)
$$

The electrical torque, T_e, is given by:

$$
T_e = \psi_{qr} i_{dr} - \psi_{dr} i_{qr},
\qquad (25.13)
$$

and the equation of the motion of the generator is:

$$
\frac{d\omega_m}{dt} = \frac{1}{2H_m}(T_m - T_e).
\qquad (25.14)
$$

The equations for active power generated, P, and the reactive power consumed, Q, are:

$$
\left.\begin{aligned}
P_s &= u_{ds} i_{ds} + u_{qs} i_{qs}, \\
Q_s &= u_{qs} i_{ds} - u_{ds} i_{qs}.
\end{aligned}\right\}
\qquad (25.15)
$$

Because only the stator winding is connected to the grid, generator and grid can exchange active and reactive power only through the stator terminals. Therefore, the rotor does not need to be taken into account. The values of the various parameters are dependent on the generator rating and can be derived from tables and graphs (Heier, 1998). Table 25.3 includes the generator parameters used in this Chapter.

Table **25.3** Simulated induction generator parameters

Generator characteristic	Value
Number of poles, p	4
Generator speed (constant-speed; rpm)	1517
Generator speed (doubly fed; rpm)	1000–1900
Mutual inductance, L_m (p.u.)	3.0
Stator leakage inductance, $L_{\sigma s}$ (p.u.)	0.10
Rotor leakage inductance, $L_{\sigma r}$ (p.u.)	0.08
Stator resistance, R_s (p.u.)	0.01
Rotor resistance, R_r (p.u.)	0.01
Compensating capacitor (constant-speed; p.u.)	0.5
Moment of inertia (s)	0.5

25.6 Model of a Wind Turbine with a Doubly fed Induction Generator

25.6.1 Model structure

Figure 25.6 depicts the general structure of a model of variable-speed wind turbines with doubly fed induction generators (Type C). It shows that a wind turbine with a doubly fed induction generator is much more complex than a constant-speed wind turbine (see Figure 25.1). Compared with a constant-speed wind turbine, a variable-speed wind turbine has additional controllers, such as the rotor speed controller and the pitch angle controller. Additionally, if it is equipped with terminal voltage control, it also has a terminal voltage controller. We will now discuss each block in Figure 25.6, with the

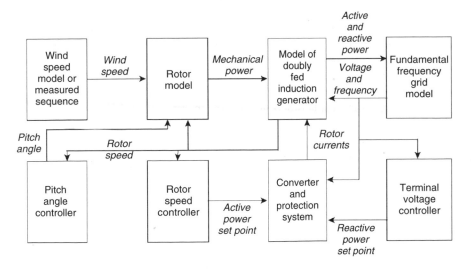

Figure **25.6** General structure of a model of a variable speed wind turbine with doubly fed induction generator (type C)

exception of the grid model and the wind speed model. The wind speed model is identical to that in the constant-speed wind turbine model, described in Section 25.5.2. The grid model is not discussed for the same reasons mentioned in Section 25.5.1.

Figure 25.6 does not include a shaft model, in contrast to Figure 25.2. The reason is that in variable-speed wind turbines the mechanical and electrical part, to a large extent, are decoupled by the power electronics. Therefore, the control approach of the power electronics converter determines how the properties and behaviour of the shaft are reflected in the terminal quantities of the generator. The mutual interdependencies between shaft, control of the power electronic converter and output power pattern are a very advanced topic which will not be treated here. If it is nevertheless desired to incorporate a shaft representation in a model of a variable-speed wind turbine with a doubly fed induction generator, the approach described in Section 25.5.4 can be used.

25.6.2 Rotor model

Again, Equation (25.5) and a numerical approximation of the $c_p(\lambda, \theta)$ curve based on Equations (25.6) and (25.7) are used to represent the rotor (see Section 25.5.3). However, it is assumed here that the variable-speed wind turbine is pitch controlled. The performance coefficient is thus dependent not only on the tip speed ratio, λ, but also on the pitch angle, θ. Therefore, a new numerical approximation for the $c_p(\lambda, \theta)$ curve has been developed, using manufacturer documentation of variable-speed wind turbines and multidimensional optimisation. Table 25.2 includes the values for the parameters in Equations (25.6) and (25.7), which are used to represent the rotor of a variable-speed wind turbine. Figure 25.7(a) depicts the resulting power curve, together with the power curves of two commercial variable-speed wind turbines. Figure 25.7(b) shows the pitch angle deviation that is necessary to limit the power to the nominal value. In this case, documentation from only one manufacturer was available.

Again, the rotor model of the simulations includes the low-pass filter depicted in Figure 25.4 in order to represent the smoothing of high-frequency wind speed components over the rotor surface. The simulations are carried out to validate the models and are described in detail in Section 25.8. This issue is less critical for variable-speed wind turbines, though, as rapid variations in wind speed are not translated into output power variations because the rotor functions as an energy buffer. For the same reason, representation of tower shadow was not included either, because in variable-speed wind turbines the tower shadow hardly affects the output power because of the decoupling of electrical and mechanical behaviour by the power electronics (Krüger and Andresen, 2001; Petru and Thiringer, 2000).

25.6.3 Generator model

The equations that describe a doubly fed induction generator are identical to those of the squirrel cage induction generator, [i.e. Equations (25.9) – (25.14); see Section 25.5.5]. The only exception is that the rotor winding is not short-circuited. Therefore, in the expressions for rotor voltages u_{dr} and u_{qr} in Equations (25.9) and (25.12), these voltages

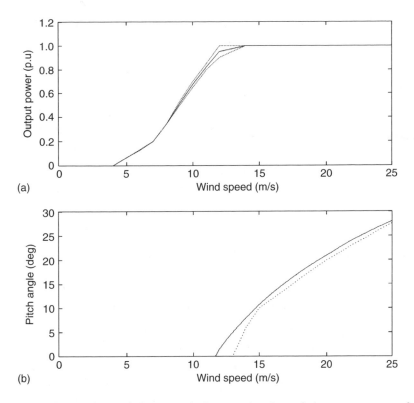

Figure   25.7 (a) Comparison of the numerical approximation of the power curve of a pitch-controlled wind turbine (solid curve) with the power curves of two existing pitch-controlled wind turbines (dotted curves). (b) Pitch angle deviation above nominal wind speed, based on a numerical approximation (solid curve) and manufacturer documentation (dotted curve)

are not equal to zero. The flux equations are identical to those of the squirrel cage induction generator, given in Equations (25.11).

If we want to obtain the voltage–current relationship using the voltage and flux linkage equations, first the stator transients must again be neglected. This way, the generator model corresponds to the assumptions used in PSDSs. This time, we also neglect the rotor transients (Fujimitsu *et al.*, 2000). To take into account the rotor transients would require detailed modelling of the converter, including the semiconductor switches and the current control loops. The result would be time constants that are significantly lower than 100 ms, the typical minimum time constant studied in PSDSs. Further, the resulting model would be much more complex and therefore difficult to use, and it would require many more parameters, which are often difficult to obtain, in practice. To avoid this, we assume that the VSCs with current control loops can be modelled as current sources, as already observed in Section 25.4. Chapter 26 presents a complete model of the doubly fed induction generator including the stator and rotor transients.

The following voltage – current relationships result in per unit quantities:

$$\left.\begin{aligned}
u_{ds} &= -R_s i_{ds} + \omega_s[(L_{s\sigma} + L_m)i_{qs} + L_m i_{qr}], \\
u_{qs} &= -R_s i_{qs} - \omega_s[(L_{s\sigma} + L_m)i_{ds} + L_m i_{dr}], \\
u_{dr} &= -R_r i_{dr} + s\omega_s[(L_{r\sigma} + L_m)i_{qr} + L_m i_{qs}], \\
u_{qr} &= -R_r i_{qr} - s\omega_s[(L_{r\sigma} + L_m)i_{dr} + L_m i_{ds}].
\end{aligned}\right\}
\qquad (25.16)$$

Note the differences between Equations (25.16) and Equations (25.12): the rotor voltages do not equal zero and the $d\psi/dt$ terms in the rotor equations have been neglected. The generator parameters are given in Table 25.3.

The equation giving the electrical torque and the equation of motion of a doubly fed induction generator are again equal to those of a squirrel cage induction generator, given in Equations (25.13) and (25.14), respectively. The equations for active and reactive power are, however, different because the rotor winding of the generator can be accessed. This leads to the incorporation of rotor quantities into these equations:

$$\left.\begin{aligned}
P &= P_s + P_r = u_{ds}i_{ds} + u_{qs}i_{qs} + u_{dr}i_{dr} + u_{qr}i_{qr}, \\
Q &= Q_s + Q_r = u_{qs}i_{ds} - u_{ds}i_{qs} + u_{qr}i_{dr} - u_{dr}i_{qr}.
\end{aligned}\right\}
\qquad (25.17)$$

It should be emphasised that the reactive power, Q, in Equation (25.17) is not necessarily equal to the reactive power fed into the grid, which is the quantity that must be used for the load-flow solution. This depends on the control strategy for the grid side of the power electronic converter that feeds the rotor winding. This does not apply to the active power. Even though the converter can generate or consume reactive power, it cannot generate, consume or store active power – at least not long enough to be of any interest here. The expression for P in Equations (25.17) gives, therefore, the total active power generated by the doubly fed induction generator, apart from the converter efficiency, which can be incorporated by multiplying the last two terms of this expression (i.e. $u_{dr}i_{dr}$ and $u_{qr}i_{qr}$) with the assumed converter efficiency. That means that all active power fed into or drawn from the rotor winding will be drawn from or fed into the grid, respectively.

25.6.4 Converter model

The converter is modelled as a fundamental frequency current source. This assumption is, however, valid only if the following conditions are fulfilled:

- the machine parameters are known;
- the controllers operate in their linear region;
- vector modulation is used;
- the terminal voltage approximately equals the nominal value.

The wind turbine manufacturer is responsible for the first two conditions and we assume that these conditions are fulfilled. The third requirement is met, because the control of the converter used in variable-speed wind turbines is nearly always based on vector modulation (Heier, 1998). The last condition is not fulfilled during grid faults. However, when a fault occurs, variable-speed wind turbines are promptly disconnected to protect the power electronic converter. Further, there are very-high-frequency phenomena when the power electronic converter responds to a voltage drop. PSDSs are unsuitable for investigating this topic. Therefore, the model incorporates a low-frequency representation of the behaviour of the converter during faults, similar to high-voltage direct-current (HVDC) converters (SPTI, 1997).

The converters are represented as current sources, and therefore the current set points equal the rotor currents. The current set points are derived from the set points for active and reactive power. The active power set point is generated by the rotor speed controller, based on the actual rotor speed value. The reactive power set point is generated by the terminal voltage or power factor controller, based on the actual value of the terminal voltage or the power factor. Both controllers will be discussed below.

When the stator resistance is neglected and it is assumed that the d-axis coincides with the maximum of the stator flux, the electrical torque is dependent on the quadrature component of the rotor current (Heier, 1998). If u_{qr} equals u_t, from the assumption that the d-axis coincides with the maximum of the stator flux, then it can be derived from Equations (25.13) and (25.16) that the following relation between i_{qr} and T_e holds.

$$T_e = \frac{L_m u_t i_{qr}}{\omega_s(L_{s\sigma} + L_m)}. \tag{25.18}$$

where u_t is the terminal voltage.

The reactive power exchanged with the grid at the stator terminals is dependent on the direct component of the rotor current. From Equations (25.16) and (25.17), neglecting the stator resistance and assuming that the d-axis coincides with the maximum of the stator flux, it can be shown that

$$Q_s = -\frac{L_m u_t i_{dr}}{L_{s\sigma} + L_m} - \frac{u_t^2}{\omega_s(L_{s\sigma+L_m})}. \tag{25.19}$$

However, the total reactive power exchanged with the grid depends not only on the control of the generator but also on the control of the grid side of the converter feeding the rotor winding. The following equations apply to the converter:

$$\left. \begin{array}{l} P_c = u_{dc}i_{dc} + u_{qc}i_{qc}, \\ Q_c = u_{qc}i_{dc} - u_{dc}i_{qc}, \end{array} \right\} \tag{25.20}$$

in which the subscript c stands for converter. In this equation, P_c is equal to the rotor power of the doubly fed induction generator P_r given in Equation (25.19). P_r may be multiplied with the converter efficiency if the converter losses are to be included. The reactive power exchanged with the grid equals the sum of Q_s from Equation (25.19) and Q_c from Equations (25.20). Q_c depends on the control strategy and the converter rating but often equals zero, which means that the grid side of the converter operates at unity power factor (see Chapter 19).

25.6.5 Protection system model

The goal of the protection system is to protect the wind turbine from damage caused by the high currents that can occur when the terminal voltage drops as a result of a short circuit in the grid. It also has the task of preventing islanding. Islanding is a situation in which a part of the system continues to be energised by distributed generators, such as wind turbines, after the system is disconnected from the main system. This situation should be prevented because it can lead to large deviations of voltage and frequency from their nominal values, resulting in damage to grid components and loads. It can also pose a serious threat to maintenance staff, who incorrectly assumes that the system is de-energised after the disconnection from the transmission system.

The thermal time constants of semiconductor components are very short. Therefore, the converter that feeds the rotor winding is easily damaged by fault currents, and overcurrent protection is essential for the converter. The generator itself is more capable of withstanding fault currents and is therefore less critical. The working principle of overcurrent protection is as follows. When a voltage drop occurs, the rotor current quickly increases. This is 'noticed' by the controller of the rotor side of the power electronics converter. Then the rotor windings are shorted using a 'crow bar' (basically turning the generator into a squirrel cage induction generator). A circuit braker between the stator and the grid is operated and the wind turbine is completely disconnected. The grid side of the converter can also 'notice' the voltage drop and the corresponding current increase, depending on the converter controls. After the voltage is restored, the wind turbine is reconnected to the system.

The anti-islanding protection of the wind turbine acts in response to voltage and/or frequency deviations or phase angle jumps. The grid side of the converter measures the grid voltage with a high sampling frequency. There are criteria implemented in the protection system for determining whether an island exists. If these criteria are met, the wind turbine is disconnected. The criteria that are applied are a trade-off between the risk of letting an island go undetected, on the one hand, and of incorrectly detecting an island when there is none, leading to 'nuisance tripping', on the other.

The response of a doubly fed induction generator to a terminal voltage drop is a high-frequency phenomenon. It cannot be modelled adequately with PSDS software. This becomes clear when we look at the assumptions on which these programs are based (see Section 25.2 and Chapter 24). Hence, if the protection system that is incorporated into a doubly fed induction generator model is used in PSDS it will react to terminal voltage and not to rotor current. It is thus a simplified representation of the actual protection system.

There are simulations in the literature that show the different responses of reduced-order and complete models (Akhmatov, 2002). However, presently there are no quantitative investigations that analyse the importance of the differences between the reduced-order model and the complete model for the interaction with the system. The latter is, however, the main point of interest in PSDS, so this topic requires further investigation. Further, it must be noted that the protection system of a model of a doubly fed induction generator that will be used in PSDSs can incorporate criteria only with respect to the amplitude and frequency of the terminal voltage. As only the effective value of fundamental harmonic components is studied, it will not be possible to detect phase jumps.

In addition to criteria for detecting whether the wind turbine should be disconnected, the protection system also includes criteria for determining whether the wind turbine can be reconnected, as well as a reconnection strategy. If a protection system of a model of a doubly fed induction generator is used in PSDS, it consists of the following elements:

- a set of criteria for deciding when to disconnect the wind turbine;
- a set of criteria regarding terminal voltage and frequency for determining when to reconnect the wind turbine;
- a reconnection strategy (e.g. a ramp rate at which the power is restored to the value determined by the actual wind speed).

25.6.6 Rotor speed controller model

The speed controller of a variable-speed wind turbine operates as follows:

- The actual rotor speed is measured with a sample frequency f_{ss} (Hz). The sample frequency is in the order of 20 Hz.
- From this value, a set point for the generated power is derived, using the characteristic for the relationship between rotor speed and power.
- Taking into account the actual generator speed, a torque set point is derived from the power set point.
- A current set point is derived from the torque set point, using Equation (25.18).

In a reduced-order model, this current set point is reached immediately as a result of the modelling approach. In practice, the current set point will be used as input to the current control loops and it will take a certain time to reach the desired value of the current, as will be discussed in Chapter 26. However, the time necessary to reach the new current is significantly below the investigated bandwidth of 10 Hz.

We will use the characteristic for the relationship between rotor speed and generator power to arrive at a set point for generated real power. In most cases, the rotor speed is controlled to achieve optimal energy capture, although sometimes other goals may be pursued, particularly noise minimisation. The solid line in Figure 25.8 depicts the relation between rotor speed and power for optimal energy capture. At low wind speeds, the rotor speed is kept at its minimum by adjusting the generator torque. At medium wind speeds, the rotor speed varies proportionally to the wind speed in order to keep the tip speed ratio, λ, at its optimum value. When the rotor speed reaches its nominal value, the generator power is kept at its nominal value as well.

Controlling the power according to this speed – power characteristic, however, causes some problems:

- The desired power is not uniquely defined at nominal and minimal rotor speed.
- If the rotor speed decreases from slightly above nominal speed to slightly below nominal speed, or from slightly above minimal speed to slightly below minimal speed, the change in generated power is very large.

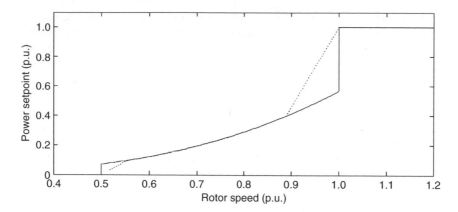

Figure 25.8 Optimal (solid curve) and practical (dotted curve) rotor speed-power characteristic of a typical variable-speed wind turbine

To solve these problems, we use a control characteristic that is similar to those that lead to optimal energy capture. The dotted line in Figure 25.8 depicts this control characteristic. Sometimes, this problem is solved by applying more advanced controller types, such as integral controllers or hysteresis loops (Bossanyi, 2000). The location of the points at which the implemented control characteristic deviates from the control characteristic leading to optimal energy capture is a design choice. If these points lie near the minimal and nominal rotor speed, the maximum amount of energy is extracted from the wind over a wide range of wind speeds, but rotor speed changes near the minimum and nominal rotor speed result in large power fluctuations. If these points lie further from the minimal and nominal rotor speed, the wind speed range in which energy capture is maximal is narrowed, but the power fluctuations near minimal and nominal rotor speed are smaller.

25.6.7 Pitch angle controller model

The pitch angle controller is active only in high wind speeds. In such circumstances, the rotor speed can no longer be controlled by increasing the generated power, as this would lead to overloading the generator and/or the converter. Therefore the blade pitch angle is changed in order to limit the aerodynamic efficiency of the rotor. This prevents the rotor speed from becoming too high, which would result in mechanical damage. The optimal pitch angle is approximately zero below the nominal wind speed. From the nominal wind speed onwards, the optimal angle increases steadily with increasing wind speed, as can be seen in Figure 25.7. Equations (25.6) and (25.7) are used to calculate the impact of the pitch angle, θ, on the performance coefficient. The resulting value can be inserted in Equation (25.5) in order to calculate the mechanical power extracted from the wind.

It should be taken into account that the pitch angle cannot change immediately, but only at a finite rate, which may be quite low because of the size of the rotor blades of modern wind turbines. Blade drives are usually as small as possible in order to save

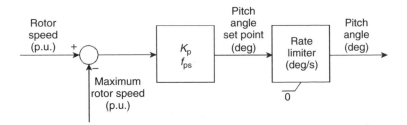

Figure 25.9 Pitch angle controller model. *Note:* K_p is a constant; f_{ps} is the sample frequency of the pitch angle controller

money. The maximum rate of change of the pitch angle is in the order of 3–10 degrees per second, depending on the size of the wind turbine. As the blade pitch angle can change only slowly, the pitch angle controller works with a sample frequency f_{ps}, which is in the order of 1–3 Hz.

Figure 25.9 depicts the pitch angle controller. This controller is a proportional (P) controller. Using this controller type implies that the rotor speed is allowed to exceed its nominal value by an amount that depends on the value chosen for the constant K_p. Nevertheless, we use a proportional controller because:

- a slight overspeeding of the rotor above its nominal value can be tolerated and does not pose any problems to the wind turbine construction;
- the system is never in steady state because of the varying wind speed. The advantage of an integral controller, which can achieve zero steady state error, would therefore be hardly noticeable.

25.6.8 Terminal voltage controller model

A variable-speed wind turbine with a doubly fed induction generator is theoretically able to participate in terminal voltage control, as discussed in Chapter 19. Equation (25.19) shows that the reactive power exchanged with the grid can be controlled, provided that the current rating of the power electronic converter is sufficiently high to circulate reactive current, even at nominal active current.

The first term on the right-hand side of Equation (25.19) determines the net reactive power exchange with the grid, which can be controlled by changing the direct component of the rotor current, i_{dr}. The second term represents the magnetisation of the stator. Equation (25.19) can be rewritten in the following way

$$Q_s = -\frac{L_m u_t (i_{dr,\ magn} + i_{dr,\ gen})}{L_{s\sigma} + L_m} - \frac{u_t^2}{\omega_s (L_{s\sigma} + L_m)}, \tag{25.21}$$

in which i_{dr} has been split in a part magnetizing the generation ($i_{dr,\ magn}$) and a part generating reactive power ($i_{dr,\ gen}$).

It can be seen that $i_{dr,\ magn}$, the rotor current required to magnetise the generator itself, is given by:

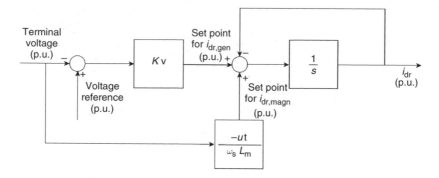

Figure 25.10 Voltage controller model for a wind turbine with a doubly fed induction generator (Type C). *Note*: K_v is the voltage controller constant; u_t is the terminal voltage; ω_s is the angular frequency of the stator; L_m is the mutual inductance; i_{dr} is the direct component of the rotor current; $i_{dr,magn}$ and $i_{dr,gen}$ are the currents to magnetise the generator and in the generator respectively

$$i_{dr,\,magn} = -\frac{u_t}{\omega_s L_m}.$$
(25.22)

The net reactive power exchange between the stator and the grid is then equal to

$$Q_s = -\frac{L_m u_t i_{dr,\,gen}}{L_{s\sigma} + L_m}.$$
(25.23)

Figure 25.10 depicts a terminal voltage controller for a doubly fed induction generator. The value of K_v determines the steady state error and the speed of response. This value should not be set too high, because this leads to stability problems with the controller. Figure 25.10 is, however, only one example of a voltage controller for a wind turbine with a doubly fed induction generator. There are also other possible controller topologies. However, any voltage controller will correspond to the basic principle that the terminal voltage is controlled by influencing the reactive power exchange with the grid.

If the value of K_v is changed to zero, it results in a controller that keeps the power factor equal to one. Currently, this is the dominant mode of operation of wind turbines with doubly fed induction generators, because it requires only small converters and reduces the risk of islanding. Islanding requires a balance between both active and reactive power consumed. If the wind turbines are prevented from generating any reactive power this situation will hardly occur.

25.7 Model of a Direct drive Wind Turbine

Figure 25.11 depicts the general structure of a model of a variable-speed wind turbine with a direct-drive synchronous generator (Type D). Only the generator model and the voltage controller model are discussed below. The wind speed model is identical to that in the case of a constant-speed wind turbine model, described in Section 25.5.2. The rotor model as well as the rotor speed and pitch angle controllers are identical to those used in the doubly fed induction generator and were described in Sections 25.6.6 and 25.6.7, respectively. The converter and the protection system of a wind turbine with a

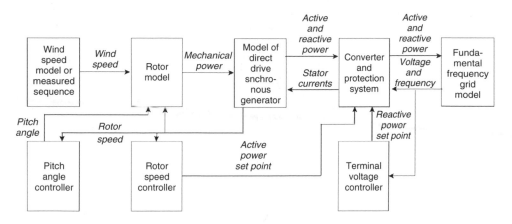

Figure 25.11 General structure of a model of a variable-speed wind turbine with a direct-drive synchronous generator (Type D)

doubly fed induction generator are different from that of a direct-drive synchronous generator. However, because of the simplifying assumptions used in models for dynamic simulations, the analyses in Sections 25.6.4 and 25.6.5 also hold for the model of the converter and the protection system of a direct-drive wind turbine. For the reasons mentioned in Section 25.5.1, we will not discuss the grid model.

25.7.1 Generator model

According to Kundur (1994), the voltage equations of a wound rotor synchronous generator in the dq reference frame, taking into account the assumptions in Section 25.4, are:

$$\left.\begin{aligned}
u_{ds} &= -R_s i_{ds} - \omega_m \psi_{qs} + \frac{d\psi_{ds}}{dt}, \\
u_{qs} &= -R_s i_{qs} + \omega_m \psi_{ds} + \frac{d\psi_{qs}}{dt}, \\
u_{fd} &= R_{fd} i_{fd} + \frac{d\psi_{fd}}{dt}.
\end{aligned}\right\} \quad (25.24)$$

The subscript fd indicates field quantities. Note that for the stator equations the generator convention is used (i.e. positive currents are outputs). The flux equations are:

$$\left.\begin{aligned}
\psi_{ds} &= -(L_{dm} + L_{\sigma s})i_{ds} + L_{dm}i_{fd}, \\
\psi_{qs} &= -(L_{qm} + L_{\sigma s})i_{qs}, \\
\psi_{fd} &= L_{fd}i_{fd}.
\end{aligned}\right\} \quad (25.25)$$

All quantities in Equations (25.24) and (25.25) are in per unit values.

In the case of a permanent magnet rotor, the expressions for u_{fd} and ψ_{fd} in Equations (25.24) and (25.25) disappear because they refer to field quantities, and expression for ψ_{ds} in Equations (25.25) becomes

$$\psi_{ds} = -(L_{ds} + L_{\sigma s})i_{ds} + \psi_{pm}, \qquad (25.26)$$

in which ψ_{pm} is the amount of flux of the permanent magnets mounted on the rotor that is coupled to the stator winding. When neglecting the $d\psi/dt$ terms in the stator voltage equations, the voltage flux relationships become:

$$\left.\begin{aligned} u_{ds} &= -R_s i_{ds} + \omega_m(L_{s\sigma} + L_{qm})i_{qs}, \\ u_{qs} &= -R_s i_{qs} - \omega_m(L_{s\sigma} + L_{dm})i_{ds}, \\ u_{fd} &= R_{fd} i_{fd} + \frac{d\psi_{fd}}{dt}. \end{aligned}\right\} \qquad (25.27)$$

The $d\psi/dt$ terms in the expressions for u_{ds} and u_{qs} are neglected because the associated time constants are small, and taking them into account would result in the need to develop a detailed representation of the power electronic converter. That would include phenomena that we are not interested in at this point. For a complete model of this wind turbine type, including the $d\psi/dt$ terms in the stator equations and a full converter model, see for instance, Chen and Spooner (2001).

The following equation gives the electromechanical torque:

$$T_e = \psi_{ds}i_{qs} - \psi_{qs}i_{ds}. \qquad (25.28)$$

Based on this equation, the set point for the stator currents can be calculated from a torque set point generated by the rotor speed controller. The equation of motion is given by Equation (24.14) (page 545). The active and reactive power of a synchronous generator are given by:

$$\left.\begin{aligned} P_s &= u_{ds}i_{ds} + u_{qs}i_{qs}, \\ Q_s &= u_{qs}i_{ds} - u_{ds}i_{qs}. \end{aligned}\right\} \qquad (25.29)$$

It has to be emphasised that the generator is fully decoupled from the grid by the power electronic converter. Therefore, the power factor of the generator does not affect the reactive power factor at the grid connection. The latter is determined by the grid side of the converter and not by the operating point of the generator. The expression for Q_s in Equations (25.29) is therefore of limited interest when studying the grid interaction but is important when dimensioning the converter. The generator parameters are given in Table 25.4.

25.7.2 Voltage controller model

The voltage controller applied in a direct-drive wind turbine is different from that in a wind turbine with a doubly fed induction generator because the generator is fully decoupled from the grid. It is therefore not the generator that generates active power and controls the terminal voltage but the power electronic converter. The generated reactive power is given by the expression for Q_c in Equations (25.20). The model of the voltage controller is depicted in Figure 25.12. It assumes that the terminal voltage u_t is equal to u_{qc} in Equations (25.20). Similar to the terminal voltage controller of a doubly

Table   25.4 Simulated synchronous generator parameters

Generator characteristic	Value
Number of poles, p	80
Generator speed (rpm)	9–19
Mutual inductance in d-axis, L_{dm} (p.u.)	1.21
Mutual inductance in q-axis, L_{qm} (p.u.)	0.606
Stator leakage inductance, $L_{\sigma s}$ (p.u.)	0.121
Stator resistance, R_s (p.u.)	0.06
Field inductance, L_{fd} (p.u.)	1.33
Field resistance, R_{fd} (p.u.)	0.0086
Inertia constant, H_m (s)	1.0

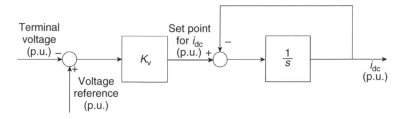

Figure   25.12 Voltage controller model for wind turbine with direct drive synchronous generator. *Note*: K_v is the voltage controller constant; i_{dc} is the direct component of the converter current;.

fed induction generator, there may be alternative controller topologies here, too. However, they are all based on the principle that the reactive power exchange with the grid has to be controlled in order to influence the terminal voltage. To simulate wind turbines operating at unity power factor, i_{dc} must be kept at zero, and the voltage controller can be removed.

25.8 Model Validation

25.8.1 Measured and simulated model response

In this section, the model's responses to a measured wind speed sequence will be analysed and then compared with measurements. The measurements were obtained from wind turbine manufacturers under a confidentiality agreement. Therefore, all values except wind speed and pitch angle are in per unit and their base values are not given. We used Matlab™ to obtain the simulation results.

Figure 25.13(a) depicts a measured wind speed sequence. In Figures 25.13(b)–25.13(d), respectively, the simulated rotor speed, pitch angle if applicable and output power are depicted for each of the wind turbine types. Figure 25.14(a) shows three measured wind speed sequences. In Figures 25.14(b) and 25.14(c), respectively, the measured rotor speed and pitch angle of a variable-speed wind turbine with a doubly

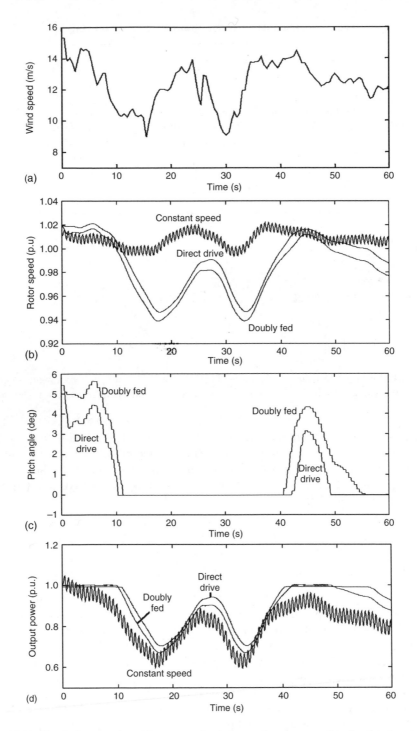

Figure 25.13 Simulation results: (a) measured wind speed sequence; (b) simulated rotor speed; (c) simulated pitch angle; and (d) simulated output power, by wind turbine type. *Note*: Constant speed = constant-speed wind turbine; Direct drive = variable-speed wind turbine with a direct-drive synchronous generator; Doubly fed = variable-speed wind turbine with a doubly fed induction generator

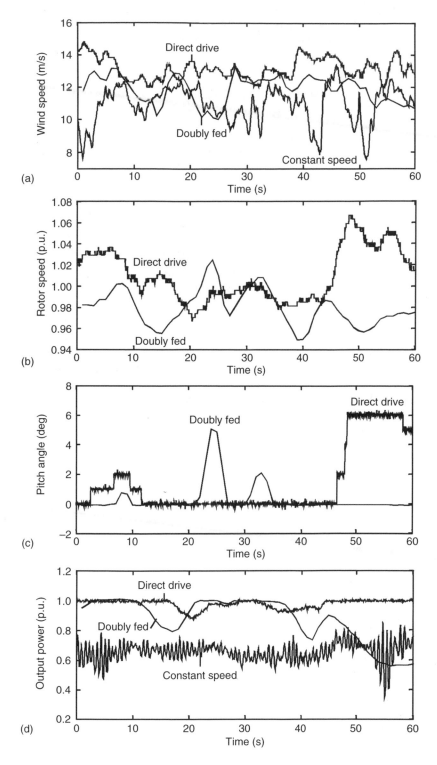

Figure **25.14** Measurement results: (a) measured wind speed sequence; (b) simulated rotor speed; (c) simulated pitch angle; and (d) simulated output power. *Note*: See Figure 25.13

Table   25.5 Simulated wind turbine parameters

Turbine characteristic	Value
Rotor speed (rpm)[a]	17
Minimum rotor speed (rpm)[b]	9
Nominal rotor speed (rpm)[b]	18
Rotor diameter (m)	75
Area swept by rotor, A_r (m^2)	4418
Nominal power (MW)	2
Nominal wind speed (m/s)[a]	15
Nominal wind speed (m/s)[b]	14
Gearbox ratio[a]	1:89
Gearbox ratio[b]	1:100
Inertia constant, H_{wr} (s)	2.5
Shaft stiffness, K_s (p.u./el. rad)[a]	0.3

[a] For a constant-speed wind turbine.
[b] For a variable-speed wind turbine.

fed induction generator and of a variable-speed wind turbine with a direct-drive synchronous generator are depicted. Figure 25.14(d) shows the measured output power of all three turbine types. The rotor speed of the constant speed wind turbine was not measured because the small rotor speed variations are difficult to measure. This quantity is therefore not included in the graph.

The characteristics of the wind turbine are given in Table 25.5. Table 25.3 includes the generator parameters of the induction generator used in the constant-speed wind turbine and the variable-speed wind turbine with a doubly fed induction generator, and Table 25.4 provides the parameters of the direct-drive synchronous generator.

25.8.2 Comparison of measurements and simulations

For two reasons, the available measurements cannot be used for a direct, quantitative validation of the models. First, the wind speed is measured with a single anemometer, whereas the rotor has a large surface. Second, the measured wind speed is severely disturbed by the rotor wake, because the anemometer is located on the nacelle. Therefore, the wind speed that is measured with a single anemometer is not an adequate measure for the wind speed acting on the rotor as a whole. It is thus not possible to feed a measured wind speed sequence into the model in order to compare measured with simulated response.

The discrepancy between the wind speed measured with a single anemometer and the aggregated wind speed acting on the rotor is clearly illustrated by the behaviour of the variable-speed wind turbine with a doubly fed induction generator after about 25 s as observed in Figure 25.14. The wind speed decreases from about 12 to 10 m/s at that time. However, rotor speed, pitch angle and generated power increase. From the fact that the observed behaviour is physically impossible, it can be concluded that the wind speed measured by the anemometer is not a good indicator of the wind speed acting on the rotor as a whole.

Thus, although it would be possible to use the wind speed sequence measured by the anemometer as the model's input, it cannot be used for a quantitative comparison of the measured with the simulated response to the wind speed sequence for validating the model. Therefore, we carry out only a qualitative comparison, which leads to the following conclusions.

The simulation results depicted in Figure 25.13 show the following.

- In particular, short-term output power fluctuations (i.e. in the range of seconds) are more severe in the case of constant-speed wind turbines than in the case of the two variable-speed wind turbine types [Figure 25.13(d)]. The reason is that in the variable-speed wind turbines the rotor functions as an energy buffer.
- The response of the variable-speed wind turbine types is similar. This is because their behaviour is, to a large extent, determined by the controllers, which are identical.

These findings also apply to the measurements depicted in Figure 25.14.

A comparison of the simulated and measured responses shows that:

- The range of the measured and simulated rotor speed fluctuations of the variable-speed turbines are similar (they fluctuate over a bandwidth of about 0.1 p.u.); see Figures 25.13(b) and 25.14(b).
- Measured and simulated pitch angle behaviour are similar with respect to the rate of change (approximately 3° to 5° per second) and the minimum (approximately 0°) and maximum value (approximately 6°); see Figures 25.13(c) and 25.14(c).
- The range of the measured and simulated output power fluctuations of constant-speed wind turbines (Type A) and that of wind turbines with doubly fed induction generators (Type C) are similar (they fluctuate over a bandwidth of approximately 0.3 to 0.4 p.u.); see Figures 25.13(d) and 25.14(d).
- The rate of change of the measured output power fluctuations of the constant-speed wind turbine [Type A; see Figure 25.14(d)] differs from that of the simulated output power fluctuations [see Figure 25.13(d)]. However, the correlation between the measured wind speed and output power is rather weak for these wind turbines. The observed discrepancies between measurements and simulation are therefore probably caused by inaccuracies in the measurements rather than by the model.
- The range of the measured output power fluctuations of wind turbines with direct-drive synchronous generators (Type D – approximately 0.2 p.u. [see Figure 25.14(d)] – differs from that of the simulated output power fluctuations – approximately 0.4 p.u. [see Figure 25.13(d)]. This is probably caused by the fact that in the measurements, the direct-drive wind turbine is exposed only to rather high wind speeds, whereas in the simulation there are also lower wind speeds.

Although a quantitative validation of the models is not possible with the available measurements, this qualitative comparison of measured and simulated responses gives confidence in the accuracy and usability of the derived models. It shows that the assumptions and simplifications applied in modelling the rotor, the generator and the controllers have rather limited consequences, and it can be assumed that they do not affect the simulation results to a larger extent than do other sources of uncertainty, such as the system load and the committed generators.

25.9 Conclusions

This chapter presented models of the three most important current wind turbine types. The models match the simplifications applied in PSDSs and include all subsystems that determine the grid interaction. Therefore, they are suitable for analysing the impact of large-scale connection of wind power on the dynamic behaviour of electrical power systems.

The response of the models to a measured wind speed sequence was investigated, and the measurements show an acceptable degree of correspondence. This gives confidence in the derived models and shows that the results of the applied simplifications are acceptable.

References

[1] Akhmatov, V. (2002) 'Variable-speed Wind Turbines with Doubly-fed Induction Generators, Part I: Modelling in Dynamic Simulation Tools', *Wind Engineering* 26(2) 85–108.

[2] Akhmatov, V., Knudsen, H., Nielsen, A. H. (2000) 'Advanced Simulation of Windmills in the Electric Power Supply', *International Journal of Electrical Power & Energy Systems* 22(6) 421–434.

[3] Anderson, P. M., Bose, A. (1983) 'Stability Simulation of Wind Turbine Systems', *IEEE Transactions on Power Apparatus and Systems* 102(12) 3791–3795.

[4] Anderson, P. M., Agrawal, B. L., Van Ness J. E. (1990) *Subsynchronous Resonance in Power Systems*, IEEE Press, New York.

[5] Bossanyi, A. A. (2000) 'The Design of Closed Loop Controllers for Wind Turbines', *Wind Energy* 3(3) 149–163.

[6] Chen, Z., Spooner, E. (2001) 'Grid Power Quality with Variable Speed Wind Turbines', *IEEE Transactions on Energy Conversion* 16(2) 148–154.

[7] Fujimitsu, M., Komatsu, T., Koyanagi, K., Hu, K., Yokoyama. R. (2000) 'Modeling of Doubly-fed Adjustable-speed Machine for Analytical Studies on Long-term Dynamics of Power System', in *Proceedings of PowerCon, Perth*, pp.25–30.

[8] Grainger, J. J., Stevenson Jr, W. D. (1994) *Power System Analysis*, McGraw-Hill, New York.

[9] Hatziargyriou, N. (Ed.) (2001) *Modeling New Forms of Generation and Storage*, Cigré Task Force 38.01.10, Paris.

[10] Heier, S. (1998) *Grid Integration of Wind Energy Conversion Systems*, John Wiley & Sons Ltd, Chicester, UK.

[11] Hinrichsen, E. N., Nolan, P. J. (1982) 'Dynamics and Stability of Wind Turbine Generators', *IEEE Transactions on Power Apparatus and Systems* 101(8) 2640–2648.

[12] Krüger, T., Andresen, B. (2001) 'Vestas OptiSpeed – Advanced Control Strategy for Variable Speed Wind Turbines', in *Proceedings of the European Wind Energy Conference, Copenhagen, Denmark*, pp. 983–986.

[13] Kundur, P. (1994) *Power System Stability and Control*, McGraw-Hill, New York.

[14] Panofsky, H. A., Dutton, J. A. (1984) *Atmospheric Turbulence: Models and Methods for Engineering Applications*, John Wiley & Sons Inc., New York.

[15] Papathanassiou, S. A., Papadopoulos, M. P. (1999) 'Dynamic Behavior of Variable Speed Wind Turbines under Stochastic Wind', *IEEE Transactions on Energy Conversion* 14(4) 1617–1623.

[16] Patel, M. R. (2000) *Wind and Solar Power Systems*, CRC Press, Boca Raton, FL.

[17] Pedersen, J. K., Akke, M., Poulsen, N. K., Pedersen, K. O. H. (2000) 'Analysis of Wind Farm Islanding Experiment', *IEEE Transactions on Energy Conversion* 15(1) 110–115.

[18] Petru, T., Thiringer, T. (2000) 'Active Flicker Reduction from a Sea-based 2.5 MW Wind Park Connected to a Weak Grid', presented at the Nordic Workshop on Power and Industrial Electronics (NORpié/2000), Aalborg, Denmark.

[19] Shinozuka, M., Jan, C.-M. (1972) 'Digital Simulation of Random Processes and its Applications', *Journal of Sound and Vibration* 25(1) 111–128.

[20] Simiu, E., Scanlan, R. H. (1986) *Wind Effects on Structures: An Introduction to Wind Engineering*, 2nd edn, John Wiley & Sons Inc., New York.

[21] Slootweg J. G., Polinder H., Kling, W. L. (2001) 'Initialization of Wind Turbine Models in Power System Dynamics Simulations', in *Proceedings of 2001 IEEE Porto Power Tech Conference, Porto*.

[22] SPTI (Shaw Power Technologies Inc.) (1997) Online documentation of PSS/E^{TM} (Power System Simulator for Engineers) 25, USPT, Schenectady, NY.

[23] Wasynczuk, O., Man, D. T., Sullivan, J. P. (1981) 'Dynamic Behavior of a Class of Wind Turbine Generators during Random Wind Fluctuations', *IEEE Transactions on Power Apparatus and Systems* 100(6) 2837–2845.

26

High-order Models of Doubly-fed Induction Generators

Eva Centeno López and Jonas Persson

26.1 Introduction

This chapter focuses on the modelling of doubly-fed induction generators (DFIGs), typically used in Type C wind turbines as defined in Section 4.2.3. This generator model is currently (2005) one of the most researched wind generator models and there is growing interest in understanding its behaviour. The term 'doubly fed' indicates that the machine is connected to the surrounding power system at two points; directly via the stator side, and also via the rotor side through a voltage source converter (VSC).

We start the chapter by presenting a model of a DFIG at the most detailed level, namely a 6th-order state-variable model including both rotor and stator electromagnetic transients. In this way, the model covers fast transient phenomena and becomes an appropriate model for studies on instantaneous value modelling of the power system. When simulating a DFIG with a reduced-order model, fast transients are neglected in the simulation. A reduced-order model can be used in transient studies that assume that all quantities vary with fundamental frequency and that neglect grid transients. This assumption may be appropriate, depending on which types of phenomena the model should be valid for.

As mentioned above, the 6th-order DFIG model is developed in detail in this chapter. The VSC of the rotor is not modelled with a detailed modulation scheme, though. It is assumed that the switching frequency is infinite. In spite of this simplification, we take into account the limitations in the voltage generation that the converter imposes due to the DC link. These limitations will be implemented as limitations to the voltage and torque controllers. The VSC is outlined in this chapter.

Wind Power in Power Systems Edited by T. Ackermann
© 2005 John Wiley & Sons, Ltd   ISBN: 0-470-85508-8 (HB)

We will describe a sequencer that includes different modes of operation of a DFIG. The need for such a sequencer arises from the fact that whenever there is a fault in the interconnected power system, the system of the DFIG must be able to handle the resulting high currents without causing any damage to the equipment.

This chapter will cover neither aerodynamic modelling of wind turbines nor gearbox representation, since these were described in Chapter 24. The DFIG model does not include any magnetic saturation, air-gap harmonics or skin effects either.

The described DFIG model has been implemented as a user model in the power system simulation software Simpow® developed by STRI (see Centeno, 2000; Fankhauser *et al.*, 1990). Simpow contains both possibilities to simulate electromagnetic transients and electromechanical transients, however, in this chapter we will simulate in electromagnetic transients mode.

26.2 Advantages of Using a Doubly-fed Induction Generator

When using conventional squirrel cage asynchronous machines (i.e. Type A wind turbines), very high magnetising currents are drawn from the power grid when recovering from a nearby fault in the power system. If the power system is weak and cannot provide a sufficient magnetising current in the postfault transient state, the squirrel cage asynchronous machine keeps on accelerating. The asynchronous machine draws as much magnetising current as possible from the grid, and a severe voltage drop takes place. The voltage drop will stop only once the protection system disconnects the wind turbine from the grid. A rapid power reduction can be used to deal with this problem (for an explanation and a demonstration, see Sections 24.4.5.1 and 29.3, respectively).

An alternative is to use an asynchronous generator with its rotor connected to a VSC via slip rings, a so-called doubly fed induction generator (DFIG; i.e. a Type C wind turbine). In this concept, the VSC is connected to a control system that determines the voltage that the VSC impresses onto the rotor of the induction generator. This concept also provides variable-speed capability, which makes it possible to optimise power production, as explained in Section 24.3.1. By controlling the rotor voltage, the VSC control can control the current that is drawn from the grid. The voltage that the VSC impresses onto the rotor is determined by speed, torque and voltage. These are the controllers of which the VSC control consists.

26.3 The Components of a Doubly-fed Induction Generator

The DFIG consists of the components depicted in Figure 26.1, which shows the two connections of the machine to the surrounding network that give rise to the term 'doubly fed'. The VSC consists of two converters with a DC link between them. The VSC connects the rotor to the network, which is labelled 'Grid' in Figure 26.1.

The components of the DFIG are:

- wind turbine;
- gearbox;
- induction generator;
- converters (VSC);

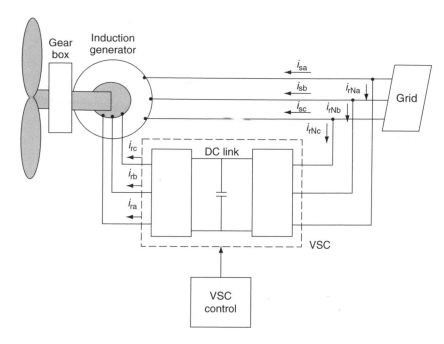

Figure 26.1 Doubly fed induction generator. *Note*: VSC = voltage source converter; i = current; subscript r indicates a rotor quantity; subscript s indicates stator quantity; subscripts *a, b* and *c* indicate phase *a, b* and *c* winding axes; subscript N indicates the grid side

- DC link (VSC);
- voltage control;
- speed control;
- torque control.

The aim of the VSC shown in Figure 26.1 is to provide the rotor voltage. The VSC control determines the rotor voltage and uses speed, torque and voltage controllers for this purpose.

Regarding the above list, we will assume here that the wind turbine provides a constant mechanical torque T_{MECH} to the generator, and we will neglect the gearbox. Chapter 24 describes the modelling of the wind turbine. The wind representation (gusts, turbulence and random variations) is discussed briefly in Sections 24.6.2.4 and 24.6.2.5 and is described in more detail in Section 25.5.2. The wind turbine representation, and consequently also the wind representation, are neglected in this chapter.

We will derive the equations of the induction generator. The converters and the DC link will not be modelled in detail, though. Instead, the VSC is represented by its output – the rotor voltage – and this is modelled as the output of the VSC control.

26.4 Machine Equations

In this section, the basic equations of the machine are presented (for the equations of induction generator, VSC and sequencer, see also Centeno, 2000). Before we actually describe these equations, we will present the vector method, which is used to represent time-dependent magnitudes, and also the reference frame that will be used throughout this chapter.

26.4.1 The vector method

The vector method is a simple but mathematically precise method for handling the transient performance of electrical machines (Ängquist, 1984). By using the vector method the number of equations required to describe the machine's behaviour is decreased. This is a function of the fact that the reactances of the machine are constant in time when representing the machine in its reference frame. Furthermore, if stator and rotor vector quantities are expressed in the same reference frame, time-varying coefficients in the differential equations are avoided. This is because of the flux linkages between stator and rotor and simplifies the solution to such a system of differential equations. For the later target (i.e. to refer stator and rotor equations to the same reference frame), the so-called Park's transformation is used.

We assume balanced voltages and nonground connection points, which means we have zero sequence free quantities. That reduces the degree of freedom to two, and it is therefore possible to work in a two-dimensional plane. Two orthogonal axes are defined, the d and q axis. The d axis is called the *direct* axis and lies collinear to the rotor phase a winding. The q axis is called the *quadrature* axis and lies perpendicular to and leading the d axis. We will work in a rotor-fixed reference frame and therefore the equations for the rotor do not have to be transformed. However, the stator equations have to be transformed. There are two reasons for choosing the rotor coordinate system. First, this applies to much lower frequencies and thus the resolution of the differential equation system will be simpler. The second reason is that the rotor circuit is connected to a VSC and its control equipment, which means that the converter needs variables in the rotor coordinate system as input. The reference frames are illustrated in Figure 26.2, including the space vector, $\overline{S}(t)$, given by:

$$\overline{S}(t) = S_d(t) + \mathrm{j}S_q(t). \tag{26.1}$$

where subscripts d and q indicate the direct and quadrature axis components of the space vector, respectively, and $\mathrm{j} = \sqrt{-1}$.

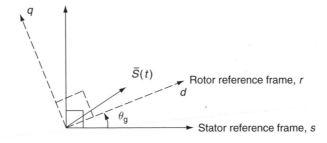

Figure 26.2 Reference frames. *Note*: $\overline{S}(t) =$ space vector; $d =$ direct axis; $q =$ quadrature axis; $\theta_\mathrm{g} =$ the angle between the two reference frames

We can introduce subscripts to refer $\overline{S}(t)$ to the two different frames as follows:

- $\overline{S}_s\,(t)$ is the space vector referred to the stator reference frame;
- $\overline{S}_r\,(t)$ is the space vector referred to the rotor reference frame.

In order to decrease the number of indices, the time-dependence (t) is included for all quantities but is not indexed in subsequent formulae.

The coordinate transformation of $\overline{S}$ from rotor to stator reference frame can be expressed as follows:

$$\overline{S}_s = \overline{S}_r e^{j\theta_g} = \left(S_{rd} + jS_{rq}\right)e^{j\theta_g}, \tag{26.2}$$

where θ_g is the angle between the two reference frames, which in our application is the rotor angle of the machine. S_{rd} is the real part of $\overline{S}_r$, and S_{rq} is the imaginary part of $\overline{S}_r$. The above transformation is necessary if the connection of the machine is to be described for the power system. The power system requires all quantities to be expressed in the stator reference frame.

Identifying real and imaginary parts in Equation (26.2), we obtain:

$$S_{sd} = S_{rd}\cos\theta_g - S_{rq}\sin\theta_g; \tag{26.3}$$
$$S_{sq} = S_{rq}\cos\theta_g + S_{rd}\sin\theta_g. \tag{26.4}$$

Furthermore, the inverse transformation is required in order to go from the stator to the rotor reference frame. The equations for this transformation are as follows:

$$\overline{S}_r = \overline{S}_s e^{-j\theta_g}; \tag{26.5}$$

identifying real and imaginary parts in Equation (26.5), we obtain:

$$S_{rd} = S_{sd}\cos\theta_g + S_{sq}\sin\theta_g; \tag{26.6}$$
$$S_{rq} = S_{sq}\cos\theta_g - S_{sd}\sin\theta_g. \tag{26.7}$$

This coordinate transformation will be carried out for all stator quantities (i.e. for the voltages, currents and fluxes) as all equations will be formulated and solved in the rotor-fixed reference frame.

The projection of the space vectors over the phase axis has to coincide with the corresponding phase quantities. This requirement results in the following relationship between the phase and dq components for the rotor:

$$S_{ra} = \mathrm{Re}[\overline{S}_r] = S_{rd}; \tag{26.8}$$

$$S_{rb} = \mathrm{Re}\left[\overline{S}_r e^{-j2\pi/3}\right] = -\frac{1}{2}S_{rd} + \frac{\sqrt{3}}{2}S_{rq}; \tag{26.9}$$

$$S_{rc} = \mathrm{Re}\left[\overline{S}_r e^{j2\pi/3}\right] = -\frac{1}{2}S_{rd} - \frac{\sqrt{3}}{2}S_{rq}. \tag{26.10}$$

It should be noted that the real axis (i.e. the d axis) is collinear to the phase a winding axis. From the equations above, it can be derived that the space vector can be expressed as a function of phase quantities. For the rotor, we arrive at the following relationship between dq and phase components:

$$S_{rd} + \mathrm{j}S_{rq} = \frac{2}{3}\left[S_{ra} + \left(-\frac{1}{2}+\mathrm{j}\frac{\sqrt{3}}{2}\right)S_{rb} + \left(-\frac{1}{2}-\mathrm{j}\frac{\sqrt{3}}{2}\right)S_{rc}\right]. \qquad (26.11)$$

Identifying real and imaginary parts in Equation (26.11), we obtain:

$$S_{rd} = \frac{1}{3}(2S_{ra} - S_{rb} - S_{rc}), \qquad (26.12)$$

$$S_{rq} = \frac{1}{\sqrt{3}}(S_{rb} - S_{rc}). \qquad (26.13)$$

From Equations (26.12) and (26.13) it can be concluded that instantaneous three-phase quantities [real-valued $S_a(t)$, $S_b(t)$ and $S_c(t)$] can be transformed into a unique space vector with real and imaginary parts $S_d(t)$ and $S_q(t)$, respectively.

Note that the complex magnitudes described above are different from the complex magnitudes used in the so-called $\mathrm{j}\omega$ method. The $\mathrm{j}\omega$ method assumes the complex magnitudes to rotate with constant speed, ω, whereas here ω is allowed to vary freely.

26.4.2 Notation of quantities

In the following, we use the subscript italic s for quantities that are expressed in the stator reference frame and the subscript italic r for quantities that are expressed in the rotor reference frame. Superscript upright s represents the index of a stator quantity and superscript upright r represents the index of a rotor quantity. Therefore, $\bar{u}_r^{\mathrm{s}}$ represents the stator voltage in the rotor reference frame, and $\bar{u}_s^{\mathrm{s}}$ represents the stator voltage in the stator reference frame. The notation $\bar{u}$ represents a complex quantity, in this case a voltage.

26.4.3 Voltage equations of the machine

The stator voltage $\bar{u}_s^{\mathrm{s}}$, referred to the stator frame is:

$$\bar{u}_s^{\mathrm{s}} = \vec{i}_s^{\mathrm{s}}r^{\mathrm{s}} + \frac{\mathrm{d}\bar{\psi}_s^{\mathrm{s}}}{\mathrm{d}t}, \qquad (26.14)$$

where $\vec{i}_s^{\mathrm{s}}$ is the complex stator current and $\bar{\psi}_s^{\mathrm{s}}$ is the complex stator flux. r^{s} is the stator winding resistance.

The coordinate transformation of any quantity between the rotor reference frame and the stator reference frame was derived in Equation (26.2). By applying this equation to the three complex quantities in Equation (26.14), we transform the equation from the stator reference frame to the rotor reference frame. The result is:

$$\vec{i}_s^{\mathrm{s}} = \vec{i}_r^{\mathrm{s}}\mathrm{e}^{\mathrm{j}\theta_{\mathrm{g}}}, \qquad (26.15)$$

$$\bar{u}_s^{\mathrm{s}} = \bar{u}_r^{\mathrm{s}}\mathrm{e}^{\mathrm{j}\theta_{\mathrm{g}}}, \qquad (26.16)$$

$$\bar{\psi}_s^{\mathrm{s}} = \bar{\psi}_r^{\mathrm{s}}\mathrm{e}^{\mathrm{j}\theta_{\mathrm{g}}}. \qquad (26.17)$$

Substituting Equations (26.15) – (26.17) into Equation (26.14), we get

$$\overline{u}_r^s e^{j\theta_g} = \overline{i}_r^s e^{j\theta_g} r^s + \frac{d\left(\overline{\psi}_r^s e^{j\theta_g}\right)}{dt}. \tag{26.18}$$

The last term of Equation (26.18) is expanded as follows:

$$
\begin{aligned}
\frac{d\left(\overline{\psi}_r^s e^{j\theta_g}\right)}{dt} &= e^{j\theta_g} \frac{d\left(\overline{\psi}_r^s\right)}{dt} + \overline{\psi}_r^s \frac{d\left(e^{j\theta_g}\right)}{dt} \\
&= e^{j\theta_g} \frac{d\left(\overline{\psi}_r^s\right)}{dt} + \overline{\psi}_r^s \frac{d\left(j\theta_g\right)}{dt} e^{j\theta_g} = e^{j\theta_g} \frac{d\left(\overline{\psi}_r^s\right)}{dt} + j\omega_g \overline{\psi}_r^s e^{j\theta_g},
\end{aligned}
\tag{26.19}
$$

where ω_g is the speed of the machine. Inserting Equation (26.19) into Equation (26.18), we obtain:

$$\overline{u}_r^s e^{j\theta_g} = \overline{i}_r^s e^{j\theta_g} r^s + e^{j\theta_g} \frac{d\left(\overline{\psi}_r^s\right)}{dt} + j\omega_g \overline{\psi}_r^s e^{j\theta_g}. \tag{26.20}$$

By multiplying both sides of Equation (26.20) by $e^{-j\theta_g}$, we arrive at the final stator voltage equation in the rotor-fixed coordinate frame:

$$\overline{u}_r^s = \overline{i}_r^s r^s + \frac{d\left(\overline{\psi}_r^s\right)}{dt} + j\omega_g \overline{\psi}_r^s. \tag{26.21}$$

The voltage equation for the rotor circuit in the rotor reference frame can be directly expressed as

$$\overline{u}_r^r = \overline{i}_r^r r_r + \frac{d\left(\overline{\psi}_r^r\right)}{dt}, \tag{26.22}$$

where $\overline{u}_r^r$ is the complex rotor voltage; $\overline{i}_r^r$ is the complex rotor current; $\overline{\psi}_r^r$ is the complex rotor flux; and r^r is the rotor winding resistance.

26.4.3.1 Per unit system of the machine

So far, all equations have been written and deduced using real values of the variables and parameters involved. In general, it is very useful to work in a per unit (p.u.) system, which will be developed in this section. In order to arrive at a per unit system, we have to define a base system first (see also Section 24.5). There is a number of base systems from which we can choose. We will, however, use peak phase voltage and peak phase current as the base values. The other base values can be calculated from these two base values, as shown in Table 26.1. It has to be stressed that the base power, S_{base}, is the three-phase power of the induction generator, S_N, (where a subscript N indicates the nominal value).

Equations (26.21) and (26.22) in p.u. of the machine base are:

$$\overline{u}_r^s = \overline{i}_r^s r^s + \frac{1}{\omega_N} \frac{d\left(\overline{\psi}_r^s\right)}{dt} + j\frac{\omega_g}{\omega_N} \overline{\psi}_r^s, \tag{26.23}$$

Table   26.1 Per unit bases

Quantity	Expression
U_{base} (kV)	$U_{base} = U_N \frac{\sqrt{2}}{\sqrt{3}}$
I_{base} (kA)	$I_{base} = \frac{S_N}{U_N} \frac{\sqrt{2}}{\sqrt{3}}$
S_{base} (MVA)	$S_{base} = S_N = \frac{2}{3} U_{base} I_{base}$
Z_{base} (Ω)	$Z_{base} = \frac{U_{base}}{I_{base}} = \frac{U_N^2}{S_N}$
ω_{base} (rad/s)	$\omega_{base} = \omega_N = 2\pi f_N$
T_{base} (MWs)	$T_{base} = \frac{P_{base}}{\omega_{base}} = \frac{S_{base}}{\omega_{base}} = \frac{S_N}{\omega_N}$
ψ_{base} (Wb)	$\psi_{base} = \frac{U_{base}}{\omega_{base}} = \frac{U_N}{\omega_N} \frac{\sqrt{2}}{\sqrt{3}}$

Note: U = voltage; I = current; S = apparent power; Z = impedance; ω = rotational speed; T = torque; ψ = flux; subscript 'base' indicates the base value for the per unit system; subscript N indicates the nominal value; U_N is the nominal root mean square (RMS) value of the phase-to-phase voltage of the machine.

$$\bar{u}_r^r = \bar{i}_r^r r^r + \frac{1}{\omega_N} \frac{d\left(\overline{\psi}_r^r\right)}{dt}. \tag{26.24}$$

Note that the speed of the machine ω_g in Equation (26.23) is in [radians per second]. In our equations, machine speed, ω_g, as well as turbine speed, ω_m, will be measured in physical units. The angular speed, ω_N, in Equations (26.23) and (26.24) is equal to $2\pi f_N$, where f_N is the power frequency (e.g. 50 Hz).

26.4.4 Flux equations of the machine

In order to solve the system of equations for the induction generator it is necessary in addition to formulate equations representing the relationship between currents and fluxes. In the following stator flux equation, the stator current and the stator flux are referred to the rotor reference frame:

$$\overline{\psi}_r^s = \bar{i}_r^s x^s + \bar{i}_r^r x_m, \tag{26.25}$$

where x^s is the stator self-reactance, equal to

$$x^s = x_l^s + x_m, \tag{26.26}$$

and where x_l^s is the leakage reactance of the stator and x_m is the mutual reactance between the stator and rotor windings. Equation (26.25) is given in p.u.

It can be shown that the values of the reactances and inductances coincide if they are given in p.u. Therefore, the reactances in Equation (26.25) can be replaced with the inductances given on the same p.u. base.

In the following rotor flux equation, the stator current is referred to the rotor frame:

$$\overline{\psi}_r^r = \overline{i}_r^s x_m + \overline{i}_r^r x^r, \tag{26.27}$$

where x_m is the mutual reactance between the stator and rotor windings and x^r is the rotor self-reactance, which is equal to

$$x^r = x_l^r + x_m, \tag{26.28}$$

and where x_l^r is the leakage reactance of the rotor.

By combining Equation (26.25) with Equation (26.27), the stator and rotor currents can be expressed as functions of stator and rotor fluxes:

$$\overline{i}_r^s = \left(\overline{\psi}_r^s - \frac{x_m}{x^r}\overline{\psi}_r^r\right) \Big/ \left[x^s\left(1 - \frac{x_m^2}{x^s x^r}\right)\right], \tag{26.29}$$

$$\overline{i}_r^r = \left(\overline{\psi}_r^r - \frac{x_m}{x^s}\overline{\psi}_r^s\right) \Big/ \left[x^r\left(1 - \frac{x_m^2}{x^s x^r}\right)\right], \tag{26.30}$$

and, by introducing the total leakage factor, σ, given by

$$\sigma = 1 - \frac{x_m^2}{x^s x^r}, \tag{26.31}$$

we can rewrite Equations (26.29) and (26.30) as

$$\overline{i}_r^s = \left(\overline{\psi}_r^s - \frac{x_m}{x^r}\overline{\psi}_r^r\right)\left(\frac{1}{x^s\sigma}\right), \tag{26.32}$$

$$\overline{i}_r^r = \left(\overline{\psi}_r^r - \frac{x_m}{x^s}\overline{\psi}_r^s\right)\left(\frac{1}{x^r\sigma}\right). \tag{26.33}$$

Equations (26.32) and (26.33) describe the currents of the machine as functions of the stator and rotor fluxes. All quantities are in the rotor reference frame.

26.4.5 Mechanical equations of the machine

There are two different mechanical bodies in the system we want to represent – the induction generator and the wind turbine. This section contains the mechanical equations of the induction generator.

The mechanical equations of the induction generator can be represented as follows:

$$\frac{d\omega_g}{dt} = \frac{\omega_N}{2H_g}(T_{EL} + T_{SHAFT}), \tag{26.34}$$

$$\frac{d\theta_g}{dt} = \omega_g, \tag{26.35}$$

where T_{EL} is the electrical torque produced by the induction generator; T_{SHAFT} is the incoming torque from the shaft connecting the induction generator with the wind turbine; and H_g is the inertia constant of the induction generator. The angle θ_g is the machine angle, and ω_g is the speed of the induction generator. All quantities, except for the machine speed, ω_g, are given in p.u.

The expression for the electrical torque in physical units is

$$T_{EL} = \frac{3}{2} \text{Im} \left[\overline{\psi}_r^{s*} \overline{i}_r^s \right]. \tag{26.36}$$

The derivation of Equation (26.36) is not included here; for further details, see Kovács, 1984. The asterisk indicates the complex conjugate of the flux, ψ. Using Equation (26.32) for the stator current in the rotor reference frame, $\overline{i}_r^s$, the equation for the electrical torque can be rewritten as

$$
\begin{aligned}
T_{EL} &= \frac{3}{2} \text{Im} \left[\frac{\overline{\psi}_r^{s*}}{x^s \sigma} \left(\overline{\psi}_r^s - \frac{x_m}{x^r} \overline{\psi}_r^r \right) \right] \\
&= \frac{3 x_m}{2 x^s x^r \sigma} (\psi_{rq}^s \psi_{rd}^r - \psi_{rd}^s \psi_{rq}^r),
\end{aligned}
\tag{26.37}
$$

where real and imaginary parts of the stator and rotor fluxes are shown in parentheses in the final expression. Their relation to the complex stator and rotor fluxes is as follows:

$$\overline{\psi}_r^s = \psi_{rd}^s + j\psi_{rq}^s, \tag{26.38}$$

$$\overline{\psi}_r^r = \psi_{rd}^r + j\psi_{rq}^r. \tag{26.39}$$

Expressed in the p.u. system, the equation for the electrical torque, Equation (26.37), is

$$T_{EL} = \frac{x_m}{x^s x^r \sigma} (\psi_{rq}^s \psi_{rd}^r - \psi_{rd}^s \psi_{rq}^r), \tag{26.40}$$

where the reactances, total leakage factor, fluxes and torque are given in p.u. of the machine ratings.

The incoming torque from the shaft to the induction generator, T_{SHAFT}, in Equation (26.34), consists of the two terms $T_{TORSION}$ and $T_{DAMPING}$:

$$T_{SHAFT} = T_{TORSION} + T_{DAMPING}. \tag{26.41}$$

Equation (26.41) represents a drive-train system, where $T_{TORSION}$ represents the elasticity of the shaft and $T_{DAMPING}$ represents the damping torque of the shaft (see Novak, Jovile and Schmidtbauer, 1994 and Akhmatov, 2002).

$T_{TORSION}$ is expressed as a function of the angle of the wind turbine, θ_m, and the angle of the machine, θ_g, as

$$T_{TORSION} = K(\theta_m - \theta_g), \tag{26.42}$$

where K is the shaft torsion constant (p.u. torque/rad); that is, it is the effective shaft stiffness.

$T_{DAMPING}$ is related to the speed of the wind turbine, ω_m, and the speed of the machine, ω_g, as follows:

$$T_{DAMPING} = D(\omega_m - \omega_g), \tag{26.43}$$

where D is the shaft damping constant [p.u. torque/(rad/sec)] and represents the damping torque in both the wind turbine and the induction generator.

It is important to note that the speed of the machine, ω_g, and the speed of the wind turbine, ω_m, are measured in radians per second and that the angle of the machine, θ_g, and the angle of the wind turbine, θ_m, are measured in radians.

26.4.6 Mechanical equations of the wind turbine

The mechanical equations of the wind turbine are represented by the following two differential equations:

$$\frac{d\omega_m}{dt} = \frac{\omega_N}{2H_m}(T_{MECH} - T_{SHAFT}),$$
(26.44)

$$\frac{d\theta_m}{dt} = \omega_m,$$
(26.45)

where T_{MECH} is the mechanical torque produced by the wind turbine; T_{SHAFT} is the torque from the shaft connecting the induction generator with the wind turbine [see Equation (26.41)]; and H_m is the inertia constant of the turbine. The angle θ_m represents the position of the wind turbine, and ω_m is the speed of the wind turbine. All quantities except ω_m and θ_m are given in p.u.

In this chapter, we assume that the mechanical torque, T_{MECH}, produced by the wind turbine is constant. However, the constant T_{MECH} can be changed to a varying T_{MECH}, as described in Section 24.3.1

The equations of the machine given in this section (Section 26.4) can also be found in Centeno, 2000.

26.5 Voltage Source Converter

It is assumed that the converters in Figure 26.1 are arranged in a VSC. In this chapter, we will not implement the detailed modulation scheme but, rather, assume that the switching frequency is infinite. If the detailed modulation scheme were implemented it would slow down the simulations significantly, as each switching operation of the converter would be handled by the simulator as a separate event. The switching operations of the converters can be neglected without significantly affecting the result.

In this chapter, the VSC and its control is condensed and lumped together in a unique module. Therefore, the voltage that the voltage source control impresses onto the rotor of the machine is represented as a direct output of the voltage source converter control, ($\bar{u}_r^r$ in Figure 26.3). In spite of this simplification, we take into account the limitations in the voltage generation that the converter imposes. These limitations will be implemented as limitations to the voltage and torque controllers (U-reg and T-reg, respectively, in Figure 26.3).

The strategy used for the converter control is field-oriented control. The input signals to the VSC are:

- the stator voltage in the stator reference frame of the machine, $\bar{u}_s^s$;
- the speed of the machine, ω_g;

- the angle of the machine, θ_g;
- the electrical torque of the machine, T_{EL}.

Figure 26.3 shows the input parameters to the converter control: the stator winding resistance, r^s; the stator self-reactance, x^s; the rotor winding resistance r^r; the rotor self-reactance, x^r; and the mutual reactance between the stator and rotor windings x_m; ω_N is equal to $2\pi f_N$, where f_N is the power frequency (e.g. 50 Hz). The input parameters x^s and x^r were defined in Equations (26.26) and (26.28), respectively. The constant σ was defined in Equation (26.31). The following two reference values are also fed into the VSC: ω_{gref} and u^s_{ref}, each set by the user to a constant value throughout the simulation. $\bar{\psi}_s^{RFC}$ in Figure 26.3 is the stator flux in the rotor flux coordinate system RFC. $\overline{V}_r^{RFC}$ is a complex quantity which real part is the output signal from the voltage controller V_{rd}^{RFC} and its imaginary part is the output signal from the torque controller V_{rq}^{RFC}. ψ_r^{RFC} in Figure 26.3 is the rotor flux in the RFC system, a real quantity, ρ is the rotor flux angle in the rotor coordinate system, $T_{Erefwlim}$ is a torque reference which is the output signal from the speed control ω-reg. In Figure 26.3 s represents the Laplace operator.

The constants τ^s and τ^r in Figure 26.3 are defined as follows:

$$\tau^s = \frac{x^s}{r^s \omega_N}, \tag{26.46}$$

$$\tau^r = \frac{x^r}{r^r \omega_N}. \tag{26.47}$$

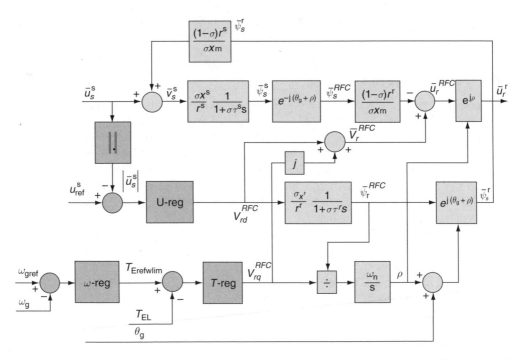

Figure 26.3 Control system configuration for the voltage source converter; for definitions of variables, see section 26.5

The speed, voltage and torque controllers – ω-reg, U-reg, and T-reg, respectively – are modelled as PI controllers (proportional–integral controllers).

26.6 Sequencer

The sequencer is a module that handles the different modes of operation of the system that consists of the DFIG, the wind turbine, the VSC and its control. The need for such a sequencer arises from the fact that whenever there is a fault in the interconnected power system, the system of the DFIG must be able to handle the resulting high currents without causing any damage to the equipment. Another reason for having different modes is that the system has to recover from a fault as soon as possible in order to remain in phase. The sequencer basically controls the rotor current level and the stator voltage level and sets the current mode of operation.

The aim of the sequencer is to protect the VSC from high currents as well as to optimise the behaviour of the system. The changes of mode are required during transient conditions. There are three different modes, with mode 0 corresponding to the mode for normal operation.

If the rotor current exceeds 2 p.u., the VSC has to be disconnected because it cannot resist such a high current. Therefore, the rotor is short-circuited during such an event. If the machine works as a short-circuited machine, the system operates in mode 1. If the machine works in mode 1 and the stator voltage magnitude has been greater than 0.3 p.u. for 100 ms, the system switches to mode 2. During mode 2, the machine is still short-circuited, but additional resistances are connected into the rotor circuit. In that case, the equivalent rotor resistance is 0.05 p.u., which forces the rotor current to decrease.

Finally, if the system works in mode 2 and the stator voltage level has exceeded 0.85 p.u. for 100 ms and the rotor current has been lower than 2 p.u. for 100 ms, the system returns to normal operation (i.e. mode 0). Once the system has returned to normal operation the machine first tries to get magnetised and then starts producing torque. That is done by setting the torque reference to 0 during 400 ms, once the disturbance has been cleared.

The left-hand side of Figure 26.6 (page 601) depicts the modes of operation during a simulation.

26.7 Simulation of the Doubly-fed Induction Generator

In this section we will test the setup of the DFIG. The machine is incorporated into a small power system where it is connected to an infinite bus via a line impedance, Z, of $0.06 + j0.60$ (Ω). In the power system, there will be a three-phase fault at $t = 36$ s at 30% of the full distance from the infinite bus (see Figure 26.4). The three-phase fault has a fault resistance, R_{fault}, of 0.30 (Ω). The power system is modelled in instantaneous value mode.

Prior to the fault, the DFIG operates at nominal speed. In Figure 26.5, the terminal voltage is shown on the left-hand side and the stator fluxes ψ_{sd} and ψ_{sq} are shown on the right-hand side.

In Figure 26.6, the rotor current and the mode of operation is shown on the left-hand side and the speed of the machine, ω_g, on the right-hand side. From Figure 26.6 we can see that the rotor is short-circuited (mode of operation equal to 1) approximately

Infinite bus   Fault   DFIG bus

Figure   26.4 Test system for the doubly-fed induction generator (DFIG)

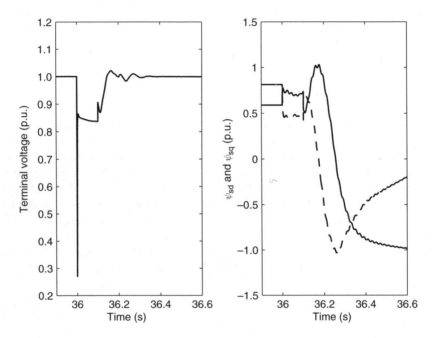

Figure   26.5 Terminal voltage and stator fluxes ψ_{sd} and ψ_{sq} of the doubly-fed induction generator (DFIG) during a three-phase fault in the power system

between $36.0 < t < 36.1$ (s). When the mode of operation is equal to 2, between $36.1 < t < 36.2$ (s), the rotor is still short-circuited but the rotor current is slowly decreasing as additional resistances are switched into the rotor circuit.

After a few seconds, the DFIG model returns to its initial state. However, this is not shown in Figures 26.5 and 26.6.

26.8 Reducing the Order of the Doubly-fed Induction Generator

Different models are available to simulate the behaviour of the DFIG under different conditions. The machine described in Section 26.4 is a 6th-order model, as the model contains the following six state variables: ψ_{rd}^s, ψ_{rq}^s, ψ_{rd}^r, ψ_{rq}^r, ω_g and θ_g. ψ_{rd}^s and ψ_{rq}^s are the real and imaginary parts of the stator flux $\overline{\psi}_r^s$, and ψ_{rd}^r and ψ_{rq}^r are the real and imaginary parts of the rotor flux $\overline{\psi}_r^r$, as shown in Equations (26.38) and (26.39).

The 6th-order model developed in this chapter is an extension of the 5th-order model described in Thiringer and Luomi (2001). The extension consists of the machine angle,

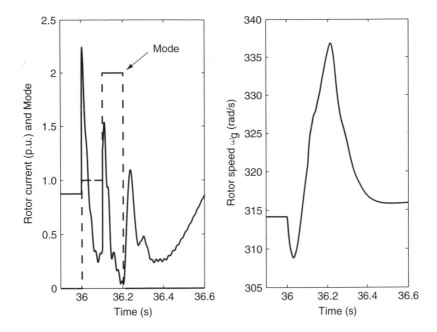

Figure   26.6 Rotor current, mode of operation, and speed of the machine ω_g during a three-phase fault in the power system

θ_g, that has to be calculated as it is used as an input signal to the control system of the VSC (see Section 26.5).

The order of the DFIG model can be reduced from six to four by omitting the stator flux transients in Equation (26.23). Then the derivatives of the stator flux linkages will be set to zero. Such a model is often referred to as the transient stability model or the neglecting stator transients model (see Thiringer and Luomi, 2001; or see Section 24.4.3.3). Equation (26.23) is then simplified to:

$$\overline{u}_r^{\text{s}} = \overline{i}_r^{\text{s}} r^{\text{s}} + \text{j}\frac{\omega_g}{\omega_N}\overline{\psi}_r^{\text{s}}. \tag{26.48}$$

In power system analysis, this simplification is commonly used in transient and small signal stability programs where the grid transients are neglected, too, since the power system is modelled with phasor models, also called fundamental frequency models, instead of instantaneous value models (Thiringer and Luomi, 2001).

26.9 Conclusions

This chapter presented a 6th-order DFIG. Special attention was paid to the used vector method and the bidirectional transformations between the stator and rotor reference frame.

The detailed modulation scheme of the VSC connected to the rotor of the induction generator was not described, as it was assumed that the switching frequency of the VSC is infinite. The VSC and its control was condensed and lumped together in a unique module.

A sequencer was described, the task of which is to protect the VSC from high currents as well as to optimise the behaviour of the system. Transient conditions require changes of mode in the sequencer. There are three different modes generated by the sequencer.

The DFIG model was tested in a small power system that was simulated in instantaneous value mode. Our model was not verified against an actual installation of a DFIG model. Finally, we described how the order of the DFIG model can be reduced.

References

[1] Akhmatov, V. (2002) 'Variable-speed Wind Turbines with Doubly-fed Induction Generators, Part I: Modelling in Dynamic Simulation Tools', *Wind Engineering* 26(2) 85–108.
[2] Ängquist, L. (1984) *Complex Vector Representation of Three-phase Quantities*. Technical report TR YTK 84-009E, ASEA Drives, Västerås, Sweden.
[3] Centeno López, E. (2000) 'Windpower Generation Using Double-fed Asynchronous Machine', B-EES-0008, masters thesis, Department of Electric Power Engineering, Royal Institute of Technology, Stockholm, Sweden, and ABB Power Systems, Västerås, Sweden.
[4] Fankhauser, H., Adielson, T., Aneros, K., Edris, A.-A., Lindkvist, L., Torseng, S. (1990) 'SIMPOW – A Digital Power System Simulator', reprint of *ABB Review* 7 pp. 27–38.
[5] Kovács, P. K. (1984) *Studies in Electrical and Electronic Engineering, 9: Transient Phenomena in Electrical Machines*, Elsevier, Amsterdam.
[6] Novak, P., Jovik, I., Schmidtbauer, B. (1994) 'Modeling and Identification of Drive-system Dynamics in a Variable-speed Wind Turbine', in *Proceedings of the Third IEEE Conference on Control Applications, August 24–26, 1994, volume 1*, Institute for Electrical and Electronic Engineers (IEEE), New York, pp. 233–238.
[7] Thiringer, T., Luomi, L. (2001) 'Comparison of Reduced-order Dynamic Models of Induction Machines', *IEEE Transactions on Power Systems* 16(1) 119–126.

27

Full-scale Verification of Dynamic Wind Turbine Models

Vladislav Akhmatov

Dedicated to my true friend C. E. Andersen

27.1 Introduction

The incorporation of wind power into electric power systems is progressing faster than predicted. In Denmark, for example, the total installed capacity was approximately 3000 MW of grid-connected wind power in mid-2003. About 2400 MW of this capacity is installed in Western Denmark (Jutland – Funen) and 600 MW in Eastern Denmark (Zealand and Lolland – Falster; for more details, see Chapter 10).

In addition, the large offshore wind farm at Rødsand in Eastern Denmark started operating in late 2003. The Rødsand wind farm comprises 72 2.3 MW fixed-speed wind turbines (Type A) from the manufacturer Bonus Energy.[1] The wind farm is connected to the transmission network of the national transmission system operator (TSO) of Denmark. Further large offshore projects have recently gone under consideration in Denmark, and incorporation of 400 MW offshore wind power due in the year 2008 in Denmark has been announced.

The majority of the electricity-producing wind turbines in Denmark are Type A wind turbines. This type is also called the Danish concept. They are scattered across the

---

[1] For definitions of wind turbine Types A–D, see Section 4.2.3.

Wind Power in Power Systems   Edited by T. Ackermann
© 2005 John Wiley & Sons, Ltd   ISBN: 0-470-85508-8 (HB)

country and are connected to the local distribution power networks. That means that the majority of wind turbines connected to the Danish power grid are based on the same concept as those at the Rødsand offshore wind farm.

Such a large penetration of wind power into electric power networks reduces the amount of electric power supplied by centralised power plants and may affect the operation of the power networks. Therefore, it is important to know what consequences the dynamic interaction between large wind farms and the power system has before the wind farm is connected to the grid. Technical documentation has to be provided and it has to show that the wind turbines comply with the technical specifications set by the power system operator. The power system operator has to grant the large wind farm permission to be connected to the grid (for more details on these issues, see Chapters 7 and 11).

It is necessary to develop and implement dynamic simulation models of wind turbines for the existing simulation software tools that are applied in the analysis of power system stability. The results of such analyses will be used for planning net-reinforcements and the incorporation of dynamic reactive compensation, for revising protective relay settings in the transmission power system and other practical arrangements, which can be complex, expensive and time-consuming. The analysis may also take into consideration economic interests of different companies and be used in the decision-making process regarding suppliers to a large wind farm project. Therefore the analysis will focus on the accuracy, credibility and documentation of the dynamic wind turbine models.

We use dynamic wind turbine models to investigate short-term voltage stability in the context of connecting large amounts of wind power to the grid. The models have to include sufficiently accurate representations of all the components of the wind turbine construction that are relevant to such investigations. We will clarify, explain and document the accuracy of the dynamic wind turbine models and their possible shortcomings by validating the models. In general, the validation process should be finished before the models are applied in the analysis of the short-term voltage stability. Thus, validation becomes an integral part of the development of dynamic wind turbine models and their implementation into dynamic simulation tools.

27.1.1 Background

The validations we will present here have their background in projects regarding the incorporation of large amounts of wind power into the East Danish power system and, more specifically, regarding the analysis of short-term voltage stability. The majority of the small wind turbine sites in Denmark and the large offshore wind farm at Rødsand consist of Type A wind turbines (Akhmatov, 2003a). Figure 27.1 illustrates this wind turbine concept. It shows a three-bladed rotor, a shaft system with a gearbox, and an induction generator with a shorted rotor-circuit. These basic components need accurate representation and validation.

We use the simulation tool Power System Simulator for Engineering (PSS/E[TM]).[2] When we carried out our analysis, the dynamic simulation tool PSS/E[TM] did not contain any

---

[2] PSS/E[TM] is a trademark of Shaw Power Technologies Inc. (PTI), New York, USA.

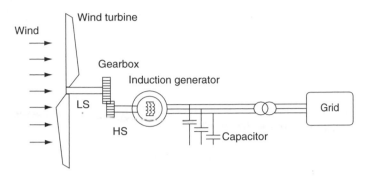

Figure   27.1 A fixed-speed wind turbine equipped with an induction generator. *Note*: LS = low speed; HS = high speed. Reprinted from Akhmatov, V. Analysis of Dynamic Behaviour of Electric Power Systems with Large Amount of Wind Power, Ph.D. thesis, Technical University of Denmark, Kgs. Lyngby, Denmark, copyright 2003, with permission from the copyright holder

model of Type A wind turbines that was sufficiently complex (Akhmatov, Knudsen and Nielsen, 2000). Therefore, it was necessary to develop a user-written dynamic model of Type A wind turbines to implement it into the PSS/E^{TM} tool and to verify it. This project was carried out at the Danish company NESA (Akhmatov, Knudsen and Nielsen, 2000). The simulation tool PSS/E^{TM} and modelling details are discussed briefly in Chapter 24.

This chapter describes the validation of the user-written dynamic model of Type A wind turbines that was developed and implemented into the simulation tool PSS/E^{TM} at NESA. It includes partial validation and an example of a full-scale validation of the dynamic wind turbine model.

27.1.2 Process of validation

When analysing short-term voltage stability, Type A wind turbines are treated as complex electromechanical systems (Akhmatov, Knudsen and Nielsen, 2000). The dynamic wind turbine model contains representations of the induction generator and the shaft system, the aerodynamic model of the turbine rotor and blade-angle control. First, representation of the individual parts of the wind turbine construction and its generator can be validated. At this point the models of the induction generator, the shaft system and the turbine rotor can be validated separately. This is called partial validation and it is necessary always to specify which part of the dynamic wind turbine model is to be validated. If the individual parts of the wind turbine model are found to be sufficiently accurate, the complete model of fixed-speed wind turbines is considered to be sufficiently accurate too. However, partial validation does not necessarily take into account the links between different parts of the model.

Second, full-scale validation can be carried out. This also focuses on validating the representations of the different parts of the wind turbine construction. In addition, it validates the interaction and links between different parts of the wind turbine construction. We will illustrate full-scale validation using an example where the interaction of the electric parameters (the machine current) and the mechanical parameters (the generator rotor speed) are verified against measurements (Raben *et al.*, 2003).

We can distinguish between two types of validation processes. First, a user-written model that is implemented into a given simulation tool can be verified against the standardised model of the same component of the other simulation tool (Knudsen and Akhmatov, 1999). This requires that the standardised model already be verified and documented by the supplier of the simulation tool. Second, the user-written model is validated against measurements. In this case, the measurements come either from planned experiments in the field (Pedersen *et al.*, 2003) or from transient events or accidents that have occurred in the power network.

Probably the easiest and cheapest way of validation is to carry it out against standardised models of another simulation tool. The validating process is then characterised by the following:

- The cases that are simulated with the two simulation tools can be set up so that they are as similar to each other as possible. Possible uncertainties and discrepancies between the two cases are minimised.
- The cases contain only those components for which representation is necessary for the validation.
- The network representation around the validated component is reduced.
- The component model can be validated during simulated operational conditions or during transient events, which are not always reasonable to apply in experiments (Raben *et al.*, 2003).
- The results of the validation need careful interpretation.

We have to keep in mind that validation is still being carried out with use of a simulation model, even though it is standardised in the given simulation tool. The standardised model may still have shortcomings regarding the model assumptions and the area of application.

An example of this is the common third-order model of induction generators, CIMTR3, which is the standardised model of the simulation tool PSS/E^{TM}. This model includes a representation of the rotor flux transients, but it does not contain the stator flux transients (see Chapter 24). The common third-order model is very suitable for simulating unbalanced, three-phased events in the power network (Pedersen *et al.*, 2003). The disconnection of a three-phased line is an example of such an unbalanced event. However, the common third-order model will predict inaccurate results in the case of balanced, three-phased, short-circuit faults with a significant voltage drop at the induction generator terminals (Knudsen and Akhmatov, 1999). Consequently, it can be very useful to include in the validating process other simulation tools and their standardised and validated models of the same component. However, this requires a careful interpretation of the results.

The validation of user-written models against measurements in the field is characterised by the following:

- Experimental work needs careful planning and preparation. Preparation includes obtaining permission from authorities, the power system controller and the wind turbine manufacturer (Raben *et al.*, 2003).
- The disturbances the wind turbines and the power network can be subjected to are often limited. It is not always possible to obtain permission, for example, to execute

experiments and subject a power system with grid-connected wind turbines to a balanced, three-phase, short-circuit fault, even though the results of such experiments would be valuable for validating the dynamic wind turbine model.

- Measurements can be affected by components, control systems and parts of the power network that are not part of the component to be validated. It may become necessary to include representations of these components in the network representation that is used for the validation. Careful planning of the experimental work can reduce this undesirable effect, though.
- The validation is likely to be complicated by uncertainties with respect to network configuration, the data of the components and the control systems affecting the measurements and so on. A careful representation of the simulation case can minimise such uncertainties.
- The simulated and the measured behaviour have to be in agreement, otherwise the model is incomplete and the interpretation of the results of the validation process is ambiguous.

Even though there may be difficulties regarding (a) planning and executing measurements in the field, (b) collecting data and (c) minimising uncertainties, the validation results reached with the use of measurements have a high credibility. The reason for this is that simulations aim at reproducing measured behaviour under similar conditions in simulations. If this is not achieved, the usefulness of the simulation will be under question.

There will, however, always be discrepancies between simulated and measured behaviour. Such discrepancies have to be minimised by accurate modelling and by discovering and explaining the possible sources of these discrepancies. If the validation process is carried out accurately, such discrepancies are small and are often caused by uncertainties in the network representation and data (Akhmatov, 2003a; Pedersen *et al.*, 2003).

27.2 Partial Validation

In the following, we will present cases with a partial validation of the dynamic wind turbine model. The cases are validated against simulations and measurements.

27.2.1 Induction generator model

Here we will describe the validation of the user-written model of induction generators that is implemented in the simulation tool PSS/E[TM] at NESA. This is a transient fifth-order model (i.e. it contains a representation of the fundamental frequency transients of the machine current). The validation is carried out against simulations with the standardised model of the tool Matlab/Simulink[TM].[3] The simulated sequence is a three-phase, short-circuit fault that lasts 100 ms. We chose to validate the user-written model in PSS/E[TM] against simulations with the standardised model of induction generators of Matlab/Simulink[TM] for the following reasons:

---

[3] Matlab/Simulink[TM] is a trademark of The MathWorks Inc; Natrick, USA.

- A balanced, three-phase, short-circuit fault is a common transient event that has been analysed in investigations on transient voltage stability (Knudsen and Akhmatov, 1999). Therefore it is necessary to ensure that the user-written model in the tool PSS/E^{TM} gives a sufficiently accurate response for this kind of fault.
- It can be difficult to obtain permission to carry out an experiment with such a disturbance in the grid because this is a serious incident.
- The tool Matlab/Simulink^{TM} is applied and recognised worldwide.
- The tool Matlab/Simulink^{TM} contains the standardised and verified model of induction generators, which is a three-phase, physical representation. The model is suitable for the simulation of balanced as well as unbalanced transient events in the power network.

The validation process is based on a simple network equivalent. It contains an induction generator connected to an infinite bus through a 0.7/10 kV transformer and a line with an impedance corresponding to the short-circuit ratio (SCR) of 6.7. Figure 27.2 shows the network equivalent. The network equivalent models are implemented both in the simulation tool PSS/E^{TM} (the positive-sequence equivalent) and the tool Matlab/Simulink^{TM} (the three-phase representation). The implementation for each simulation tool is as similar as possible. The induction generator works at the rated operation point and supplies 2 MW. Table 27.1 includes the data.

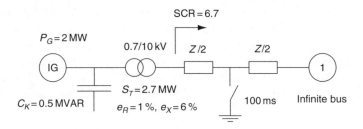

Figure   27.2 Model of a network equivalent with an induction generator (IG). *Note*: SCR = short-circuit ratio; P_G = generator electric power; C_K = no-load compensation; S_T = transformer power capacity; e_R = real part of transformer impedance; e_x = imaginary part of transformer impedance. Reprinted from Akhmatov, V. Analysis of Dynamic Behaviour of Electric Power Systems with Large Amount of Wind Power, Ph.D. thesis, Technical University of Denmark, Kgs. Lyngby, Denmark, copyright 2003, with permission from the copyright holder

Table   27.1 Data of a 2 MW induction generator. Reprinted from Akhmatov, V. Analysis of Dynamic Behaviour of Electric Power Systems with Large Amount of Wind Power, Ph.D. thesis, Technical University of Denmark, kgs. Lyngby, Denmark, copyright 2003, with permission from the copyright holder

Parameter	Value	Parameter	Value
Rated power (MW)	2	Stator resistance (p.u.)	0.048
Rated voltage (v)	690	Stator reactance (p.u.)	0.075
Rated frequency (Hz)	50	Magnetising reactance (p.u.)	3.80
Rated slip	0.02	Rotor resistance (p.u.)	0.018
Generator rotor inertia (s)	0.5	Rotor reactance (p.u.)	0.12

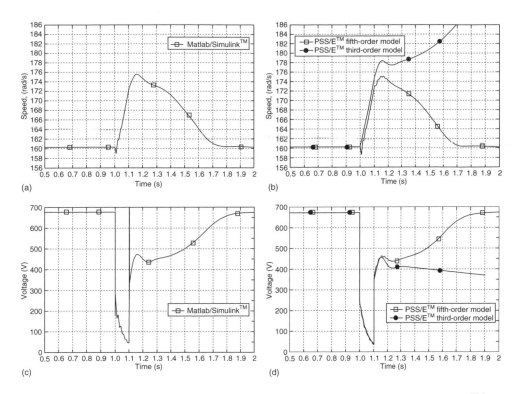

Figure   27.3 Simulated behaviour of the generator rotor speed: (a) using Matlab/Simulink[TM] and (b) using PSS/E[TM]. The terminal (RMS phase–phase) voltage: using (c) Matlab/Simulink[TM], and (d) using PSS/E[TM]. Reprinted from Akhmatov, V. Analysis of Dynamic Behaviour of Electric Power Systems with Large Amount of Wind Power, Ph.D. thesis, Technical University of Denmark, Kgs. Lyngby, Denmark, copyright 2003, with permission from the copyright holder

Figure 27.3 illustrates the simulated behaviour of the voltage and the generator rotor speed, which is applied to validate the transient fifth-order model of induction genera-tors. For a comparison, the behaviour computed with the use of a common third-order model in the tool PSS/E[TM] (neglecting fundamental frequency transients in the machine current) is plotted in the same figure.

As expected (Knudsen and Akhmatov, 1999), the simulated results of the transient fifth-order model of induction generators in the tool PSS/E[TM] are in agreement with the results of the standardised model of the tool Matlab/Simulink[TM]. The common third-order model of induction generators predicts overpessimistic results with respect to maintaining short-term voltage stability, for the following reasons:

- The common third-order model predicts more overspeeding of the induction genera-tor during the grid fault, compared with the results of the fifth-order model. An explanation of this phenomenon is given in Chapter 24.
- The mechanical parameter, which is the generator slip, and the electric parameters of the induction generator are strongly coupled. If the prediction of the overspeeding is too high, the prediction of the reactive absorption will be too high as well.

- Consequently, the common third-order model of induction generators predicts a slower voltage re-establishing after the grid fault than does the transient fifth-order model.
- In this particular case, the common third-order model predicts voltage instability, whereas according to the transient fifth-order model of induction generators the voltage is re-established.

Knudsen and I reached the same conclusion when validating against the standardised induction generator model of the Alternative Transient Program (ATP) simulation tool (Knudsen and Akhmatov, 1999). Chapter 24 briefly describes the ATP tool. This was also the first time that the validity and accuracy of the common third-order model of induction generators were questioned as to whether the model should be used in the analysis of short-term voltage stability.

Wind turbine generators are equipped with protective relays. This means that several electrical and mechanical parameters are monitored, and if one of the monitored parameters exceeds its respective relay settings the wind turbine generators will be disconnected. One of the monitored parameters is the machine current. Figure 27.4 shows the computed behaviour of the machine current during the grid fault. As can be seen, the machine currents computed with the transient fifth-order model in the tool PSS/E^{TM} and with the standardised model of the tool Matlab/Simulink^{TM} are in agreement. The machine current computed with the common third-order model in the tool PSS/E^{TM} does not coincide with the result of the tool Matlab/Simulink^{TM}.

This needs further explanation. The transient fifth-order model takes the following into account when computing the behaviour of the machine current:

- A three-phased, short-circuit fault is a balanced transient event, meaning that the fault occurs at the same moment in all three phases. The phase current can be characterised by the DC offset, and the current phasor contains the fundamental frequency transients. The computed current behaviour during the faulting time shows this.
- The short-circuit fault is cleared separately in each phase of the faulted three-phased line. This happens when the phase-currents pass zero in the respective phases. The fault clearance is therefore an unbalanced event. It does not initiate the DC offset in the phase currents, and the fundamental frequency transients in the current phasor are eliminated after the fault is cleared.
- The standardised model of induction generators in the tool Matlab/Simulink^{TM} automatically takes this behaviour into account because Matlab/Simulink^{TM} operates with three-phased representations of electric machines and power networks.
- The transient fifth-order model of induction generators in the tool PSS/E^{TM} is adapted to this behaviour. This is necessary because the simulation tool PSS/E^{TM} operates with positive-sequence equivalents of power network models.
- Additionally, the phase current behaviour is plotted. This phase current is modelled in Matlab/Simulink^{TM} and is included here in order to demonstrate that the current phasor follows the behaviour of the magnitude of the phase current with developed DC offset. This observation is important for the validating process described in Section 27.3.

According to the current behaviour shown in Figure 27.4, the common third-order model of induction generators underpredicts values of the machine current during the faulting time. There are no fundamental frequency transients. The insufficient representation of the machine current with the common third-order model can be misleading with respect to the prediction of the protective relay action during grid faults.

The transient fifth-order model of induction generators is a part of the dynamic model of Type A wind turbines, which was implemented in the simulation tool PSS/E^{TM} at NESA. Until this implementation, the simulation tool PSS/E^{TM} contained only the common third-order model of induction generators, in this case the model CIMTR3. PTI is currently working on a standardised higher-order model of induction generators to be implemented in the tool PSS/E^{TM}. This model also includes the representation of the fundamental frequency transients in the machine current (Kazachkov, Feltes and Zavadil, 2003).

27.2.2 Shaft system model

Regarding the modelling of Type A wind turbines, it is common knowledge that this type has a relatively soft coupling between the turbine rotor and the induction generator rotor (Hinrichsen and Nolan, 1982). This is because the turbine rotor and the induction

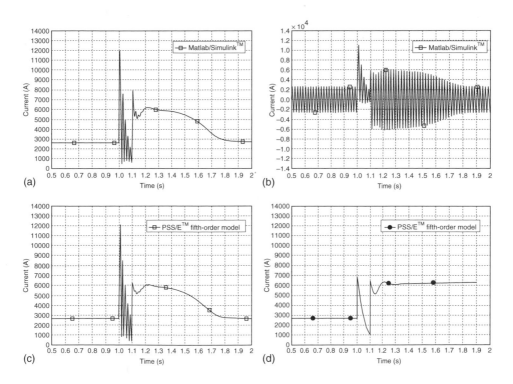

Figure   27.4 Simulated behaviour of the (peak phase) machine current: (a) the current phasor in Matlab/Simulink^{TM}, (b) the phase current in Matlab/Simulink^{TM}, (c) the current phasor with the fifth-order model in PSS/E^{TM}, and (d) the current phasor with the third-order model in PSS/E^{TM}. Reprinted from Akhmatov, V. Analysis of Dynamic Behaviour of Electric Power Systems with Large Amount of Wind Power, Ph.D. thesis, Technical University of Denmark, kgs. Lyngby, Denmark, copyright 2003, with permission from the copyright holder

generator rotor are connected through a shaft system with a relatively low stiffness. When using a typical measure of shaft stiffness (i.e. p.u./el. rad) the shaft stiffness of Type A wind turbines is between 30 and 100 times lower than the shaft stiffness of conventional power plant units (Hinrichsen and Nolan, 1982). Therefore, the two-mass model is used for the shaft systems of Type A wind turbines (Akhmatov, Knudsen and Nielsen, 2000). The two masses represent the turbine rotor and the generator rotor, respectively. Since its inception, the shaft representation in the two-mass model has been the most disputed part of the dynamic wind turbine model that is applied in the analysis of short-term voltage stability (Raben et al., 2003).

Before that time, wind turbine shafts were assumed to be stiff, so that turbine inertia and generator rotor inertia were 'lumped together' – the lumped-mass model. When the shaft system is represented by the two-mass model and relatively low values of shaft stiffness are applied, the wind turbine model predicts the following phenomena (Akhmatov, Knudsen and Nielsen, 2000):

- It predicts oscillations of the voltage, the machine current, the electric and the reactive power and the speed of the wind turbine induction generators when subjected to a transient fault: the natural frequency of such oscillations is that of the shaft torsional mode and is in the range of a few hertz, depending on the shaft construction and data.
- It predicts the slower re-establishment of voltage in the power grid after the grid fault is cleared. The two-mass model predicts larger demands for dynamic reactive compensation than does the lumped-mass model.

That more dynamic reactive compensation is predicted for the same amount of grid-connected wind power appears to be an even more important issue than the risk of voltage oscillations lasting a few seconds after a grid fault. The reason for this is that a higher demand for dynamic reactive compensation makes the incorporation of wind power generally more costly.

The validation of the shaft system representation is therefore very important. One of the first experiments that indicated the soft coupling between the turbine rotor and the generator rotor was the islanding experiment carried out at the Rejsby Hede wind farm (Pedersen et al., 2000). The wind farm comprises 40 600 kW fixed-speed wind turbines equipped with induction generators. It is situated in Western Denmark, in the area of the transmission system operator Eltra.

During the experiment, the wind farm and a part of the local distribution network were disconnected from the transmission power network. This islanded operation continued for less than 1 s before the wind turbine generators were tripped and the wind turbines were stopped (Pedersen et al., 2000). The wind speed was about 10 m/s and the induction generators were 80 % reactive compensated (Pedersen et al., 2000).[4] During the experiment, the phase currents, the phase voltages and the electric frequency from a section of the wind farm were measured. Later, the results of the islanding experiment at

---

[4] This implies that 80 % of the reactive power absorption of the induction generators was covered by the capacitor banks and the charging from the wind farm cable network, whereas 20 % was covered by the reactive power absorption from the entire grid.

the Rejsby Hede wind farm were used for the partial validation of the dynamic model of fixed-speed wind turbines (Pedersen *et al.*, 2003).

During islanded operation, the electric frequency showed an oscillating behaviour (Pedersen *et al.*, 2003; see also Figure 27.5). The natural frequency of these oscillations is 1.7 Hz and corresponds to the shaft torsional mode (Akhmatov *et al.*, 2003).

Figure 27.6 shows the behaviour of the electric frequency that was simulated under similar conditions to those during the islanding experiment. The two-mass model and the lumped-mass model, respectively, were used to simulate the behaviour.

The figures show that the measured behaviour and the behaviour simulated with the two-mass model are in agreement. The behaviour predicted with the lumped-mass model does not contain any oscillations but follows the average line sketched on the measured frequency curve in Figure 27.5. The lumped-mass model gives an oversimplified representation of the wind turbine shaft system.

The measurements of the machine current during this experiment cannot be used to validate the transient fifth-order model of induction generators. The reason is that the disconnection of the wind farm is an unbalanced transient event. As explained in Section 27.2.1, the fundamental frequency transients in the machine current are eliminated during unbalanced events. The validation described in Pedersen *et al.* (2003) uses the common third-order model of induction generators for the validation.

The way in which the parameters of the two-mass model can be evaluated by using measurements of the oscillating behaviour of the electric frequency has been described elsewhere (Akhmatov *et al.*, 2003). This approach can be useful if the data of the shaft system model are unknown.

27.2.3 Aerodynamic rotor model

The aerodynamic model of the wind turbine rotor introduces a feedback between mechanical power and rotational speed (Akhmatov, Knudsen and Nielsen 2000). In

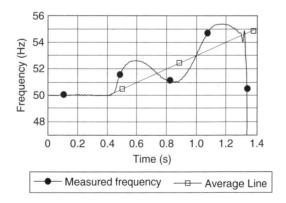

Figure   27.5 Measured electric frequency reproduced Reprinted from the *International Journal of Electrical Power and Energy Systems*, volume 22, issue 6, V. Akhmatov, H. Knudsen and A. H. Nielsen, 'Advanced Simulation of Windmills in the Electrical Power Supply', pp. 421–431 copyright 2000, with permission from Elsevier

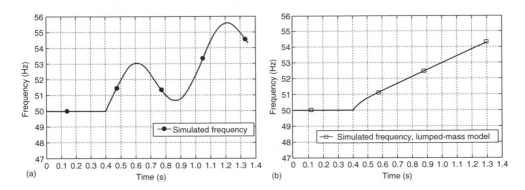

Figure 27.6 Simulated electric frequency: (a) two-mass model and (b) lumped-mass model Part (a) reproduced Reprinted from the *International Journal of Electrical Power and Energy Systems*, volume 22, issue 6, V. Akhmatov, H. Knudsen and A. H. Nielsen, 'Advanced Simulation of Windmills in the Electrical Power Supply', pp. 421–434 copyright 2000, with permission from Elsevier

addition, the aerodynamic rotor model is required to represent the mechanical power control with blade-angle control (pitch or active-stall). The aerodynamic rotor model has to be accurate if the blade-angle control is to be applied to improve short-term voltage stability (Akhmatov, 2001). The use of blade-angle control for stabilising the operation of large wind farms is discussed in Chapter 29, for instance.

The wind turbine rotor can be modelled in several ways, depending on the purpose of the analysis. If short-term voltage stability is to be analysed, the wind turbine rotor is commonly represented by using $c_P-\lambda-\beta$ curves. The $c_P-\lambda-\beta$ curves are often precalculated and provided by the wind turbine manufacturer. However, it may be necessary to have more detailed aerodynamic representations in order to achieve a better accuracy. In the following, aerodynamic rotor models of different complexity will be validated:

- The wind turbine is modelled with the use of its $c_P-\lambda-\beta$ curves (see Chapter 24).
- A model with unsteady inflow phenomena is applied (Øye, 1986).
- The reduced aeroelastic code (AEC) is applied. The AEC is reduced because only oscillations of the wind turbine blades are included, and the tower and the foundation are considered to be stiff (Akhmatov, 2003b).

My colleagues and I implemented the aerodynamic rotor models of different complexity as user-written models into the simulation tool PSS/E™ at NESA (Akhmatov, 2003b). The aerodynamic rotor models were validated against the measurements in Snel and Schepers (1995). These measurements came from a 2 MW fixed-speed, pitch-controlled wind turbine equipped with an induction generator. The rotor diameter was 61 m. The study also included the complete dataset of the wind turbine. In the experiment, the wind turbine was subjected to the 'step' of the pitch angle. This 'pitch step' was achieved by a sudden change of the pitch reference. The pitch angle was then adjusted by the pitch servo with a servo time constant of 0.25 s and a pitch rate limit of 7.2° per second.

Figure 27.7(a) shows the measured curve of the wind turbine's mechanical torque during the selected study case. The selected study case corresponds to the operational point at a wind speed of 8.7 m/s and an initial pitch angle of 0°. At the time $t = 1$ s, the pitch reference was set to $+3.7°$, and at the time $t = 30$ s, the pitch reference was reset to $0°$. It can be seen that the measured torque shows noticeable overshoots during the pitching.

Figures 27.7(b)–27.7(d) give the simulated curves of the wind turbine's mechanical torque. The following observations can be made:

- The model that uses $c_P–\lambda–\beta$ curves does not produce any overshoot during pitching. The reason is that this model describes the wind turbine rotor in equilibrium, whereas overshoot corresponds to the transition between the two states of equilibrium. When applying $c_P–\lambda–\beta$ curves, the transition process is neglected and the wind turbine passes immediately from one state of equilibrium to another. Consequently, the $c_P–\lambda–\beta$ curves give stationary operational points of the wind turbine.
- The model with unsteady inflow phenomena produces a sufficiently accurate representation of the behaviour of the mechanical torque, including overshoots during pitching. In terms of the engineering model suggested by Øye (1986), such overshoots are explained by lags in the induced velocities around the wind turbine blades. The characteristic time constant of such lags, τ_v, is in the order of $2R/V$, where R is blade length and V is the incoming wind velocity. According to Øye (1986), such lags describe unsteady inflow phenomena occurring around the wind turbine – the transition process. This model is often called the Øye model and it is applied in this computation.
- Finally, the reduced AEC predicts overshooting and also oscillations in the mechanical torque. These oscillations are caused by oscillations of the wind turbine blades due to the pitching. The AEC takes into account the effect of the blades moving in the same direction as the incoming wind or in the direction opposite to the relative wind velocity acting on the blades. This and other phenomena, which are not described here, produce the feedback between the blade oscillations and the mechanical torque.

The behaviour that was simulated with the model with unsteady inflow phenomena and with the reduced AEC both agree with the measurements. This also indicates a sufficiently accurate implementation of these models into the dynamic simulation tool.

The aerodynamic rotor model that uses $c_P–\lambda–\beta$ curves predicts accurately the stationary operational points of the wind turbine rotor. However, this model does not seem to provide an entirely accurate representation of the transition between stationary operational points during pitching. The dynamic wind turbine model will be applied to analyse transient voltage stability rather than for aerodynamic simulations, though. In a number of other situations, it may nevertheless be acceptable to use reasonable simplifications to represent the aerodynamics of the rotor.

The model with unsteady inflow phenomena does not require more data than the model that uses $c_P–\lambda–\beta$ curves. However, the model with unsteady inflow phenomena is more complex and requires more computational resources than does the model that uses (precalculated) $c_P–\lambda–\beta$ curves. The reduced AEC requires both more data and significantly more computational resources than do the other two models.

The reduced AEC model is the most accurate among the aerodynamic rotor models that we have compared. The discrepancies between this model and the model with unsteady inflow phenomena are only marginal when studying transient voltage stability. Therefore, we do not include the reduced AEC into the final discussion below on how detailed the representation of the aerodynamic rotor model has to be for the analysis of short-term voltage stability.

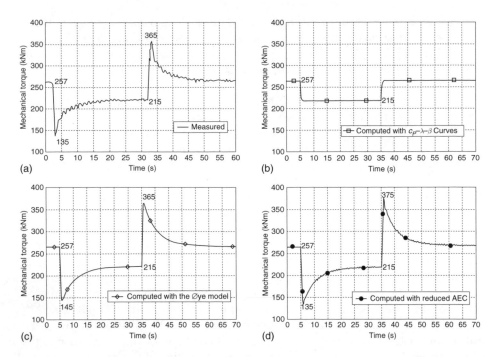

Figure 27.7 Wind turbine torque during pitching: (a) measured, (b) computed with use of $c_P–\lambda–\beta$ curves, (c) computed with use of the Oye (1986) model with unsteady inflow phenomena, and (d) computed with the reduced aeroelastic code (AEC). Part (a) reproduced from H. Snel and J. G. Schepers (Eds), 1995, 'Joint Investigation of Dynamic Inflow Effects and Implementation of an Engineering Method', by permission of the Netherlands Energy Research Foundation ECN. Part (d) reproduced from V. Akhmatov, 2003, 'On Mechanical Excitation of Electricity-producing Wind Turbines at Grid Faults', *Wind Engineering*, volume 27, issue 4, pp. 257–272, by permission of Multi-Science Publishing Co. Ltd

It is also possible to apply the power ramp (reduction of wind farm power output of up to 20 % of the rated power in less than 2 s) in order to stabilise a large wind farm during grid faults (Eltra, 2000). The power ramp can be carried out using either pitch or active-stall control modes. Before the power ramp is set to 20 % of the rated power, the wind turbine is at rated operation. Figure 27.8 shows the simulation results for both control modes.

Also, this can be used to validate the wind turbine model that uses $c_P–\lambda–\beta$ curves. In this case, the model is validated against simulations that use the already validated aerodynamic rotor model – the model with unsteady inflow phenomena. In pitch control mode there is a noticeable discrepancy due to overshoots in the mechanical torque predicted by the model with unsteady inflow phenomena. In the pitch control mode, the blades are pitched away from the incoming wind, which leads to a relatively significant change of wind profile around the turbine rotor. Consequently, the unsteady inflow phenomena become significant. The presence of overshoots in the mechanical torque of the wind turbine rotor during pitching has been confirmed by measurements (Øye, 1986; Snel and Shepers, 1995).

In active-stall control mode, overshoots in the mechanical torque are almost eliminated. The reason is that the blades are pitched across the incoming wind and the wind profile

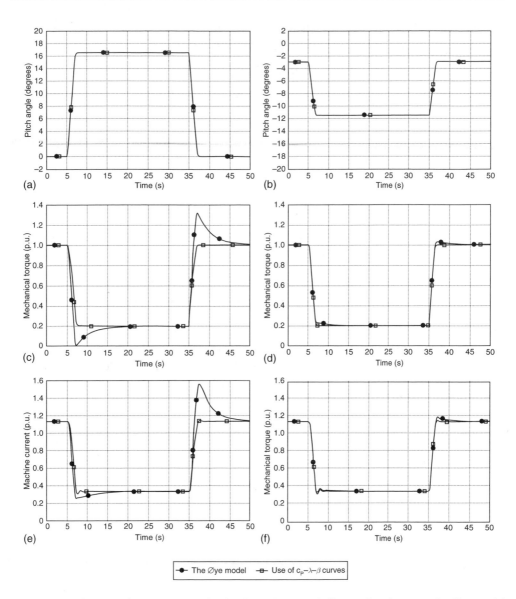

Figure 27.8 Computed power ramp using blade-angle control: Comparison between the Øye model (Øye, 1986) and the model that uses $c_P-\lambda-\beta$ curves. Computed pitch angle for (a) pitch control and (b) active stall. Computed mechanical torque for (c) pitch control and (d) active-stall. Computed machine current for (e) pitch control and (f) active-stall. Reprinted from Akhmatov, V. Analysis of Dynamic Behaviour of Electric Power Systems with Large Amount of Wind Power, Ph.D. thesis, Technical University of Denmark, Kgs. Lyngby, Denmark, copyright 2003, with permission from the copyright holder

formed around the rotor does not change significantly. So far, the following can be stated regarding the aerodynamic rotor modelling for the analysis of transient voltage stability:

- Fixed-pitch and active-stall controlled wind turbines can always be modelled with the use of their $c_P - \lambda$ curves and $c_P - \lambda - \beta$ curves, respectively, without reducing accuracy, because blade pitching does not produce any significant overshoots in the mechanical torque in this case.
- Pitch-controlled wind turbines should be modelled with use of the aerodynamic model with unsteady inflow phenomena (i.e. the Øye model), because there are overshoots in the mechanical torque during (relatively fast) pitching (for a detailed review of several aerodynamic rotor models with unsteady inflow phenomena, see Snel and Schepers, 1995). The simplified model with the $c_P - \lambda - \beta$ curves does not predict such overshoots.
- The behaviour shown in Figure 27.8 is computed at the rated operation, that is, at a wind speed of 15 m/s and for a wind turbine with a rotor diameter of 61 m. However, the overshoots will be stronger for lower wind speeds and larger diameters of wind turbine rotors (Akhmatov, 2003a). Current wind turbines have rotor diameters of around 80–90 m and there is a tendency towards even larger diameters.
- Pitch-controlled wind turbines can be represented by using $c_P - \lambda - \beta$ curves in situations where the accuracy of the mechanical torque computation is not critical to the results of the investigation.
- During sufficiently slow pitching, the overshoots are eliminated and the wind turbine model that uses $c_P - \lambda - \beta$–curves can be applied in any case without losing accuracy. I have demonstrated that this applies for pitch rates in the range of 2° per second or below for the given wind turbine (Akhmatov, 2003a).

Why should overshoots in the mechanical torque during pitching be taken into account? First, this indicates that the rate of the pitch control is restricted in order to reduce overshoots in the torque applied to the low-speed shaft of the wind turbine. Excessive overshoots may affect the gearbox of the shaft system. Second, the overshoot will transfer to the electrical and mechanical parameters of the wind turbine. In the case of Type A wind turbines, overshoots in the mechanical torque lead to similar behaviour in the generator current, the electric power and the reactive consumption and may affect terminal voltage. It is important to be aware of this when applying simplified models that use $c_P - \lambda - \beta$ curves to represent pitch-controlled wind turbines.

Similarly, there are restrictions on the rate of active stall control. However, such restrictions are caused by the stronger sensitivity of the mechanical torque to the blade angle position. The simulated curves in Figure 27.8 illustrate this. When using pitch control, the pitch angle has to be changed by 16.5° (positive) in order to reduce power output to 20 % of the rated power. When using active stall control, the pitch angle is changed by around 8° (negative) in order to reach the same power level.

27.2.4 Summary of partial validation

The parts of the dynamic model of Type A wind turbines have been validated; that is, the model of induction generators, the shaft system model and the aerodynamic rotor model. The models representing the parts of the dynamic model of fixed-speed wind turbines are in agreement with the measurements and with the results of other validated models. Thereby, the dynamic wind turbine model is also validated and can be considered sufficiently accurate for an analysis of transient voltage stability.

27.3 Full-scale Validation

Full-scale validation is necessary to ensure that the links between several component models are also sufficiently accurate. We will use an example to illustrate the full-scale validation of the dynamic wind turbine model. For this, we take measurements from a tripping–reconnection experiment executed at the Danish wind farm at Nøjsomheds Odde (Raben *et al.*, 2003). The experiment was planned and carried out by the Danish power distribution company SEAS in order to clarify details regarding the complexity of the dynamic wind turbine model. The wind farm consists of 24 fixed-speed, active stall controlled wind turbines equipped with induction generators (Raben *et al.*, 2003). The results of this work have practical implications:

- The dynamic wind turbine model is applied in the analysis of the transient voltage stability regarding the incorporation of a large amount of wind power into the Eastern Danish power system (Akhmatov, Knudsen and Nielsen, 2000).
- The wind turbines at the Nøjsomheds Odde wind farm are of the same wind turbine concept and delivered from the same manufacturer, Bonus Energy, as for the Rødsand offshore wind farm (Raben *et al.*, 2003), except they are 1 MW wind turbines.

The experiment was a cooperation between the consulting company Hansen-Henneberg (Copenhagen, Denmark), which carried out the measurements; the manufacturer Bonus Energy, which contributed its expertise regarding wind turbines and provided exact data; and the power company NESA, which planned the experiment and did the simulations and validation (Raben *et al.*, 2003).

27.3.1 Experiment outline

The internal power network of the wind farm consists of four sections with six wind turbines in each section (Raben *et al.*, 2003). Through 0.7/10 kV transformers, the wind turbines are connected to the internal power network of the wind farm. The wind farm is connected to the local 50 kV distribution power system through a 10/50 kV transformer and through a 50/132 kV transformer to the transmission power network. The short-circuit capacity at the connection point of the transmission power network (132 kV voltage level) is 350 MVA (Raben *et al.*, 2003).

In the selected section, five wind turbine generators were temporarily disconnected from the internal power network and stopped. The induction motors (IMs) of these five disconnected wind turbine generators were still in operation for maintaining cooling of the disconnected generators. The reason is that the experiment took a very limited time. These five wind turbine generators were disconnected just before the experiment started and were to be reconnected again after the experiment was finished (Raben *et al.*, 2003).

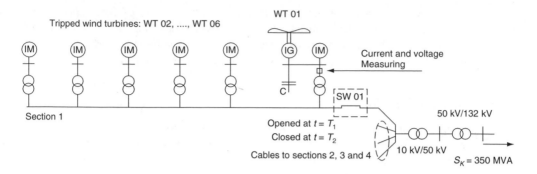

Figure 27.9 Schematic representation of the selected section with grid-connected wind turbine WT 01 during the experiment. *Note*: IM = induction motor; IG = induction generator; WT = wind turbine; SW = switch; t = time; S_k = short-circuit power Reproduced from N. Raben, M. H. Donovan, E. Jørgensen, J. Thirsted and V. Akhmatov, 2003, 'Grid Tripping and Re-connection: Full-scale Experimental Validation of a Dynamic Wind Turbine Model', *Wind Engineering*, volume 27, issue 3, pp. 205–213, by permission of Multi-Science Publishing Co. Ltd

Figure 27.9 represents the selected section schematically, including the wind turbine WT 01 and the induction motors of the five disconnected wind turbine generators.

The tripping–reconnection experiment was carried out in moderate winds (Raben *et al.*, 2003). These conditions were chosen in order successfully to reconnect the wind turbine generators and provide measurements for the validation. In high winds (i.e. close to 70–80 % of rated power) the wind turbine was tripped shortly after the reconnection was attempted. In this case, no measurements were available for the validation. Computations showed that the machine current transients would be around 10 times the rated current if reconnection were attempted at rated operation. Such excessive current transients should be the reason of disconnection. Therefore, the experiments were carried out in moderate winds.

Table 27.2 includes the main data of the wind turbine that was studied. The wind turbine operates with a 200 kW generator in moderate winds and shifts to a 1 MW generator in high winds.

At the time of tripping, T_1, switch SW 01 was opened and wind turbine WT 01 was in islanded operation with the induction motors of section 1 and the internal power network of this section given by the impedance of the transformers and the cable charging. The islanded operation lasted about 500 ms (Akhmatov, 2003a).

Table 27.2 Main wind turbine data.

Parameter	Value	Parameter	Value
Rated Power	1 MW/200 kW	Gear ratio	1:69
Synchronous rotor speed	22 rpm/15 rpm	Rated voltage	690 V
Rotor Diameter	54.2 m	Rated frequency	50 Hz

Source: Raben *et al.*, 2003. (Reproduced by permission of Multi-Science Publishing Co. Ltd.)

At the time of reconnection, T_2, switch SW 01 was closed and section 1 with wind turbine WT 01 was reconnected to the entire power system. During the experiment, the voltage and the current at the low-voltage side of the 0.7/10 kV transformer of wind turbine WT 01 were measured, as shown in Figure 27.9. The sampling frequency of the measuring equipment was 1 kHz (Raben *et al.*, 2003) and any delay in the measured signals produced by the equipment can be neglected.

During islanded operation, the no-load capacitor of wind turbine generator WT 01 was kept grid-connected. This was done to reduce the possibility of a voltage drop at the terminals of wind turbine WT 01 during islanded operation (Raben *et al.*, 2003).

27.3.2 Measured behaviour

Figure 27.10 presents the measured behaviour of the phase current, I_L, and the phase-to-phase voltage, U_{LL}. During islanded operation, the current of the no-load compensated induction generator of wind turbine WT 01 was not zero, because the induction motors in section 1 absorbed a certain amount of electric power produced by the induction generator of wind turbine WT 01; also, some reactive power was exchanged via the internal power network of section 1, as there was the no-load impedance of the transformers and the cable charging.

During islanded operation, and shortly after reconnection, higher harmonics of the fundamental frequency of 50 Hz were seen in the measured current. The higher harmonics were presumably caused by the induction machines. This coincides with the results of previous studies (Pedersen *et al.*, 2000). The measured voltage showed no (significant) drop during islanded operation.

At tripping, $t = T_1$, there was no DC offset in the measured phase current. However, there was a noticeable DC offset in the measured phase current at reconnection, $t = T_2$. There was no DC offset during tripping because the opening of switch SW 01 was an unbalanced three-phased event (Section 27.2.1 explains the elimination of the DC offset in the phase current during tripping.) Switch SW 01 was closed at the same time, $t = T_2$,

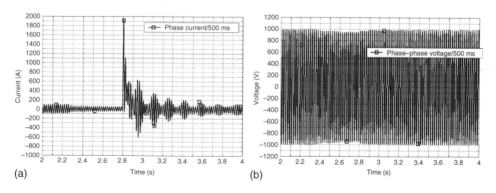

Figure 27.10 Measured behaviour of (a) phase-current and (b) phase-to-phase voltage. Reprinted from Akhmatov, V. *Analysis of Dynamic Behaviour of Electric Power Systems with Large Amount of Wind Power*, Ph.D. thesis, Technical University of Denmark, Kgs. Lyngby, Denmark, copyright 2003, with permission from the copyright holder

in all the three phases. The reconnection was therefore a balanced transient event initiating the DC offset in the phase current.

During and after islanded operation, the voltage magnitude was almost unchanged. The measured current showed a fluctuating behaviour after reconnecting. The natural frequency of the current fluctuations was about 7 Hz and could not be explained by the dynamic behaviour of the induction generator only. The value of the natural frequency indicated that the current fluctuations were related to the torsional oscillations of the shaft system of wind turbine WT 01.

27.3.3 Modelling case

The positive-sequence equivalent of the experimental network is implemented in the tool PSS/E^{TM} that is used for the validation. It is essential that the network representation and its load flow solution are in agreement with the factual conditions of the experiment. SEAS provided the data of the internal power network of the wind farm, including for the cables and the transformers, and the manufacturer Bonus Energy supplied the data of the fixed-speed wind turbines (Akhmatov, 2003a).

The values of the phase current and phase-to-phase voltage, which were measured just before opening switch SW 01, are used for initialising wind turbine generator WT 01. The numeric value of the apparent power, S, of the no-load compensated induction generator of wind turbine WT 01 is computed from:

$$S = \sqrt{P^2 + Q^2} = \sqrt{3}U_{LL}I_L, \qquad (27.1)$$

where P denotes the electric power of the induction generator of wind turbine WT 01 (P_G), and $Q = Q_G - Q_C$, where Q_G is the reactive power absorbed by the induction generator and Q_C is the reactive power of the no-load capacitor.

From the measured behaviour of the current and the voltage shown in Figure 27.10, the numeric value of the apparent power of the no-load compensated induction generator is estimated to be 80 kVA. This is the value before the islanded operation. Using this information and the data of the generator and the no-load capacitor, and applying iterations, we arrive at the initial operational point of the induction generator of wind turbine WT 01. Assuming a regular wind distribution over the wind farm, other grid-connected wind turbines (in three other sections) were initialised to the same operational points.

The operational points of the induction motors of the disconnected wind turbine generators are not given but are estimated from the measured current and voltage during islanded operation. During islanded operation, the measured value of the phase current peak is in the range of 60–65 A, and the voltage magnitude does not change significantly. This current represents mainly the electric power absorbed by the induction motors in section 1 of the wind farm. Using the above-described procedure for initialising wind turbine generator WT 01, the operational points of the induction motors are estimated. The induction motors absorb approximately 10 kW per (disconnected) wind turbine.

Despite our efforts carefully to represent the experimental conditions in the simulation case, there will always be a small number of uncertainties and missing data. The manufacturer did not provide the damping coefficients that are required for the two-mass model equations, for instance. The representation is simplified and the damping coefficients are set to zero in the simulations.

In dynamic simulations, the wind turbine generator is computed with the transient fifth-order model. For the induction motors, however, the standardised PSS/E^{TM} model CIMTR4 is used – that is, the common third-order model of induction motors (i.e. the current transients in the induction motors during reconnection are neglected). The no-load capacitor of wind turbine WT 01 is modelled in accordance with the dynamic interface of the tool PSS/E^{TM}. Therefore, the transients in the capacitor current are also neglected, even though there are probably such current transients in the measurements (Larsson and Thiringer, 1995).

27.3.4 Model validation

The main target is to validate the user-written dynamic model of Type A wind turbines, which was implemented in the tool PSS/E^{TM}. This model contains the fifth-order model of induction generators and a shaft system representation with the two-mass model. For reasons of comparison, the same simulations are carried out using of the following models:

- A model with the fifth-order model of induction generators and the lumped-mass model (i.e. assuming a very stiff shaft system);
- a model containing the common third-order model of induction generators and the two-mass model of the shaft system.

The simulation tool PSS/E^{TM} operates with positive-sequence equivalents of the network models and computes the phasor values of voltages and currents. Figure 27.11 shows the simulated behaviour, in phasor values, of the voltage and the current. For the validation, we therefore compare the measured phase values with the computed phasor values. As demonstrated in Section 27.2.1, the current phasor follows the magnitude behaviour of the phase current with developed DC offset. The behaviour simulated with the dynamic wind turbine model is in agreement with the measured behaviour. The simulated current phasor is correctly initiated, drops to around 60 A (peak-phase) during islanded operation, contains the fundamental frequency transients during reconnection and shows an oscillating behaviour with the natural frequency of around 7 Hz.

The wind turbine model containing the lumped-mass model of the mechanical system underpredicts values for the current transients at reconnection. Furthermore, it does not predict oscillations of the current after reconnecting.

The wind turbine model with the common third-order model of induction generators underpredicts values of the current at reconnection, because the fundamental frequency transients in the machine current are neglected. This model does, however, predict current oscillations with the natural frequency of around 7 Hz.

The oscillating behaviour of the current phasor is related to the shaft torsional mode and to the natural frequency of the generator itself. The simulated generator rotor speed shown in Figure 27.12 illustrates this. When representing the shaft system with the

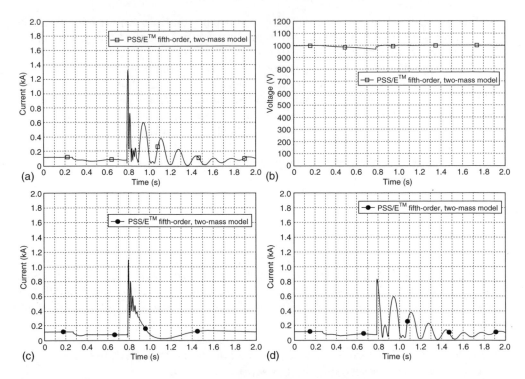

Figure 27.11 Computed behaviour of (a) current phasor and (b) voltage phasor, both computed with the dynamic wind turbine model containing the fifth-order generator model and the two-mass shaft model. Current-phasor computed with the model containing (c) the fifth-order generator model and the lumped-mass model of the mechanical system and (d) the third-order generator model and the two-mass shaft model. Reprinted from Akhmatov, V. Analysis of Dynamic Behaviour of Electric Power Systems with Large Amount of Wind Power, Ph.D. thesis, Technical University of Denmark, Kgs. Lyngby, Denmark, copyright 2003, with permission from the copyright holder

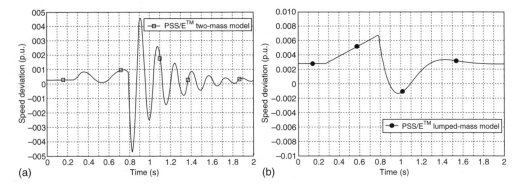

Figure 27.12 Computed behaviour of the generator rotor speed deviation (minus slip) for (a) the two-mass model and (b) the lumped-mass model. *Note*: The curves are to different scales. Reprinted from Akhmatov, V. Analysis of Dynamic Behaviour of Electric Power Systems with Large Amount of Wind Power, Ph.D. thesis, Technical University of Denmark, Kgs. Lyngby, Denmark, copyright 2003, with permission from the copyright holder

two-mass model, the behaviour of the generator rotor speed shows oscillations with the natural frequency of around 7 Hz. This is similar to the current phasor behaviour. The reason is the strong coupling between electrical and mechanical parameters in induction generators (Akhmatov, Knudsen and Nielsen 2000).

When using the lumped-mass model, the generator rotor speed does not show such oscillating behaviour. Consequently, there are no oscillations in the simulated current phasor.

To summarise this discussion, only the dynamic wind turbine model containing the fifth-order model of induction generators and the two-mass representation of the shaft system gives sufficiently accurate results. This full-scale validation also demonstrates that the links between the various parts of the dynamic wind turbine model have been accurately implemented.

27.3.5 Discrepancies between model and measurements

Having shown that the dynamic wind turbine model with the fifth-order induction generator model and the two-mass model of the shaft system gives sufficiently accurate results, we will now evaluate and explain possible discrepancies between the simulations carried out with this model and the measurements.

First, the simulated results do not include higher harmonics of the fundamental electric frequency, which are there in the measurements. The reason is that this user-written model is implemented in the tool PSS/E^{TM}, which is a fundamental frequency tool. This is one of the restrictions of this dynamic wind turbine model and of the simulation tool.

Second, during reconnection, the fundamental frequency transients of the simulated current phasor are somewhat lower than in the measured phase current. This can be explained by the fact that the induction motors and the no-load capacitor are modelled according to the interface of the tool PSS/E^{TM}. That means that during balanced transient events the fundamental frequency transients are neglected in the simulated current. However, such transients have probably influenced the measured phase current during reconnection.

Last, the damping of the current oscillations is lower in the simulated behaviour than it is in the measured behaviour. The reason is that the simulations are carried out with the damping coefficients set to zero. However, there is always finite damping in the real wind turbine construction and its generator.

This validation shows only relatively small discrepancies between simulated and measured behaviour. The discrepancies were minimised by accurately implementing the network model equivalent into the simulation tool and by accurate modelling.

27.4 Conclusions

Validation is an important and indispensable part of modelling electricity-producing wind turbines in dynamic simulation tools. The dynamic wind turbine models are to be used for analysing power system stability in the context of connecting large amounts of wind power to the grid. Therefore, the dynamic wind turbine models have to be based on sufficiently accurate assumptions and to predict accurate responses of the wind farms to transient events in power networks. There is no point in applying models that have not been validated and that are inaccurate.

The results presented here demonstrate that it is possible to develop and implement in existing simulation tools dynamic wind turbine models of sufficient accuracy and complexity. The focus here was on the validation of the user-written dynamic model of Type A wind turbines implemented in the simulation tool PSS/E^{TM}, for use by the Danish power company NESA. Type A wind turbines may be considered the simplest among the existing wind turbine concepts. Even for this concept, though, modelling and validation has been challenging.

The dynamic wind turbine model contains (a) the transient fifth-order model of induction generators, (b) the two-mass representation of the shaft system and (c) the aerodynamic rotor model and the model of blade-angle control. The common third-order model of induction generators should not be used to investigate balanced transient events with a significant voltage drop. The lumped-mass model of the shaft system does not predict accurately the interaction between the wind turbines and the electric power network during grid disturbances. It is also important to choose the details of the aerodynamic rotor model in accordance with the target of the investigation.

The dynamic wind turbine model was validated against measurements and also against simulations. Here, the partial validation has demonstrated that the main parts of the dynamic wind turbine model are developed and implemented in the simulation tool with sufficient accuracy. The full-scale validation showed that all the parts of the complete model and also the links between them are represented with sufficient accuracy.

There were only relatively small discrepancies between the results of the dynamic wind turbine model and measurements. There will always be such discrepancies because of uncertainties in the data of the wind turbine and the power network and restrictions of the simulation tool and the dynamic wind turbine model. These discrepancies can be minimised by careful modelling.

Dynamic wind turbine models (of different wind turbine concepts) should be implemented, in the near future, in the existing simulation tools that are applied to analyse power system stability. Individual validations are expensive and therefore simulation tool suppliers should commercialise and standardise the dynamic wind turbine models whenever possible. In this way such models would become accessible to every user, which would be generally advantageous. These standardised models should be validated in cooperation with wind turbine manufacturers, organisations involved in large wind farm projects or independent research organisations; the validation reports should also be published.

References

[1] Akhmatov, V. (2001) 'Note Concerning the Mutual Effects of Grid and Wind Turbine Voltage Stability Control', *Wind Engineering* 25(6) 367–371.
[2] Akhmatov, V. (2003a) *Analysis of Dynamic Behaviour of Electric Power Systems with Large Amount of Wind Power*, PhD thesis, Technical University of Denmark, Kgs. Lyngby, Denmark, available at http://www.oersted.dtu.dk/eltek/res/phd/00-05/20030403-va.html.

[3] Akhmatov, V. (2003b) 'On Mechanical Excitation of Electricity-producing Wind Turbines at Grid Faults', *Wind Engineering* 27(4) 257–272.

[4] Akhmatov, V., Knudsen, H., Nielsen, A. H. (2000) 'Advanced Simulation of Windmills in the Electrical Power Supply' *International Journal of Electrical Power and Energy Systems* 22(6) 421–434.

[5] Akhmatov, V., Knudsen, H., Nielsen, A. H., Pedersen, J. K., Poulsen, N. K. (2003) 'Modelling and Transient Stability of Large Wind Farms', *Electrical Power and Energy Systems*, 25(2) 123–144.

[6] Eltra (2000) 'Specifications for Connecting Wind Farms to the Transmission Network', ELT1999-411a, Eltra Transmission System Planning, Denmark.

[7] Hinrichsen, E. N., Nolan, P. J. (1982) 'Dynamics and Stability of Wind Turbine Generators', *IEEE Transactions on Power Apparatus and Systems*, 101(8) 2640–2648.

[8] Kazachkov, Y. A., Feltes, J. W., Zavadil, R. (2003) 'Modeling Wind Farms for Power System Stability Studies' presented at IEEE Power Engineering Society General Meeting, Toronto, Canada.

[9] Knudsen, H., Akhmatov, V. (1999) 'Induction Generator Models in Dynamic Simulation Tools', *International Conference on Power System Transients IPST'99, Budapest, Hungary*, pp. 253–259.

[10] Larsson, A., Thiringer, T. (1995) 'Measurements on and Modelling of Capacitor-connecting Transients on a Low-voltage Grid Equipped with Two Wind Turbines', *International Conference on Power System Transients IPST'95, Lisbon, Portugal*, pp. 184–188.

[11] Øye, S. (1986) 'Unsteady Wake Effects Caused by Pitch-angle Changes', in *IEA R&D WECS Joint Action on Aerodynamics of Wind Turbines, 1st Symposium, London, U.K.*, pp. 58–79.

[12] Pedersen, J. K., Akke, M., Poulsen, N. K., Pedersen, K. O. H. (2000) 'Analysis of Wind Farm Islanding Experiment', *IEEE Transactions on Energy Conversion* 15(1) 110–115.

[13] Pedersen, J. K., Pedersen, K. O. H. Poulsen, N. K., Akhmatov, V., Nielsen, A. H. (2003) 'Contribution to a Dynamic Wind Turbine Model Validation from a Wind Farm Islanding Experiment', *Electric Power Systems Research* 64(2) 41–51.

[14] Raben, N., Donovan, M. H., Jørgensen, E., Thisted, J., Akhmatov, V. (2003) 'Grid Tripping and Re-connection: Full-scale Experimental Validation of a Dynamic Wind Turbine Model', *Wind Engineering* 27(3) 205–213.

[15] Snel, H., Schepers, J. G. (Eds) (1995) 'Joint Investigation of Dynamic Inflow Effects and Implementation of an Engineering Method', ECN-C-94-107, Netherlands Energy Research Foundation, ECN, Petten, The Netherlands.

28

Impacts of Wind Power on Power System Dynamics

J. G. Slootweg and W. L. Kling

28.1 Introduction

In most countries, the amount of wind power generation integrated into large-scale electrical power systems covers only a small part of the total power system load. However, the amount of electricity generated by wind turbines is increasing continuously. Therefore, wind power penetration in electrical power systems will increase in the future and will start to replace the output of conventional synchronous generators. As a result, it may also begin to influence overall power system behaviour. The impact of wind power on the dynamics of power systems should therefore be studied thoroughly in order to identify potential problems and to develop measures to mitigate those problems.

The dynamic behaviour of a power system is determined mainly by the generators. Until now, nearly all power has been generated with conventional directly grid-coupled synchronous generators. The behaviour of the grid-coupled synchronous generator under various circumstances has been studied for decades and much of what is to be known is known. However, although this generator type used to be applied in wind turbines in the past, this is no longer the case. Instead, wind turbines use other types of generators, such as squirrel cage induction generators or generators that are grid-coupled via power electronic converters. The interaction of these generator types with the power system is different from that of a conventional synchronous generator. As a consequence, wind turbines affect the dynamic behaviour of the power system in a way that might be different from synchronous generators. Further, there are also differences in the interaction with the power system between the various wind turbine types presently applied, so that the various wind turbine types must be treated separately. This also applies to the various wind park connection schemes that can be found discussed in the literature.

Wind Power in Power Systems   Edited by T. Ackermann
© 2005 John Wiley & Sons, Ltd   ISBN: 0-470-85508-8 (HB)

This chapter discusses the impact of wind power on power system dynamics. First, we present the concepts of power system dynamics and of transient and small signal stability, as well as the most important currently used wind turbine types. Then, we will study the impact of these wind turbine types and of various wind farm interconnection schemes on the transient stability of the power system. For this, we will analyse the response to disturbances, such as voltage and frequency changes. Finally, we will deal with the impact of wind power on the small signal stability of power systems and use eigenvalue analysis for this purpose.

28.2 Power System Dynamics

Power system dynamics investigates how a power system responds to disturbances that change the system's operating point. Examples of such disturbances are frequency changes because a generator trips or a load is switched in or disconnected; voltage drops due to a fault; changes in prime mover mechanical power or exciter voltage, and so on. A disturbance triggers a response in the power system, which means that various properties of the power system, such as node voltages, branch currents, machine speeds and so on, start to change. The power system is considered stable if the system reaches a new steady state and all generators and loads that were connected to the system before the disturbance are still connected. The original power system is considered unstable if, in the new steady state, loads or generators are disconnected.

Two remarks must be made at this point. First, when a system is stable, the new steady state can either be identical to or different from the steady state in which the system resided before the disturbance occurred. This depends on the type of disturbance, the topology of the system and the controllers of the generators. Second, that a power system is unstable does not necessarily mean that a disturbance leads to a complete blackout of the system. Rather, the system's topology is changed by protection devices that disconnect branches, loads and/or generators during the transient phenomenon, in order to protect these. In most cases the changed system will be able to reach a new steady state, thus preventing a complete blackout. However, although the 'new' system that results after the changes is stable, the 'old' system was unstable and stability has been regained by changing the system's topology.

There are two different methods to investigate the dynamics of a power system in order to determine whether the system is stable or not. The first method is time domain analysis. The type of time domain analysis that we use here is also referred to as dynamics simulation, fundamental frequency simulation or electromechanical transient simulation (see also Chapter 24). This approach subjects the system to a disturbance after which its response (i.e. the quantitative evolution of the system's properties over time) is simulated. In this way, it can be decided whether the system is stable or not. In the case of instability, strategies can be designed to change the system's topology in such a way that stability is regained with minimum consequences to loads and generators.

The second method is frequency domain analysis, also referred to as an analysis of the small signal properties of the system or as eigenvalue analysis. Frequency domain analysis studies a linearised representation of the power system in a certain state. The linearised representation makes it possible to draw conclusions as to how the power system will respond in the analysed state to incremental changes in the state variables. The results of the frequency domain analysis consist of the following:

- An overview of the system's eigenvalues; in case of complex eigenvalues, the damping and frequency of the corresponding oscillation can be calculated from the eigenvalue, whereas in case of real eigenvalues the associated time constant can be calculated.
- An overview of the participation factors for each of the eigenvalues or at least those eigenvalues that are considered of importance, because their damping is low, for instance; the participation factors contain information on the relationship between the calculated eigenvalues and the system's state variables. This information can be used to identify the state variables that can be used to affect a certain eigenvalue.

For more detailed information on the frequency domain analysis of a power system, including the calculation and the use of participation factors, see, for example, Kundur (1994).

The two methods mentioned are complementary. Dynamics simulation can analyse the complete response of a system to a disturbance. However, the simulation contains information on the system's response only to the specific disturbance that is studied. For other disturbances new simulations have to be carried out. Frequency domain analysis, in contrast, yields a complete overview of the response to an incremental change in any of the system's state variables for the system in its current operating state. However, owing to the nonlinearity of a power system, the results are valid only for the investigated power system topology and for small changes in the state variables. If the topology of the power system is changed or the state variables change significantly (e.g. by connecting or disconnecting a load or a line or by changing the operating points of the generators) different eigenvalues will result, because if the changes in state variables are large the linearised representation of the system will no longer be valid.

28.3 Actual Wind Turbine Types

The vast majority of wind turbines that are currently being installed use one of three main types of electromechanical conversion system. The first type is known as the Danish concept. In Chapter 4, the Danish concept is introduced as Type A.[1] An (asynchronous) squirrel cage induction generator is used to convert the mechanical energy into electricity. Owing to the different operating speeds of the wind turbine rotor and the generator, a gearbox is necessary to match these speeds. The generator slip slightly varies with the amount of generated power and is therefore not entirely constant. However, because these speed variations are in the order of 1%, this wind turbine type is normally referred to as a constant-speed or fixed-speed wind turbine. The Danish, or constant-speed, design is nowadays nearly always combined with stall control of the aerodynamic power, although pitch-controlled constant-speed wind turbine types have been built, too.

---

[1] For definitions of wind turbine Types A–D, see Section 4.2.3.

The second type uses a doubly fed induction generator instead of a squirrel cage induction generator, introduced in Chapter 4 as Type C. Similar to the previous type, it needs a gearbox. The stator winding of the generator is coupled to the grid and the rotor winding to a power electronic converter, nowadays usually a back-to-back voltage source converter with current control loops. In this way, the electrical and mechanical rotor frequencies are decoupled, because the power electronic converter compensates the difference between mechanical and electrical frequency by injecting a rotor current with variable frequency. Variable-speed operation becomes possible. That means that the mechanical rotor speed can be controlled according to a certain goal function, such as energy yield maximisation or noise minimisation. The rotor speed is controlled by changing the generator power in such a way that it equals the value derived from the goal function. In this type of conversion system, the control of the aerodynamic power is usually performed by pitch control.

The third type is called the direct-drive wind turbine because it does not need a gearbox. It corresponds to Type D in Chapter 4. A low-speed multipole synchronous ring generator with the same rotational speed as the wind turbine rotor converts the mechanical energy into electricity. The generator can have a wound rotor or a rotor with permanent magnets. The stator is not coupled directly to the grid but to a power electronic converter. This may consist of a back-to-back voltage source converter or a diode rectifier with a single voltage source converter. The electronic converter makes it possible to operate the wind turbine at variable speed. Similar to the case for Type C wind turbines, pitch control limits the mechanical power input. The three main wind turbine types are illustrated in Figure 19.2 (page 421).

Apart from the wind turbine types depicted in Figure 19.2, other types have been developed and used. They include wind turbines with directly grid-coupled synchronous generators and with conventional synchronous as well as squirrel cage induction generators combined with a gearbox and coupled to the grid with a full-scale power electronic converter. Currently, there are hardly any of these wind turbine types on the market, though, and we will not take them into consideration here.

28.4 Impact of Wind Power on Transient Stability

28.4.1 Dynamic behaviour of wind turbine types

28.4.1.1 Constant-speed wind turbines

Constant-speed wind turbines (Type A) use a directly grid-coupled squirrel cage induction generator to convert mechanical into electrical power. The behaviour of a constant-speed wind turbine is determined by the intrinsic relationship between active power, reactive power, terminal voltage and the rotor speed of the squirrel cage induction generator. This relationship can be studied using the network equivalent, depicted in Figure 28.1 (Kundur, 1994). In this figure, U is the voltage, I is current, s the slip, R the resistance and L the reactance. The indices σ, t, s, m and r stand for leakage, terminal, stator, mutual and rotor, respectively. The values of the generator parameters are given in Table 25.3 (page 567).

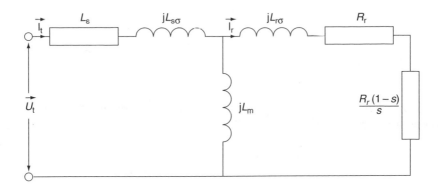

Figure 28.1 Network equivalent of squirrel cage induction generator. *Note*: $U =$ voltage; $I =$ current; $s =$ slip; $j = \sqrt{-1}$; $L =$ reactance; $R =$ resistance; subscripts σ, t, s, m and r refer to leakage, terminal, stator, mutual and rotor, respectively

Squirrel cage induction generators are likely to become unstable after a voltage drop (Van Cutsem and Vournas, 1998). The explanation for this is as follows. Figure 19.3 (page 422) illustrates the relationship between active power output and rotor slip as well as between reactive power consumption and rotor slip with the terminal voltage U_t as a parameter. It shows that:

- The lower the terminal voltage, the larger the absolute value of the rotor slip that corresponds to a certain amount of active power generation.
- The larger the rotor slip, the larger the reactive power consumption.

If the generator terminal voltage drops (e.g. because of a fault) only a small amount of electrical power can be fed into the grid because the generated electrical power is proportional to the terminal voltage. However, the wind continues to supply mechanical power. Owing to the resulting imbalance between supplied mechanical power and generated electrical power, the generator speeds up. This results in a decreasing slip. Once the fault is cleared, the squirrel cage induction generator draws a large amount of reactive power from the grid because of its high rotational speed, as can be seen in Figure 19.3(b). Owing to this reactive power consumption, it can happen that the terminal voltage recovers only relatively slowly after the fault is cleared.

However, if the generator terminal voltage is low, the electrical power generated at a given slip is lower than that at nominal terminal voltage, as shown in Figure 19.3(a) (page 422). If the rotor accelerates faster than the terminal voltage is restored, the reactive power consumption continues to increase. This leads to a decrease in the terminal voltage and thus to a further deterioration of the balance between mechanical and electrical power and to a further acceleration of the rotor. Eventually, the voltage at the wind turbine will collapse. It may then be necessary to disconnect the turbine from the grid to allow the grid voltage to restore.

Depending on the design and the settings of its protection system, the wind turbine will either be stopped by its undervoltage protection or accelerate further and be stopped and disconnected by its overspeed protection. It can only be reconnected after restoration of the grid voltage in the affected parts of the network. That may take several minutes, particularly if other protection systems were activated during the disturbance too. The exact quantitative behaviour of the terminal voltage and the required restoration time depend on the actual wind speed, wind turbine characteristics, network topology and protection system settings.

Wherever possible, a fault should be removed from the system before the wind turbine becomes unstable because of the mechanism pointed out above. Otherwise, a large amount of generation may be lost. The fault should be cleared quickly in order to limit the overspeeding and, consequently, to limit the amount of reactive power that is consumed for restoring the voltage. The time that is available to clear the fault before it leads to voltage and rotor speed instability is called the critical clearing time.

Akhmatov *et al.* (2003) propose a number of countermeasures to prevent instability of constant speed wind turbines (see also Chapter 29):

- Constant-speed wind turbines, which are usually stall-controlled, can be equipped with pitch drives that quickly increase the pitch angle when an acceleration of the rotor is detected. This reduces the mechanical power and consequently limits the rotor speed and the reactive power consumption after the fault, thus reducing the risk of instability.
- The wind turbines can be equipped with a controllable source of reactive power [e.g. a static condenser (STATCON) or static VAR compensator (SVC)] to deliver the reactive power required to increase the speed at which the voltage is restored.
- Mechanical and/or electrical parameters of the wind turbine and the generator can be changed, but this often has the disadvantage of increased cost, reduced efficiency and a more complicated mechanical construction.

These measures aim either at reducing the amount of overspeeding during the fault or at supplying reactive power to accelerate voltage restoration after the fault. Although the measures mitigate the problem, they do not completely solve it: it originates from the working principle of an induction generator, which is not principally changed by the above measures.

It should be noted that constant-speed wind turbines may become unstable at times other than after a fault. The above sequence of events may also be initiated by a relatively small drop in terminal voltage. This may be the result of a nearby synchronous generator tripping or a highly inductive load switching in, for instance. When the wind turbine delivers its nominal power and the terminal voltage drops slightly, rotor speed will increase, because a larger slip is required to deliver nominal power at a terminal voltage below nominal. This leads to an increase in the reactive power consumption, which in turn results in a further decrease in terminal voltage. This can lead to a voltage collapse that is not preceded by a short circuit. This is an example of voltage instability.

The response of a squirrel cage induction generator to changes in grid frequency is similar to that of synchronous generators. The frequency of the stator field is identical to the grid frequency divided by the number of pole pairs of the generator. If this frequency changes, the mechanical rotor frequency changes as well. The change in energy stored in the rotating mass, which is caused by the rotor speed change is either fed into the system (in the case of a drop in grid frequency) or drawn from the system (in the case of an increase in grid frequency).

However, there is an important difference in the responses of synchronous generators in power plants and of squirrel cage induction generators that are used in constant-speed wind turbines. In conventional power plants, a controllable prime mover is used. If there is a frequency drop, the mechanical power applied to the generator can be adjusted in order to counteract this frequency drop. When the frequency increases, the prime mover power is reduced, whereas when the frequency decreases, the prime mover power is increased. In wind turbines, this is not possible because the wind cannot be controlled. Thus, although constant-speed wind turbines tend to damp frequency deviations by either releasing energy from or storing energy in the rotating mass, the effect is weaker than in the case of synchronous generators in power plants. We would like to stress here that the different responses are not due to the different generator types but to the fact that in power plants the prime mover can be adjusted to counteract frequency changes. This is normally not possible in wind turbines.

28.4.1.2 Variable-speed wind turbines

The dynamic behaviour of variable-speed wind turbines is fundamentally different from that of constant-speed wind turbines. Variable-speed wind turbines use a power electronic converter to decouple mechanical frequency and electrical grid frequency. This decoupling takes place not only during normal operation but also during and after disturbances.

Power electronic components are very sensitive to overcurrents because of their very short thermal time constants, as mentioned in Chapter 25. When a drop in terminal voltage occurs, the current through the semiconductors increases very quickly. The controller of the power electronic converter samples many quantities, such as terminal voltage, converter current, grid frequency and so on, at a high sampling frequency, in the order of kilohertz. A fault is therefore noticed instantly by the power electronic converter. The variable-speed wind turbine is then quickly disconnected in order to prevent damage to the converter.

If a power system has a high penetration of wind power from variable-speed turbines, this effect is naturally undesirable. If the variable-speed wind turbines are disconnected at a relatively small voltage drop, a large amount of generation might be lost. Such a situation may arise when a fault in the high-voltage transmission grid causes a voltage drop in a large geographical region. This would lead to severe problems with the power balance in the associated control area or even on the system level.

Therefore, grid companies with large amounts of wind power presently revise their connection requirements for wind turbines, and are starting to require wind turbines to remain connected to the grid during a fault (E.On Netz, 2001; see also Chapter 7). This does not seem to be a major problem, though. The literature presents approaches where the semiconductors are controlled in such a way that during voltage drops the current is limited to the nominal value (Petterson, 2003; Saccomando, Svensson and Sannino, 2002). Presently, most variable-speed wind turbines are not equipped with current controllers for a continued operation during voltage drops. However, if system interaction were to require variable-speed wind turbines to have current controllers this would not be a very complicated issue.

When the fault is cleared, variable-speed wind turbines must resume normal operation. In contrast to the case for directly grid-coupled generators, for variable-speed turbines there are various degrees of freedom to switch them back to normal operation. They are not governed by the intrinsic behaviour of the generator. It would be possible, for instance, to generate extra reactive power when the voltage starts to increase again, in order to accelerate voltage restoration.

Another question that must be answered is what to do with the energy that is stored in the rotating mass because of the imbalance between mechanical power supplied and electrical power generated during the fault. One possibility is to feed the energy into the system. This would mean that shortly after the fault the constant-speed wind turbines would generate more power than before the fault and more than would be possible given the actual wind speed. After a while, the wind turbines would return to normal operation where generated power was in line with wind speed. The exact course of the transition would depend on the design of the controllers. Another possibility would be to let the turbine generate its prefault amount of power while using the pitch controller to slow down the rotor. The tuning of the fault response of a variable-speed wind turbine is similar to that of a high-voltage direct-current (HVDC) link, which is described extensively in the literature (Kundur, 1994).

The decoupling of mechanical rotor frequency and electrical grid frequency in variable-speed wind turbines also affects the response of variable-speed wind turbines to changes in grid frequency. If the grid frequency changes because of a mismatch between generation and load, the mechanical frequency of a variable-speed wind turbine does not change. Thus, no energy is stored in or withdrawn from the rotating mass and drawn from or supplied to the system, as would be the case with a constant-speed wind turbine.

Rather, the controllers of the power electronic converters compensate changes in grid frequency, and the mechanical rotor frequency is not affected. It would be possible, however, to equip a variable-speed wind turbine with additional controllers that change the active power based on the measured value of the grid frequency. The response would then be similar to the intrinsic behaviour of directly grid-coupled generators, within the limits imposed by the actual value of the wind speed, though.

28.4.2 Dynamic behaviour of wind farms

Increasingly, wind turbines are being or will be grouped in wind farms, either onshore or offshore. General reasons behind this are the desire to use good wind locations effectively and to concentrate the visual impact of wind turbines to a limited area. Modern wind turbines can reach a total height of 150 m if one of the blades is in a vertical position. Wind farms tend to be located offshore because the turbulence intensity is lower, wind speeds are higher, noise problems are less severe and the visual impact is even further reduced if the wind farm is located far away from the coastline.

There is a number of wind farm configurations that are feasible (Bauer *et al.*, 2000). All possible configurations share certain characteristics regarding the interaction between the wind farm and the grid, namely those that are inherently associated with using wind turbines for power generation. Examples of such characteristics are fluctuating output power and poor controllability and predictability of generated power. However, the response of the wind farm to voltage and frequency disturbances strongly depends on the wind farm's configuration. Therefore, we will discuss the response to a terminal voltage drop associated with a fault and to a frequency change separately for the various wind farm configurations.

If conventional AC links and transformers are used for implementing both the infrastructure within the wind farm and the grid connection, the response of the wind farm to disturbances is determined by the wind turbines, because the connections themselves are passive elements. In this case, the response to voltage and frequency changes depends on the wind turbine type used. The main distinction here is between constant-speed and variable-speed turbines, as discussed in Section 28.4.1. If a DC link is used to connect the wind farm to the grid, the wind turbines are electrically decoupled from the analysed system. The response of the wind farm is governed by the technology used for implementing the DC connection rather than by the wind turbine concept. Thus, the differences between the various wind turbine types, pointed out in Section 28.4.1, become largely irrelevant.

The DC connection can be of a current source type or of a voltage source type. Conventional HVDC technology based on thyristors is a current source type of DC connection (see also Chapters 7 and 22). Power electronic AC/DC converters based on thyristors always have a lagging current and thus consume reactive power. Capacitors can be installed to compensate for this reactive power. With this type of DC link, additional technological components become necessary in order to control reactive power, similar to the case of constant-speed wind turbines.

If there is a voltage drop caused by a fault in the power system the response of a current source HVDC connection can be summarised as follows (Kundur, 1994). During a voltage drop, even in the case of only small dips, commutation failures are likely to occur. That means that the current does not transfer from one semiconductor switch to another. If the voltage stays below nominal, but increases sufficiently to clear the commutation failure, the system may continue to operate at a lower DC voltage, thus transferring less power. If the voltage stays low and commutation failures continue to occur, the inverter is bypassed by shorting its input and blocking its output. Once the voltage is restored, the inverter is reconnected. The time of recovery is in the range of 100 ms to several seconds, depending on the control strategy and the characteristics of the grid to which the inverter is connected. It should be stressed that the disconnection and reconnection of a HVDC link leads to transient phenomena in the wind farm and in the power system to which it is connected.

In a voltage source type of DC connection, integrated gate bipolar transistor (IGBT) or metal oxide semiconductor field effect transistor (MOSFET) switches are used. Such technology is often referred to as HVDC Light[TM] or HVDC Plus, depending on the manufacturer of the system.[2] The interaction of a voltage source HVDC connection with the grid is similar to that of a variable-speed wind turbine with a direct-drive generator. The technology of both is essentially identical. This means that the converter sets the limits to the reactive power control. It is possible to limit the converter current

---

[2] HVDC Light[R] is produced by ABB, HVDC Plus is produced by Siemens.

during faults to the nominal value in order to keep the converter grid connected. For this, the semiconductors have to be controlled accordingly.

With respect to the factors that determine the response of a wind farm to frequency changes, the conclusions that can be drawn are similar to those regarding individual wind turbines, presented in Section 28.4.1. First, in case of a wind farm consisting completely of passive AC components, the response to frequency changes is determined by the turbines. The response of constant-speed and variable-speed wind turbines to frequency changes was described in Section 28.4.1. If a DC link is used to connect the wind farm to the grid, the wind turbines in the wind farm will not 'notice' changes in grid frequency, because the HVDC connection decouples the internal frequency of the wind farm and the grid frequency. It depends, therefore, on the control of the converter at the grid side of this connection whether the wind farm responds to a change in grid frequency by changing the amount of generated power or not.

28.4.3 Simulation results

In this section, simulations are used to provide a quantitative illustration of the qualitative analysis of the impact that wind power has on the transient behaviour and stability of a power system. The widely known power system dynamics simulation program PSS/E[TM] was used for these simulations.[3] The New England test system, consisting of 39 buses and 10 generators, was used as the test system to which wind turbine models were connected (Pai, 1989). The reasons for using a test system, rather than a model of a real power system, are:

- Models of real power systems are not very well documented and the data are partly confidential. This tends to shift the focus from using the model to investigate certain phenomena towards improving the model itself. In contrast, all parameters of test systems are given in the literature, which makes them convenient to use.
- Models of real power systems tend to be very big, which makes the development and calculation of numerous scenarios cumbersome and time-consuming and complicates the identification of general trends.
- The results obtained with models of real systems are less generic than are those obtained with general purpose test systems, and such test systems can be more easily validated by and compared with the results of other investigations.

A one line diagram of the New England test system is depicted in Figure 28.2.

An aggregated model was used to represent the wind farms in order to avoid modelling each wind turbine individually and including the internal infrastructure of the wind farm (Akhmatov and Knudsen, 2002; Slootweg et al., 2002; see also Chapter 29). Further, both types of variable-speed wind turbine were represented with a single model, a general variable-speed wind turbine model (Slootweg et al., 2003b). As their interaction

---

[3] PSS/E[TM] (Power System Simulator for Engineers) is a registered trademark of Shaw Power Technologies Inc. (PTI), New York.

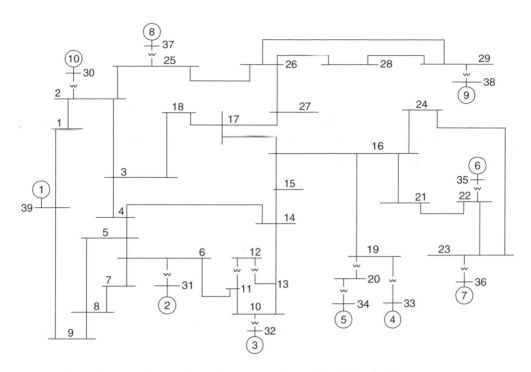

Figure   28.2 The New England test system. *Note*: 1–39 = buses; 1–10 = generators

with the power system is governed mainly by the controllers of the power electronic converter, both types can be assumed to show very similar behaviour if the control approaches are identical.

28.4.3.1 Response to a drop in terminal voltage

This section looks at the mechanisms that lead to voltage and rotor speed instability. First, the synchronous generator at bus 32 of the test system is replaced with a wind farm with constant-speed wind turbines, and a 150 ms fault occurs at bus 11. Figures 28.3(a.i) and 28.3(a.ii) depict the rotor speed and the voltage at bus 32 of the constant-speed wind turbines, respectively. Figure 28.3(a.ii) shows that the voltage does not return to its predisturbance value. Instead, the voltage oscillates. The oscillation is caused by the relatively soft shaft of the wind turbine. The shaft softness causes a large angular displacement between the two shaft ends and a significant amount of energy is stored in the shaft. When the fault occurs, this energy is released and rotor speed increases quickly. Although the voltage is restored, the shaft causes the oscillation that can be seen in Figure 28.3(a.ii)

Then, the synchronous generator at bus 32 is replaced with a wind farm with variable-speed wind turbines, either with (dotted line) or without (solid line) terminal voltage or reactive power control. The same fault is applied. Figures 28.3(b.i) and 28.3(b.ii) depict the results. Figure 28.3(b.ii) show that the terminal voltage behaves more favourably for

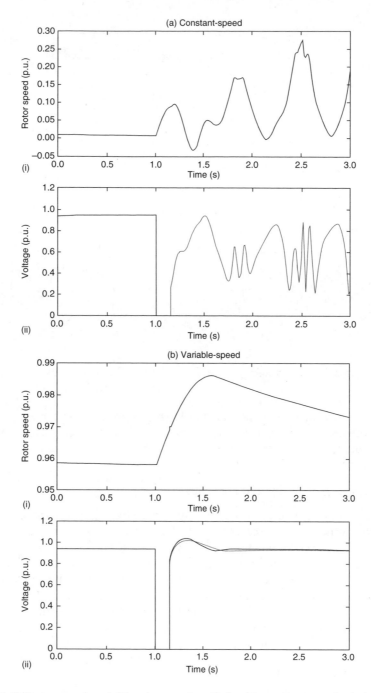

Figure 28.3 (i) Rotor speed and (ii) voltage at bus 32 for (a) constant-speed wind turbines and (b) variable-speed wind turbines after a 150 ms fault at bus 32 (see Figure 28.2). In part (b.ii), the dotted and solid lines correspond to variable-speed wind turbines operating at unity power factor or in terminal voltage control mode, respectively

variable-speed wind turbines than for constant-speed wind turbines [Figure 28.3(a.ii)], particularly if they are equipped with terminal voltage control. Further, it can be seen that the rotor speed of the variable-speed turbine behaves more smoothly than that of a constant-speed wind turbine. Variable-speed wind turbines do not have to be resynchronised by the system, as is the case with constant-speed turbines. Rather, the controllers of the variable-speed turbine control the rotor speed independent of the grid frequency. Constant-speed turbines do not have such controllers. As mentioned earlier, the exact behaviour of the rotor speed depends on the control strategy.

Voltage and rotor speed instability are not triggered only by faults. The tripping of a synchronous generator from the grid may also result in a voltage drop that causes voltage and rotor speed instability. Figures 28.4(a.i) and 28.4(a.ii) depict the rotor speed of a wind farm with constant-speed turbines and the voltage at bus 32 when the synchronous generator at bus 31 trips. Figures 28.4(b.i) and 28.4(b.ii) present the rotor speed and the voltage at bus 32 for variable-speed turbines with (dotted line) and without (solid line) terminal voltage or reactive power control. Figure 28.4 shows that constant-speed wind turbines become unstable as a result of bus voltage decreases, whereas variable-speed wind turbines stay connected. Further, it can be seen that wind turbines with terminal voltage control behave more favourably than do wind turbines without voltage control, because they attempt to bring the voltage back to its predisturbance value.

28.4.3.2 Response to a change in grid frequency

As discussed above, the mechanical and electrical behaviour of variable-speed wind turbines is decoupled by power electronic converters. As a result, mechanical quantities, such as rotor speed and mechanical power, are largely independent of electrical quantities, such as active and reactive power and generator terminal voltage and frequency. Therefore, the mechanical frequency of variable-speed wind turbines does not react to a change in electrical grid frequency. The energy stored in the rotating mass of variable-speed wind turbines is not released when the grid frequency drops. In practice, it is much more likely that the frequency decreases than that it increases. To illustrate this effect, the generators at buses 32, 36 and 37, generating 1750 MW, are replaced by wind farms. This corresponds to a wind power penetration of about 28.5 %.

If wind turbines replace synchronous generators on such a large scale, the power from the wind turbines has not only to replace the active power from the synchronous generators but also to take on the other tasks carried out by the synchronous generators. This refers mainly to the generation of reactive power and voltage control, as discussed in Chapter 19. Otherwise, it will not be possible to keep the voltage at each bus within the allowable deviation from its nominal value. The reason is that there are only limited options for controlling the voltage if wind turbines replace conventional synchronous generators. Therefore, this section studies only constant-speed wind turbines with SVCs, and variable-speed wind turbines with terminal voltage controllers. Constant-speed wind turbines without controllable reactive power source, and variable-speed wind turbines running at unity power factor, are not taken into account: it is unreasonable to look at the dynamic behaviour of a system that cannot be operated because of a lack of voltage control possibilities.

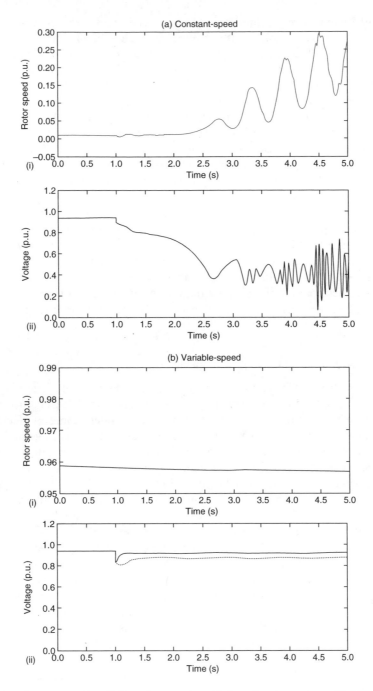

Figure   28.4 (i) Rotor speed and (ii) voltage at bus 32 for (a) constant-speed wind turbines and (b) variable-speed wind turbines after the tripping of the synchronous generator at bus 31 (Figure 28.2). In part (b.ii), the dotted and solid lines correspond to variable-speed wind turbines operating at unity power factor or in terminal voltage control mode, respectively

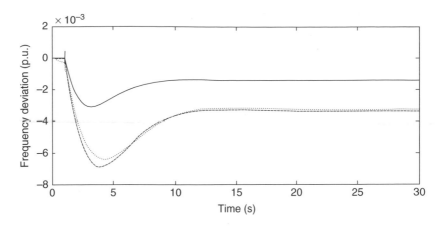

Figure 28.5 Frequency drop after the tripping of the synchronous generator at bus 30 (Figure 28.2): base case (solid line) compared with wind farms with variable-speed wind turbines (dashed line) and constant-speed wind turbines (dotted line) at buses 32, 36 and 37 (Figure 28.2), corresponding to a wind power penetration level of 28.5 %

Figure 28.5 shows the simulation results for the case in which the synchronous generator at bus 30, delivering 570 MW, trips. The solid line corresponds to the base case, the dashed line to the case with variable-speed wind turbines, and the dotted line to the case with constant-speed wind turbines. Figure 28.5 illustrates that the frequency drop is both deeper and lasts longer for wind turbines than for synchronous generators. As discussed above, the reason is that the wind turbines are not equipped with governors. This is not possible either, because their prime mover is uncontrollable. Figure 28.5 also shows that the frequency drop is deepest in the case of variable-speed wind turbines because of the decoupling of electrical and mechanical quantities.

28.4.3.3 Rotor speed oscillations of synchronous generators

Now, we want to analyse the third effect of increasing wind turbine penetration in electrical power systems; that is, possible changes in the damping of the rotor speed oscillations of the remaining synchronous generators after a fault. Again, only constant-speed wind turbines with SVCs, and variable-speed wind turbines with terminal voltage controllers, are included in the analysis, because we assume a high wind energy penetration. It would be unreasonable to include here constant-speed wind turbines without controllable reactive power source and variable-speed wind turbines at unity power factor.

A 150 ms second fault is applied to bus 1. Figure 28.6 depicts the results (i.e. the rotor speed of the synchronous generators at buses 30, 31, 35 and 38). The solid line corresponds to the base case, the dotted line to constant-speed wind turbines and the dashed line to variable-speed wind turbines. This figure shows that the presence of the wind turbines does not significantly affect the time constant of the damping of the oscillations that occur after the fault. Although the oscillations have rather different shapes, the

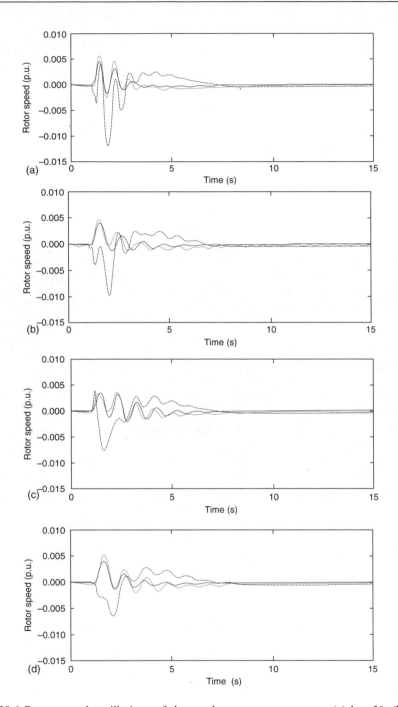

Figure   28.6 Rotor speed oscillations of the synchronous generators at (a) bus 30, (b) bus 31, (c) bus 35 and (d) bus 38 at a wind power penetration level of 28.5 %. The solid line corresponds to the base case, the dotted line to constant-speed wind turbines and the dashed line to variable-speed wind turbines

system remains stable and has in all cases returned to steady state after about 10 s. Figure 28.6 also shows the following:

- The shape of the oscillations changes when constant-speed wind turbines are connected. This is probably caused by the shaft of the turbines. As already pointed out, the shaft is relatively soft, resulting in a large angular displacement between the shaft ends and in a significant amount of energy being stored in the shaft. When the fault occurs, this energy is released and rotor speed increases quickly. While the voltage is restored, the shaft causes an oscillation, as can be seen in Figure 28.6.
- With variable-speed wind turbines, the rotor speed of some of the generators drops instead of increasing when a fault occurs. This is particularly true for remote wind turbines. The reason is that the variable-speed wind turbines generate hardly any power during the fault and they need time to bring their power back to the prefault value. In our simulation, this took 0.5 s after the terminal voltage was restored to 80 % of the nominal value. During the fault, there is a generation shortage, which causes the rotors of some of the synchronous generators to slow down rather than to speed up, which usually is the case during a fault.

28.5 Impact of Wind Power on Small Signal Stability

28.5.1 Eigenvalue–frequency domain analysis

When analysing the small signal stability of a power system, the nonlinear equations describing its behaviour are linearised. This means that the nonlinear equations are approximated with a polynomial of which only the first-order (linear) terms are taken into account, while neglecting the higher-order terms. The resulting equations can be written in matrix notation and the eigenvalues of one of the matrices, the so-called state matrix, give important information on the system's dynamics. The reader is referred to textbooks on this topic for a more formal, mathematical treatment of the linearisation of a power system (Kundur, 1994).

As mentioned in Section 28.2, the information that the eigenvalues contain is valid only for the actual state of the system and when the disturbance that elicits a response of the system is small. This can be explained as follows. Because a power system is nonlinear, its response to a certain disturbance depends on its operating point. This also applies to the eigenvalues, which describe the system's response and will change accordingly when the system's response changes. Thus, the results of the linearisation depend on the initial operating point: a linearisation of another operating point would yield different eigenvalues. Further, if a large disturbance is applied to the power system, the contribution of the nonlinearities that have been neglected in the linearisation becomes substantial and the assumptions on which linearisation is based are no longer valid. The study of the response of the power system to large disturbances, such as faults or unit trips, requires time domain simulations rather than frequency domain simulations.

28.5.2 Analysis of the impact of wind power on small signal stability

Power system oscillations have been studied in many contexts. Examples are the impact of HVDC links and their controllers on power system oscillations, the impact of long-distance power transmission and the damping of power system oscillations by means of flexible AC transmission systems (FACTS). However, in the literature there are no studies that investigate the impact of wind power on power system oscillations. Therefore, at this stage it is not possible to arrive at detailed conclusions regarding the impact of wind power on power systems. We will therefore present a qualitative analysis and some preliminary conclusions. It should be stressed, though, that further research is necessary.

In synchronous generators, the electrical torque depends mainly on the angle between rotor and stator flux. This angle is the integral of the rotational speed difference between these two fluxes which in turn depends on the difference between electrical and mechanical torque. This makes the mechanical part of the synchronous machine a second-order system that intrinsically shows oscillatory behaviour. Further, small changes in rotor speed are unlikely to affect the electrical torque developed by the machine, as they lead to hardly any change in rotor angle. Therefore, the damping of rotor speed oscillations must come from other sources, such as damper windings, the exciter and the rest of the power system.

Relatively weak links and large concentrations of synchronous generators contribute to the risk of weakly damped or undamped oscillations (Kundur, 1994; Rogers, 1996). The reason is that if a synchronous generator is large compared with the scale of the system as a whole and/or if it is weakly coupled the contribution of the rest of the system to the damping torque diminishes and thus the damping of oscillations deteriorates. The oscillation of a generator that is large compared with the system will also affect other generators, thus spreading the oscillation through the system and giving rise to power system oscillations that comprise a number of generators that oscillate against each other. The lower the frequency, the less damping provided by the damper windings. Power system oscillations can have frequencies of about 1 Hz and lower, so that hardly any damping is provided by the damper windings.

The generator types used in wind turbines rarely if even engage in power system oscillations. Squirrel cage induction generators used in constant-speed wind turbines show a correlation between rotor slip (i.e. rotor speed) and electrical torque instead of between rotor angle and electrical torque, as in the case of synchronous generators. The mechanical part is therefore of first order and does not show oscillatory behaviour, as opposed to that of a synchronous generator. Although an oscillation can be noticed when including the rotor transients in the model, as this increases the model order, this oscillation is small and well damped. Thus, squirrel cage induction generators are intrinsically better damped and rely less than do synchronous generators on the power system to provide damping and thus rarely lead to power system oscillations.

The generator types used in variable-speed wind turbines are decoupled from the power system by power electronic converters that control the rotor speed and electrical power and damp any rotor speed oscillations that may occur. Variable-speed wind turbines do not react to any oscillations in the power system either. The generator does not 'notice' them, as they are not transferred through the converter.

This analysis shows that if wind power is assumed to replace the power generated by synchronous generators, these synchronous generators contribute less to cover the power demand. The topology of the system remains unchanged, though. Thus, the synchronous generators become smaller in proportion to the impedances of the grid. This strengthens the mutual coupling, which in most cases improves the damping of any oscillation between the synchronous generators. Hence, replacing synchronous generators with wind turbines can be expected to improve the damping of power system oscillations.

28.5.3 Simulation results

This section shows some of the simulation results. A small test system was developed in order to analyse the impact of wind power on the eigenvalues of a power system. Figure 28.7 depicts the test system. It consists of two areas, one with a large, strongly coupled system, represented by an infinite bus, and the other one consisting of two synchronous generators. The impedances are in per unit (p.u.) on a 2500 MVA base and the loads are modelled as constant MVA. This test system shows an oscillation of a group of generators against a strong system and an intra-area oscillation. The generators at buses 3 and 4 oscillate against the strong system, and generators 3 and 4 oscillate against each other. The shapes of these oscillatory modes are also depicted in Figure 28.7. In order to analyse the impact of an inter-area power flow, the loads were increased to 3000 MW without changing the output power of the generators.

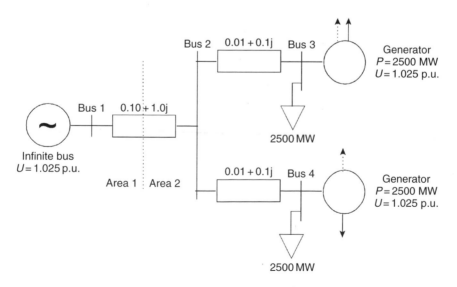

Figure   28.7 Test system with two generators, the oscillatory modes are indicated by arrows: oscillation of a group of generators (dotted-line arrows), and intra-area oscillation (solid-line arrows). *Note*: $U =$ voltage; $P =$ active power

The load flow scenarios used to study the three oscillation types are developed in the following way:

- One or more buses to which a synchronous generator involved in the oscillatory mode to be studied is connected are selected. The buses are selected according to the scenario that is being studied. The analysis includes scenarios in which generators in only one of the two swing nodes are gradually replaced by wind power as well as scenarios in which generators in both swing nodes are gradually replaced by wind power.
- An aggregated model of a wind farm with either constant-speed or variable-speed wind turbines is connected to the selected buses.
- The active power, reactive power capability and MVA rating of the selected synchronous generator(s) are gradually reduced. The reduction in active power is compensated by increasing the power generated by the wind farm, the MVA of which rating is increased accordingly. The reduction in reactive power generation, if any, is not compensated for.

In order to calculate the eigenvalues for each of the scenarios, the load flow cases are solved and dynamic models of the synchronous generators and the wind farm(s) are attached. The resulting dynamic model of the investigated scenario is then linearised and the eigenvalues of the state matrix are calculated as described above. In this way, we can show the trajectory of the eigenvalues in the complex plane with changing wind power penetration. This provides information on the oscillatory behaviour.

Figure 28.8 depicts the results. In both Figure 28.8(a) and Figure 28.8(b) the damping ratio is indicated on the horizontal axis and the oscillation frequency on the vertical axis. The upper graphs Figures 28.8(a.i) and 28.8(b.i) show the eigenvalues that correspond to one wind farm, at bus 3. The lower graphs Figures 28.8(a.ii) and 28.8(b.ii) depict the eigenvalues that correspond to two wind farms, at buses 3 and 4. The direction in which the eigenvalues move with increasing wind power penetration is indicated by an arrow.

28.5.4 Preliminary conclusions

The analysis and the simulation results presented in Sections 28.5.2 and 28.5.3 show that replacing synchronous generators with wind power tends to improve the damping of oscillations of a (group of coherent) generator(s) against a strong system and the damping of inter-area oscillations, particularly if constant-speed wind turbines are used. The effect on intra-area oscillations seems to be insignificant.

The reason for the damping effect of wind power is that the remaining synchronous generators connected to the system become proportionally smaller and therefore more strongly coupled. However, wind power itself does not induce new oscillatory modes, because the generator types used in wind turbines do not engage in power system oscillations. Oscillations in squirrel cage induction generators that are used in constant-speed wind turbines are intrinsically better damped, and the generators of variable-speed wind turbines are decoupled from the power system by a power electronic converter, which controls the power flow and prevents them from engaging in power system oscillations.

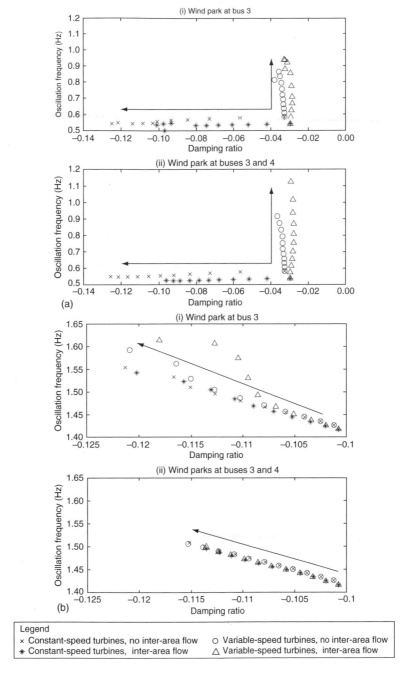

Figure 28.8 (a) Impact of increasing wind power penetration on the oscillation of the generators at buses 3 and 4 against the infinite bus: (i) the eigenvalues with a wind farm at bus 3; (ii) the eigenvalues with two wind farms, at buses 3 and 4 (b) Impact of increasing wind power penetration on the intra-area oscillation of the generators at buses 3 and 4 against each other: (i) the eigenvalues with a wind farm at bus 3; (ii) the eigenvalues with two wind farms, at buses 3 and 4. The direction of the arrows indicates increasing wind power penetration; for the test system, see Figure 28.7

When a very large part of the synchronous generation capacity in a swing node is replaced with wind power, the results become ambiguous in some cases. This is probably caused by the fact that the mode shape changes, which can also change the oscillation type.

We have used theoretical insights regarding the origin and mitigation of power system oscillations to explain our observations. More detailed analyses of other power systems can be expected to yield similar results. The results presented here are mainly qualitative. Computations with other power systems will be required to confirm the obtained results and determine whether it is possible to quantify the impact of wind power on power system oscillations.

28.6 Conclusions

This chapter discussed the impact of wind power on the dynamics of power systems. The reasons for investigating this topic are that the dynamics of a power system are governed mainly by the generators. Thus, if conventional power generation with synchronous generators is on a large scale replaced with wind turbines that use either asynchronous squirrel cage induction generators or variable-speed generation systems with power electronics, the dynamics of the power system will at some point be affected, and perhaps its stability, too.

The main conclusion of this chapter is that although wind turbines indeed affect the transient and small signal dynamics of a power system, power system dynamics and stability are not a principal obstacle to increasing the penetration of wind power. By taking adequate measures, the stability of a power system can be maintained while increasing the wind power penetration.

In the case of constant-speed wind turbines, measures must be taken to prevent voltage and rotor speed instability in order to maintain transient stability. This can be done by equipping them with pitch controllers in order to reduce the amount of overspeeding that occurs during a fault; by combining them with a source of reactive power to supply the large amount of reactive power consumed by a squirrel cage generator after a fault, such as a STATCON or SVC; or by changing the mechanical and/or electrical parameters of the turbine. In the case of variable-speed wind turbines, the sensitivity of the power electronic converter to overcurrents will have to be counteracted in another way than is presently done, namely, by switching off the turbine. However, the literature seems to indicate that there may indeed be other options.

The small signal stability was studied here using the eigenvalues obtained from the linear analysis. Increased levels of wind power penetration do not seem to require any additional measures in order to maintain small signal stability. The generator type that is most likely to engage into power system oscillations is the synchronous generator, and that generator type is not used in the wind turbine types that are presently on the market. Therefore, replacing synchronous generation seems to have either a negligible or a favourable impact on power system oscillations. However, the impact of wind power on the small signal stability of power systems is a rather recent research subject. Hence, results are still very limited. The conclusions we have presented here should therefore be considered preliminary and be used prudently.

References

[1] Akhmatov, V., Knudsen, H. (2002) 'An Aggregate Model of a Grid-connected, Large-scale, Offshore Wind Farm for Power Stability Investigations – Importance of Windmill Mechanical System', *International Journal of Electrical Power and Energy Systems* 24(9) 709–717.

[2] Akhmatov, V., Knudsen, H., Nielsen, A. H., Pedersen, J. K., Poulsen, N. J. (2003) 'Modelling and Transient Stability of Large Wind Farms', *International Journal of Electrical Power and Energy Systems*, 25(2) 123–144.

[3] Bauer, P., De Haan, S. W. H., Meyl, C. R. and Pierik, J. T. G. (2000) 'Evaluation of Electrical Systems for Offshore Windfarms', presented at IAS Annual Meeting and World Conference on Industry Applications of Electrical Energy, Rome, Italy.

[4] E.On Netz (2001) 'Ergänzende Netzanschlussregeln für Windenergieanlagen', E.ON Netz. Bayreuth, Germany.

[5] Kundur, P. (1994) *Power System Stability and Control*, McGraw-Hill Inc., New York.

[6] Pai, M. A. (1989) *Energy Function Analysis for Power System Stability*, Kluwer Academic Publishers, Boston, MA.

[7] Petterson, A. (2003) 'Analysis, Modeling and Control of Doubly-fed Induction Generators for Wind Turbines', licentiate thesis, Department of Electric Power Engineering, Chalmers University of Technology, Göteborg, Sweden.

[8] Rogers, G. (1996) 'Demystifying Power System Oscillations', *IEEE Computer Applications in Power* 9(3) 30–35.

[9] Saccomando, G., Svensson, J., Sannino, A. (2002) 'Improving Voltage Disturbance Rejection for Variable-speed Wind Turbines', *IEEE Transactions on Energy Conversion* 17(3) p. 422–428.

[10] Slootweg J. G., de Haan S. W. H., Polinder H., Kling W. L. (2003a) 'Aggregated modeling of Wind Farms with Variable Speed Wind Turbines in Power System Dynamics Simulations', in *Proceedings of the 14th Power Systems Computation Conference, Sevilla, Spain*.

[11] Slootweg, J. G., de Haan, S. W. H., Polinder, H., Kling, W. L. (2003b) 'General Model for Representing Variable Speed Wind Turbines in Power System Dynamics Simulations', *IEEE Transactions on Power Systems* 18(1) 144–151.

[12] Van Custem, T., Vournas, C. (1998) *Voltage Stability of Electric Power Systems*, Kluwer Academic Publishers, Boston, MA.

29

Aggregated Modelling and Short-term Voltage Stability of Large Wind Farms

Vladislav Akhmatov

Dedicated to my true friend C. E. Andersen

29.1 Introduction

In aggregated modelling of large wind farms, a distinction is made between two broad issues. The first issue is concerned with the detailed representation of the electricity-producing wind turbines, the internal network of the farm and its grid connection to the entire power system. The individual responses of many wind turbines and their possible mutual interaction are studied. Therefore, an aggregated model represents a large number of grid-connected wind turbines within a large wind farm (Akhmatov *et al.*, 2003a). In this chapter, we will look mainly at this issue.

The transmission power network on the opposite side of the connection point can be given by a simplified equivalent. The equivalent may consist of one or a small number of transmission lines between the connection point of the wind farm and the swing bus generator. The impedance of the transmission lines is related to the short-circuit capacity, S_K, of the transmission network at the connection point of the wind farm. The generator equivalent is based on the lumped power capacity and the lumped inertia, corresponding to the power generation units of the external power system. According to Edström (1985), this simplified representation is valid for a group of coherent synchronous generators of conventional power plant units.

Wind Power in Power Systems   Edited by T. Ackermann
© 2005 John Wiley & Sons, Ltd   ISBN: 0-470-85508-8 (HB)

The second issue relates to reduced models of large wind farms that are implemented in detailed models of large power systems and are used to analyse power system stability. In this case, the focus is on the collective impact of a large wind farm on a large power system (Slootweg *et al.*, 2002). This implies that a wind farm can be represented by the single machine equivalent (Akhmatov and Knudsen, 2002). This means that the wind farm is given by one model of the wind turbine of the given concept with rescaled power capacity, power supply and reactive power exchanged between the wind farm and the entire transmission network at the connection point.

A power system may even include a number of small and large wind farms. In this case, the wind farms are represented by their respective single machine equivalents. Knudsen and I give examples of flicker investigations in a power system with more than 60 wind farms (in many decentralised sites; see Knudsen and Akhmatov, 2001) and in Akhmatov (2002) I report the results of a voltage stability analysis with 60 decentralised wind turbine sites and two large offshore wind farms.

29.1.1 Main outline

At the beginning of this chapter, I wish to discuss the application of aggregated models of wind farms to the analysis of voltage stability. Section 29.2 describes the model of a large offshore wind farm consisting of 80 2 MW wind turbines. I will discuss the simulation results based on the wind turbine types according to the classification in Chapter 4 (see Section 4.2.3):

- Type A2 wind turbines (i.e. fixed-speed, active-stall-controlled wind turbines equipped with no-load compensated induction generators; see Section 29.3), among the largest manufacturers that produce this type of wind turbine are Vestas Wind Systems, Bonus Energy and Nordex.
- Type B1 wind turbines (i.e. pitch-controlled wind turbines equipped with induction generators with variable rotor resistance (VRR); see Section 29.4). The Danish manufacturer Vestas Wind Systems produces OptiSlip^{TM} wind turbines.[1]
- Type C1 wind turbines (i.e. variable-speed, pitch-controlled wind turbines with doubly fed induction generators (DFIGs) and partial-load frequency converters; see Section 29.5), largest manufacturers are Vestas Wind Systems, GE Energy, Nordex and Gamesa Eólica.
- Type D1 wind turbines (i.e. variable-speed, pitch-controlled wind turbines with full-load converters and synchronous generators excited by permanent magnets; see Section 29.6). The Dutch manufacturer Lagerwey the Windmaster produced 2 MW wind turbines equipped with permanent magnet generators (PMGs). The German manufacturer Enercon produces wind turbines with multipole, synchronous generators with electrical excitation and frequency converters.

Finally, I will discuss the use of single machine equivalents of large wind farms in the analysis of voltage stability (Section 29.7) and present some conclusions (Section 29.8).

---

[1] OptiSlip^{TM} is a registered trademark of Vestas Wind Systems.

29.1.2 Area of application

Many wonder why and where to use aggregated models of large wind farms in the analysis of short-term voltage stability and under what conditions the wind farm representation can be reduced to a single machine equivalent.

Aggregated models of large wind farms are needed to answer another common question – whether there is a risk of mutual interaction between the electricity-producing wind turbines when clustered together in a large wind farm (Akhmatov *et al.*, 2003a). Such interaction between many wind turbines could be triggered by a grid disturbance in the external network. This is likely to lead to self-excitation and, subsequently, a group of wind turbines would be disconnected. This question is most frequently asked in the case of wind turbines equipped with converter-controlled generators regarding the risk of mutual interaction between the converter control systems of various wind turbines. There is also concern in the case of wind turbines equipped with induction generators and blade-angle control regarding a possible risk of mutual interaction.

Another broad area where aggregated modelling of wind farms is applied is the design of the farm's internal network and its (static) reactive compensation (Jenkins, 1993). In the case of Type A wind turbines, the focus is on the use of power factor correction capacitors in order to reduce reactive power demands and prevent dangerous over-voltages during the isolated operation of a section of the wind farm.

Aggregated models can also be used for development and validation of the control coordination between a large number of wind turbines with similar or different control systems (Kristoffersen and Christiansen, 2003). An analysis of the robustness of the control principles can be included in such a study. The robustness of the control principles can be evaluated from simulations where the control system fails in a number of wind turbines of a large wind farm (Akhmatov *et al.*, 2003a). Such an evaluation will look at the maximum number of individual wind turbines that may have a control system failure without jeopardising the desired outcome of the study (i.e. maintaining short-term voltage stability during a grid disturbance; see Akhmatov *et al.*, 2003a). Aggregated models of large wind farms can also be used to evaluate relay settings during short-circuit faults or other disturbances in the internal network of a wind farm.

To my knowledge, wind farm owners and power system controllers have asked wind turbine manufacturers to present aggregated models of large wind farms with a representation of more than 200 wind turbines. This illustrates that the development of aggregated models of large wind farms can be of both practical and commercial interest. It is, however, important to know that aggregated models of large wind farms can require relatively large computational resources. This is especially true for those cases where the detailed representation of advanced control systems of wind turbines is required for reaching accurate simulation results (Akhmatov, 2002). One should also take into account that there are concerns regarding the computational resources and time that the setting up of aggregated models may take.

29.1.3 Additional requirements

If the required computational resources become too large there will be restrictions regarding the practical application of aggregated models of large wind farms. It is only reasonable that users require the computational time [i.e. the time during which the user has (passively) to wait for the results of a single simulation] to be as short as possible. And a whole analysis may include a significant number of simulations.

The analysis of short-term voltage stability usually takes a few seconds. In this case, the computational time will be expected not to exceed a few seconds. Therefore, the wind turbine models and the entire power grid model need to have very efficient codes and numerical algorithms. This means that an appropriate simulation tool has to be chosen that is able to treat models of large power systems with many generation units. To reach the results presented in this chapter I have applied user-written models of electricity-producing wind turbines implemented in the simulation tool Power System Simulator for Engineering (PSS/E^{TM}) from the manufacturer Shaw Power Technologies Inc. (PTI), New York, USA.[2] The computational time lies within an acceptable range. For other simulation tools that also could have been applied for the same purpose, see Chapter 24.

29.2 Large Wind Farm Model

An aggregated model of a large offshore wind farm consisting of 80 2 MW wind turbines was set up. Figure 29.2 represents the wind farm model schematically.

In this large wind farm model, single wind turbines are represented by the simulation models that correspond to their wind turbine concepts. Through the 0.7 kV/30 kV transformers the wind turbines are connected to the internal network of the large wind farm. The internal network is organised into 8 30 kV sea-cable sections (8 rows) with 10 electricity-producing wind turbines per section. Within the individual sections, the wind turbines are connected through the 30 kV sea-cables. The distance between two neighbouring wind turbines in the same section is 500 m, and the distance between two neighbouring sections is 850 m. The electrical connection between the wind turbines is organised in a way that the disconnection of a single wind turbine in a section does not cause isolation of the other wind turbines in the same section.

The 8 sea-cable sections are connected through the 30 kV sea-cables to the offshore platform with a 30 kV/30 kV/132 kV tertiary transformer and then through the 132 kV, 20 km sea-cable/underground-cable to the connection point at the transmission power network onshore.

The aggregated model allows simulations of the wind farm assuming an irregular wind distribution. This assumption is realistic because the wind turbine rotors shadow each other from the incoming wind. Further, the area of the wind farm is $4.5 \times 6 \, km^2$, and it is reasonable to expect an irregular wind distribution over such a large area. Figure 29.1 shows the direction of the incoming wind and the power production pattern of the wind farm. The efficiency of the wind farm at the given wind distribution is 93%, and the wind farm supplies approximately 150 MW to the transmission power network.

---

[2] PSS/E^{TM} is a registered trademark of PTI.

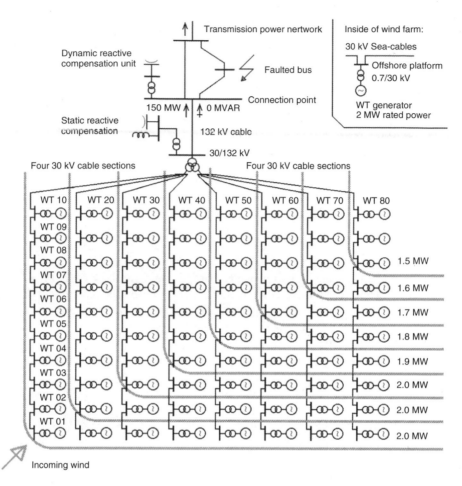

Figure 29.1 Aggregated model of large offshore wind farm with 80 wind turbines (WTs) Reprinted from *Electrical Power and Energy Systems*, volume 25, issue 1, V. Akhmatov, H. Knudsen, A. H. Nielsen, J. K. Pederson and N. K. Poulsen, 'Modelling and Transient Stability of Large Wind Farms', pp. 123–144, copyright 2003, with permission from Elsevier

29.2.1 Reactive power conditions

Type A and B wind turbines are equipped with no-load compensated induction generators. For excitation, the induction generators absorb reactive power from the wind farm's internal network and from the capacitors (Saad-Saoud and Jenkins, 1995). The reactive power absorbed by the induction generators depends on the generator parameters and its operational point (i.e. the generated electric power, the terminal voltage magnitude and the slip). The reactive power of all the generators in the wind farm has to be adjusted before the transient simulations are started (Feijóo and Cidrás, 2000).

Type C wind turbines with DFIGs are exited from the rotor circuits by the rotor converters (Akhmatov, 2002). That means that it is not necessary to excite the DFIG from the power grid. Type D wind turbines with PMGs are excited by permanent magnets and can additionally be excited by generator-side converters. The PMGs are grid-connected through the frequency converters. This means that the PMGs and the power grid are separated by the DC link. This is the reason why there is no reactive power exchange between the PMG and the power grid. The power system regards wind turbines of this concept as grid-side converters supplying electric power. During normal operation, Type C and D wind turbines can be set not to exchange reactive power with the power grid.

According to the Danish specifications for connecting wind farms to transmission networks (Eltra, 2000), large wind farms have to be reactive-neutral towards the transmission power network at the connection point. This means that the reactive power exchanged between the wind farm and the transmission system must be around zero, independent of the operational point of the wind farm. This may require the incorporation of a bank of capacitors and reactors to impose static control of the reactive power of the large wind farm. These capacitor and reactor banks can be incorporated either at the connection point onshore or on the offshore platform (with the tertiary transformer). In this study, the static compensation units are placed offshore. In the case of Type A and B wind turbines, it will be necessary to incorporate capacitors for covering reactive absorption of the induction generators. Type C and D wind turbines will, rather, require reactors to absorb reactive power produced by the sea-cables (charging).

29.2.2 Faulting conditions

The short-circuit capacity, S_K, of the transmission power network at the connection point of the wind farm is 1800 MVA. In all simulations, the disturbance is a three-phased, short-circuit fault applied to the faulted node of the transmission power system. The fault lasts 150 ms. The fault is cleared by tripping the lines that are connected to the faulted node. When the fault is cleared, S_K is reduced to 1000 MVA. The line tripping itself does not lead to voltage instability. This ensures that the short-circuit fault can be the reason for a possible voltage instability.

The large wind farm is designed according to Danish specifications (Eltra, 2000), which require the voltage to reestablish after the short-circuit fault in the transmission power network without any subsequent disconnection of the wind turbines.

29.3 Fixed-speed Wind Turbines

In this analysis, the wind farm consists of 80 wind turbines of Type A2. Table 27.1 (page 608) gives the main generator data of the wind turbines and Table 29.1 includes the shaft system data. The shaft system data have been estimated from the validating experiment described in Section 27.2.2.

Table 29.1 Shaft system of a 2 MW Type A wind turbine

Parameter	Value	Parameter	Value
Rated power (MW)	2	Turbine rotor inertia (s)	2.5
Generator rotor inertia (s)	0.5	Shaft stiffness (p.u./el.rad)	0.3

A regular control system of active stall is applied in order to optimise the power output in moderate winds and in order to keep the wind turbines at rated operation during wind speeds that are above the rated wind speed. The active-stall control system compares the electric power of the generator with its reference value, which is defined in accordance with the incoming wind, and sets the blade angle to minimise the error signal (Hinrichsen, 1984).

During a short-circuit fault there will be a voltage drop, and the electric power is reduced as a consequence of this voltage drop. The regular control system of active stall will interpret this as a lack of power output and keep the optimised value of the blade angle. That means that active-stall wind turbines will operate with fixed blade angles during the grid fault (i.e. as stall-controlled wind turbines; see Akhmatov, 2001). For the analysis of voltage stability under such operational conditions active-stall wind turbines can therefore be represented as fixed-pitch wind turbines.

If no dynamic reactive compensation is applied, the grid fault will result in voltage instability. Figure 29.2 presents the simulated behaviour of voltage, electric and reactive power and generator rotor speed of selected wind turbines. When the wind turbines are grid-connected, the generator rotor speed, terminal voltage and other electrical and

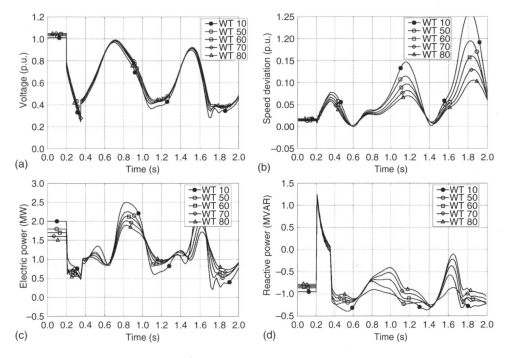

Figure 29.2 Response of a large wind farm to a short-circuit fault: (a) terminal voltage, (b) generator rotor speed, (c) electric power and (d) reactive power of selected wind turbines for the case with no dynamic reactive compensation and the occurrence of voltage instability. *Note*: For the layout of wind turbines WT 01–WT 80, see Figure 29.2 Parts (a) and (b) reprinted from *Electrical Power and Energy Systems*, volume 25, issue 1, V. Akhmatov, H. Knudsen, A. H. Nielsen, J. K. Pederson and N. K. Poulsen, 'Modelling and Transient Stability of Large Wind Farms', pp. 123–144, copyright 2003, with permission from Elsevier

mechanical parameters of the individual wind turbines in the wind farm show a coherent fluctuating behaviour. The wind turbines do not oscillate against each other. The frequency of the fluctuations is the shaft torsional mode, which is 1.7 Hz in this particular case.

Note that voltage instability will not necessarily cause a voltage collapse. The reason is that the wind turbines will be tripped by their protective relays if an uncontrollable voltage decay is registered (Akhmatov *et al.*, 2001). After this trip, the voltage is reestablished, but immediate power reserves of approximately 150 MW will be required for this (Akhmatov *et al.*, 2001). A risk of tripping indicates that the solution does not comply with the specifications (Eltra, 2000).

If a static VAR compensator (SVC) with a rated capacity of 100 MVAR is applied, the voltage will be reestablished. Figure 29.3 shows the simulated curves of voltage, electric power and generator rotor speed. Fluctuations of voltage, electric power, speed and other parameters are damped quickly and efficiently if an SVC unit is applied. There is no risk of a self-excitation of the large wind farm.

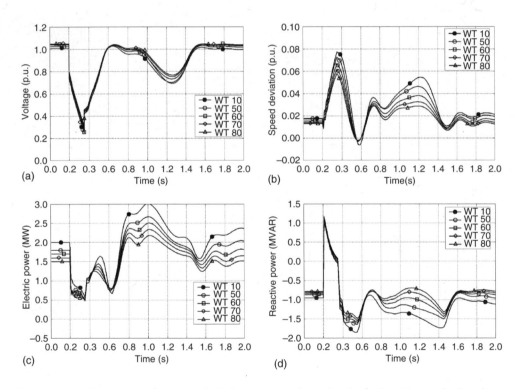

Figure 29.3 Response of large wind farm to a short-circuit fault: (a) terminal voltage, (b) generator rotor speed, (c) electric power and (d) reactive power of selected wind turbines for the case with a 100 MVAR static VAR compensation (SVC) unit and voltage reestablishment. *Note*: For the layout of wind turbines WT 01–WT 80, see Figure 29.2 parts (a) and (b) reprinted from *Electrical Power and Energy Systems*, volume 25, issue 1, V. Akhmatov, H. Knudsen, A. H. Nielsen, J. K. Pederson and N. K. Poulsen, 'Modelling and Transient Stability of Large Wind Farms', pp. 123–144, copyright 2003, with permission from Elsevier

I have used the model of the SVC unit that was developed and implemented as a user-written model in the simulation tool PSS/E™ at the Danish company NESA in cooperation with the manufacturer ABB Power Systems (Noroozian, Knudsen and Bruntt, 1999). NESA kindly permitted use of the SVC model.

29.3.1 Wind turbine parameters

The demand for dynamic reactive compensation is dependent on the parameters of fixed-speed wind turbines, which are both the induction generator parameters and the mechanical system parameters. In Akhmatov *et al.* (2003) my colleagues and I showed that demands for dynamic reactive compensation can be significantly reduced if:

- stator resistance, stator reactance, magnetising reactance and rotor reactance are reduced;
- rotor resistance is increased (however, this will also increase power losses in the rotor circuit during normal operation);
- mechanical construction is reinforced in terms of increasing turbine rotor inertia and increasing shaft stiffness.

By reducing the demand for dynamic reactive compensation one improves short-term voltage stability and the ride-through capability. Table 29.2 illustrates the effect of reinforcing the wind turbine's mechanical construction in order to reduce the demand for dynamic reactive compensation. The generator and shaft system data show that the only reasonable comparison is that for different Type A wind turbines. Therefore, an analysis of short-term voltage stability could be started by comparing such data for Type A wind turbines produced by different manufacturers. The controllability of Type A wind turbines also has to be taken into account.

29.3.2 Stabilisation through power ramp

According to Danish specifications (Eltra, 2000), large wind farms have to be able to reduce their power supply on request. The power may have to be reduced from any

Table 29.2 Mechanical construction in relation to the static VAR compensation (SVC) unit capacity needed for voltage reestablishment.

Mechanical construction parameter		Capacity of SVC unit (MVAR)
Turbine rotor inertia (s)	Shaft stiffness (p.u./el.rad)	
2.5	0.30	100[a]
2.5	0.15	125
2.5	0.60	50
4.5	0.30	50

[a] Default case.

Source: Akhmatov *et al.*, 2003a. From *Electrical Power and Energy Systems*, volume 25, issue 1, V. Akhmatov, H. Knudsen, A. H. Nielsen, J. K. Pederson and N. K. Poulsen, 'Modelling and Transient Stability of Large Wind Farms', pp. 123–144, copyright 2003, with permission from Elsevier.

operational point to an operational point of less than 20 % of rated power in less than 2 s. This procedure is called power ramping. It has been demonstrated that a power ramp can stabilise large wind farms at a short-circuit fault (Akhmatov, 2001). The power ramp can effectively be achieved by using the active-stall control of fixed-speed wind turbines (Akhmatov *et al.*, 2001). In the case of Type A wind turbines, excessive overspeeding of the wind turbines initiates voltage instability (Akhmatov, 2001). Power ramp operation means that the wind turbines are decelerated during a grid fault, and in this way excessive overspeeding is prevented. An external system gives an external signal that requests a power ramp. The signal is given when abnormal operation at the connection point of the wind farm is registered, such as a sudden voltage drop.

There is a delay between the moment the grid fault occurs and the moment the external signal is given to the wind farm. This delay is about 200–300 ms (Akhmatov *et al.*, 2003). This means that the request comes only after the grid fault is cleared.

Before the power ramp is requested, the wind turbines are controlled by the regular control systems of active stall (i.e. by the power optimisation algorithm). If the external signal requests the power ramp, the regular control systems are switched off and replaced by the control strategy of the power ramp. This means that the reference blade angle of each wind turbine in the farm is set to the predefined value corresponding to an output power of 20 % of rated power (Akhmatov *et al.*, 2003). The reference blade angle during the power ramp is computed based on the incoming wind and the rotational speed of the wind turbine rotor. Using the blade servo, the reference value of the blade angle will be reached. However, there are restrictions on the pitching rate (see Section 27.3.3).

Figure 29.4 shows the computed behaviour of a modern 2 MW wind turbine using the power ramp during a grid fault. The curves correspond to selected wind turbines in large wind farms. It can be seen that the mechanical power can be reduced from rated operation to 20 % of rated operation in less than 2 s. Operation with reduced power output lasts only a few seconds. This is necessary to prevent fatal overspeeding and it contributes to voltage reestablishment after a grid fault and to the fault ride through capability of the wind farm. In this particular case, the voltage in the power system is reestablished without use of dynamic reactive compensation (i.e. it is reestablished only through the power ramp achieved by active-stall control) and the wind turbines ride through the fault.

When the power ramp mode is cancelled, the regular control system of active stall is restarted and becomes operational. The power output will be optimised according to the incoming wind and the rotational speed of the turbine rotor.

If the grid fault is cleared by tripping a number of transmission lines, the short-circuit capacity of the transmission network is changed. The voltage does not reestablish itself to the same level as before the grid fault. The rotational speed will also be slightly different from the value during prefault operation. The result is that the operational conditions of the wind turbine rotors differ slightly from the those during prefault conditions. It is argued here that the blade angles of individual wind turbines will not necessarily be reestablished to the positions prior to the grid fault. The regular control systems of active stall will find other optimised blade-angle positions to reach the desired power outputs. The curves in Figure 29.5 illustrate this. If the power ramp is used to stabilise a large wind farm, this will not trigger interaction between the wind turbines. The simulation results show that the wind turbines will show a coherent

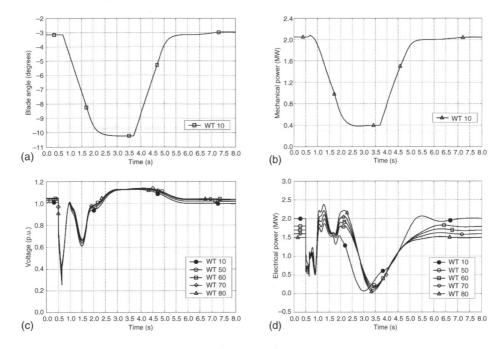

Figure   29.4 Response of a large wind farm to a short-circuit fault, with use of a power ramp: (a) blade angle of a selected wind turbine, (b) mechanical power of a selected wind turbine, (c) voltage and (d) electric power of selected wind turbines. Voltage is reestablished without using dynamic reactive compensation. *Note*: for the layout of wind turbines WT 01–WT 80, see Figure 29.1 Reprinted from *Electrical Power and Energy Systems*, volume 25, issue 1, V. Akhmatov, H. Knudsen, A. H. Nielsen, J. K. Pederson and N. K. Poulsen, 'Modelling and Transient Stability of Large Wind Farms', pp. 123–144, copyright 2003, with permission from Elsevier

response during the grid disturbance. Once the operation of the wind farm is reestablished, there are no further fluctuations of voltage or electric power.

In general, use of blade-angle control to stabilise large wind farms equipped with Type A wind turbines is a useful tool for maintaining transient voltage stability (Akhmatov *et al.*, 2003). The Danish offshore wind farm at Rødsand applies a similar technical solution. The rated power of the Rødsand offshore wind farm is 165 MW, and the farm was taken into operation in 2003. The Rødsand offshore wind farm comprises 72 Type A2 wind turbines from the manufacturer Bonus Energy.

29.4 Wind Turbines with Variable Rotor Resistance

The feature of variable rotor resistance (VRR) is designed as follows: the converter is connected to the rotor circuit of an induction generator through the slip rings. The operation of the converter means that an external resistance is added to the impedance of the rotor circuit. For an analysis of voltage stability, this is reflected in a simplified representation of the converter control as a dynamically controlled external rotor

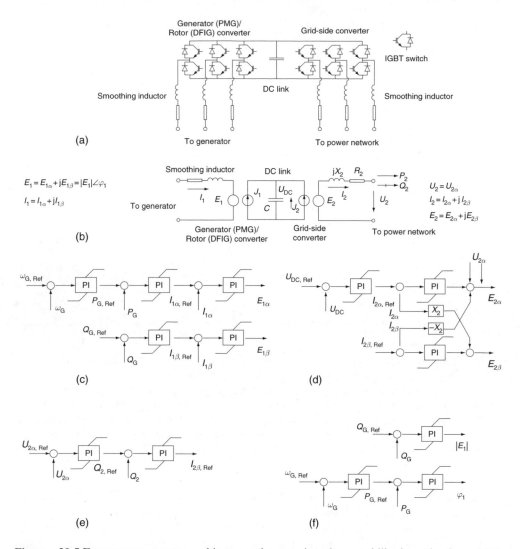

Figure 29.5 Frequency converter and its control system in voltage stability investigations: (a) the main components and integrated gate bipolar transistor (IGBT) switches, (b) generic electric scheme, (c) generic control of rotor converter of the doubly fed induction generator (DFIG), (d) generic control of the grid-side converter, (e) supplementary control of reactive power from the grid-side converter, and (f) generic control of the generator converter of the permanent magnet generator (PMG). *Note*: PI = proportional–integral controller; $c = \ldots$; E_1 = voltage source of DFIG rotor converter on PMG converter; $|E_1|$ = magnitude of E_1; $E_{1\alpha}$ = active component of E_1; $E_{1\beta}$ = reactive component of E_1; E_2 = voltage source of grid-side converter; $E_{2\alpha}$ = active component of E_2; $E_{2\beta}$ = reactive component of E_2; I_1 = rotor current of DFIG or generator current of PMG; $I_{1\alpha}$ = active component of I_1; $I_{1\beta}$ = reactive component of I_1; $I_{1\alpha, \text{Ref}}$ = desired value of $I_{1\alpha}$; $I_{1\beta, \text{Ref}}$ = desired value of $I_{1\beta}$; I_2 = current of grid-side converter; $I_{2\alpha}$ = active component of I_2; $I_{2\beta}$ = reactive component of I_2; $I_{2\alpha, \text{Ref}}$ = desired value of $I_{2\alpha}$; $I_{2\beta, \text{Ref}}$ = desired value of $I_{2\beta}$; J_1 = charging DC current; J_2 = discharging DC current;

resistance, VRR. In Akhmatov *et al.* (2003) my colleagues and I describe the model of Type B wind turbines with a generic control system of VRR with a proportional integral (PI) controller. The rotor current magnitude controls the value of the external resistance.

The VRR feature is commonly applied in combination with pitch control in order to reduce flicker (i.e. to improve power quality). In Akhmatov *et al.* (2003) we demonstrated that this feature can also be used to improve short-term voltage stability. We also included the results of an analysis of short-term voltage stability carried out with a wind farm model with 80 Type B wind turbines. Grid faults did not trigger any mutual oscillations between the Type B wind turbines, and the electric and mechanical parameters of the wind turbines show a coherent response.

Use of VRR means that the demand for dynamic reactive compensation in order to reestablish voltage can be reduced significantly. Furthermore, if pitch control is applied to prevent excessive overspeeding of wind turbines during grid faults, this will also contribute to improved short-term voltage stability and the fault-ride-through capability.

VRR (i.e. the converters) cannot be used to control reactive power or cover the static reactive demands of the induction generators because the converter is applied together with induction generators that absorb reactive power from the power grid; it is not designed to control the excitation of such generators. During transient events in the electric power networks converter protection also has to be taken into account. The converters may, for example, block during excessive machine current transients (Akhmatov, 2003a).

Note: the results presented in this section have been discussed with the Danish manufacturer Vestas Wind Systems, which produces the Opti-Slip^{TM} wind turbines.

29.5 Variable-speed Wind Turbines with Doubly-fed Induction Generators

I now wish to present the simulation results for 80 variable-speed wind turbines equipped with DFIGs. Partial-load frequency converters use the integrated gate bipolar transistor (IGBT) switches to control the DFIGs. Figure 29.5 shows schematically the

Figure 29.5 (*continued*) P_2 = electric power of grid-side converter; P_G = generator electric power; $P_{G, Ref}$ = desired value of P_G; Q_2 = reactive power of grid-side converter; $Q_{2, Ref}$ = desired esired value of Q_2; Q_G = generator reactive power; $Q_{G, Ref}$ = desired value of Q_G; R_2 = resistance of smoothing inductor; U_2 = terminal voltage; $U_{2\alpha}$ = magnitude of terminal voltage; $U_{2\alpha, Ref}$ = desired value of $U_{2\alpha}$; U_{DC} = DC link voltage; $U_{DC,Ref}$ = desired value of U_{DC}; X_2 = reactance of smoothing inductor; φ_1 = phase angle; ω_G = generator rotor speed; $\omega_{G, Ref}$ = desired value of ω_G; Parts (c)–(e) reprinted from *Wind Engineering*, volume 27, issue 2, V. Akhmatov, 'Variable-speed Wind Turbines with Doubly-fed Induction Generators, Part III: Model with the Back-to-back Converters', pp. 79–91, copyright 2003, with permission from Multi-Science Publishing Co. Ltd, Part (f) reprinted from *Wind Engineering*, volume 27, issue 6, V. Akhmatov, A. H. Nielsen, J. K. Pedersen and O. Nymann, 'Variable-speed Wind Turbines with Multi-preSynchronous Permanent Magnet Generators and Frequency Converters, Part I: 'Modelling in Dynamic Simulation Tools', pp 531–548, copyright 2003, with permission from multi-science Publishing Co. Ltd

converter and its control. In Akhmatov (2002) I provide the modelling details regarding the DFIG, shaft system, turbine rotor and generic pitch-control system. In Akhmatov (2003b) I explain the generic model of the partial-load frequency converter and its control systems applied in the analysis of voltage stability. It is, however, necessary to present briefly the converter model in order to understand better the results presented in this section.

The accuracy of the simulation results depends on a variety of factors. One is the representation of the partial-load frequency converter and its control. Several studies on the analysis of voltage stability (Pena *et al.*, 2000; Røstøen, Undeland and Gjengedal, 2002) assume that the rotor converter representation is sufficient for such an analysis and therefore they neglect the grid-side converter. The reason is that the grid-side converter has a small power capacity, and it is assumed that this converter can always follow its references, such as DC link voltage and reactive current.

Unless the terminal voltage changes significantly, this assumption is correct. However, if there is a significant voltage drop during a short-circuit fault, the grid-side converter is not able to follow its references (Akhmatov, 2003b). If one neglects the converter one introduces inaccuracy with respect to the converter's response during grid disturbances.

First, this inaccuracy shows when the predictions of transients of the machine current are excessively high (Akhmatov, 2003b). This is unacceptable, because, basically, the rotor converter blocks in order to protect against overcurrents (Akhmatov, 2002). This means that the oversimplified converter model may predict a too frequent blocking of the converters.

Second, the transient behaviour of the DC link voltage is not available if the grid-side converter is neglected. When the grid voltage drops, the grid-side converter cannot supply electric power and the DC link voltage starts to fluctuate (Akhmatov, 2003b). The converter's protective system monitors the DC link voltage and orders the converter to block if the DC voltage exceeds a given range.

Third, the damping characteristics of the torsional oscillations excited in the shaft system may be predicted incorrectly when one applies the oversimplified converter representation (Akhmatov, 2003a). This may result in misleading conclusions with respect to the intensity of shaft oscillations and the predicted load on the shaft gear.

It is also important to mention that the grid-side converter can be set to control reactive power. Through its restricted power capacity, this reactive power control may have an effect on the voltage recovery rate and may contribute to the successive converter restart, if the rotor converter has been blocked during the grid fault (Akhmatov, 2003a). This important behaviour is omitted if the grid-side converter is neglected in the model.

Details regarding the complexity of the converter models in the case of DFIGs and their partial-load converters were discussed with the manufacturer Vestas Wind Systems, which produces the Opti-Speed^{TM} wind turbines. Vestas Wind Systems agreed that an analysis of short-term voltage stability should also include a converter model with representations of the grid-side converter and the DC link rather than only of the rotor converter. It is necessary to predict with sufficient accuracy the electric parameters that have an effect on the converter blocking (and restart) during transient events in the power grid.

Vestas Wind Systems kindly provided support regarding the tuning of the parameters of the generic model of a partial-load frequency converter and its control system.

Consequently, the converters of all the wind turbines in the farm are modelled with representation of the rotor converters as well as the grid-side converters and their respective control systems, in order to reach a higher accuracy. Figures 29.5(c) and 29.5(d) give the generic control systems of the rotor and the grid-side converters of the DFIG. The rotor converter controls the generator in a synchronously rotating (α, β)-reference frame with the α-axis oriented along the terminal voltage vector. Using this control, the electric and reactive power of the DFIG are controlled independent of each other (Yamamoto and Motoyoshi, 1991).

The grid-side converter control is similar to that of a Statcom (Akhmatov, 2003a; Schauder and Mehta, 1999). The grid-side converter is controlled in the same reference frame as the rotor converter (Akhmatov, 2003b). The grid-side converter control is designed with independent control of the DC link voltage and the reactive current. Voltage compensation is achieved by cross-coupling. The switching dynamics in both converters is neglected, since the rotor converter and the grid-side converter are able to follow their respective reference values for the induced voltage sources at any time (Akhmatov, 2003b).

Figure 29.6(e) shows the supplementary system to control reactive power from the grid-side converter. Here, control coordination between the rotor and the grid-side converters is necessary (Akhmatov, 2003b).

29.5.1 Blocking and restart of converter

The wind turbines have to operate without interruption during grid faults that is the fault-ride-through. During such an event, the voltage drops. This causes transients in the machine and the grid-side converter currents. There are also fluctuations in the DC link voltage (Akhmatov, 2003b). The converter's protective system monitors currents in the rotor circuit and the grid-side converter, the DC link voltage, the terminal voltage, the grid frequency and so on. The converter will block if one or more monitored values exceed their respective relay settings. The characteristic blocking time is in the range of a few milliseconds (Akhmatov, 2002). The rotor converter stops switching and trips. The converter blocking may lead to disconnection of the wind turbine (Akhmatov, 2002).

In Akhmatov (2002) I suggest a feature to maintain the ride-through capability with a fast restart of the converter after the fault. This feature was then validated by simulations (Akhmatov, 2003a). The simulated curves shown in Figure 29.6 illustrate the fault-ride-through capability with a fast restart of the converter. The curves correspond to wind turbine WTG 01. From the start of the simulation to time $t = T_1$, the power system is in normal operation. At time T_1, the transmission power network is subjected to a short-circuit fault.

At time $t = T_2$, the rotor converter blocks by overcurrent (transients) in the rotor circuit. The rotor circuit is short-circuited through an external resistor (Akhmatov, 2002). When the rotor converter blocks, the wind turbine operates as a Type A1 wind turbine with an increased rotor resistance. Pitch control protects against excessive overspeeding of the wind turbine, as explained in Section 29.3.2. The grid-side converter operates as a Statcom, controlling DC link voltage and reactive power. This contributes to a faster reestablishment of the terminal voltage. This controllability is, however, restricted by the power capacity of the grid-side converter.

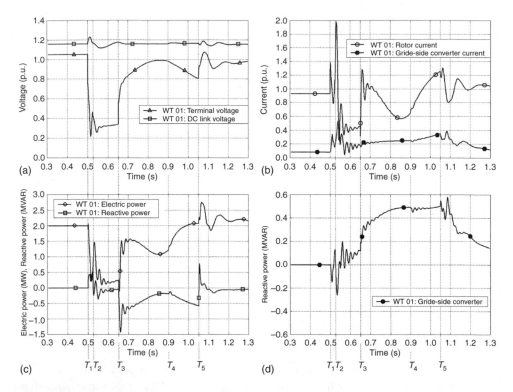

Figure   29.6 Uninterrupted operation feature with fast restart of the rotor converter: (a) terminal and DC link voltages, (b) rotor and grid-side converter current, (c) electric and reactive power, and (d) reactive power of grid-side converter. Reprinted from Akhmatov, V., Analysis of Dynamic Behaviour of Electric Power Systems with Large Amount of Wind Power, Ph.D dissertation, Technical University of Denmark, Kgs. Lyngby, Denmark, copyright 2003, with permission from the copyright holder

The grid fault is cleared at time $t = T_3$. Once the voltage and grid frequency are reestablished within their respective ranges, the rotor converter synchronisation is started. The IGBT of the rotor converter starts switching and the external resistance is disrupted from the rotor circuit. During synchronisation, the rotor converter prepares to restart. Synchronisation begins at time $t = T_4$ and protects the rotor converter against blocking during the restart sequence, which could be caused by excessive transients in the rotor current or unacceptably large fluctuations of the DC link voltage.

At time $t = T_5$, the rotor converter has restarted. Shortly after, normal operation of the wind turbine is reestablished. The feature with a fast restart of the rotor converter is designed for a ride-through operation of wind turbines in a large wind farm during grid disturbances.

29.5.2 Response of a large wind farm

One of the main concerns regarding wind farms is the risk of mutual interaction between the converter control systems of a large number of Type C wind turbines. Such concerns are reinforced during the following situations: (a) fast-acting partial-load frequency converters of the DFIG and (b) when the rotor converters of many wind turbines execute blocking and restarting sequences during a short time interval at a grid fault. Voltage stability is maintained without dynamic reactive compensation. The grid-side converters of the DFIG will control reactive power and voltage during the transient event (Akhmatov, 2003a). Figure 29.7 includes the simulated curves for the selected wind turbines operating at different operational points. The simulation results do not indicate any risk of mutual interaction between the converters of the different wind turbines. The Type C wind turbines show a coherent response during the grid fault.

Properly tuned converters are not expected to cause mutual interaction between wind turbines in large wind farms (Akhmatov, 2003a). This result is, however, based on the given control strategy of the variable-speed wind turbines and the given power network.

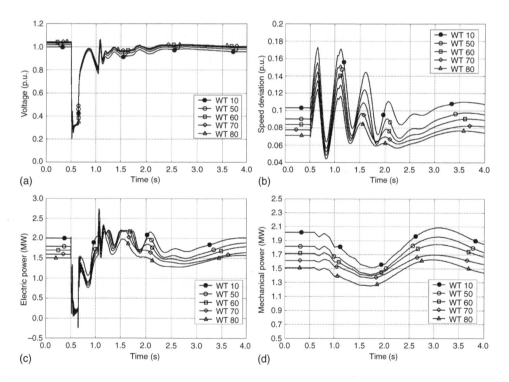

Figure 29.7 Response of a large wind farm with doubly fed induction generators (DFIGs) to a short-circuit fault using: (a) terminal voltage, (b) generator rotor speed, (c) electric power, and (d) mechanical power (illustration of pitching) of selected wind turbines; voltage is reestablished without dynamic reactive compensation. *Note*: For the layout of wind turbines WT 01–WT 80, see Figure 29.1. Reprinted from Akhmatov, V., Analysis of Dynamic Behaviour of Electric Power Systems with Large Amount of Wind Power, Ph.D dissertation, Technical University of Denmark, Kgs. Lyngby, Denmark, copyright 2003, with permission from the copyright holder

29.6 Variable-speed Wind Turbines with Permanent Magnet Generators

In this Section I present the simulation results based on 80 variable-speed wind turbines equipped with PMG and frequency converters. Table 29.3 shows the data for the wind turbine. This wind turbine type has no gearbox and therefore the generator is direct-driven by the wind turbine through the low-speed shaft (Grauers, 1996). The generator has a large number of poles and a relatively large value of reactance (Spooner, Williamson and Catto, 1996).

The generator consists of two sections of slightly above 1 MW placed on one rotor shaft. Figures 29.5(d)–29.5(f) illustrate the generic control system of the frequency converter. The generic control system of the grid-side converter is similar to the control system applied in the case of a DFIG. The PMG is controlled by the generator converter, where the electric power is controlled by the phase angle, and its reactive power (kept reactive-neutral) is controlled by the voltage magnitude.

Again, the main concern is to achieve a ride-through operation of the Type D wind turbines with PMGs. The protective system of the frequency converter monitors the machine current, the current in the grid-side converter, the DC link voltage, the terminal voltage, the grid frequency and so on; the converter will block if one or more of the monitored parameters exceed their relay settings. This may lead to disconnection and stopping of the wind turbine. In the case of a PMG, the machine current magnitude is also among the critical parameters because an excessive machine current may demagnetise the permanent magnets.

These concerns have to be taken into account when presenting a ride-through feature with blocking and fast restart of the converter during a grid fault. Figure 29.8 shows the dynamic behaviour of Type D wind turbines with PMGs. The short-circuit fault occurs at time $t = T_1$. At time $t = T_2$, the converter's protective system registers an abnormal operation and requests a blocking of the generator converter. The IGBT switches stop switching and open. Then the DC link capacitor will be charged through the diode

Table 29.3 Data for a 2 MW wind turbine equipped with a permanent magnet generator (PMG)

Generator	Value	Grid-side converter	Value
Rated power (MW)	1.01×2	Rated power (MW)	2
No-load voltage (p.u.)	1.40^a	Rated voltage (p.u.)	1.0^b
Rotational speed (rpm)	10.5–24.5	Rated frequency (Hz)	50
Number of poles	64	Rated DC link voltage	1.16^c
Lumped inertia (s)	4.8	DC link capacitor (p.u.)	0.1
Resistance (p.u.)	0.042	Mains resistance (p.u.)	0.014
Reactance X_D (p.u.)/X_Q (p.u.)	1.05/0.75	Mains reactance (p.u.)	0.175

[a] 966 V.
[b] 690 V.
[c] 800 V.

Source: Akhmatov *et al.*, 2003. From Wind Engineering, volume 27, issue 6, V. Akhmatov, A. H. Nielsen, J. K. Pedersen and O. Nymann, 'Variable-speed wind turbines with multi-pole synchronous permanent magnet generators. Part 1: Modelling in dynamic simulation tools', pp. 531–548, copyright 2003, with permission from Multi-Science Publishing Co.

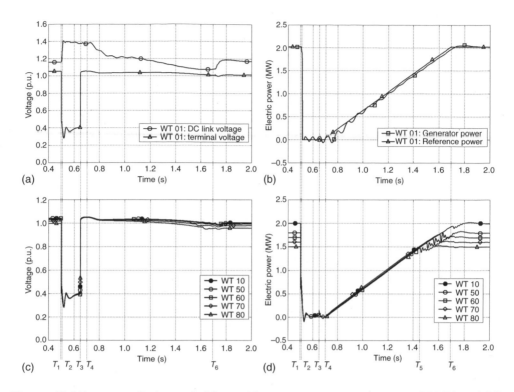

Figure   29.8 Response of a large wind farm with permanent magnet generators (PMGs) and full-load converters to a short-circuit fault: (a) terminal voltage of grid-side converter and DC link voltage, (b) generator power and its reference, (c) terminal voltage of grid-side converters, and (d) electric power supplied by grid-side converters of selected wind turbines; voltage is reestablished without use of dynamic reactive compensation. *Note*: For the Layout of wind turbines WT 01–WT 80, see Figure 29.2. Reprinted from Akhmatov. V., Analysis of Dynamic Behaviour of Electric Power Systems with Large Amount of Wind Power, Ph.D dissertation, Technical University of Denmark, Kgs. Lyngby, Denmark, copyright 2003, with permission from the copyright holder

bridges up to the voltage value that is equal to the value of the line-to-line voltage magnitude of the PMG at no-load, which is 966 V. This takes only a few milliseconds and, during this time, the machine current is reduced to zero. That the DC link capacitor is charged means that the machine current going through the generator reactance cannot be disrupted instantly. The magnetic energy accumulated in the generator reactance at the given machine current, before the converter blocking is started, is transformed into the electric energy of the charged DC link capacitor (Akhmatov, 2002).

The grid-side converter does not supply electric power to the network, but controls reactive power and voltage. This is similar to the operation of a Statcom. There are restrictions on this controllability because of the limited power capacity of the grid-side converter.

At time $t = T_3$, the fault is cleared. Shortly after, at $t = T_4$, the generator converter is synchronised and restarted. After a short while, the wind turbine operation will be reestablished. In our example, this takes less than one second, when t is between T_5 and T_6. However, there are no general requirements regarding how fast the wind turbines have to reestablish operation after the restart of the converter. In general, it may take up to several seconds to reestablish operation of a large wind farm.

Voltage is reestablished without using dynamic reactive compensation units. One of the reasons for this is that the reactive controllability of the frequency converters is used during the transient event. Figure 29.8 shows the simulated behaviour of the terminal voltage and the generator power of the selected wind turbines operating at different operational points. The simulation results do not indicate that there is any risk of mutual interaction between the frequency converters of the different wind turbines. The variable-speed wind turbines equipped with PMGs and full-load converters show a coherent response during the grid fault.

29.7 A Single Machine Equivalent

The simulated behaviour of a wind turbine operating at the rated operational point will be representative of the collective response of the large wind farm at rated operation. The reason is that the wind turbines in the large wind farm show a coherent response when subjected to a transient event in the power system. Also, the rated operation corresponds to the worst case with respect to maintaining voltage stability of Type A and B wind turbines. During rated operation, Type A and B wind turbines are closest to the level of excessive overspeeding, and the induction generators absorb most reactive power.

The rated operation will also be the worst case regarding the risk of a converter blocking in the case of Type C and D wind turbines. This is because the machine current and the grid-side converter current are both closest to their respective relay settings during rated operation (Akhmatov, 2003a). It is likely that there is no risk of mutual interaction between the converter control systems of variable-speed wind turbines. This is what is commonly expected regarding properly tuned control systems of converters.

In this case, a large wind farm can simply be represented in the analysis of voltage stability by a single machine equivalent. A single machine equivalent means that a single wind turbine model represents a large wind farm. The following assumptions apply here:

- The power capacity of the single machine equivalent is the sum of the power capacities of the wind turbines in the wind farm (Akhmatov and Knudsen, 2002).
- The power supplied by the single machine equivalent is the sum of the power of the wind turbines in the wind farm (Akhmatov and Knudsen, 2002).
- The reactive power of the single machine equivalent at the connection point is zero according to Danish specifications (Eltra, 2000).
- The mechanism of accumulation of potential energy by the shaft systems of the wind turbines and the shaft relaxation process at the grid fault have to be taken into account in the case of Type A and B wind turbines (Akhmatov and Knudsen, 2002). This has an effect on the behaviour of the generator rotor speed. Voltage instability relates to excessive overspeeding of fixed-speed wind turbines (Akhmatov et al., 2003).
- The mechanism relating to shaft twisting and relaxing mentioned in the item above is not relevant in the case of Type C wind turbines. Electric and reactive power are controlled independent of each other. Therefore, dynamic behaviour of the generator speed and the voltage are decoupled in this type (Akhmatov, 2002).

- The risk of mutual interaction between the converter control systems of Type C and D wind turbines is eliminated by efficient tuning of the converter control (Slootweg and Kling, 2002).
- Similar arguments also apply to Type D wind turbines, except in special situations (Westlake, Bumby and Spooner, 1996).

Using single machine equivalents instead of detailed models of wind farms with representations of a large number of wind turbines can reduce the complexity of the analysis of voltage stability. The focus is on the incorporation of large wind farms into large power systems. This simplification is also reasonable because conventional power plant units are commonly represented in the analysis of voltage stability by their lumped, single machine equivalents. It is also an advantage if the power system model is already sufficiently complex and contains a number of power plants, including their control, consumption centres, parts of the transmission and distribution networks, their static and dynamic reactive compensation units and so on (Akhmatov, 2002, 2003a).

29.8 Conclusions

It is possible to implement aggregated models of large wind farms with a large number of electricity-producing wind turbines and their control systems into simulation tools for the analysis of (short-term) voltage stability. In this chapter this has been illustrated with an example that includes a detailed representation of an offshore wind farm consisting of 80 wind turbines. The analysis of voltage stability has included the representations of the four main concepts of modern wind turbines: Type A2 fixed-speed, active-stall wind turbines equipped with induction generators, Type B1 pitch-controlled wind turbines equipped with induction generators with VRR, Type C1 variable-speed, pitch-controlled wind turbines equipped with DFIG and partial-load frequency converters, and Type D1 variable-speed wind turbines equipped with PMG and frequency converters.

It has been demonstrated that there is practically no risk of mutual interaction between the wind turbines and their control systems (the converters) in large wind farms. The wind turbines do not oscillate against each other. The wind turbines show a coherent response when the power system is subjected to a transient short-circuit fault.

For the analysis of short-term voltage stability in connection with the incorporation of large wind farms into large power systems the wind farms can be represented by single machine equivalents. This means that the wind farm is represented by a single wind turbine model of the chosen concept and with rescaled rated power capacity. Commonly, the worst case will arise when the wind farm is at rated operation.

References

[1] Akhmatov, V. (2001) 'Note Concerning the Mutual Effects of Grid and Wind Turbine Voltage Stability Control', *Wind Engineering* 25(6) 367–371.
[2] Akhmatov, V. (2002) 'Modelling of Variable-speed Wind Turbines with Doubly-fed Induction Generators in Short-term Stability Investigations', presented at 3rd International Workshop on Transmission Networks for Offshore Wind Farms, Stockholm, Sweden.

[3] Akhmatov, V. (2003a) *Analysis of Dynamic Behaviour of Electric Power Systems with Large Amount of Wind Power*, PhD dissertation, Technical University of Denmark, Kgs. Lyngby, Denmark, available at www.oersted.dtu.dk/eltek/res/phd/00-05/20030403-va.html.

[4] Akhmatov, V. (2003b) 'Variable-speed Wind Turbines with Doubly-fed Induction Generators, Part III: Model with the Back-to-back Converters', *Wind Engineering* 27(2), 79–91.

[5] Akhmatov, V. (2004) 'An Aggregated Model of a Large Wind Farm with Variable-speed Wind Turbines Equipped with Doubly-fed Induction Generators', *Wind Engineering* in press.

[6] Akhmatov, V., Knudsen, H. (2002) 'An Aggregated Model of a Grid-connected, Large-scale, Offshore Wind Farm for Power Stability Investigations – Importance of Windmill Mechanical System', *Electrical Power and Energy Systems* 24(9) 709–717.

[7] Akhmatov, V., Knudsen, H., Nielsen, A. H., Poulsen, N. K., Pedersen, J. K. (2001) 'Short-term Stability of Large Wind Farms', presented at European Wind Energy Conference EWEC-2001, Copenhagen, Denmark, paper PG3.56, 1182-6.

[8] Akhmatov, V., Knudsen, H., Nielsen, A. H., Pedersen, J. K. Poulsen, N. K. (2003) 'Modelling and Transient Stability of Large Wind Farms', *Electrical Power and Energy Systems* 25(1) 123–144.

[9] Akhmatov, V., Nielsen, A. H., Pedersen, J. K., Nymann, O. (2003b) 'Variable-speed Wind Turbines with Multi-pole Synchronous Permanent Magnet Generators and Frequency Converters, Part I: Modelling in Dynamic Simulation Tools', *Wind Engineering* 27(6) 531–548.

[10] Edström, A. (1985) 'Dynamiska ekvivalenter för stabilitetsstudier', Vattenfall, Systemteknik, Sweden.

[11] Eltra. (2000) 'Specifications for Connecting Wind Farms to the Transmission Network', ELT1999-411a, Eltra Transmission System Planning, Denmark.

[12] Feijóo, A. E., Cidrás, J. (2000) 'Modelling of Wind Farms in the Load Flow Analysis', *IEEE Transactions on Power Systems* 15(1) 110–115.

[13] Grauers, A. (1996) 'Design of Direct-driven Permanent-magnet Generators for Wind Turbines', Technical Report 292, School of Electrical and Computer Engineering, Chalmers University of Technology, Göteborg, Sweden.

[14] Hinrichsen, E. N. (1984) 'Controls of Variable Pitch Wind Turbine Generators', *IEEE Transactions on Power Apparatus and Systems* 103(4) 886–892.

[15] Jenkins, N. (1993) 'Engineering Wind Farms', *Power Engineering Journal* 4 (April) 53–60.

[16] Knudsen, H., Akhmatov, V. (2001) 'Evaluation of Flicker Level in a T&D Network with Large Amount of Dispersed Windmills', presented at the International Conference CIRED-2001, Amsterdam, The Netherlands, Paper 226.

[17] Kristoffersen, J. R., Christiansen, P. (2003) 'Horns Rev Offshore Wind Farm: Its Main Controller and Remote Control System', *Wind Engineering* 27(5) 351–360.

[18] Noroozian, M., Knudsen, H., Bruntt, M. (1999) 'Improving a Wind Farm Performance by Reactive Power Compensation', presented at IEEE PES Summer Meeting, Singapore.

[19] Pena, R. S., Cardenas, R. J., Asher, G. M., Clare, J. C. (2000) 'Vector Controlled Induction Machines for Stand-alone Wind Energy Applications', in *Conference Record of the 2000 IEEE Industry Applications Conference, Volume 3*, Institute of Electrical and Electronic Engineers, New York, pp. 1409–1415.

[20] Røstøen, H. Ø., Undeland, T. M., Gjengedal, T. (2002) 'Doubly-fed Induction Generator in a Wind Turbine', presented at the IEEE/CIGRE Workshop on Wind Power and the Impacts on Power Systems, Oslo, Norway.

[21] Saad-Saoud, Z., Jenkins, N. (1995) 'Simple Wind Farm Model', *IEE Proceedings on Generation, Transmission and Distribution* 142(5) 545–548.

[22] Schauder, C., Mehta, H. (1999) 'Vector Analysis and Control of Advanced Static VAR Compensators', *IEE Proceedings-C* 140(4) 299–306.

[23] Slootweg, J. G., Kling, W. (2002) 'Modelling of Large Wind Farms in Power System Simulations', in *IEEE Power Engineering Society Summer Meeting, Chicago, U.S.A., Volume 1*, Institute of Electrical and Electronic Engineers, New York, pp. 503–508.

[24] Slootweg J. G., de Haan, S. W. H., Polinder, H., Kling W. (2002) 'Aggregated Modelling of Wind Parks with Variable Speed Wind Turbines in Power System Dynamics Simulations', presented at the 14th Power Systems Computation Conference, Seville, Spain.

[25] Spooner, E., Williamson, A. C. (1996) 'Direct Coupled, Permanent Magnet Generators for Wind Turbine Applications', *IEE Proceedings on Electrical Power Applications* 143(1) 1–8.

[26] Spooner, E., Williamson, A. C., Catto, G. (1996) 'Modular Design of Permanent-magnet Generators for Wind Turbines', *IEE Proceedings on Electrical Power Applications* 143(5) 388–395.

[27] Westlake, A. J. G., Bumby, J. R., Spooner, E. (1996) 'Damping the Power-angle Oscillations of a Permanent-magnet Synchronous Generator with Particular Reference to Wind Turbine Applications', *IEE Proceedings on Electrical Power Applications* 143(3) 269–280.

[28] Yamamoto, M., Motoyoshi, O. (1991) 'Active and Reactive Power Control for Doubly-fed Wound Rotor Induction Generators', *IEEE Transactions Power Electronics* 6(4) 624–629.

Index

Wind Power in Power Systems   Edited by T. Ackermann
© 2005 John Wiley & Sons, Ltd   ISBN: 0-470-85508-8 (HB)

Royal Institute of Technology

One of the leading European Technical Universities in Power Engineering

- Research
- Undergraduate Education
- Graduate Education
- International Master programs
- Expertise in Wind Power
- High level course in Wind Power including Power System integration

For more information:
http://www.ets.kth.se/ees/
lennart.soder@ets.kth.se